COLLINS GUIDE TO THE MUSHROOMS
OF BRITAIN AND EUROPE

To W
Fungi (and lady) friend.
000 error
Happy foraging
Robert.

COLLINS WILDLIFE TRUST GUIDES

The Wildlife Trusts are working for a UK richer in wildlife. Their extensive network covers all parts of the countryside and reaches into cities as far apart as Belfast and Ipswich. They care for more than 2,300 nature reserves: from rugged coastline to urban oases, remote islands to flower-rich meadows.

With more than a third of a million members, The Wildlife Trusts provide an unrivalled source of expertise on wildlife and conservation issues, offering advice that helps local landowners and communities care for their local wildlife. This work also includes the implementation of rescue plans for nationally rare species, including the red squirrel, the dormouse and its ancient woodland home, the marsh fritillary butterfly and the otter.

The Wildlife Trusts also campaign for better protection for wildlife at the local, national and international level, and recognise that people are a vital part of the solution to the challenges they face.

Joint projects like this one between The Wildlife Trusts and Collins Natural History are important, not just because they help raise vital funds for wildlife in the UK (for each copy of this book sold 30p will be donated to The Wildlife Trusts), but also because they show people the wonders of the natural world through the medium of first-class photographs, which encourage their interest and commitment to help protect it.

COLLINS WILDLIFE TRUST GUIDE
MUSHROOMS

A photographic guide to the mushrooms of Britain and Europe

Régis Courtecuisse

HarperCollins*Publishers*

HarperCollins*Publishers*
77–85 Fulham Palace Road
London
W6 8JB

The HarperCollins website address is:
www.fireandwater.com

First published 1999

2 4 6 8 10 9 7 5 3 1

99 01 03 04 02 00

Titles in this series include:

Birds of Britain and Europe, John Gooders

Butterflies of Britain and Europe, Michael Chinery

Trees of Britain and Europe, Keith Rushforth

Night Sky, Mark Chartrand

© in the text: HarperCollins*Publishers*

The author asserts his moral right
to be identified as the author of this work
All rights reserved

ISBN 0 00 220012 0

Edit and page make-up by D & N Publishing, UK

Translation by Jo Weightman

Illustrations by the author

Colour origination by Colourscan, Singapore
Printed and bound by Rotolito Lombarda SpA, Milan, Italy

Contents

- 6 The Author
- 6 Tribute to Jacques Vast (1921–1995)
- 7 Introduction
- 9 Fungal Morphology
- 13 Mycology
- 15 Toxicology
- 24 Physiology and Fungal Biology
- 25 Classification Used in this Book
- 28 Geographical Area Covered
- 28 The Guide: General Layout

Part I Colour Plates
- 34 General Key Using Visual Index
- 37 Pyrenoid Species
- 41 Pezizoid Species
- 55 Corticioid Species
- 61 Hydnoid Species
- 65 Polyporoid Species
- 73 Clavarioid Species
- 79 Cantharelloid Species
- 95 Species with Gills
- 227 Boletoid Species
- 237 Gasteroid Species

Part II Family and Species Descriptions
45–882

Part III Appendices
- 885 Glossary
- 892 Picture Credits
- 894 Index

The Author

Régis Courtecuisse was born in northern France in 1956. He is senior lecturer in the Department of Biological and Pharmaceutical Sciences at the University of Lille II. He has been studying fungi for almost twenty years and is one of the few professional mycologists in France. He is president of the Société Mycologique du Nord and committee member of various societies and several French and international scientific boards. His work is mainly concerned with the taxonomy and systematics of the agarics, both in Europe and in the tropics (especially in the mountains of Guiana and the West Indies). In 1991, he set up a national programme for listing and mapping the fungi in France. The long-term objective is to produce a national red data list (a provisional list on these lines has already been drawn up for the Nord-Pas-de-Calais region), which can be used to formulate appropriate measures to conserve prime habitats for rare or endangered fungi. He is the author of more than one hundred scientific publications, papers and mycological works. These include R. Courtecuisse & B. Duhem, 1995 *Mushrooms & Toadstools of Britain & Europe* (HarperCollins*Publishers*).

Author's note: the arrangement of the species within the colour section of this guide has been altered from its original proposed order. As a result, some transitions between morphological groups, initially intended to be between pages, may now occur within a single page. For this reason, for example, some coloured species may appear on pages including white species. I do hope that such slight anomalies will not disturb the reader in his or her understanding of the structure of this guide.

Tribute to Jacques Vast (1921–1995)

Jacques Vast, who was born in Alsace in 1921, had his roots deep in northern France and Picardy. He was a mycologist, naturalist and talented photographer, with a slide collection of great scientific importance. Production of this book gave us the opportunity to work together, pooling our skills. His untimely death prevented him from seeing the conclusion of a guide over which he had worked for such a long time. I completed the preparation of this work alone, in sadness at his loss. Without the splendid illustrations which form its core, without Jacques Vast, I would not have written this book. I dedicate it to him.

Régis Courtecuisse, 1999

INTRODUCTION

Fungi have always troubled man and aroused a variety of emotions in him: fear, wonder, longing, etc. The somewhat mysterious relationship between our ancestors and these organisms gave rise to numerous beliefs and a wide variety of uses. Sometimes, these uses were daily or even decisive in human progress and occasionally they played a major role in religious ceremonies. Now, a tasty meal is often the first thing to come to mind when fungi are mentioned. The recent surge of interest in edible fungi is probably associated with the 'back to the earth' movement generated by the current interest in ecology, whose consequences are not always necessarily beneficial. In some countries, however, real economic problems force the population to exploit natural resources. This guide will, of course, deal with questions of edibility but also those of toxicity. While a few species of wild fungi are delicious, more are dangerous: some are even deadly. Therefore, it is absolutely essential, before collecting fungi to eat, even before trying to recognize a few edible species, to learn carefully what these public enemies look like. Apart from these delicious or deadly species, the majority of fungi are more or less edible or poisonous. In fact, many fungi which are held to be edible, may cause various adverse reactions in some people. It is the duty of every mycologist to alert those who enjoy eating fungi to the very real dangers associated with pollution from various sources. Fungi can be very efficient at accumulating chemicals discharged into the atmosphere or into the ground. Industry and even traffic on the roads emit frightening quantities of toxic pollution. The quantity absorbed by repeated or frequent consumption may greatly exceed the norms recommended by the World Health Organization (WHO), seriously endangering the health of the consumer. The occurrence of more or less illegal discharges of doubtful substances and the nuclear catastrophe at Chernobyl are examples that should make every potential collector think twice.

More than ever before, it can be said that fungi do not exist to make a meal. They have diverse and extremely important roles to play in the wild. As well as being fundamental agents in nature, their roles include garbage collection, population control and plant auxiliaries (for example mycorrhizal species). Their beauty is diverse and sometimes surprising. We should respect these living organisms as members of the biosphere and ecosystems we share with them and other living animal and vegetable forms. Learning to recognize them in all their multiplicity and diversity will help us to enjoy and respect them. This is the aim of this guide which has superb photographs accompanied by detailed descriptions. With it, the reader can identify a certain number of common woodland and grassland species and rarer ones characteristic of certain particular habitats. They should always bear in mind that they may have collected a fungus not dealt with in this book: several thousand species are listed in Europe and no book can include them all (indeed, new discoveries are still being made). If close attention is paid to physical characters, it is possible to avoid those pitfalls, common in the fungal world, that are associated with similar species and variation within a species. Only with patience and perseverance will the reader achieve a good understanding of these capricious, erratic but fascinating organisms.

What is a fungus? Its place among living things: the traditional division of living forms into two kingdoms, animal and plant, is now out of date. Five kingdoms are now recognized although there could eventually be more. The Prokaryotae (or Monera) kingdom includes the prokaryotes or bacteria. These are relatively simple organisms: they have no nucleus (DNA, the carrier of genetic information, is present in the cell in the form of a single circular chromosome) and cell division is by a simple binary fission process.

The other kingdoms (Protista, Plantae, Animalia, Fungi) comprise the eukaryotes, whose cells have a true nucleus within which DNA is present in chromosomes of varying numbers. Cell division is by mitosis (a complex process which ensures distribution of genetic material in the daughter cells).

The Protists are a heterogeneous group of single-celled organisms. They are studied by mycologists as well as by botanists and zoologists. Some groups of 'fungi' are included in this group.

The Plantae (or plants) are organisms capable of photosynthesis (able to manufacture their own organic food from sunlight, CO_2 and water, owing to the presence of pigments such as chlorophyll). Conversely, the Animalia (or animals) feed on previously formed organic matter.

The Fungi are not chlorophyll-based and differ from the animal kingdom in several respects: fungi are usually decomposers of organic substances; they are distinguished by the presence of special sugars that are rare elsewhere (trehalose and mannitol) and by other particular chemical characters, by the diversity and complexity of their life cycles (in the extreme, the total disappearance of sexual differentiation, duration of the $2 \times n$ chromosome phase), nutrition by absorption (by photosynthesis in plants, by digestion in animals) and the ability to reproduce independently by asexual reproduction (very rare elsewhere). Fungi are placed in two kingdoms (Protista and Fungi), and their definition is based on common features in their way of life.
B. Kendrick wrote, 'Fungi are eukaryotic, heterotrophic, absorptive organisms, that develop a rather diffuse, branched, tubular body and reproduce by means of spores' (*The Fifth Kingdom* 2nd Edition, 1992).

FUNGAL MORPHOLOGY

A fungus is composed of long filaments which, under normal conditions, are invisible, these very fine threads being

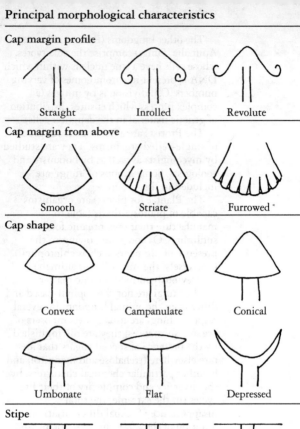

Cap surface

Smooth

Granular

Scaly

Fibrillose

Erect (hairs or fine scales)

Fleecy

Gill attachment

Adnate

Notched

Free

Decurrent

Gill shape

Horizontal

Ventricose

Arching

Sinuate

almost always buried in the depths of the substrate they colonize. In a few rare cases, filamentous masses or mats may appear on the surface. These filaments can be septate (the septa being the wall which separates each cell from its neighbour) and are then called hyphae or they may be nonseptate (the juxtaposed cells form a long tube: a siphon enclosing numerous nuclei). This latter type is limited to relatively lower groups. Special cases include fungi whose vegetative structure is unicellular (like yeasts). Generally, hyphae combine to form a mycelium, which is a more or less dense mass of cobwebby threads. In the mycelial state, a fungus cannot be identified by traditional morphological methods.

A fungus can survive and even flourish for a while in the vegetative state, but like all living forms it tends to reproduce when conditions are right. It will then produce spores. Several mechanisms may be deployed and there are many kinds of spores. Reproduction may be asexual (the thallus may simply fragment into more or less differentiated parts) or sexual (when fusion of cellular nuclei cause a genetic mix). Spores may be produced directly on the mycelium or, at an opportune moment, on differentiated bodies, the sporophores or carpophores (a term mistakenly used since 'karpos' from the Greek means 'fruit' and these nonflowering organisms do not have fruit). These sporophores are infinitely varied in form, size and complexity. When you are in the woods gathering morels or chanterelles, you are only actually collecting the sporophore of an underground organism. While the sporophores of the 'higher' fungi are the best known (they are, indeed, the actual subject of this book), other groups produce less obvious bodies although these may be complex or have distinctive forms. Fungi are identified by, and a large part of the classification system is based on, the distinctive morphological, anatomical and structural characters of the sporophores.

Mycology

First Beginnings

The first mycologists, towards the end of the eighteenth century and early in the nineteenth, were general naturalists and their work was essentially descriptive. The state of learning and the techniques of the period only introduced them to the morphological characters of sporophores, hence their tendency to group together fungi which are phylogenetically distant. Thus, the one genus *Boletus* included all species whose fertile surface was exposed to the outside world through pores (now called boletes and polypores).

With the expansion of microscopy, the world of the lower fungi in all their vast complexity was revealed. It led to the discovery of different types of reproductive cell and made it possible to differentiate between the large groups recognized today (for example Ascomycetes and Basidiomycetes). In the case of the macrofungi, which concern us most in this book, significant new ideas arose once the anatomical structure of different parts of sporophores were revealed. Certain modern taxa are partly defined by the character of their cuticle, such as *Bolbitiaceae* (Cortinariales), which have a hymeniform cap cuticle (*see* p.396). As mycologists described new species (taxonomy), they began to focus their attention on understanding the way the fungal organism works (biology). The gradual integration of new biological data and increasingly sophisticated techniques led to advances in setting up a system of classification (systematics) and also a theoretical plan of the relationship between different recognized groups (phylogeny).

Contemporary Mycology

Because of the high cost of some diseases affecting food plants or tree species, there were early developments in the field of phytopathology. Applications in forestry also generated large research programmes, especially in connection with mycorrhizas. The study of conditions under which fungi live in the wild has led to the development of research projects on their ecology and ecophysiology. There is also interest in fungal communities, in their fundamental aspects (qualitative composition: mycocoenology and mycosociology) and applied or quantitative aspects (population biology, ecological role, etc.).

When the ecological requirements of a species is known, much information about the site and area in question is gained. Fungi can then be considered as bio-indicators. Such work is fairly recent with regard to fungi in the strict sense but is more advanced in lichenology (lichens are full members of the fungal kingdom; *see* p.26). Certain lichens have proved to be excellent bio-indicators for atmospheric pollution. Conversely, as fungi accumulate certain contaminants (trace elements, radioactive elements, etc.), they can be useful indicators of the condition of their habitats. The discovery of various properties in fungi, in cultivation and in the wild, has led to consideration of the possibility that they might be used in the biological control of certain parasitic diseases or that certain species considered harmful could be destroyed by virulent fungal parasites. We are touching here upon the field of biotechnology where fungi are used as auxiliaries in chemical reactions or in various biological processes. Medicinal semi-syntheses, the food industry and research into the biodegradation of hydrocarbons, plastics and certain organic wastes are examples of fields in which fungi could be used. We should not forget that fungi are indispensable in the manufacture of cheese (moulds), wine, bread (yeasts) and that this has been known to Man since classical times. The cultivation of edible fungi also has some economic significance.

Other aspects of mycology have medicinal implications. It is known that fungi have been used in traditional pharmacopoeia since ancient times. The anti-cancer properties of certain polypores have been proven by epidemiological studies in the former USSR. The historical (and accidental) discovery of antibiotics by Sir Alexander Fleming from *Penicillium* cultures is famous and undoubtedly prompted the surge of many research projects in this field. The therapeutic uses of fungi are currently under investigation in many different areas of medicine, including acquired immune-deficiency syndrome (AIDS) research, since interesting immunostimulant properties appear to be present in *Lentinula edodes* (shiitake). However, fungi sometimes have harmful effects on our health: dermatoses attack the skin, hair and nails; forms of candidiasis are very invasive (white mouth in infants, thrush of the mouth or genital areas in an adult); and some pulmonary and visceral endomycoses are much more serious and sometimes fatal.

The role played by fungi in the life of Man is therefore understood. We can, moreover, observe that certain fungi are declining seriously. In the current ecological context, we must envisage protecting fungi in the same way that we protect other living forms or particular habitats. This brief overview concludes by stressing that the mycologist, in addition to describing species, is interested in fungi that are useful to man and those that may harm his activities or economy.

TOXICOLOGY

Two types of poisoning resulting from ingesting fungi (intentionally or otherwise) can be distinguished: mycetismus (eating so-called 'higher' fungi), and mycotoxicoses (ingestion of products contaminated by toxins secreted by fungi, often by 'lower' groups).

Mycetismus Syndromes: Syndromes with a Short Incubation Period

DIGESTIVE SYNDROMES

These are caused by two very different mechanisms. The first concerns the adverse reactions of some individuals to fungi that other people can eat without ill effects. Some humans can have a deficiency of certain enzymes which makes them incapable of digesting complex molecules that are particularly frequent in fungi (trehalose, for example). Other molecules composing fungi or secreted by them can be difficult to digest (e.g. chitin), may provoke excessive intestinal osmotic pressure (e.g. mannitol) or generate allergic reactions (e.g. antibiotics). Excessive consumption of fungi carries the risk of inducing reactions that may be very severe and care should be taken not to over-indulge in any one meal. The second is caused by a toxin or laxative component as, for example, in *Ramaria formosa*, *Ramaria pallida* and other club fungi and in the genus *Suillus* among the boletes. The upset, which is often minor, appears within a few minutes or hours, and ceases after treatment of the symptoms.

RESINOID POISONING

This is a gastrointestinal syndrome which appears fairly rapidly (in a few minutes or hours), producing violent pain, nausea, vomiting and diarrhoea. The fungi responsible are *Tricholoma pardinum* (the cause of most incidents in mountainous districts), *Omphalotus olearius* and *Entoloma lividum*. Also responsible are some other *Entoloma* species, certain boletes (especially the *Boletus satanas* group) and a few indigestible fungi.

The symptoms can be treated (with antispasmodic drugs) but weak or sick individuals may suffer quite serious complications.

PSILOCYBINE POISONING

This is caused by molecules in the LSD family. Tropical fungi are largely responsible for this hallucinogenic syndrome. In Europe, only a few species have sufficient active constituents. It is illegal to gather, store, transport or sell them.

MUSCARINE OR SUDORINE POISONING

The active principle responsible is muscarine (a poison of the autonomic nervous system), present above all in *Inocybe* species (*Inocybe patouillardii* is considered extremely poisonous, even deadly), in the small white *Clitocybe* species (especially *Clitocybe dealbata*, *C. candicans*, *C. cerussata*, *C. phyllophila*) and in *Mycena pura* and related species (*M. rosea* in particular). The syndrome is rarely fatal but often serious and very alarming to the patient, causing vasodilatation and slowing of the cardiac rate, with serious lowering of blood pressure, myosis, increased intestinal movements, diarrhoea and stimulation of the secretory glands causing sweating and excessive salivation, sometimes accompanied by nausea and vomiting. Individuals suffering from a heart condition may die. The antidote is atropine or tincture of belladonna.

PANTHERINE POISONING

Clinical indications are dominated by the action of ibotenic acid, muscimol and muscazone which induce vasoconstriction and increased cardiac rate, with hyperactivity, drying out of the mucous membranes, etc. This group of symptoms is complicated by the action of other stimulatory, hallucinogenic or aphrodisiac toxins. *Amanita muscaria* is the typical example of the agents causing this syndrome. Although it is not dangerous in small amounts, it may have a much more serious effect on individuals with cardiovascular problems. *Amanita gemmata* is also sometimes associated with this type of poisoning. *Amanita pantherina* is, however, the most dangerous in this group. Fatal effects have been reported. Treatment is based on sedatives, barbiturates, etc.

HAEMOLYTIC POISONING

Some fungi which are edible when cooked, are more or less poisonous raw or undercooked. This applies to certain amanitas and many species in the Ascomycotina (morels, *Helvella* species, etc.) whose thermolabile poisons are usually destroyed if well cooked. Symptoms are often of a digestive nature (nausea, vomiting) but the underlying cause, the destruction of red blood cells, may have serious consequences.

Syndromes with a Long Incubation Period

These syndromes, which have a reduced survival or recovery rate, are much more dangerous than those described above because the poison has time to cause damage before the victim is alerted by the first symptoms.

PHALLOIDINE POISONING

Symptoms appear several hours after consumption (6–12 h). Species responsible are the deadly *Amanita* species (*A. phalloides*, *A. virosa*, *A. verna* and closely related taxa), other species more recently associated with this syndrome (small *Lepiota* species: *L. helveola*, *L. josserandii*, *L. brunneoincarnata*, *L. brunneolilacina*, *L. pseudohelveola*, *L. subincarnata*, *L. helveoloides*) and species close to *Galerina marginata*, *G. autumnalis*, etc. (lookalikes of *Kuehneromyces mutabilis*). The Death Cap, *Amanita phalloides*, is responsible for 95% of poisonings.

The first symptoms are respiratory difficulty, vertigo and a general feeling of ill health. The next stage of acute gastroenteritis, violent and painful vomiting, severe foul-smelling liquid cholera-like diarrhoea and signs of intense dehydration may cause the victim to die at this point from cardiovascular failure. This stage lasts for three or four days after eating the fungus. It is succeeded by a period of apparent but deceptive remission where the patient may resume normal activities. The next stage is liver failure, which actually began in the first

24 h and can be detected for several days. Signs of this attack (present very early on and then weakening) can be revealed by biochemical analysis (e.g. by the enzymes transaminase and lactic dehydrogenase). Sometimes, in serious cases, the patient experiences pain and a hepatomegaly may occur. The final stage does not arrive until at least 6 days after ingestion. The outcome will be favourable if only a small amount of the poison was eaten or if treatment was undertaken very early. In this case, there will be a complete cure. However, severe liver poisoning usually leads to death.

The toxins responsible are very complex polypeptides and treatment is problematic. To be successful, it is most important to act swiftly. It is possible to use 'Dr Bastien's treatment' which consists of intravenous injections of vitamin C and administration of intestinal antiseptics (©Ercefuryl) and an antibiotic (neomycin), with the possible addition of an anti-emetic to ease the vomiting. Other treatments are often difficult to set up and appear only to work sometimes (but the main difficulty, once again, arises from the fact that the patient is often condemned because diagnosis is frequently delayed). Hospital treatment is essential and a liver graft may be necessary.

ORELLANINE POISONING

This is caused by species of *Cortinarius* (*C. orellanus*, *C. rubellus* = *C. speciosissimus*) and related species (and perhaps others). It attacks the kidneys, sometimes very severely. The outcome is death or kidney failure, which means lifelong dependence on a kidney machine. The first symptoms often appear a long time after consumption, occasionally a week later; diagnosis is difficult as the patient does not necessarily make the connection between his symptoms and the fungus. Treatment is symptomatic and usually involves dialysis.

GYROMITRINE POISONING

This is primarily caused by *Gyromitra esculenta*. Other Ascomycotina contain the same toxin (*Gyromitra* spp., *Helvella* spp., *Cudonia circinans*, *Spathularia flavida*, etc.). The incubation period varies from 5 to 48 h. Early symptoms are gastroenteritis accompanied by weakness, nausea and vomiting, sometimes with violent diarrhoea and fever (this is rare in fungal poisoning). The patient may recover fully after 2–6 days, but in some cases there is a second stage in which the liver, kidneys and nervous system are affected. Hepatitis occurs, sometimes accompanied by haemolysis and kidney failure, neurological disorders, delirium, cramps and generalized muscular convulsions. The patient may recover but the hepatitis could prove fatal. Treatment is primarily symptomatic (stomach pump, dialysis, pain killers, vitamin B_6).

Syndromes Associated with Particular Species

COPRINUS POISONING

This syndrome is linked with the simultaneous consumption of alcohol and certain species of *Coprinus* (*C. atramentarius* in particular, occasionally *C. micaceus* and related species). Symptoms are similar to those experienced by individuals taking Antabuse as a cure for alcoholism (flushing, palpitations and great discomfort). As the effects linger, drinking alcohol several days after eating the fungus may provoke this reaction.

PAXILLUS POISONING

This syndrome, which is caused by *Paxillus involutus* only, has been recently recognized. This species, which is eaten in huge quantities by some peoples (especially in central Europe), has been responsible for cases of serious poisoning, leading even to death. The reaction is of an immuno-allergic type, arising after several meals have been consumed without after-effects. First contact with the fungus causes the body to form antibodies which, after the decisive meal, act with antigens in the fungus and set off an immunological reaction which leads to a large-scale destruction of the red corpuscles. The symptoms, which appear rapidly 1–2 h after the fatal meal, are hypertension, jaundice (liver failure), haemoglobinuria, oliguria then anuria (kidney failure) and disseminated intravascular coagulation. The outcome may be death. Treatment is primarily symptomatic and aims to counterbalance failing or lost renal function. *Paxillus involutus* should therefore be considered a deadly species.

Poisoning from Secondary Sources

Such poisoning is due, not to the fungus itself, but to toxic products it has accumulated. The main problems of this type of poisoning are connected with two principal causes.

TRACE ELEMENTS (HEAVY METALS)

Eating fungi collected near roads and motorways, industrial sites, etc., may lead to consumption of huge amounts of contaminants. Some edible species are able to accumulate these elements or molecules and levels of lead and mercury far in excess of levels permitted by the WHO have been recorded. This type of pollution is particularly insidious and may occur quite far from a direct source of pollution. Current advice therefore is not to eat fungi-based meals repeatedly. Some sources recommend limiting oneself to two or three meals per year!

RADIOACTIVE ELEMENTS

Pollution by radioactive elements is a serious problem. The consequences of the Chernobyl disaster, nuclear tests and numerous sources of low level doses of ambient radioactivity can accumulate in fungi. Here also, concentrations far in excess of the permissible levels laid down by the WHO have been detected. A few

Poisoning Attributable to Mycotoxicoses

Mycotoxicoses are caused by certain toxins, secreted by microfungi contaminating food stocks and present in edible foods. It is almost impossible to avoid at least occasional contact with these formidable toxins.

ERGOTISM

Ergotism was very widespread in the Middle Ages, when flour was sometimes contaminated with the toxins secreted by ergot (a parasite in the ears of cereal crops). The active agents are lysergic acid derivatives which induce extremely severe vasoconstriction, possibly leading to gangrene of the extremities, with loss of fingers or limbs, and often to convulsions (this latter condition is also known as St Antony's Fire – a case was reported in Ethiopia in 1979).

AFLATOXINS

These are produced by species of *Aspergillus* (*A. flavus*, *A. parasiticus* and *A. niger*), but other genera are equally capable of producing them. Products liable to be contaminated are oil-bearing seeds (groundnuts, cotton, soya, sesame, etc., and cattlecake made from them), cereals (especially maize and wheat) and their derivatives (flour, pastas and bread, etc.) and dairy products. Aflatoxins are carcinogenic and are responsible for primary liver cancer in Africa. Teratogenic and mutagenic effects have been demonstrated experimentally. Other syndromes or illnesses are also attributable to aflatoxins (Reyes syndrome, infantile cirrhosis, etc.).

OCHRATOXINS

These toxins are produced by species of *Aspergillus* and *Penicillium*. They may contaminate cereals (and their derivatives, such as beer) and both dry and oil-bearing fruit. They mainly attack the kidneys and are largely responsible for certain types of nephropathy (the nephropathy endemic in the Balkans, for example, is interstitial tubal nephritis, which is often associated with cancers of the urinary tract). Effects which are immunosuppressive, teratogenic, poisonous to genes and carcinogenic have also been cited. The level of ochratoxins in foodstuffs is increasing. That it is currently present in the blood of Europeans has been amply demonstrated.

TRICHOTHECINES

These molecules, which are produced by species of *Fusarium*, are immunosuppressant and necrotizing. Some are carcinogenic, cytotoxic or adversely affect the cardiovascular or nervous system. Several conditions linked with these toxins have been described, including alimentary toxic aleukia, a serious haemorrhagic condition. Cereals are the main contaminated foodstuff.

FUMONISINS

Fusarium moniliforme is responsible for producing these toxins which have mutagenic, carcinogenic and neurological effects. Some are the cause of equine leukoencephalopathy (a condition of horses, causing death within a few days by destruction of the brain; large cavities are revealed by autopsy) and others are implicated in cancers of the oesophagus in humans.

OTHER TOXINS

These include citreoviridin (from *Penicillium citreoviride* which contaminates rice, causing beriberi), citrinin (from *Penicillium citrinum* on cereal crops) which is strongly synergetic with ochratoxins and aflatoxins in their carcinogenic effects, patulin (from various species of *Aspergillus* and *Penicillium* on apples and their derivatives) which is carcinogenic and zearaleones (from *Fusarium* and *Gibberella* on cereals) which interfere with the transmission of genes and are carcinogenic to man.

PHYSIOLOGY AND FUNGAL BIOLOGY

Physiology

Since fungi are heterotrophic with respect to carbon, they require a ready-formed source of organic carbon in order to feed (like animals). They also need oxygen (although there are some exceptions), hydrogen, various trace elements, vitamins and water, at least during part of their life cycle. Temperature and light also play a part in their physiology.

Because fungi are rich in enzymes, they have great metabolic potential and are good at adaptation. Secondary metabolites (natural antibiotics, for example) play an important role in the life of a fungus, especially in competition between organisms for the colonization of various substrates.

Despite these often decisive advantages, fungi suffer from a major constraint, linked to their heterotrophy in respect of carbon.

- **Saprotrophs** exploit dead or inert organic matter. Such species occur on leaf litter, dead plant and animal debris and on dung.
- **Parasites** exploit living organic matter, to its detriment (the host may be an animal, plant or fungus). Such species are phytopathogens or agents of parasitic disease in man or animals.
- **Symbionts** join with another organism (an autotroph), forming a mutually beneficial relationship. The best examples are lichens (alga–fungus association) and mycorrhizal species (plant autotroph–fungus). Over and above these characteristics and requirements, fungi have a certain number of original biological characteristics which vary from one group to another.

Growth

When the spore (the basic reproductive element) germinates, a primary mycelial filament grows. This is composed of haploid cells (consisting of a nucleus with a set of chromosomes, n). When conditions are suitable, reproduction takes place. This may be asexual in nature, in which case either the mycelium breaks up, or spores, which are often called conidia, are formed directly and may or not be borne on special bodies called conidiophores.

Reproduction may also be sexual. In this case, two mycelia of complementary 'polarity' – we could say 'sex' – must come together to form a secondary mycelium by plasmogamy (fusion of cytoplasms: many fungi are tetrapolar, that is they have four different 'sexes', in compatible pairs). The secondary mycelium has cells with two nuclei which are not, for the moment, fused. It is usually on this secondary mycelium that the sporophores develop. These are structures bearing the fertile cells (meiosporangia: asci or basidia, for example) within which fertilization occurs (fusion of the nuclei forms one nucleus with $2n$ chromosomes) and this is immediately followed by a series of three divisions which redistribute in the spores a haploid nuclear set.

Many species can reproduce either asexually or sexually, depending on prevailing conditions, the phase of their cycle or other determining factors. In this alternating situation, the fungus in the asexual phase (mitotic) is called an anamorph and the sexual phase (meiotic) is called a teleomorph.

CLASSIFICATION USED IN THIS BOOK

It would not be appropriate or helpful, in a field guide of this kind, to set out the full systematic picture, drawing on all the latest publications and research. I have chosen a middle way, adopting some new thinking and rejecting archaic or mistaken concepts, even if they have been hallowed by time.

Protista or 'Pseudofungi'

These organisms that, until recently, were still thought to be fungi never form a mycelium but have a vegetative structure without cellular septa. In most cases there is a plasmodium (a kind of giant amoeba) or a coenocytic siphon. They are close to fungi in that they produce spores, which are sometimes flagellate, in their reproductive phases. Nutrition is often by ingestion (phagocytosis).

Among those recently excluded are the former Myxomycetes, Acrasiomycetes, Labyrinthulomycetes and Plasmodiophoromycetes, the last of these being plant pathogens of great economic significance as agents of club root in the cabbage family (*Plasmodiophora brassicae*), potato powdery scab (*Spongospora subterranea*), seedling damp-off (*Pythium* spp.), potato blight (*Phytophthora infestans*) and vine downy mildew (*Plasmopara viticola*).

'True Fungi' (Fungal Kingdom in the Strict Sense)

ANAMORPHS AND TELEOMORPHS

Many fungi, in the course of their life cycle, alternate between an asexual stage (anamorph) and a sexual stage (teleomorph). Although these forms are different stages of just one species, they often have different Latin names and were, for a long time, placed in different groups. All forms with asexual reproduction were known as 'fungi imperfecti' and placed in Adelomycetes or Deuteromycetes (Deuteromycota), a placing which is no longer valid. Alternation between anamorph and teleomorph is rare in the most evolved groups.

DIVISION: ZYGOMYCOTA

This division includes fungi with a siphonic thallus (coenocytic structure), reproducing by fusion of gametangia, thus producing a zygospore. These are usually very unobtrusive microscopic fungi. Examples include: the Mucorales, some of which are used in the chemical and pharmaceutical industries while others are plant pathogens and occasionally parasites of man, causing zygomycoses; the Entomophthorales, which are plant and animal parasites sometimes used in the biological offensive against insects considered noxious (vectors of parasitic diseases, phytophages etc.); and the Glomales, endomycorrhizal fungi which are abundant in the soil and play an important part in the natural balance of certain habitats.

DIVISION: DIKARYOMYCOTA

This guide will actually be devoted to the most spectacular representatives of this division, which includes most species known to the general public (morels, truffles, boletes, toadstools, etc.).

These fungi have a septate thallus, each cell being separated from its neighbour by a dividing wall. The cell wall itself contains chitin. The definition of this group is confirmed by a cytological argument: the nuclei do not fuse immediately after plasmogamy (fusion of cytoplasm). This fusion takes place later in the meiosporangium, a particular cell in which karyogamy occurs, followed rapidly by chromatid reduction and distribution of the nuclear contents into haploid spores.

There are two subdivisions: Ascomycotina and Basidiomycotina. Their sporophores are called ascoma (once called ascocarps) and basidioma (formerly basidiocarps) respectively.

The Ascomycotina produce their meiospores (sexual reproductive spores) within the meiosporangium (fertile cell), which is an ascus (Fig. 1). These meiospores are ascospores and, when ripe, they are all expelled together from the ascus, which acts like a spore gun. A very large number of fungi are placed in this subdivision.

The Basidiomycotina produce their meiospores outside the fertile cell which is called a basidium (Fig. 2). This generally clavate cell bears the spores on the tip of small points, the sterigma. The basidiospores, which are often actively ejected, have a kind of scar, near the point of attachment, called an apiculus.

GEOGRAPHICAL AREA COVERED

The guide concentrates on fungi in Europe. The limits used are fairly traditional, not posing any problems in the north (Scandinavia up to the North Cape), west (British Isles and Iceland) or south (Mediterranean basin), but the eastern limits are more difficult to define since the area covered could be extended to include the former Eastern Bloc countries as far as the Caucasus and Urals. Nearly all the illustrations are of collections made in France and Britain (with some in Belgium and Italy). However, the species described are mostly widespread in northern, southern and eastern Europe.

THE GUIDE: GENERAL LAYOUT

The introduction on the previous pages indicates the general approach and aims of the book.

Part I (p.33) contains the colour plates consisting of photographs of over 900 species, most of which were taken in their habitat. The layout of this part will be explained below. Part II (p.245) contains all the species descriptions. The layout of this part will also be explained below. Part III (p.883) contains the appendices: the glossary and index which enable the reader to make the best use of the guide.

Colour Plates

VISUAL INDEX
In the illustrated section, species are divided into fifteen groups. These are introduced diagrammatically in the visual

key on p.34. In this morphological arrangement, unrelated species with similar shapes are grouped together. The reader can leaf through the photographic section, stopping (using the thumb-tab) at the drawing matching the group in question. It is a good idea to make the first determination using the visual key, where details of similar species in the basic type are given. Although simplified, this little morphological key with its analytical approach will assist the collector to avoid some serious pitfalls.

MORPHOLOGICAL ARRANGEMENT OF SPECIES
The fifteen groups are presented in the following order (as in the visual key): pyrenoid (one group), pezizoid (one group), corticio-stereoid (one group), hydnoid (one group), polyporoid (one group), clavarioid (one group), cantharelloid (one group), gilled fungi (different silhouettes: six groups), boletoid (one group) and gasteroid (one group).

In each of these groups, fungi are arranged in order of structural complexity: resupinate species, then those with a cap, followed by those with cap and stipe; species with cap, stipe and gills are arranged in increasing robustness of silhouette, and/or by the cap colour (or sporophore colour – ascomata and basidiomata – for species lacking a cap). For this last character, cap colour, species are arranged in the following sequence: white, yellow, orange, red, beige to dark brown with all shades in between, grey to black, blue, purple and other rare colours. However, certain species are distinctly polychromic or variable macroscopically. The reader will therefore understand that it has been very difficult to arrange certain groups. They should not be surprised by unusual groupings or if the chromatic sequence (colour grading) is repeated within the same group (as defined by the single thumb-tab). For example, *Mycena*- and *Collybia*-like silhouettes include so many species, that they had to be placed in several subgroups determined by increasing order of size (which cannot be done with absolute accuracy). Where there was ambiguity, I chose statistical criteria, as far as possible, in order to place a species in a given group. A few taxonomic groups, which are very well defined and easy to recognize in the field (*Inocybe*, *Pluteus*, *Amanita*, *Russula*, etc.), have been left intact. This way, the reader can quickly benefit from field experience which they cannot fail to acquire.

CAPTIONS UNDER THE PHOTOGRAPHS
Captions under the photographs give the following information: Latin name for the species (genus and specific epithet with variety where appropriate) and page number of the description.

Layout of the Descriptions

The descriptions follow the colour plates. They are set out in the same simplified order of the major systematic groups as in the section on Classification (*see* p.25).

This guide, which is only concerned with the so-called 'higher' fungi, begins with the Ascomycetes: first the Pyrenomycetes (*see* p.247), then the Discomycetes (*see* p.256). Within this group, inoperculate species are followed by the operculate species, then by the Tuberales. Basidiomycetes follow. The order in this group is: Phragmo-basidiomycetes (*see* pp.302–6 for auriculario-tremelloid then caloceroid species), Homobasidiomycetes including *Aphyllophoromycetideae* (*see* p.307, corticio-stereoids, polypores, hydnoids, clavarioids, chanterelles), then *Agaricomycetideae* (*see* p.363, species with gills) and finally *Gasteromycetideae* (*see* p.865). In the gilled fungi, the group with most species represented, the major families are listed in alphabetic order (*see* p.365).

Each species description follows the same layout, except where variations are imposed by morphological differences between groups. As an example, here is the layout and content of a description of a gilled fungus.

p.000 (cross reference to photograph page); **common English name of the species;** *Latin binomial* followed by author(s) and perhaps by useful synonyms; family; order.

Description: *Summary of main characters of the species described (general shape, appearance, colour, etc.).*

Cap Minimum, maximum size, shape and silhouette when young and when mature, other diagnostic features of the cap. Shape and appearance of the margin. Description of the surface. Colour.

Gills Insertion, shape, closeness, colour when young and when mature. Description and diagnostic features of the gill edge.

Stipe Minimum–maximum size, consistency, texture, shape, surface features, colour. Shape of the base, where appropriate.

Flesh Thickness, texture, colour. Smell. Taste.

Chemical tests Description of reactions to particular chemical tests (reagents, oxidization when cut, etc.).

Microscopic features spores (minimum and maximum length × minimum and maximum width, shape, ornamentation, germ pore, apiculus, thickness of the spore wall, etc.; colour and appearance of the spore, possible chemical tests). Basidia (number of sterigmata, size, clamps, other features). Cystidia (facial then marginal with size and shape, possible chemical reactions, etc.). Gill trama (possibly). Cap cuticle (structure, disposition and size of hyphae, pigmentation). Clamps.

Season: Outer limits of usual time of appearance indicated by two months.

Habitat: Description of the usual habitat for the species (associated vegetation, type of habitat, soil or bioclimatic preference, etc.).

Distribution: Indication of the frequency and geographical distribution of the species, and possibly a list of the countries where it is known to occur.

Remarks on possible difficulties of identification and information about closely related or similar species.

Special Note
In order to make the descriptions less cumbersome, some characteristics which are obvious in the photographs are not repeated every time within a homogeneous group. The most frequently observed characters do not need to be written each time, for example: margin not striate, cap surface dry or smooth, gills quite close to close, stipe solid, cylindrical or smooth, flesh firm, taste mild, spores smooth non-amyloid, lacking a germ pore, thin-walled, apiculus ordinary, basidia four-spored, with clamps, trama regular. If nothing is written under a heading then the particular feature does not merit discussion for the group in question.

The following abbreviations have been used in the microscopical descriptions:
Bas: basidium(a) CC: cap cuticle
Caulo: caulocystidium(a)
Cheilo: cheilocystidium(a)
Cyst: cystidium(a) (if cheilo = pleuro!)
Frb: fruit body GP: germ pore
MC: marginal cell(s)
Pleuro: pleurocystidium(a)
p.p.: Latin *pro parte*, 'in part'
SC: stipe cuticle
SP: suprahilar plage Sp: spores

Line drawings in the margin beside each species description illustrate the main microscopical characters. All are drawn by the author and are taken from his work files and personal notes and, in certain cases, also from the literature. They are mostly drawings of spores and, where necessary, of various cystidia and cuticles. When several cells are shown, letter(s) indicate(s) which cell is which:
S: spore
C: cheilocystidium
P: pleurocystidium
CC: cap cuticle
SC: stipe cuticle

PART I

COLOUR PLATES

GENERAL KEY USING VISUAL INDEX

A. Fertile surface smooth or with alveolate folds, on the upper side of the sporophore or turned upwards. If it is internal, the sporophore is very small (fertile cell an ascus).

    A1. Pyrenoid type, p.37 (size <2 mm (0.08 in)):

        simple (dispersed singly) (1)
        compound (aggregated on a stroma)

    A2. Pezizoid type, p.41: simple, disc-shaped forms derived from the basic disc (2):

        helvelloid 1 (3)
        helvelloid 2 (4)
        morchelloid (5)

B. Fertile surface smooth, folded, alveolate, with spines, tubular/poroid or lamellate, on the underside of the sporophore or upturned. If it is internal, the sporophore is rarely very small (fertile cell a basidium).

  B1. Hymenophore smooth. Usually lignicolous:

        Corticioid type, p.55:
        corticioid (1)
        stereoid (2)
        tremelloid (3)

  B2. Hymenophore with spines:

        Hydnoid type, p.61:
        resupinate
        stipitate

  B3. Hymenophore tubular/poroid, texture woody or leathery:

        Polyporoid type, p.65:
        resupinate (1)
        pileate or dimidiate (2)
        stipitate (3)

GENERAL KEY USING VISUAL INDEX

B4. Hymenophore amphigenous, smooth:

> Clavarioid type, p.73:
> simple (1)
> compound and
> branching (2)

1

2

B5. Hymenophore fairly smooth to wrinkled:

> Cantharelloid type, p.79:
> cantharelloid (1)
> craterelloid (2)

1

2

B6. Hymenophore lamellate: C
Hymenophore tubular-poroid
(and texture fleshy):   D
Hymenophore internal: E

C. Hymenophore lamellate, p.95.

C1. Stipe absent or clearly eccentric:

> Pleurotoid type

C2. Gills decurrent, texture fibrous:

> Omphalinoid-clitocyboid type:
> omphalinoid (1)
> clitocyboid (2)

 1

 2

C3. Gills adnate. Silhouette slender
to medium, texture fibrous:

> Mycenoid-collybioid type:
> mycenoid (1)
> marasmioid (2)
> collybioid (3)

1

2

3

General Key Using Visual Index

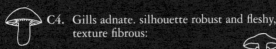

C4. Gills adnate. silhouette robust and fleshy, texture fibrous:

Tricholomatoid type

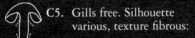

C5. Gills free. Silhouette various, texture fibrous:

Amanitoid type:
amanitoid (1)
pluteoid (2)

C6. Gills adnate to decurrent, texture granular and brittle:

Russuloid type:
russuloid (1)
lactarioid (2)

D. Hymenophore tubular-poroid and texture fleshy.

Boletoid type, p.227:
boletoid (1)
xerocomoid (2)
leccinoid, etc. (3)

E. Hymenophore internal (at least when young).

Gasteroid type, p.237:
lycoperdioid (1)
calvatioid (1)
geastroid (2)
nidularioid (3)

F. Phalloid type.

Pyrenoid Species

All the species in this group are Ascomycetes (*see* p 247), whose fertile cells (asci) occur within more or less spherical perithecia (shown in the sketch), which are always tiny (less than 2 mm diameter). These perithecia either occur singly (species are then difficult to see unless they have a distinctive colour) or in groups on various kinds of stroma when they are easier to see with the naked eye.

38 Pyrenoid species

Leptosphaeria acuta p.248

Lasiosphaeria ovina p.248

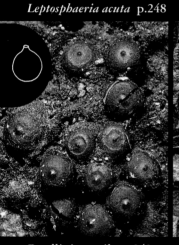
Rosellinia aquila p.249

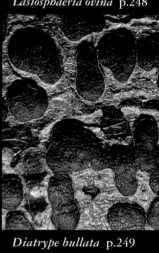
Diatrype bullata p.249

Hypoxylon multiforme p.250

Hypoxylon fragiforme p.250

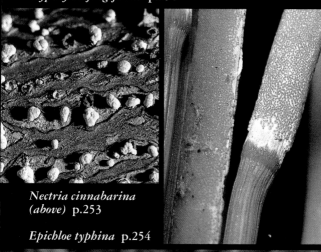

Nectria cinnabarina (above) p.253

Epichloe typhina p.254

Hypocreopsis lichenoides p.251

40 Pyrenoid species

Xylaria polymorpha p.253

Xylaria hypoxylon p.252

Daldinia concentrica (above) p.251

Claviceps purpurea p.254

Claviceps purpurea p.254

Cordyceps militaris p.255

PEZIZOID SPECIES

The basic fruitbody of these Ascomycetes is an apothecia (as shown in the sketch). Simple apothecia are typical of the cup fungi, which are either large or small and occupy a range of substrates; while many species are difficult to see, others are remarkably spectacular and easily spotted. When the apothecia form only a part of the fruit body, because they are raised on a more or less well-developed stalk, the species are referred to as helvellas or morels, the latter among the most highly prized of the edible fungi (must be well cooked!)

42 Pezizoid species

Rhytisma acerinum p.258

Callorina fusarioides p.261

Pyronema omphalodes p.272

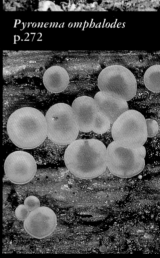

Bisporella citrina p.263

Orbilia xanthostigma p.262

Aleuria bicucullata
p.283

Poculum firmum
p.266

Bulgaria inquinans p.269

Neobulgaria pura p.268

Ciboria batschiana p.265

44 Pezizoid species

Lanzia luteovirescens p.266

Rustroemia maritima p.267

Saccobolus versicolor p.274

Iodophanus carneus p.273

Peziza merdae p.289

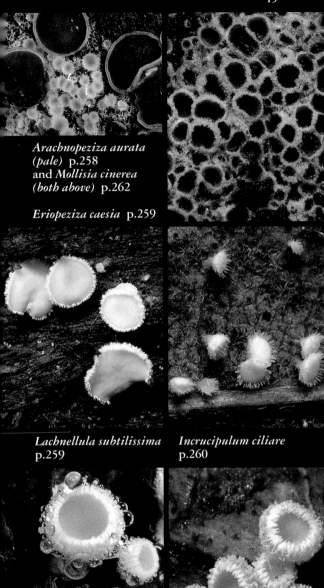

Arachnopeziza aurata (*pale*) p.258 and *Mollisia cinerea* (*both above*) p.262

Eriopeziza caesia p.259

Lachnellula subtilissima p.259

Incrucipulum ciliare p.260

Lachnum virgineum p.261

Capitotricha bicolor var. *rubi* p.260

46 Pezizoid species

Lasiobolus cuniculi p.276

Coprobia granulata p.277

Anthracobia maurilabra *(above)* p.276

Scutellinia setosa p.281

Scutellinia kerguelensis p.281

Cheilymenia stercorea p.277

48 Pezizoid species

Sarcoscypha coccinea p.291

Ascobolus furfuraceus
p.273

Calyptella capula
p.640

Cudoniella aciculare p.264

Aleuria aurantia p.283

Melastiza chateri p.282

Octospora rutilans p.275

Heyderia cucullata p.270

Bryoglossum gracile p.271

Leotia lubrica p.269

Mitrula paludosa p.270

50 Pezizoid species

Otidea onotica p.284

Otidea bufonia p.285

Aleurodiscus amorphus p.314

Peziza fimeti p.289

Chlorociboria aeruginascens p.265

51

Peziza ampelina p.288

Peziza petersii p.287

Peziza gerardii p.287

Peziza cerea p.288

Peziza succosa p.290

Peziza ammophila p.290

52 Pezizoid species

Peziza phyllogena p.286

Marcelleina benkertii p.274

Peziza badia p.286

Rhizina undulata p.292

Disciotis venosa p.296

Helvella unicolor p.295

Helvella leucomelas p.296

Helvella elastica p.294

Helvella macropus p.295

Helvella crispa (above) p.293

Helvella sulcata p.294

54 Pezizoid species

Mitrophora semilibera p.297

Gyromitra esculenta p.292   *Morchella elata* p.298

Morchella vulgaris p.298

CORTICIOID SPECIES

The simplest species in this group form crusts on wood or sometimes even on the ground. The very simple body shape does not lend itself to much macroscopic variety, so identification in the field is often delicate, if not impossible. Other species become detached from the host, forming a small lateral bracket, or even a more complex form like an ear or a brain.

56 Corticioid species

Lyomyces sambuci p.308

Hyphoderma roseocremeum p.310

Bulbillomyces farinosus p.309

Trechispora farinacea p.311

58 Corticioid species

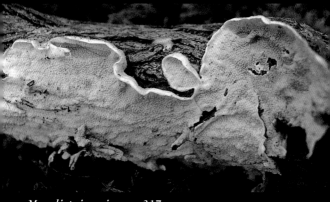

Meruliopsis corium p.317

Merulius tremellosus p.318

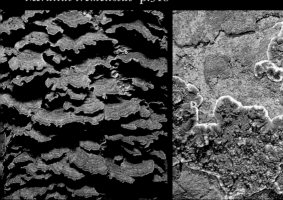

Hymenochaete tabacina p.316

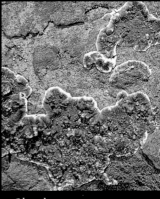

Chondrostereum purpureum p.316

Stereum hirsutum p.314 *Stereum insignitum* p.315

Auricularia mesenterica p.302

Auricularia auricula-judae p.302

60 Corticioid species

Exidia glandulosa
p.304

Tremella mesenterica
p.303

Tremella foliacea p.303

Tremiscus helvelloides
p.305

HYDNOID SPECIES

These species have a very characteristic fertile surface covered in small spines, but they do not form a homogeneous group. Among these fungi are species closely attached to the woody host on which they are decomposing, while others form lateral brackets and others are even terrestrial species with a cap and stipe living symbiotically with higher plants.

62 Hydnoid species

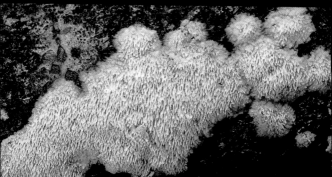

Mycoacia uda p.319

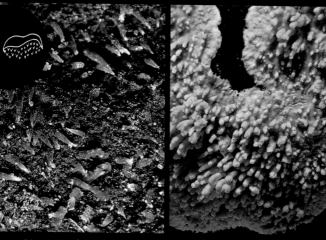

Mycoacia fuscoatra p.319

Hyphoderma radulum p.310

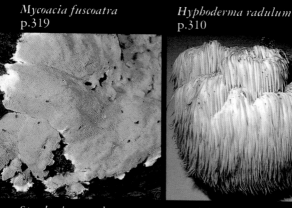

Steccherinum ochraceum

Hericium erinaceum

Hericium clathroides
p.323

Pseudohydnum gelatinosum p.304

Hydnellum peckii p.321

Hydnellum peckii
p.321

Hydnellum concrescens
p.321

64 Hydnoid species

***Phellodon niger** (above)*
p.322
***Sarcodon imbricatum**
(right)* p.322

Auriscalpium vulgare
p.323

Hydnum repandum
p.325

Hydnum rufescens p.325

POLYPOROID SPECIES

Unlike the boletes, which also have a tubular-poroid fertile surface, polypores in the broad sense usually have a leathery or even woody consistency. Confusion only occurs with those polypores that have a cap and stipe, but the vast majority of species in this group have a more simple body shape; sometimes they remain totally attached to the substrate but may perhaps form small lateral brackets or grow into a hoof shape in a position perpendicular to the woody host.

66 Polyporoid species

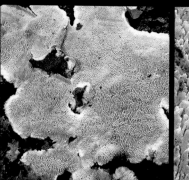

Junghuhnia nitida
p.326

Schizopora paradoxa
p.327

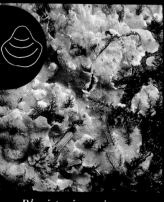

Physisporinus vitraeus
p.326

Bjerkandera adusta
p.342

Trametes pubescens p.342

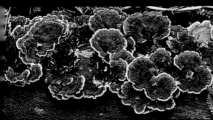

Trametes versicolor p.341

Hapalopilus rutilans (below) p.334

Oligoporus caesius p.343

Abortiporus biennis p.334

Pycnoporus cinnabarinus p.339

68 Polyporoid species

Inonotus hispidus p.328

Fistulina hepatica p.331

Inonotus dryadeus p.329

Fomes fomentarius p.335   ***Fomitopsis pinicola*** p.336

Polyporoid species

Ganoderma lipsiense p.330

Grifola frondosa p.332

Meripilus giganteus p.333

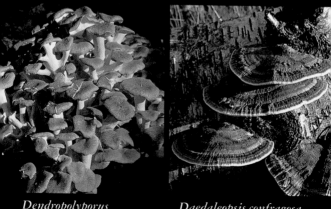

Dendropolyporus umbellatus p.332

Daedaleopsis confragosa p.337

Daedaleopsis tricolor p.338

Lenzites betulinus
p.336

Daedalea quercina
p.355

*Gloeophyllum
saepiarium* p.338

Scenidium nitidum
p.343

Polyporus durus p.345

72 Polyporoid species

Polyporus lepideus p.345

Coltricia perennis p.327

Bondarzewia montana (*above*) p.346

Boletopsis leucomelaena p.347

Albatrellus pes-caprae p.346

CLAVARIOID SPECIES

These species are typically club-shaped or tree-like, more or less branching Basidiomycetes (*see* p.301), sometimes with a thick, fleshy base. Almost their entire surface is fertile (sometimes there is a more or less well-developed sterile base).
However, some Ascomycetes have a very similar, club-like shape and these, (especially the earth tongues), have also been placed in this morphological group.

74 Clavarioid species

Geoglossum cookeianum p.272

Microglossum viride p.271

Spathularia neesii p.267

Calocera cornea p.305

Clavaria falcata p.352

Typhula quisquilaris p.353

Macrotyphula fistulosa
p.354

Clavariadelphus pistillaris
p.354

Clavulinopsis fusiformis
p.353

Calocera viscosa p.306

Clavulinopsis corniculata p.352

76 Clavarioid species

Clavulina cristata p.351

Ramariopsis kuntzei p.355

Clavaria zollingeri p.352

Ramaria stricta p.356

Ramaria gracilis p.355

Ramaria ochraceovirens p.356

Ramaria formosa p.356

Ramaria botrytis (above) p.357

Ramaria largentii p.357

Sparassis crispa p.351

Pterula multifida p.350

78 Clavarioid species

Artomyces pyxidatus p.350

Thelephora penicillata p.349

Thelephora palmata p.349

Thelephora terrestris p.348

Cotylidia pannosa p.348

CANTHARELLOID SPECIES

Species of *Craterellus* and *Cantharellus* have a fruitbody comprising a stipe and often funnel-shaped cap; the fertile surface beneath the cap is sometimes smooth or may be more or less folded. Most of these species are good edible mushrooms (for example, Chanterelle and Horn of Plenty); however, there are others with a similar shape but far less pleasant taste.

80 Cantharelloid species

Craterellus cornucopioides p.358

Pseudocraterellus undulatus p.359

Cantharellus lutescens p.360

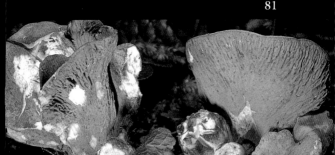
Gomphus clavatus p.362

Cantharellus melanoxeros p.361

Cantharellus tubiformis p.360

Cantharellus cibarius p.359

Cantharellus cinereus p.361

Faerberia carbonaria p.807

SPECIES WITH GILLS

For practical purposes, this is the largest group. To help reach the right place in this section, several morphological sub-groups are proposed, taking account of shape (with or without stipe; delicate or fleshy type), gill attachment (decurrent, adnate or free), presence of veils (as indicated by a ring or volva, for example) and the texture of the flesh (fibrous or brittle). In an attempt to present these sub-groups in a logical evolutionary order, they are arranged as follows:

Pleurotoid species
pp.84–8

Plitocyboid species
pp.89–105

Collybioid-mycenoid species
pp.106–60

Tricholomatoid species
pp.161–75

Species with free gills
pp.176–201

Species with granular texture
pp.202–26

84 Species with gills

Plicaturopsis crispa
p.318

Schizophyllum commune
p.362

Resupinatus applicatus p.806

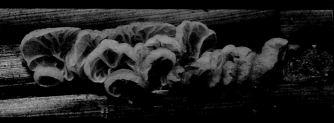

Campanella inquilina p.630

Campanella caesia p.630

Arrhenia spathulata p.809

Clitopilus hobsonii var. *daamsii* p.556

Crepidotus variabilis p.522

Crepidotus mollis var. *calolepis* p.522

Paxillus panuoides p.683

86 Species with gills

Melanotus phillipsii p.804

Hohenbuehelia algida p.807

Arrhenia lobata p.809

Arrhenia auriscalpium

Arrhenia acerosa p.808

Panellus stypticus
p.686

Panellus serotinus
p.686

Pleurotus ostreatus
(above) p.687

Lentinellus micheneri
p.684

Lentinellus cochleatus p.684

88 Species with gills

Pleurotus cornucopiae p.688

Paxillus atrotomentosus p.682

Pleurotus eryngii p.688

Lentinus tigrinus (right) p.685

Delicatula integrella (below) p.669

Entoloma sericellum p.550

Mycena rorida p.669

Rhodocybe fallax (above) p.558

Clitocybe dealbata (right) p.823

Clitocybe phyllophila (below) p.823

90 Species with gills

Cuphophyllus borealis p.564

Cuphophyllus cereopallidus p.566

Cuphophyllus virgineus p.565

Cuphophyllus russocoriaceus p.566

Cuphophyllus berkeleyi p.564

Cuphophyllus colemannianus p.567

Hygrophorus eburneus p.586

Hygrophorus penarius p.581

Hygrophorus fagi p.582

Clitopilus prunulus p.555

92 Species with gills

Hygrophorus chrysodon p.590

Leucopaxillus paradoxus p.860

Clitocybe geotropa p.819

Clitocybe inornata p.821

Hygrophorus agathosmus p.585

94 Species with gills

Mycena belliae p.668

Camarophyllopsis foetens p.675

Phaeotellus rickenii p.810

Entoloma rusticoides p.555

Rickenella swartzii p.816

Rickenella fibula
p.816

Marasmiellus omphaliformis
p.629

Clitocybe graminicola
p.824

Omphalina barbularum
p.814

Cuphophyllus pratensis p.563

96 Species with gills

Omphalina galericolor
p.811

Omphalina pyxidata
p.812

Omphalina rivulicola
p.813

Omphalina hepatica
p.812

Omphalina pseudomuralis p.814

Omphalina oniscus p.815

Phaeotellus griseopallidus p.811

Clitocybe ditopa p.825

Entoloma undatum p.554

Phytoconis ericetorum p.815

98 Species with gills

Cuphophyllus grossulus p.563

Gerronema chrysophyllum p.817

Hydropus marginellus p.671

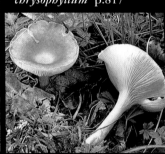

Clitocybe gibba p.819

Clitocybe decembris p.826

Clitocybe phaeophthalma p.825

Clitocybe umbilicata p.826

Lepista inversa p.832

Clitocybe clavipes p.818

Clitocybe odora p.822

100 Species with gills

Rhodocybe gemina p.557

Clitocybe nebularis p.821

Rhodocybe popinalis p.557

Hygrophorus leucophaeus
p.583

Hygrophorus lindtneri
p.587

Hygrophorus arbustivus
p.582

Hygrophorus pudorinus
p.587

Hygrophorus hypothejus p.591

Hygrophorus latitabundus p.590

Hygrophorus persoonii
p.589

102 Species with gills

Hygrophorus russula p.583

Hygrophorus camarophyllus p.584

Hygrophorus olivaceoalbus p.588

Clitocybe lateritia p.820

Cantharellula umbonata p.817

Armillaria tabescens (above) p.827

Gomphidius roseus p.561

Omphalotus illudens
p.681

Catathelasma imperiale
p.860

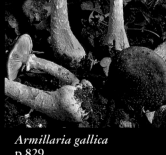

Armillaria gallica
p.829

104 Species with gills

Armillaria mellea p.827

Armillaria mellea p.827

Tubaria autochtona p.532

Tubaria romagnesiana p.534

Tubaria conspersa p.533

Chroogomphus rutilus p.561

106 *Species with gills*

Hygrocybe reidii p.572

Hygrocybe coccinea p.569

Hygrocybe punicea p.567

Hygrocybe splendidissima p.568

Hygrocybe aurantiosplendens p.568

Hygrocybe insipida p.571

Hygrocybe chlorophana p.576

Hygrocybe coccineocrenata p.573

108 Species with gills

Hygrocybe konradii p.580

Hygrocybe quieta p.570

Hygrocybe calyptriformis p.580

Hygrocybe fornicata var. *streptopus* p.576

Hygrocybe psittacina p.574

Hygrocybe unguinosa p.575

Hygrocybe conicoides
p.578

Hygrocybe cinereifolia
p.578

Hygrocybe olivaceonigra
p.579

Hygrocybe conica
p.577

Marasmius androsaceus p.632

Marasmius epiphylloides p.633

Crinipellis stipitarius p.631

Mycena tenerrima p.665   *Mycena stylobates* p.664

112 Species with gills

Mycena corynephorus
p.665

Hemimycena tortuosa
p.666

Marasmiellus candidus p.628

Hemimycena lactea
p.667

Oudemansiella mucida
p.673

Psathyrella cotonea p.446

Anellaria semiovata p.407

Coprinus niveus p.440

Bolbitius tener p.399

Inocybe geophylla p.507

114 Species with gills

Nyctalis parasitica p.865

Collybia cookei p.621

Collybia racemosa p.622

Collybia confluens p.626

Marasmius wynneae p.638

115

Entoloma rhodopolium fo. *nidorosum* p.540

Mycena galericulata p.641

Mycena polygramma p.641

Mycena inclinata p.642

Mycena maculata p.643

116 Species with gills

Marasmius cohaerens p.637 *Marasmius oreades* p.639

Marasmiellus ramealis p.627 *Psathyrella candolleana* p.450

Mycena capillaripes p.646

Inocybe kuehneri p.512

Collybia peronata p.627

Cystoderma carcharias p.676

Nyctalis agaricoides p.864

Mycena flavoalba p.660

118 Species with gills

Mycena filopes p.657

Mycena arcangeliana p.650

Mycena clavicularis p.656

Mycena leptocephala p.658

Mycena erubescens p.655

Mycena vitilis p.659

Mycena amicta p.656

Mycenella bryophila p.640

Mycena epipterygia p.654

Mycena galopus p.651

120 Species with gills

Alnicola escharoides p.485

Hypholoma ericaeum p.801

Entoloma cetratum p.551

Inocybe petiginosa p.518

Panaeolina foenisecii p.407

Panaeolus rickenii p.403

Galerina subfusispora p.527

Flammulaster carpophilus var. *autochtonoides* p.523

Flammulaster ferruginea p.524

Mycena pterigena p.648

122 Species with gills

Tephrocybe palustris p.863   *Entoloma hebes* p.551

Psathyrella conopilus p.448

Psathyrella coprobia p.449   *Psathyrella artemisiae* p.453

Psathyrella badiophylla var. *neglecta* p.454

Psathyrella gracilis
p.447

Psathyrella lutensis
p.452

Coprinus kuehneri
p.436

Coprinus auricomus
p.435

124 Species with gills

Coprinus xanthothrix p.440

Coprinus disseminatus p.437

Coprinus saccharinus p.439

Coprinus micaceus p.438

Coprinus narcoticus p.437

Coprinus lagopides
p.443

Coprinus ammophilae
p.443

Coprinus picaceus p.444

Coprinus atramentarius
p.442

Coprinus comatus
p.441

126 Species with gills

Marasmius undatus p.636

Cortinarius acutus p.473

Alnicola scolecina p.485

Phaeocollybia arduennensis (above) p.529

Galerina paludosa p.526

Macrocystidia cucumis p.560

Mycena chlorantha (above) p.649

Galerina laevis p.525

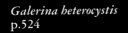

Galerina heterocystis p.524

Conocybe dunensis p.401

128 Species with gills

Conocybe tenera
p.400

Conocybe inocybeoides
p.401

Bolbitius vitellinus p.399

Mycena speirea var. *camptophylla* p.667

Entoloma icterinum p.549

Mycena aurantiomarginata p.650

Entoloma incanum
p.549

Hypholoma elongatum
p.800

Pholiotina arrhenii
(above) p.402

Pholiotina vestita p.403

130 Species with gills

Psilocybe squamosa p.802

Mycena meliigena p.662

Entoloma roseum p.653

Mycena crocata p.548

Mycena rosella p.647

Mycena sanguinolenta p.652

Mycena haematopus p.652

Mycena renatii p.648

Mycena adonis p.662

Mycena acicula p.661

132 Species with gills

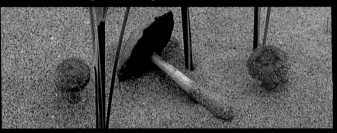

Psathyrella ammophila p.448

Entoloma aranaeosum p.552

Rhodocybe caelata p.558   *Mycena aetites* p.660

Hydropus scabripes p.671

Tephrocybe rancida p.864

Laccaria laccata p.830

Laccaria tortilis p.831

Mycena seynii p.645

Mycena rosea p.644

134 Species with gills

Mycena pura p.643

Mycena pelianthina p.645

Psilocybe chionophila p.803

Psilocybe pratensis p.803

Panaeolus dunensis p.406

Panaeolus ater p.405

Panaeolus obliquoporus p.405

Panaeolus sphinctrinus p.404

Entoloma excentricum p.541

Melanoleuca iris p.858

136 Species with gills

Entoloma sericeum p.538

Pholiota lenta p.795

Entoloma ameides p.542

Entoloma alpicola p.539

Collybia dryophila p.625

Collybia distorta p.623

Collybia butyracea p.622

Oudemansiella radicata p.672

Marasmius alliaceus p.635

Entoloma sordidulum p.539

138 Species with gills

Cortinarius uliginosus
p.466

Cortinarius pholideus
p.465

Cortinarius orellanus
p.459

Cortinarius polaris
p.467

Cortinarius epsomiensis var. *alpicola* p.464

140 *Species with gills*

Cortinarius hinnuleus
p.470

Dermoloma atrocinereum
p.674

Cortinarius armillatus p.468

Psathyrella lacrymabunda
p.444

Psathyrella pyrotricha
p.445

Entoloma porphyrophaeum p.537

Entoloma clypeatum p.536

Rugosomyces carneus p.862

Cortinarius sphagneti p.468

Cortinarius cinnamomeus p.466

142 Species with gills

Cortinarius humicola
p.460

Pholiota graminis p.793

Cortinarius melanotus
p.458

Cortinarius saniosus
p.459

Agrocybe vervacti p.398

Agrocybe praecox
p.397

Agrocybe pediades
p.398

Stropharia coronilla p.798

Cortinarius delibutus
p.482

Cystoderma amianthinum
p.676

144 Species with gills

Cystoderma terreyi
p.677

Cortinarius puniceus
p.465

Cortinarius bolaris p.461

Lepista sordida
p.836

Rugosomyces ionides
p.862

Cortinarius paleifer
p.472

Cortinarius bicolor
p.469

Cortinarius croceocaeruleus p.481

Laccaria amethystina
p.829

Inocybe geophylla var.
lilacina p.508

146 Species with gills

Entoloma lazulinum p.544   *Entoloma chalybaeum* p.543

Entoloma euchroum p.543

Entoloma tjallingiorum p.545   *Entoloma lampropus* p.547

Entoloma querquedula p.546

Entoloma mougeotii p.545

Entoloma poliopus p.547

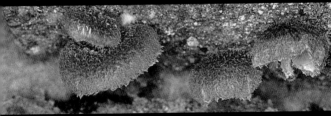

Phaeomarasmius erinaceus p.531

Pholiota curvipes p.792

148 *Species with gills*

Collybia fusipes
p.623

Psathyrella leucotephra
p.450

Ramicola centunculus p.532

Ramicola centunculus
fo. *filopes* p.532

Agrocybe aegerita
p.396

Psathyrella maculata p.446

Psathyrella spadicea p.452

Psathyrella piluliformis p.451

Hypholoma fasciculare p.799

Hypholoma sublateritium p.800

150 Species with gills

Hemipholiota populnea
p.795

Rhodotus palmatus
p.559

Tricholomopsis decora
p.856

Kuehneromyces mutabilis p.789

Galerina marginata p.526

Flammulina velutipes p.674

Pholiota squarrosa p.790

Pholiota jahnii p.791

Pholiota flammans p.792

Gymnopilus spectabilis p.528

152 Species with gills

Pholiota alnicola
p.794

Gymnopilus penetrans
p.529

Inocybe calamistrata p.500

Inocybe hystrix p.504

Inocybe calospora p.518

Inocybe obscura p.506 *Inocybe squarrosa* p.505

Inocybe vulpinella p.512

Inocybe nitidiuscula p.510

Inocybe gausapata p.510

154 Species with gills

Inocybe pholiotinoides p.513

Inocybe cookei p.499

Inocybe hirtella p.515

Inocybe lacera p.507

Inocybe dulcamara p.493

Inocybe heimii p.490

Inocybe squarrosoannulata p.492

Inocybe terrigena p.491

Inocybe paludosa p.492

Inocybe umbrinofusca p.494

Inocybe asterospora p.520

Inocybe fastigiata p.496

Inocybe fastigiella p.498

Inocybe praetervisa p.519

Phaeocollybia lugubris p.530

158 Species with gills

Inocybe jurana
(above) p.495

Inocybe maculata
p.497

Inocybe psammophila
(below) p.508

Inocybe arenicola p.497

Inocybe agardhii p.490

Inocybe splendentoides
p.511

Inocybe piriodora
p.502

Inocybe patouillardii p.494

160 Species with gills

Inocybe lanuginosa p.517

Cortinarius anomalus p.463

Inocybe bongardii (left) p.501

Inocybe corydalina (right) p.503

Inocybe pruinosa p.509

161

Stropharia caerulea
p.796

Stropharia ochrocyanea
p.797

Tricholoma saponaceum var. ***squamosum*** p.844

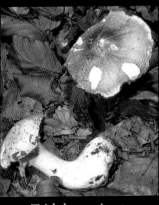
Tricholoma sejunctum
var. *fagetorum* p.845

Tricholoma viridilutescens
p.846

162 *Species with gills*

Tricholoma columbetta
p.838

Tricholoma pseudoalbum
p.837

Tricholoma album p.836

Collybia maculata p.624

Tricholoma acerbum p.838

Calocybe gambosa
p.863

Entoloma lividum
p.535

Hebeloma crustuliniforme p.437

Hebeloma edurum p.488

Lepista glaucocana p.834

165

Megacollybia platyphylla p.625

Lepista panaeola p.834

Cortinarius herculeus p.474

Phaeolepiota aurea p.679

Rozites caperatus p.602

166 Species with gills

Tricholoma aurantium p.855

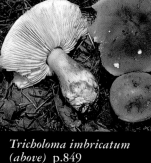

Tricholoma imbricatum (above) p.849

Tricholoma vaccinum p.850

Tricholoma fulvum (above) p.851

Cortinarius triumphans (right) p.475

Cortinarius splendens (above) p.480

Cortinarius elegantissimus p.479

Tricholoma auratum (below) p.848

Tricholoma flavovirens p.847

Tricholoma sulfureum p.848

168 Species with gills

Cortinarius trivialis p.483

Tricholoma ustale p.853

Cortinarius stillatitius p.484

Tricholoma pessundatum p.852

Tricholoma ustaloides p.853

Phylloporus rhodoxanthus p.419

Melanoleuca grammopodia (above two) p.858

Tricholoma psammopus (above) p.851

Cortinarius subtortus (right) p.476

Cortinarius caninus (below) p.463

170 Species with gills

Tricholoma caligatum p.856

Tricholoma fracticum p.854

Tricholoma pardinum p.844

Tricholoma virgatum p.839

Tricholoma terreum p.841

Tricholoma scalpturatum p.842

Tricholoma cingulatum p.843

Tricholoma portentosum p.840

172 Species with gills

Tricholoma atrosquamosum
p.841

Hygrophorus marzuolus
p.585

Lyophyllum fumosum p.861

Melanoleuca cinereifolia var. *maritima* p.859

Entoloma inopiliforme p.536

Entoloma lividoalbum p.541

Tricholomopsis rutilans p.857

174 Species with gills

Lepista nuda p.835

Cortinarius camphoratus p.462

Cortinarius alboviolaceus (above) p.461

Cortinarius traganus p.462

Cortinarius torvus (below) p.470

Cortinarius praestans p.475

Cortinarius rufoolivaceus (right) p.481

Cortinarius terpsichores p.477

Cortinarius purpurascens p.478 (right)

Cortinarius caligatus p.476 *Cortinarius violaceus* p.457

176 Species with gills

Agaricus bisporus p.367

Agaricus campestris p.366

Agaricus devoniensis p.369

Agaricus bernardii p.370

Agaricus arvensis p.373   *Agaricus essettei* p.374

Agaricus xanthoderma p.375

1/8 Species with gills

Agaricus menieri p.377

Agaricus bitorquis p.369

Agaricus praeclaresquamosus p.376

Agaricus romagnesii p.371

Agaricus subperonatus p.368

Agaricus spissicaulis p.371

Agaricus porphyrrhizon p.375

Agaricus koelerionensis p.372

180 Species with gills

Agaricus cupreobrunneus p.366

Agaricus silvaticus p.373

Agaricus bohusii p.367

Pluteus petasatus p.690

Pluteus cervinus p.689

Pluteus phlebophorus *(above)* p.694

Pluteus thomsonii *(right)* p.695

Pluteus romellii p.694

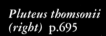

Pluteus umbrosus p.692

Volvariella hypopithys p.696

Volvariella gloiocephala p.698

Amanita vaginata p.383

Amanita mairei p.384

184 Species with gills

Amanita submembranacea p.380

Amanita crocea p.383

Amanita pseudofriabilis p.381

Amanita malleata p.382

Amanita ceciliae p.380

Amanita oreina p.384

Amanita muscaria p.386

Amanita muscaria ssp. *americana* p.386

186 Species with gills

Amanita pantherina p.387

Amanita caesarea p.385

Amanita virosa p.394

Amanita ovoidea p.392

Amanita phalloides p.393

Amanita dunensis p.393

Amanita phalloides p.393

Amanita echinocephala p.390

188 *Species with gills*

Amanita strobiliformis p.389

Amanita singeri p.391

Amanita rubescens p.388   *Amanita franchetii* p.389

Amanita spissa p.388

Amanita beillei p.391   *Amanita citrina* p.395

Amanita porphyria p.395

190 Species with gills

Macrolepiota rickenii p.617

Macrolepiota fuliginosa p.616

Macrolepiota rhacodes p.618

Limacella guttata p.379

Limacella illinita
p.379

Chamaemyces fracidus
p.678

Leucoagaricus macrorrhizus p.610

Leucoagaricus pinguipes p.611

192 Species with gills

Leucoagaricus littoralis p.612

Leucoagaricus leucothites p.615

Lepiota ochraceodisca p.597

Lepiota laevigata p.596

Leucoagaricus badhamii p.614

Leucoagaricus badhamii p.614

194 Species with gills

Lepiota ignivolvata (above) p.596

Cystolepiota seminuda p.594

Sericeomyces sericatellus p.609

Cystolepiota adulterina p.593

Cystolepiota hetieri p.594

Melanophyllum haematospermum p.619

Melanophyllum eyrei p.620

Leucocoprinus brebissonii p.616

196 Species with gills

Lepiota felina p.607

Lepiota pseudofelina p.602

Lepiota clypeolaria p.597

Lepiota ventriosospora p.598

Lepiota ochraceofulva p.608

Echinoderma perplexum p.593

Echinoderma echinaceum p.592

198 Species with gills

Lepiota audreae p.600

Lepiota josserandii p.604

Lepiota cristata p.599

Leucoagaricus gauguei p.611

Lepiota rhodorrhiza (right) p.606

Lepiota ignicolor p.602

Lepiota pseudohelveola var. *sabulosa* p.606

Lepiota grangei p.603

Cystolepiota bucknallii p.595

Leucoagaricus purpureorimosus p.613

Lepiota brunneoincarnata p.604

Lepiota fuscovinacea p.608

Russula delica (above) p.701

Russula chloroides p.702

Russula nigricans (below) p.700

Russula amoenolens p.707

Russula lepida var. *ochroleucoides* p.712

Russula acrifolia p.700   *Russula farinipes* p.703

Russula subfoetens p.704

Russula foetens p.703

Russula laurocerasi p.706

Russula illota p.705

Russula mustelina p.711

Russula heterophylla p.709

Russula violeipes fo. *citrina* p.715

Russula aeruginea p.715

Russula claroflava p.724

Russula ochroleuca p.732

Russula fellea p.733

Russula virescens p.710

Russula cutefracta p.708   *Russula parazurea* p.716

Russula medullata p.717

Russula exalbicans (above) p.744

Russula gracillima p.743

208 *Species with gills*

Russula luteotacta p.734

Russula betularum p.735

Russula emetica p.734

Russula fageticola p.736

Russula nana p.737

Russula lepida p.712   *Russula aurora* p.713

Russula vesca p.710

Russula veternosa p.751   *Russula sanguinaria* p.746

210 Species with gills

Russula xerampelina
p.719

Russula subrubens p.720

Russula velenovskyi p.721

Russula mesospora p.752

Russula aurea p.723

Russula laeta p.725

Russula risigallina p.729

Russula decolorans p.723

Russula nitida p.722

Russula turci p.730

212 Species with gills

Russula unicolor p.718

Russula integra p.727

Russula puellaris (above) p.718

Russula melitodes p.726

Russula torulosa (below) p.748

Russula queletii p.746

Russula drimeia p.747

Russula fuscorubra p.749

Russula cessans p.728

Russula badia p.750

214 Species with gills

Russula cyanoxantha p.707

Russula amoena p.714

Russula cavipes
p.742

Russula amarissima
p.753

Russula olivacea p.731

Russula artesiana (above) p.745

Russula amara p.730

Russula krombholzii p.739

Russula norvegica p.740

Russula fragilis p.738

216 Species with gills

Russula alnetorum p.740

Russula pelargonia p.741

Lactarius vellereus p.753

Lactarius piperatus
p.754

Lactarius controversus
p.755

Lactarius torminosus p.755

Lactarius zonarius p.757

Lactarius necator p.756

Lactarius pyrogalus p.759

Lactarius circellatus p.759

218 Species with gills

Lactarius blennius p.762

Lactarius chrysorrheus p.776

Lactarius vietus p.761

Lactarius volemus p.775

Lactarius trivialis p.760   *Lactarius nanus* p.760

Lactarius robertianus p.784

Lactarius uvidus p.782

220 Species with gills

Lactarius luridus p.783

Lactarius uvidus var. *candidulus* p.783

Lactarius lilacinus p.765

Lactarius spinosulus p.765

Lactarius fuscus p.764

Lactarius hysginus p.762   *Lactarius cimicarius* p.766

Lactarius badiosanguineus p.774

222 Species with gills

Lactarius tabidus p.772

Lactarius quietus p.768

Lactarius camphoratus p.768

Lactarius subdulcis p.769

Lactarius hepaticus p.773

Lactarius brunneohepaticus p.773

Lactarius lacunarum p.771

Lactarius decipiens p.776

Lactarius mitissimus p.769

Lactarius atlanticus p.767

224 Species with gills

Lactarius fulvissimus p.770   **Lactarius rufus** p.763

Lactarius sanguifluus *(above)* p.785

Lactarius deliciosus p.786   **Lactarius quieticolor** p.786

Lactarius deterrimus p.787

Lactarius salmonicolor p.788

Lactarius bresadolanus p.758

226 Species with gills

Lactarius pterosporus p.778

Lactarius acris p.777

Lactarius ruginosus
(above) p.779

Lactarius romagnesii
(right) p.779

Lactarius picinus p.780

Lactarius lignyotus p.781

BOLETOID SPECIES

Boletes are easy to recognize from their shape (cap and stipe) and the fertile surface of tubes opening as pores beneath the cap. They often have a fleshy texture, although in some cases it is brittle, but it is never leathery or woody as in some polypores. Practically all species have a mycorrhizal relationship (symbiosis) with higher plants.

228 Boletoid species

Strobilomyces floccopus p.408

Porphyrellus porphyrosporus p.409

Gyrodon lividus p.411

Gyroporus castaneus p.410

Gyroporus cyanescens p.410

Boletus pulverulentus p.420

Suillus grevillei p.413

Boletinus cavipes p.412

Suillus luteus (above) p.412 *Suillus collinitus (right)* p.414

230 Boletoid species

Suillus bovinus p.415

Suillus variegatus p.415

Xerocomus badius p.418

Xerocomus rubellus p.418

Xerocomus armeniacus
p.416

Xerocomus subtomentosus
p.416

Xerocomus subtomentosus
(above) p.416

Xerocomus chrysenteron
p.417

Tylopilus felleus
p.429

232 Boletoid species

Boletus edulis p.422

Boletus aereus p.424

Boletus pinophilus p.423

Boletus depilatus p.424

Boletus appendiculatus p.421

Boletus radicans p.420   *Boletus satanas* p.428

Boletus calopus p.421

Boletus luteocupreus p.427

234 Boletoid species

Boletus luridus p.427

Boletus regius p.422

Boletus dupainii p.426

Boletus erythropus p.425

Boletus queletii p.425

Leccinum duriusculum p.430

Leccinum crocipodium (left) p.433

Leccinum lepidum p.434

Leccinum scabrum p.431

236 Boletoid species

Leccinum aurantiacum p.429

Leccinum carpini p.433

Leccinum variicolor p.432

Leccinum brunneogriseolum p.432

Xerocomus parasiticus p.418

GASTEROID SPECIES

This final group brings together those Basidiomycetes with an internal fertile part, at least when young. The simplest body type is spherical (but this sphere is always large, i.e. macroscopic, unlike the perithecia of the pyrenoid Ascomycetes appearing at the beginning of the colour section); in some instances, complex adaptation has led to highly original adult fruitbody shapes. The truffles which, however, belong in the Ascomycetes, have been included in this group as they have converged morphologically as a result of their underground way of life.

238 Gasteroid species

Vascellum pratense
p.875

Lycoperdon perlatum
p.874

Lycoperdon piriforme
p.873

Lycoperdon mammiforme
p.874

Lycoperdon echinatum
p.875

Calvatia excipuliforme
p.873

Bovista limosa p.876

Pisolithus arrhizus p.869

Pisolithus arrhizus p.869

Scleroderma citrinum p.866

240 Gasteroid species

Scleroderma bovista p.867

Scleroderma geaster p.867

Mycenastrum corium p.876

Geastrum sessile p.871

Geastrum triplex p.870

Geastrum morganii p.872

Astraeus hygrometricus p.869

Myriostoma coliforme p.872

Tulostoma brumale p.868

242 Gasteroid species

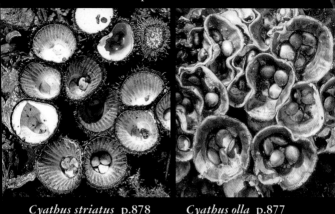

Cyathus striatus p.878

Cyathus olla p.877

Crucibulum laeve p.877

Nidularia farcta p.878

Sphaerobolus stellatus p.870

Phallus impudicus p.879

Phallus hadriani p.879

Battarraea phalloides p.868

Mutinus caninus p.880   *Clathrus archeri* p.881

244 Gasteroid species

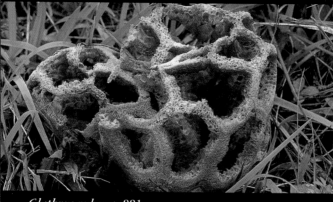

Clathrus ruber p.881

Geopora clausa p.280

Balsamia vulgaris p.300   *Tuber uncinatum* p.299

PART II

SPECIES DESCRIPTIONS

The number preceding each species description in the following pages is the page number in the colour section on which that species can be found.

Ascomycetes (Subdivision of the Ascomycotina)

Although the general public is mainly acquainted with the most easily seen or most delicious Ascomycetes (truffles, morels), the group includes thousands of small, sometimes microscopic species of considerable economic importance, with implications in human and plant toxicology. The systematics are complicated, depending upon many characters which are mostly undetectable without sophisticated methods of study. This work includes only a few examples of the two most representative subclasses, *Pyrenomycetideae* (more recently known as Pyrenomycetes) and *Pezizomycetideae* (commonly called Discomycetes). Both belong to the class Hymenoascomycetes which characteristically have asci disposed regularly within a hymenium (a unicellular layer of fertile asci lying beside each other and forming a regular palisade lining a surface of given shape). Two kinds of structure bearing this hymenium can be distinguished: a sphere (in the Pyrenomycetes) and a disc (in the Discomycetes).

I. *Pyrenomycetes*

This group includes all the forms with asci enclosed in a spherical ascoma or fruit body. Typically, this is a perithecium, which has an opening (an ostiole) through which spores can be dispersed into the environment. In many species, the perithecia occur singly, arising directly in or on the substrate, sometimes on a subiculum (a kind of mat of hyphae) covering the host. In others, they share a common base, a structure called a stroma which varies greatly in size, shape and colour. Species with elementary ascomata (perithecia) on a stroma are more striking (because the stroma can be quite large or brightly coloured) than those with single perithecia, which are nearly always overlooked on account of their small size.

The Pyrenomycetes contain many orders and families whose characters are difficult to assess in the field.

---

p.38 *Leptosphaeria acuta* (Hoffm.:Fr.) P. Karsten
Leptosphaeriaceae, Dothideales

Description: *Very small, pointed, black.*
Up to 0.5 mm tall (0.02 in), black, shiny, conical with a rather swollen base and very pointed top. *Microscopic features* Sp 35–50 × 5–6 µm, yellowish, fusiform, with 6–10 septa, slightly constricted at this point, in two rows in the asci. Asci 140–160 × 8–12 µm, no reaction with iodine. Paraphyses forked.
Season: January–April.
Habitat: Gregarious or in troops on dead nettle stems (*Urtica dioica*).
Distribution: Very common wherever the host occurs.

Easy to recognize by its morphological and ecological characters. The genus *Leptosphaeria* (in the broad sense) includes hundreds of species colonizing all kinds of herbaceous plants. Species in this group are known to damage certain economically important crops.

---

p.38 *Lasiosphaeria ovina* (Pers.:Fr.) Ces. & de Not.
Lasiosphaeriaceae, Sordariales

Description: *Small white sphere with a black ostiole.*
Perithecia up to 0.5 mm (0.02 in) diameter, spherical, with a whitish felty coating. Black ostiole on top. *Microscopic features* Sp 35–45 × 4–5 µm, pale to yellowish, curved, nonseptate, in two rows in the asci. Asci 160–220 × 15–22 µm, no iodine reaction.
Season: All year round.
Habitat: On dead wood, especially broad-leaved.
Distribution: Throughout Europe.

Easy to recognize, despite its small size, on account of the white or very pale covering

p.38 *Rosellinia aquila* (Fr.:Fr.) de Notaris
Xylariaceae, Sphaeriales

Description: *Frb hemispherical with a prominent black ostiole, on an often abundant subiculum.*
Perithecia up to 1.5 mm (0.06 in) diameter, hemispherical or slightly flattened, smooth, brown-black, with a nipple-like ostiole. Consistency brittle, like coal. *Microscopic features* Sp 20–27 × 7–8 µm, fusoid with one side less curved, brown-black, having a longitudinal ventral slit and a small rounded hyaline appendage at each end (sometimes absent). Asci 150–200 × 8–10 µm, apex clearly stained blue with iodine.

Season: All year round, especially in spring.
Habitat: Branches of broad-leaved trees, living or dead. Gregarious, the ascomata are borne on a rather thick, brown-black subiculum.
Distribution: Throughout Europe.

The fairly numerous species of *Rosellinia* are distinguished mainly by the spores and the asci. Ecological preferences may help identification (on conifers for example), but no species are host specific.

p.38 *Diatrype bullata* (Hoffm.:Fr.) Fr.
Diatrypaceae, Diatrypales

Description: *Perithecia borne on an erumpent black discoid stroma.*
Perithecia with a barely projecting ostiole, borne on a black to brown, flat or slightly rounded, circular or slightly elliptic to angular, erumpent stroma. *Microscopic features* Sp 5–8 × 1.5–2 µm, hyaline or nearly so, sausage-shaped, smooth, often with two drops, in one or two rows in the ascus. Asci 40–50 × 5–6 µm, with long stalks, no iodine reaction.

Season: All year round.

Habitat: Especially dead willow branches, sometimes on poplar or alder.
Distribution: Throughout Europe.

The very similar *D. disciformis* (Hoffm.:Fr.) Fr., colonizes beech. *D. stigma* (Hoffm.:Fr.) Fr., forms long erumpent black patches, which may cover entire branches.

---

p.39 *Hypoxylon fragiforme* (Pers.:Fr.) Kickx
Xylariaceae, Sphaeriales

Description: *Stroma rounded, with a prominent ostiole. Brick-red to red-brown.*
Perithecia with a prominent ostiole, borne on a globular stroma, up to 1 cm (0.4 in) diameter, brick-red to brown-black and finally black, with the appearance of a small strawberry. *Microscopic features* Sp 10–14 × 5–7 µm, elliptic to subfusiform, flattened on one side, dark brown, with a colourless germination slit on the flat side. Asci 120–150 × 8–9 µm, apex staining faintly blue in iodine.
Season: All year round.
Habitat: Beech bark. Gregarious.
Distribution: Throughout Europe. Very common.

The imperfect stage (anamorph), which is particularly frequent in France, is easily found. Small rufous tubercles, surrounded by brownish felty and branching filaments, are arranged in a star-like formation, preceding the appearance of the fertile stromata. *H. fuscum* (Pers.:Fr.) Fr., which is also very common, has a smaller black stroma which is significantly less regular in shape and has less visible ostioles; found on hazel and alder.

---

p.38 *Hypoxylon multiforme* (Fr.:Fr.) Fr.
Xylariaceae, Sphaeriales

Description: *Stroma irregular, slightly elongate, black, quite thick.*
Perithecia with a rather prominent ostiole, borne on a thick, erumpent, irregular

stroma that is initially ovoid then elliptic or misshapen. Stromata confluent, forming irregular mounds. *Microscopic features* Sp 9–11 × 4–5 µm, dark brown, elliptic with a flattened or depressed ventral side, and a germination slit. Asci 100–140 × 6–7 µm, apical pore coloured blue in iodine.

Season: All year round.
Habitat: On the bark of birch trees; gregarious.
Distribution: Throughout Europe. Very common.

H. serpens (Pers.:Fr.) Fr., which occurs on a range of hosts, has thinner stroma. Most species in this genus are saprotrophic but, like the *Rosellinia*, some are parasitic, for example, *H. mammatum* (Wahl.) P. Karst., on poplars.

---

p.39 *Hypocreopsis lichenoides* (Tode:Fr.) Seaver
Hypocreaceae, Hypocreales

Description: *Stroma a circular orange-brown crust, strongly radially wrinkled.*
Perithecia with barely visible ostioles, borne on a rather thick orange to orange-brown stroma, with branches radiating out from a folded or wrinkled central zone, up to 10 cm (4 in) diameter. *Microscopic features* Sp 25–30 × 8–9 µm, fusiform, with one septum, and constricted at this point, hyaline. Asci 150–180 × 8–10 µm, no reaction with iodine.

Season: All year round.
Habitat: Various broad-leaved trees.
Distribution: A disappearing species. Recorded in Belgium, Denmark, Finland, France, Germany, Holland, Norway, Sweden and the UK. Appears to be absent in southern parts of Europe.

Could not be overlooked.

---

p.40 **King Alfred's Cakes, Cramp Balls**
Daldinia concentrica (Bolt.:Fr.) Ces. & Not.
Xylariaceae, Sphaeriales

Description: *Stroma large, black, with shiny concentric internal layers.*

Perithecia with tiny ostioles, borne on a hemispherical stroma up to 7 cm (2.8 in) diameter, reddish brown then black, smooth to finely punctate. When cut open, shiny concentric zones of dark-brown and silver-grey can be seen. The current year's perithecia are borne in the outer layer and are easily seen in section. *Microscopic features* Sp 13–17 × 6–9 µm, elliptic or fusiform with one flatter side and a germination slit, black. Asci 180–210 × 11–12 µm, with an apical ring, coloured blue in iodine. Paraphyses septate.

Season: All year round.
Habitat: On ash especially but also on birch.
Distribution: Widespread.

An easy genus to identify. *D. vernicosa* (Schw.) Ces. & de Not., with very thick, pale zones (up to six times the thickness of the dark zones, unique in the genus) is smaller (up to 2 or 2.5 cm (0.8 to 1 in)), substipitate, slightly hollow, usually in burnt areas (rare).

---

p.40 **Candle Snuff**
Xylaria hypoxylon (L.:Fr.) Greville
Xylariaceae, Sphaeriales

Description: *Stroma upright like a small tree, white at first at the top (conidia) then wholly black.*
Perithecia with a prominent ostiole, on an erect stroma, more or less branched like a small tree, up to 8 cm (3.1 in) tall, black. The imperfect state is velvety at the base and white powdery above. *Microscopic features* Sp 11–15 × 5–6 µm, slightly kidney-shaped, black, with a germination slit. Asci 100–140 × 8 µm, apical ring coloured blue in iodine.

Season: All year round, but forms bearing asci occur mainly in the autumn.
Habitat: Stumps and dead wood, especially of broad-leaved trees.
Distribution: Throughout Europe. Very common.

There are quite a number of species in this genus. The most characteristic include:

X. filiformis (Alb. & Schw.:Fr.) Fr., very slender, with distant perithecia, occurs on wood. *X. carpophila* (Pers.) Fr., also very slender grows on beech mast. Two more robust species occur frequently (*see* below).

p.40 Dead Man's Fingers
Xylaria polymorpha (Pers.:Fr.) Greville
Xylariaceae, Sphaeriales

Description: *Stroma clavate or irregularly lobed, black, uneven or slightly warted.*
Perithecia with minutely projecting ostioles, on a stroma 3–10 × 1–3 cm (1.2–4 × 0.4–1.2 in), club-like, irregular, sometimes flattened or slightly lobed at the apex, somewhat shortly stalked, black and warted, matt. Flesh of the stroma fibrous, almost white. *Microscopic features* Sp 20–32 × 5–9 µm, elliptic or almost flat on the ventral side, almost kidney-shaped, brown, with a straight, rather short germination slit. Asci 130–200 × 7–10 µm, with an apical ring like *X. hypoxylon* (*see* p.252).

Season: August–April.
Habitat: Dead wood of broad-leaved trees, especially beech.
Distribution: Widespread, common.

X. longipes Nits. is similar but more slender and more distinctly stalked. The surface is slightly more cracked.

p.39 Coral Spot
Nectria cinnabarina (Tode:Fr.) Fr.
Hypocreaceae, Hypocreales

Description: *Perithecia in clusters or densely crowded, bright red.*
Perithecia up to 1–1.5 mm (0.04–0.06 in) diameter, bright red, with a slightly darker ostiole, shining slightly under the lens. Stroma erumpent or immersed in the substrate. *Microscopic features* Sp 15–20 × 4–8 µm, cylindro-elliptic, hyaline, with a transverse septum, slightly constricted, in two irregular rows in the asci. Asci

70–90 × 8–12 µm, not reacting with iodine. Paraphyses branching.
Season: All year round.
Habitat: Gregarious, on twigs, dead branches or trunks. Generally on broad-leaved trees.
Distribution: Widespread. Very common.

This very common species is generally associated with an anamorph (*Tubercularia vulgaris* Tode:Fr.) which forms pale pink cushions the same size. The genus *Nectria* has many species which are generally saprotrophic but sometimes also parasitic, such as *N. galligena* Bres. which causes canker in some fruit trees, ash and willow.

---

p.39 *Epichloe typhina* (Pers.:Fr.) Tulasne
Clavicipitaceae, Hypocreales

Description: *Stroma forming a sheath on stalks of grasses, yellow.*
Perithecia with a rather prominent ostiole, on a stroma forming a sheath towards the bottom of grass stems. *Microscopic features* Sp 80–100 × 1.5–2 µm, filiform, multiseptate when mature, parallel in the asci. Asci 80–100 × 6–8 µm.
Season: June–September.
Habitat: Stalks of living grasses.
Distribution: Uneven. Rare or very rare.

Appearance, colour and habitat make this an easy taxon to identify.

---

p.40 **Ergot**
Claviceps purpurea (Fr.) Tulasne
Clavicipitaceae, Hypocreales

Description: *Stroma capitate, yellow buff, with a long stalk. On black, elliptic sclerotia (ergots).*
Perithecia with a prominent ostiole on a spherical or ovoid stroma, yellow-brown to orange-yellow or mauve-pink, at the end of a thread-like violaceous, red-brown then yellow-brown stipe. This structure develops on a long black sclerotium fallen, in the previous winter, from the heads of

grasses on which it had been growing during the summer and autumn.
Microscopic features Sp 90–120 × 1 μm, filiform, multiseptate at maturity, parallel in the asci. Asci 120–150 × 4–5 μm, not reacting with iodine.

Season: April–July (sexual form).
August–November (sclerotia).

Habitat: The sclerotia occur in the inflorescences of various grasses.

Distribution: Widespread, in most of Europe (the sexual stage, with perithecia, is very rarely collected).

Well known because it is responsible for ergot poisoning, which was very widespread in the Middle Ages when it was known as St Anthony's Fire (*see* p.22), and has now died out. *C. microcephala* Wallr. is sometimes distinguished by the small size of the sclerotia (ergots) and the habitat (grasses in damp places).

---

p.40 **Scarlet Caterpillar Fungus**
Cordyceps militaris (L.:Fr.) Link
Clavicipitaceae, Hypocreales

Description: *Stroma with a yellow stipe and orange club.* Perithecia with a sharp ostiole, on the fertile area of a complex stroma, which has a scurfy, yellow or yellowish, sterile base up to 4 × 0.5 cm (1.6 × 0.5 in) tall and a confluent fertile head, which is cylindric, blunt, up to 2 × 0.8 cm (0.8 × 0.3 in) tall, bright orange and punctate. *Microscopic features* Sp filiform, very long but breaking up into numerous cylindric segments, 4–7 × 1–1.5 μm. Asci initially eight-spored, very long, up to 400 × 4 μm, not reacting with iodine. Apex thick-walled.

Season: August–November.

Habitat: Parasitic on buried insect larva. The stroma develops in mosses (woods and damp banks).

Distribution: Widespread. Rather common.

Avoid confusion with species of *Onygena*, which are slightly similar but have a very specific habitat (on feathers for *O. corvina* Alb. & Schw.:Fr., on hoof and horn for *O. equina* (Willd.:Fr.) Pers.). The genus *Cordyceps* is readily distinguished by the structure of its stromas and its habitat, on buried insect larvae or on *Elaphomyces* (false truffles).

C. sinensis (Berk.) Sacc., is included in the pharmacopoeia in China where it is sold in small bundles, by herbalists and in markets.

II. Discomycetes ('Cup Fungi')

There is considerable diversity within this group, even though the sporophore type is fairly homogeneous: the apothecia are discoid or cup-shaped, lined on the upper side with a hymenium formed by a palisade of asci. There are several differences which give rise to the numerous orders and families: the nature and chemistry of the ascus, especially at the apex; spore shape; complexity of the sporophore (simple apothecia grouped on a subiculum or on a common complex sporophore, the latter sometimes being stalked); nature and anatomy of the outer layer of the apothecia. Three groups are traditionally recognized.

INOPERCULATE DISCOMYCETES

The diagnostic feature of this group is an ascus without a 'lid'. The ripe spores must force their way through a narrow pore. The group includes a large number of microfungi (these are very small species whose life cycle often involves imperfect forms (asexual, vegetative reproduction), some of which are practically invisible to the naked eye) and perfect forms (sexual) which may themselves be more or less visible. This guide illustrates only a few examples of perfect forms. Quite a large number of families have been created on the basis of their life style, the morphology and structure of the ascomata or fruit bodies (texture, anatomy of the outer surface of the apothecia) and other characters.

OPERCULATE DISCOMYCETES

Species in this group, which occupies a halfway position between microfungi and macrofungi, are usually considered to be more evolved than the inoperculate species, perhaps because imperfect (asexual) forms are much less frequent. They have asci with an apical lid, which enables the ripe spores to disperse more easily. This guide illustrates only a few examples of certain families, since this group cannot be studied properly without a microscope. They are defined by their lifestyle, the morphology and structure of the ascomata or fruit bodies (texture, anatomy of the outer surface of the apothecia) and by other characters.

TUBERALES

It is merely for convenience that this traditional group is retained in the book since it derives from operculate Discomycetes that have adapted to life underground. *Geopora clausa* (p.280) forms one evolutionary link that can be summed up in the following way: the apothecia develops into a closed sphere and folds develop on the inside of this closed ascoma or fruit body until a densely tortuous, brain-like internal structure is formed (the hymenium then lines the veins of the gleba formed in this way or may even disappear completely as the folds break up); an underground way of life is adopted when ascomata are produced below the surface of the soil. The spores are dispersed by the action of small burrowing mammals or by simple disintegration of the envelope (peridium). For this reason, the function and morphology of the asci are profoundly altered: they become elliptic or spherical and lose the capacity to open by an apical operculus. This group includes the truffles, which are highly prized edible species, but also others which are uninteresting to eat or sometimes even unsuitable for consumption. There is no danger of poisoning from a member of this group but there are also frauds. To check that the product sold is really what it claims to be, it should be examined under the microscope as the spores are usually diagnostic for each species.

---

p.42 **Tar Spot**
Rhytisma acerinum (Pers.:Fr.) Fr.
Rhytismataceae, Rhytismatales

Description: *Black spots on leaves of maples.*
Apothecia elliptic, on a more or less circular stroma, black, up to 2 cm (0.8 in) diameter, forming a blister-like spot on the upper side of the leaves. *Microscopic features* Sp 60–80 × 1.5–2.5 µm, filiform or slightly clavate at one end, hyaline, bunched together in the asci. Asci clavate, 100–130 × 8–10 µm, not reacting with iodine. Paraphyses curved at the tip, sometimes forked.

Season: Spring (conidial stage usually from July onwards).
Habitat: On sycamore leaves (rarely on field maple) that fell the year before.
Distribution: Throughout Europe. Very common.

The imperfect stage, *Melasmia acerina* Lév., is best known. It can easily be seen from late summer onwards on leaves still on the tree.

---

p.45 *Arachnopeziza aurata* Fuckel
Hyaloscyphaceae, Leotiales

Description: *Minute white to pale yellow cup fungus, in groups on a cobwebby subiculum.*
Apothecia up to 0.7 mm (0.03 in) diameter, cup-shaped, white to pale yellow, ciliate, with marginal hairs, resting on a concolorous cobwebby subiculum. *Microscopic features* Sp 50–75 × 1.5–3 µm, narrowly cylindric, hyaline, pluriseptate, parallel in the asci. Asci cylindro-clavate, 90–100 × 10–12 µm, the apex colouring blue in iodine. Paraphyses forked. Excipulum hairs more or less encrusted.
Season: Usually spring and autumn.

Habitat: Damp rotten wood of broad-leaved trees, especially birch.
Distribution: Throughout Europe. Common.

A minute species, easy to recognize to genus.

---

p.45 *Eriopeziza caesia* (Pers.:Fr.) Rehm
Hyaloscyphaceae, Leotiales

Description: *Minute bluish-grey cup fungus, in dense groups on a cobwebby white subiculum.*
Apothecia up to 0.6 mm (0.02 in) diameter, slightly cup-shaped, lead-grey with brownish or olivaceous tones, sometimes more bluish, the margin fringed with white hairs, resting on a very dense, whitish subiculum. *Microscopic features* Sp 5–6 × 1–2 µm, narrowly fusiform, hyaline, in two irregular rows in the ascus. Asci cylindro-clavate, 30–38 × 4–5 µm, the apex colours blue in iodine. Hairs filiform, encrusted.
Season: Usually late autumn and winter.
Habitat: Decorticated, rotten wood, especially oak.
Distribution: Throughout Europe. Very common.

Particularly easy to confuse with certain 'Cyphellaceae' (especially in the genus *Lachnella*) in the Basidiomycotina with a very reduced sporophore.

---

p.45 *Lachnellula subtilissima* (Cooke) Dennis
Hyaloscyphaceae, Leotiales

Description: *Small orange cup-fungus with white marginal hairs.*
Apothecia up to 5 mm (0.2 in) diameter, cup-shaped then spreading, shortly stalked, bright orange; margin and excipulum with short, white, dense hair, forming a scurfy felty coating. *Microscopic features* Sp 6–10 × 2–2.5 µm, fusiform to narrowly elliptic, hyaline, in one or two rows in the asci. Asci 45–50 × 4–5 µm, not reacting with iodine. Paraphyses up to 60 µm, septate. Hairs on the outside cylindric, narrow, hyaline, finely encrusted, thick-walled.

Season: All year round.
Habitat: Conifers, especially pine and spruce.
Distribution: Especially in mountainous districts.

There are several species, all on conifers.

---

p.45 *Capitotricha bicolor* var. *rubi* (Bres.) Courtec.
Hyaloscyphaceae, Leotiales

Description: *Small orange cup-fungus with white hairs on the margin.*
Apothecia up to 2 mm (0.1 in) diameter, top or cup-shaped, orange. Margin and excipulum closely covered with fairly short, white hairs, giving a slightly tomentose, frosted appearance. *Microscopic features* Sp $6-8 \times 1.5-2$ μm, fusiform, hyaline, in two irregular rows in the asci. Asci $50-60 \times 4-5$ μm, pointed at the top, with a pore plug, colouring blue in iodine. Paraphyses projecting beyond the asci, lanceolate. Hairs cylindric, hyaline, septate, with fine granular encrustation, apex with a tuft of crystals.
Season: March–July.
Habitat: Dead raspberry canes.
Distribution: Widespread. Probably overlooked.

The type species, *C. bicolor* (Bull.:Fr.) Baral, which is very similar (larger spores), occurs on small branches of broad-leaved trees, especially oak. The genus *Capitotricha* is one of many close to *Lachnum*, the current name for the genus *Dasyscyphus*.

---

p.45 *Incrucipulum ciliare* (Schrad.:Fr.) Baral
Hyaloscyphaceae, Leotiales

Description: *Tiny whitish cup-fungus with white hairs on the margin.*
Apothecia up to 1 mm (0.04 in) diameter, stalked, discoid or slightly cup-shaped, whitish, the margin edged with stiff whitish hairs. *Microscopic features* Sp $14-20 \times 2-3$ μm, fusiform or slightly curved, hyaline, in two irregular rows in the ascus. Asci pointed at the top, with a pore plug, colouring blue in iodine.

Paraphyses projecting beyond the asci, lanceolate, sometimes branched. Hairs cylindric, tapering towards the tip, thick-walled, septate, with apical crystals.

Season: August–October.
Habitat: Dead oak leaves.
Distribution: Not known.

A tiny species, indistinguishable from a few related species which share the same specialized habitat.

---

p.45 *Lachnum virgineum* (Batsch:Fr.) P. Karsten
Hyaloscyphaceae, Leotiales

Description: *Small stalked white cup-fungus, fringed with white hairs.*
Apothecia up to 1 mm (0.04 in) diameter, often distinctly stalked, hymenium cup-shaped, white to ochraceous cream, margin and excipulum covered with pure white hairs, which hold droplets of dew.
Microscopic features Sp 6–10 × 1.5–2 µm, fusiform or slightly clavate, hyaline. Asci 45–55 × 3.5–5 µm, blunt-tipped, with a pore plug, colouring blue in iodine. Paraphyses projecting beyond the asci, fusoid, pointed, septate. Hairs thickened at the tip, blunt, hyaline, thin-walled, finely granular, septate.

Season: Mainly in the spring.
Habitat: Gregarious, on twigs, branches, stumps, etc.
Distribution: Throughout Europe. Very common.

Other white *Hyaloscyphaceae* are very similar. *L. virgineum* is the most common, before *Dasyscyphella nivea* (Hedw.:Fr.) Raitviir (terminal part of the hairs smooth and more inflated), on wood of oak, sometimes beech, often in interstices of the bark or dark cavities.

---

p.42 *Callorina fusarioides* (Berk.) Korf
Dermateaceae, Leotiales

Description: *Apothecia small, bright orange, on dead nettles.*
Apothecia up to 1 mm (0.04 in) diameter,

often elongated or discoid, sessile and erumpent, waxy or fleshy, reddish orange. Margin plain. *Microscopic features* Sp 9–15 × 3–4 μm, cylindro-elliptic, hyaline, uniseptate, in two rows in the ascus. Asci 70–95 × 7–10 μm, cylindro-clavate, no reaction with iodine. Paraphyses very slender, forked, the tip capitate, up to 4 μm.

Season: February–May.
Habitat: On dead nettle stems (*Urtica dioica*).
Distribution: Very common.

Often occurs with *Leptosphaeria acuta* (p.248), and frequently in its very similar, imperfect stage (*Cylindrocolla urticae* (Pers.) Bonorden)).

---

p.42 *Orbilia xanthostigma* (Fr.:Fr.) Fr.
Orbiliaceae, Leotiales

Description: *Apothecia tiny, orange, translucent.*
Apothecia up to 1 mm (0.04 in) diameter, circular, translucent, golden-yellow to orange, with a very short central stalk. Margin very thin. *Microscopic features* Sp 3–4 × 1–1.5 μm, strongly curved, hyaline, in two irregular rows in the ascus. Asci 30–40 × 4 μm, clavate, no reaction with iodine. Paraphyses projecting beyond the asci, the tip distinctly inflated capitate, up to 4 μm.

Season: All year round.
Habitat: On damp wood and bark of broad-leaved trees.
Distribution: Throughout Europe.

The genus is easy to recognize by its translucent appearance and often bright colours (but there are many species).

---

p.45 *Mollisia cinerea* (Batsch:Fr.) P. Karsten
Dermateaceae, Leotiales

Description: *Apothecia very small, blue-grey or blue, with a slightly thick margin.*
Apothecia up to 2–3 mm (0.08–0.1 in)

diameter, cup-shaped then flattened, sometimes with wavy lobes, bluish grey with a slightly thick and often paler or greyish margin. *Microscopic features* Sp 7–10 × 2–2.5 µm, cylindric, more or less curved, hyaline. Asci 55–70 × 5–6 µm, apical pore colours blue in iodine. Paraphyses cylindric, blunt-tipped. Margin and excipulum formed by dark globular cells.

Season: All year round.
Habitat: Damp wood and bark of broad-leaved trees.
Distribution: Throughout Europe. Very common.

Easy to recognize as a genus but the numerous species are difficult to separate. They are distinguished by the host (herbaceous or woody) and microscopic details.

---

p.42 *Bisporella citrina* (Batsch:Fr.) Korf & Carpenter
Leotiaceae, Leotiales

Description: *Apothecia small, gregarious, with a short slender stalk, bright yellow.*
Apothecia up to 3 mm (0.1 inch) diameter, cup-shaped or flat, with a distinct very slender stalk, uniformly egg-yellow to pale orange-yellow. Margin glabrous and excipulum sometimes a little paler.
Microscopic features Sp 8–13 × 3–4 µm, elliptic, hyaline, with oil drops, uniseptate when mature, in two irregular rows in the ascus. Asci 100–135 × 7–10 µm, the apex slightly pointed, apical pore plug colours blue in iodine. Paraphyses slightly thickened above, with numerous yellow oil drops.

Season: July–November.
Habitat: Forming large troops on often decorticated wood of various broad-leaved trees; usually on beech.
Distribution: Throughout Europe. Very common.

This species can sometimes be confused with *Hymenoscyphus umbilicatus* (Le Gal)

Dumont, which is top-shaped (like a truncated cone) with a small umbilicus in the centre, has a thick stalk, an unpleasant bitter taste, and can grow bigger. *B. sulfurina* (Quélet) Carpenter, which is slightly smaller, bright sulphur-yellow, and often found on old Pyrenomycetes (black traces on the wood), is equally well-known.

---

p.48 *Cudoniella acicularis* (Bull.:Fr.) Schroeter
Leotiaceae, Leotiales

Description: *Fruit bodies gregarious, small, white, on long stalks, like pinheads.*
Apothecia up to 4 mm (0.15 in) diameter, cup-shaped, stalked (stalk up to 2 cm (0.8 in)), sometimes more grey to almost blackish at the base of the stalk.
Microscopic features Sp 15–20 × 4–5 µm, fusiform, sometimes slightly asymmetrical, hyaline, sometimes uniseptate at maturity, in two irregular rows in the asci. Asci 100–120 × 10–13 µm, cylindro-clavate, the apex slightly pointed, no reaction with iodine. Paraphyses slightly thickened above.
Season: August–December.
Habitat: Gregarious, on stumps and trunks of broad-leaved trees, especially of oak.
Distribution: Throughout Europe. Common.

C. clavus (Alb. & Schw.:Fr.) Dennis, found in damp sites, is larger, up to 1.5 cm (0.6 in); the stipe is thick or like a truncated cone.

p.50 **Green Wood Cup**
Chlorociboria aeruginascens (Nylander) Kanouse ex Ramamurthi, Korf & Batra
Leotiaceae, Leotiales

Description: *Cup blue-green, stalked. Mycelium staining the wood the same colour.*
Apothecia up to 1 cm (0.4 in) diameter, cup-shaped but soon flat, circular or irregularly lobed to wavy, stalked, intensely blue-green. *Microscopic features* Sp 6–10 × 1.5–2 µm, narrowly fusiform, hyaline, with oil drops, in one or two irregular rows in the ascus. Asci 60–70 × 5 µm, cylindro-clavate, apical spore colours blue in iodine. Paraphyses cylindric, branched at the base.
Season: April–December.
Habitat: Gregarious, on wood of broad-leaved trees.
Distribution: Throughout Europe. Very common.

The wood, which is characteristically stained blue-green by the mycelium, has been used in marquetry. *C. aeruginosum* (Pers.:Fr.) Seaver ex Ram., Korf & Batra, with much larger spores (9–15 × 1.5–2.5 µm), may also be found, but much more rarely.

---

p.43 *Ciboria batschiana* (Zopf) Buchwald
Sclerotiniaceae, Leotiales

Description: *Cup rather small, brown, stalked.*
Apothecia up to 1.5 cm (0.6 in) diameter, stalked (stalk up to 1 cm (0.4 in)), cup-shaped, rather uniform brown, rather leathery. *Microscopic features* Sp 6–10 × 4–6 µm, elliptic, hyaline. Asci 130–150 × 6–8 µm, cylindro-clavate, apical pore colours blue in iodine.
Season: August–November.
Habitat: Often in a group, on fallen acorns.
Distribution: Uneven. Rather rare.

The *Sclerotiniaceae* usually occur on sclerotia (very hard, perennating forms, often dark, visible or immersed in the host) or on mummified fruits. They are

266 Discomycetes

often strictly host specific. This is a difficult to identify family, in which the anamorphs (asexual forms) sometimes have economic implications: *Monilia* (anamorphs of the genus *Monilinia*) attack certain fruits (apples) and may ruin the harvest; *Botrytis* (anamorphs of the genus *Botryotinia*), on grapes, form 'noble rot', essential for some very distinctive wines such as Sauternes.

---

p.43 *Poculum firmum* (Pers.:Fr.) Dumont
(= *Rutstroemia f.*)
Sclerotiniaceae, Leotiales

Description: *Cup rather small, brown with a short stalk.* Apothecia up to 1.5 cm (0.6 in) diameter, stalked (stalk up to 5 mm (0.2 in)), cup-shaped then flat, warm brown with yellowish or reddish tones, the hymenium sometimes olivaceous, rather leathery. Margin thin, irregular or easily torn. *Microscopic features* Sp 13–19 × 3.5–6 µm, cylindro-elliptic or asymmetrical, with one side flatter, initially with oil drops, 1–5 septate at maturity, budding small globular secondary spores at the apex, hyaline. Asci 130–150 × 8–12 µm, the apical pore colours blue in iodine. Paraphyses septate, forked.
Season: April–December.
Habitat: Gregarious, on fallen branches of broad-leaved trees, especially oak.
Distribution: Widespread, but rarely abundant.

Apothecia arising from a black stromatic base, buried in the wood.

---

p.44 *Lanzia luteovirescens* (Rob.) Dum & Korf
(= *Rutstroemia l.*)
Sclerotiniaceae, Leotiales

Description: *Cup small, olive yellow with a long stalk.* Apothecia up to 4 mm (0.16 in) diameter, with a long stalk (stalk up to 1 cm (0.4 in)), cup-shaped then discoid, olive yellow, rather leathery. *Microscopic features*

Sp 12–15 × 5–7 µm, elliptic subfusiform, hyaline, often with two oil drops. Asci 130–150 × 10–12 µm, apical pore colours blue in iodine. Paraphyses slightly clavate at the tip.

Season: August–November.
Habitat: In small groups on the petioles of sycamore leaves.
Distribution: Widespread.

The apothecia arise from black stromatic patches, visible on the petioles. Other species occur on leaves of alder (*Ciboria conformata* (P. Karsten) Svrcek); oak and beech (*Poculum petiolorum* (Rob.) Dumont); oak (*P. sydowianum* (Rehm) Dumont).

---

p.44 *Rutstroemia maritima* (Rob.) Dennis
Sclerotiniaceae, Leotiales

Description: *Apothecia very small, top-shaped, pink.*
Apothecia up to 2 mm (0.08 in) diameter, top-shaped or like a truncated cone, stalk not very distinct, hymenium cup-shaped then discoid or even convex, pinkish brown or pink, rather tough. *Microscopic features* Sp 12–15 × 4–5 µm, elliptic subfusiform, hyaline, initially with two oil drops. Asci 140–170 × 8–12 µm, apical pore colours blue in iodine.

Season: August–November.
Habitat: Coastal, on the remains of grasses and sedges and also on rabbit droppings.
Distribution: Little known on account of its habitat and small size. Has been recorded in France and UK.

---

p.74 *Spathularia neesii* Bresadola
Geoglossaceae, Leotiales

Description: *Frb an upright, spathulate, pale yellow tongue, with a sterile stalk tapering at the base, somewhat rufous.*
Apothecia not discoid, up to 8 cm (3.2 in) tall and 2 cm (0.8 in) diameter, shaped like a spatula, tongue or fan, the fertile head

pale yellow or ochraceous yellow, often wrinkled. Sterile base rufous brown.
Microscopic features Sp 50–80 × 2–2.5 µm, long, cylindro-fusoid, hyaline, with fairly numerous oil drops, and with many septa at maturity, lying parallel in the ascus. Asci 120–150 × 10–14 µm, clavate, no reaction with iodine. Paraphyses forked, inrolled or twisting at the tip.

Season: August–October.
Habitat: Under spruce. Usually lower slopes of mountains or continental.
Distribution: Quite widespread.

Of the related species, *S. flavida* Pers.:Fr. is the easiest to distinguish being up to 10 cm (4 in) tall with colours quite bright and clear (yellow head and white stalk) and having shorter spores; it is found particularly under larch. *S. alpestris* (Rehm) E. Rahm is smaller, all one colour, yellowish ochraceous cream and found under spruce and fir. *S. rufa* Sw. is wholly rufous brown and *S. pectinata* E. Rahm, has small spores up to 35 µm, an ochraceous head and dark red-brown stalk and is found under spruce.

---

p.43 *Neobulgaria pura* (Pers.:Fr.) Petrak
Leotiaceae, Leotiales

Description: *Apothecia small to medium, top-shaped, translucent, violaceous pink.*
Apothecia up to 2 cm (0.8 in) diameter, top-shaped, the hymenium flat or slightly depressed, a pretty violaceous pink, more or less translucent and gelatinous. *Microscopic features* Sp 6–9 × 3–4.5 µm, elliptic, hyaline. Asci 70–90 × 5–9 µm, the apical pore colours blue in iodine.
Season: July–December.
Habitat: In crowded clusters, erumpent on fallen trunks of broad-leaved trees, especially beech.
Distribution: Throughout Europe. Common.

The variety *foliacea* (Bres.) Dennis & Gamundi, with numerous confluent apothecia, forms a brain-like or leafy mass (up to 5 cm (2 in) diameter). *Ascotremella faginea* (Peck) Seaver (on beech) is similar but has striate spores.

---

p.43 **Black Bulgar**
Bulgaria inquinans (Pers.:Fr.) Fr.
Leotiaceae, Leotiales

Description: *Apothecia soft, top-shaped, brown to black.* Apothecia fleshy, up to 5 cm (2 in) diameter, hemispherical then like a truncated cone, finally flatter, the margin and outer surface brown and scurfy and the hymenium black. *Microscopic features* Sp dimorphic, the four upper ones black or almost so, 10–16 × 6–7 µm, elliptic or asymmetrical, with a flatter side, the four lower ones hyaline, smaller and similarly shaped. Asci cylindro-clavate, 120–180 × 8–10 µm. Paraphyses forked near the base, the tip wavy, brown.
Season: September–April.
Habitat: Bark of fallen oak, sometimes also sweet chestnut, birch, hornbeam and elm.
Distribution: Very common throughout Europe.

Very easy to recognize (especially as the spores stain the fingers, which does not happen in similar species, such as *Exidia truncata* Fr.:Fr., Basidiomycete).

---

p.49 **Jelly Babies**
Leotia lubrica (Scop.:Fr.) Pers.
Leotiaceae, Leotiales

Description: *Frb stalked, the head almost globular, gelatinous, orange-yellow or green.* Apothecia not discoid, up to 6 cm (2.4 in) tall, the head (up to 1.5 cm (0.6 in) diameter) subglobular, orange-yellow, yellowish buff, greenish or dark olive and the stalk cylindric or tapering, orange-yellow to ochraceous yellow, sometimes dotted with olivaceous to dark-green granules. Consistency gelatinous. *Microscopic features* Sp 20–25 × 5–6 µm, fusiform to cylindro-fusiform or slightly curved, hyaline, 5–7-septate at maturity, in one (sometimes two) rows in the ascus. Asci 130–160 × 10–12 µm, blunt-tipped, no reaction with iodine. Paraphyses septate, branched, the tip slightly thickened.

Season: July–November.
Habitat: Gregarious to tufted, on the ground (especially in damp places) under broad-leaved trees.
Distribution: Throughout Europe. Common.

Forms with dark-green head and stalk are sometimes called *L. atrovirens* Pers.

---

p.49 *Heyderia cucullata* (Batsch:Fr.) Boudier
(= *H. abietis*)
Leotiaceae, Leotiales

Description: *Frb very small, with a cylindric head and very slender stalk.*
Apothecia not discoid, up to 3 cm (1.2 in) tall, the head distinctly separated from the stalk, cylindric or long cone-shaped, up to 7 mm (0.3 in) long, yellowish buff to yellowish brown, the stalk filiform, concolorous or darker below. Consistency slightly gelatinous. *Microscopic features* Sp 10–15 × 2–3 µm, cylindro-fusoid or slightly curved, hyaline, with oil drops, not septate, in one or two rows in the ascus. Asci 60–80 × 5–7 µm, the apical pore colours blue in iodine. Paraphyses septate, slightly thickened at the tip (up to 4 µm), slightly brownish.

Season: August–November.
Habitat: On fallen conifer needles.
Distribution: Not fully understood.

---

p.49 *Mitrula paludosa* (Fr.:Fr.)
Geoglossaceae, Leotiales

Description: *Frb small to medium with a long bright orange head and white stalk.*
Apothecia not discoid, up to 5 cm (2 in) tall, the head elongated, cylindro-elliptic or fusoid, bright orange, viscid, the stipe

cylindric, pure white or slightly greyish, often translucent. Consistency gelatinous. *Microscopic features* Sp 10–15 × 2.5–3 µm, cylindric or elliptic, blunt-tipped, rarely uniseptate at maturity, hyaline, in two irregular rows in the ascus. Asci 100–150 × 8–9 µm, the apical pore blue in iodine.

Season: March–October.
Habitat: Plant remains in wet places, ditches, slow-moving streams.
Distribution: Throughout Europe. Not very common, although sometimes locally abundant.

---

p.49 *Bryoglossum gracile* (P. Karsten) Redhead
Geoglossaceae, Leotiales

Description: *Frb small, with an orange head and whitish stipe.*
Apothecia not discoid, up to 2 cm (0.8 in) tall, the head small, almost globular or shortly ovoid to elliptic, orange-yellow, the stipe cylindric, whitish, slightly scurfy. *Microscopic features* Sp 11–12 × 3–5 µm, cylindro-elliptic or slightly curved, hyaline, faintly punctate and uniseptate when ripe. Asci 70–90 × 5–6 µm, the apical pore colours blue in iodine. Paraphyses forked, septate.

Season: July–September.
Habitat: Very wet moss in marshy ground.
Distribution: A rare species of mountain regions.

The habitat is specific.

---

p.74 *Microglossum viride* (Schrad.:Fr.) Gillet
Geoglossaceae, Leotiales

Description: *Frb club-shaped, green.*
Apothecia not discoid, up to 6 cm (2.4 in) tall, like a narrow club, head more or less distinct from the slightly tapering or grooved stalk. Uniformly olive green. Stalk slightly scurfy. *Microscopic features* Sp 15–20 × 5–6 µm, cylindro-fusoid to slightly curved or sinuous, hyaline, with oil drops then 1–3-septate at maturity, in one or two rows in the ascus. Asci 120–160 × 8–10 µm, the apical pore colours blue in iodine.

Paraphyses forked below, slightly capitate.
Season: August–November.
Habitat: On the ground or in moss, sometimes on debris, in damp broad-leaved woodland.
Distribution: Not fully understood. Always rather rare.

M. fuscorubens Boudier (very rare) has reddish tones. Among the clavate to spathulate *Geoglossaceae* (lacking a distinct globular head), the *Microglossum* spp. are distinguished by their white spores.

---

p.74 **Earth Tongue**
Geoglossum cookeianum Nannfeldt
Geoglossaceae, Leotiales

Description: *Frb spathulate or tongue-shaped, black.* Apothecia not discoid, up to 10 cm (4 in) tall, black, matt, with a slender, scurfy or smooth stalk, topped by a rather wider, smoother head, like a narrow club, or flattened and grooved. *Microscopic features* Sp 50–90 × 5–7 µm, fusiform, curved, with tapering blunt ends, brown, 7-septate at maturity, parallel in the ascus. Asci 150–180 × 15–20 µm, cylindro-clavate, the apical pore colours blue in iodine. Paraphyses numerous, brown, septate, the terminal elements very short or subglobular.
Season: September–December.
Habitat: Mossy lawns, especially on sandy soils.
Distribution: Atlantic in distribution (more frequent in coastal turf).

G. cookeianum is the commonest earth tongue, at least in temperate and coastal areas. The many other black species cannot be identified without a microscope. *Trichoglossum hirsutum* (Pers.:Fr.) Boud. is similar but has erect, finely velvety hairs, (usually in sphagnum).

---

p.42 *Pyronema omphalodes* (Bull.:Fr.) Fuckel
Pyronemataceae, Pezizales

Description: *Apothecia very small, tender pink to salmon, in confluent clusters.*

Apothecia up to 1 mm (0.04 in) diameter, cushion-shaped, gregarious, confluent, making large mounds on a white felty mat of mycelium. *Microscopic features* Sp 10–15 × 6.5–8.5 µm, elliptic, hyaline. Asci 120–150 × 12–15 µm, no reaction with iodine. Paraphyses slightly inflated at the tip, up to 4 µm.

Season: All year round.
Habitat: Bonfire sites and burned areas.
Distribution: Widespread. Common.

P. domesticum (Sow.:Fr.) Sacc. has a less developed subiculum (spores and asci larger).

---

p.44 *Iodophanus carneus* (Pers.:Fr.) Korf
Ascobolaceae, Pezizales

Description: *Apothecia tiny, flesh pink to salmon, top-shaped.* Apothecia up to 1.5 mm (0.06 in) diameter, top-shaped then disc-like, flesh pink to salmon, the hymenium smooth then punctate. Margin slightly frosted. *Microscopic features* Sp 18–25 × 10–15 µm, elliptic, hyaline, very finely punctate at maturity, in one or two irregular rows in the ascus. Asci 200–230 × 25–30 µm, broadly clavate, entirely blue in iodine. Paraphyses septate, thickened at the tip, up to 10 µm.

Season: All year round.
Habitat: On dung (rabbit, etc.) but also rotting vegetation.
Distribution: Widespread. Common.

---

p.48 *Ascobolus furfuraceus* Pers.:Fr.
Ascobolaceae, Pezizales

Description: *Apothecia small, brown, with scurfy margin.* Apothecia up to 5 mm (0.2 in) diameter, top-shaped then discoid, yellow-brown then dotted black, the margin scurfy. *Microscopic features* Sp 20–30 × 10–14 µm, elliptic, hyaline then purple, finally purple-brown, with longitudinal anastomosing striations, in two rows or irregularly disposed in the ascus. Asci 180–250 × 25–30 µm, clavate, the wall colours blue in iodine. Paraphyses septate, sometimes forked, inflated at the tip which may be capitate, embedded in a greenish-yellow mucilage.

Season: All year round.
Habitat: On dung, sometimes rotting vegetation. In groups.
Distribution: Cosmopolitan. Common.

This genus, which has many species, has very striking spores. The ascomata are sometimes highly coloured (yellow, green).

---

p.44 *Saccobolus versicolor* (P. Karst.) P. Karst.
Ascobolaceae, Pezizales

Description: *Apothecia very small, rather top-shaped, dark violaceous brown.*
Apothecia small, up to 1 mm (0.04 in) diameter, cushion to lens-shaped, more or less dark purple, dotted when ripe. Margin distinct. *Microscopic features* Sp 13–22 × 6.5–10 µm, fusoid-elliptic, hyaline then violaceous to very dark violaceous crimson, arranged in the ascus in two rows of three spores on which lies one row of two oblique spores. Asci 100–145 × 22–35 µm, clavate, the wall colours blue in iodine. Paraphyses septate, forked, sometimes inflated at the tip where they are held in a violaceous mucilage.
Season: All year round.
Habitat: Dung, sometimes rotting plant remains.
Distribution: Cosmopolitan.

In this family, the ripe asci turn towards the light before shooting off their spores. The discharge is more likely to be successful because the spores have little chance of meeting any obstacles before being borne away by the wind.

---

p.52 *Marcelleina benkertii* J. Moravec
Otideaceae, Pezizales

Description: *Apothecia small, cup-shaped or flat, violaceous black species.*

Apothecia small, up to 1 cm (0.4 in) diameter, cup-shaped then flat, the margin thin, sometimes irregular or torn, violaceous black to almost black. Excipulum becoming paler as it dries.
Microscopic features Sp 9–11.5 µm, globose, hyaline. Asci 190–250 × 12–15 µm, no reaction with iodine. Paraphyses forked below, septate, the tip slightly inflated, up to 4.5 µm, curved or branched with a purplish intracellular pigmentation and embedded in brownish matter.

Season: September–November.
Habitat: Prefers damp places on the ground.
Distribution: Very rare. Recorded in Bohemia, Germany and France.

Black or very dark cup fungi are difficult to distinguish and fall into several genera.

---

p.49 *Octospora rutilans* (Fr.:Fr.) Denn. & Itz.
Otideaceae, Pezizales

Description: *Apothecia small, orange with a slightly fringed margin.*
Apothecia up to 1.5 cm (0.6 in) diameter, concave to flat, sometimes slightly top-shaped or slightly stalked, circular, reddish orange, the margin slightly felty or rough from the hyaline or whitish hairs.
Microscopic features Sp 22–25 × 13–15 µm, broadly elliptic, with oil drops, hyaline and ornamented with a colourless network. Asci 250–300 × 17–20 µm, no reaction with iodine. Paraphyses forked, septate, slightly inflated at the tip, with orange intracellular granules, turning green in iodine. Excipulum hairs hyaline to yellowish, thick-walled, septate, blunt-tipped.

Season: September–January.
Habitat: On sandy soils, near or in contact with moss.
Distribution: Widespread.

One of a complex of small Discomycetes more or less associated with moss.

p.46 *Anthracobia maurilabra* (Cooke) Boudier
Otideaceae, Pezizales

Description: *Apothecia small, gregarious, dull orange-yellow, with a scurfy margin.*
Apothecia up to 5 mm (0.2 in) diameter, cushion- then cup-shaped, often misshapen because they are so crowded, dull or often pale orange-yellow, the margin and excipulum slightly scurfy, brownish. *Microscopic features* Sp 19–22 × 8–9 µm, elliptic, hyaline, with two oil drops. Asci 170–200 × 15–20 µm, no reaction with iodine. Paraphyses septate, forked, the tip inflated up to 5 µm. Marginal hairs 50–130 × 7–20 µm, cylindric, brown, septate, slightly thick-walled, blunt-tipped.
Season: June–November.
Habitat: Bonfire sites and burned ground.
Distribution: Widespread (not fully understood).

The three main species (same habitat) can only be distinguished by looking at them under the microscope.

---

p.46 *Lasiobolus cuniculi* Velenovsky
Otideaceae, Pezizales

Description: *Apothecia tiny, top-shaped with long, stiff, hyaline hairs.*
Apothecia up to 0.7 mm (0.03 in) diameter and 0.4 mm (0.02 in) tall, top-shaped with a flat hymenium, yellowish to yellowish pink, edged with stiff erect hairs up to 0.5 mm (0.02 in). *Microscopic features* Sp 18–25 × 12–14 µm, elliptic, hyaline, in two irregular rows in the ascus. Asci 150–220 × 20–36 µm, broadly clavate, no reaction with iodine. Paraphyses septate, sometimes forked, slightly inflated and sometimes curved at the tip. Hairs 100–500 × 10–25 µm, straight, the base inflated and the point long tapering, stiff.
Season: All year round.
Habitat: Rabbit droppings (and other dung).
Distribution: Rather rare.

L. ciliatus (Schm.:Fr.) Boudier is the most widespread taxon: Apothecia up to 1 mm (0.04 in), asci 15–30 µm and hairs 20–42 µm.

---

p.46 *Coprobia granulata* (Bull.:Fr.) Boudier
Otideaceae, Pezizales

Description: *Apothecia very small, reddish orange, with a granular margin.*
Apothecia up to 2.5 mm (0.1 in) diameter, discoid, flat, yellowish orange then orange-red, the margin scurfy-granular. *Microscopic features* Sp 15–18 × 7–8 µm, elliptic, hyaline, lacking oil drops. Asci 170–190 × 10–15 µm, no reaction with iodine. Paraphyses septate, the tip capitate, up to 15 µm, with orange granules turning green in iodine. Margin and excipulum covered with large rounded cells, up to 100 µm diameter.
Season: April–December.
Habitat: Cow dung, in meadows.
Distribution: Widespread. Very common.

Various, rather similar, species of *Cheilymenia*, have hairs which can be seen with a hand lens (*see C. stercorea below*).

---

p.46 *Cheilymenia stercorea* (Wigg.:Fr.) Boud.
Otideaceae, Pezizales

Description: *Apothecia very small, orange, urn-shaped or discoid, with brownish hairs.*
Apothecia up to 4 mm (0.16 in) diameter, top-shaped then opening a little, orange, with a flat hymenium and sparse, rather stiff, brown hairs on the margin. *Microscopic features* Sp 17–20 × 8–10 µm, elliptic, hyaline. Asci 200–230 × 12–18 µm, no reaction with iodine. Paraphyses forked, septate, the tip slightly clavate, with internal orange granules, turning green in iodine. Hairs (margin and excipulum) 300–700 × 10–30 µm, brown, thick-walled (up to 3 µm), bristly and pointed, the base

forked. Hairs at the base of the apothecia branching or starry.
Season: All year round.
Habitat: Especially on cow dung.
Distribution: Widespread. Common.

Often growing together with the preceding species. Other *Cheilymenia* (bright yellow, orange yellow, orange) occur on various hosts (dung, bare ground, rotting vegetation).

---

p.47 *Sphaerosporella brunnea* (Alb. & Schw.:Fr.) Svrcek & Kubicka
Otideaceae, Pezizales

Description: *Frb small, rather rufous brown with a slightly scurfy margin.*
Apothecia up to 5 mm (0.2 in) diameter, shallow cup-shaped then almost flat, sometimes lobed, rather rufous brown, the margin thin, looking slightly scurfy, like the excipulum, from the tufts of very short hairs set in more or less radial lines. *Microscopic features* Sp globose, 13–15 µm, hyaline, with oil drops, in one row in the ascus. Asci 150–200 × 15–20 µm, blunt at the apex, no reaction with iodine. Paraphyses septate, slightly clavate at the tip up to 6 µm, with reddish brown contents. Marginal and excipulum hairs in tufts, 50–150 × 10–12 µm, pointed, 1–2-septate, brown, thick-walled.
Season: July–November.
Habitat: Bonfire sites, sometimes in short mosses.
Distribution: Probably widespread like most of the bonfire species. Rather rare but easily overlooked.

S. hinnulea (Berk. & Br.) Rifai is a pale species on damp ground.

---

p.47 *Humaria hemisphaerica* (Wigg.:Fr.) Fuck.
Otideaceae, Pezizales

Description: *Apothecia deeply cup-shaped, blue-grey with black-brown hairs.*

Apothecia up to 3 cm (1.2 in) diameter, initially closed and almost buried then hemispherical and finally wide open, the hymenium bluish grey, the margin and excipulum with dark brown hairs.
Microscopic features Sp 20–25 × 10–13 μm, elliptic, hyaline, with two oil drops, irregularly warted when ripe. Asci 250–300 × 17–22 μm, blunt-ended, no reaction with iodine. Paraphyses septate, with a long clavate tip, up to 7–8 μm, hyaline, not greening in iodine. Marginal and excipulum hairs 200–1000 × 15–20 μm, straight or wavy, drawn out to a more or less sharp point, brown, septate, thick-walled.

Season: July–November.
Habitat: Bare ground on banks, the sides of ditches, usually on light or sandy soils, in cool, damp places. Sometimes found on very rotten wood.
Distribution: Widespread. Common or very common.

Easy to recognize by its large size, but small forms may resemble other species.

---

p.47 *Geopora arenicola* (Lév.) Kers
Otideaceae, Pezizales

Description: *Cup medium-sized, very deep, cream, the margin and excipulum covered with brown hairs.*
Apothecia up to 3 cm (1.2 in) diameter, urn-shaped, half buried, then opening cup-shaped, the margin sometimes starry. Hymenium blue-grey to cream. Excipulum and margin with short brown hairs.
Microscopic features Sp 20–30 × 10–16 μm, elliptic, hyaline, with oil drops. Asci 200–b250 × 15–20 μm, no reaction with iodine. Paraphyses septate, slightly thickened at the tip, up to 6 μm. Marginal and excipulum hairs 50–300 × 6–10 μm, wavy, brown, septate, thick-walled, blunt-tipped, sometimes branched.

Season: All year round.
Habitat: Sandy or stony substrates, dunes, banks of rivers and ponds.
Distribution: Widespread.

p.47 *Geopora sumneriana* (Cooke) De la Torre
Otideaceae, Pezizales

Description: *Cup large, initially half buried, bluish cream. Excipulum with long brown hairs.*
Apothecia up to 7 cm (2.8 in) diameter, bladder-like and subterranean then emerging and opening in a star-like manner, revealing the light ochraceous grey or pale-cream hymenium. Margin and excipulum set with long brown hairs. *Microscopic features* Sp 30–35 × 14–15 µm, elliptic, hyaline, with oil drops. Asci 330–380 × 16–20 µm, no reaction with iodine. Paraphyses septate, forked below, slightly clavate at the tip. Marginal and excipulum hairs 200–2000 × 10–15 µm, wavy, branched, brown, septate, thick-walled and often encrusted with crystals.
Season: January–May.
Habitat: Under cedar, more rarely under yew.
Distribution: Widespread. Usually found in warm situations.

Very easy to recognize (size, habitat and season).

---

p.244 *Geopora clausa* (Tulasne) Burdsall
Otideaceae, Pezizales

Description: *Frb globular or irregularly shaped, hollow, buff. Hypogeous or semihypogeous.*
Apothecia a closed sphere, up to 3 cm (1.2 in) diameter, often deformed to brain-like, light-buff to brownish, finely granular scurfy. The inside white then somewhat rufous, composed of a cavity often containing several irregular chambers. *Microscopic features* Sp 15–25 × 13–19 µm, elliptic, hyaline. Asci 200–300 × 18–24 µm, no reaction with iodine. Paraphyses septate, the sections sometimes slightly inflated. Excipulum hairs of varying length, brown, septate, sometimes branched, thick-walled.
Season: August–November.
Habitat: Sandy soil.
Distribution: Requires further study. A rare species,

occurring in warm sites or in Atlantic areas, recorded from Spain, France and Italy.

This species forms a link between the cup fungi (cup-shaped, epigeous) and the truffles (hypogeous, spherical, with the hymenium arranged on internal veins).

---

p.46 *Scutellinia kerguelensis* (Berk.) O. Kuntze
Otideaceae, Pezizales

Description: *Cup medium-sized, bright orange-red, fringed with quite long brown hairs.*
Apothecia up to 1.5 cm (0.6 in) diameter, sessile, discoid, rather bright orange-red, the margin and excipulum covered with rather short black-brown hairs, up to 0.5 mm (0.02 in). *Microscopic features* Sp 22–28 × 14–19 µm, broadly elliptic, hyaline, finely warted when ripe. Asci 250–320 × 18–26 µm, blunt, no reaction with iodine. Paraphyses septate, the terminal cell inflated up to 14 µm, with internal orange granules, turning green in iodine. Hairs (margin and excipulum) 100–450 × 10–30 µm, brown, septate, thick-walled, slightly wavy, the base tapering and rooting, not or very little forked.
Season: June–December.
Habitat: Gregarious. Very damp places, on the ground or on rotting vegetation.
Distribution: Widespread. Fairly common, usually in arctic and alpine areas.

It depends on the region as to which *Scutellinia* is most common. In the field, it is customary to call most of them *S. scutellata*. In fact, in the lowlands of western and Atlantic Europe, *S. crinita* (Bull.:Fr.) Lambotte appears to be the most frequent.

---

p.46 *Scutellinia setosa* (Nees:Fr.) O. Kuntze
Otideaceae, Pezizales

Description: *Cup small, brownish orange, the margin fringed with black hairs.*

Apothecia up to 5 mm (0.2 in) diameter, cup-shaped then discoid, reddish to dull brownish-orange, the margin and excipulum covered with black-brown or paler hairs, up to 1 mm (0.04 in) long at the margin. *Microscopic features* Sp 18–20 × 10–12.5 μm, elliptic, hyaline, almost smooth to very finely warted. Asci 200–230 × 11–15 μm, blunt, no reaction with iodine. Paraphyses septate, the terminal cell slightly inflated, up to 7 μm. Hairs (margin and excipulum) 450–900 × 15–30 μm, dark brown to almost hyaline, septate, thick-walled, the rooting base forking several times.

Season: June–November.
Habitat: On rotten wood and vegetation, in woods. Damp places.
Distribution: Recorded from Germany, France, Russia and former Czechoslovakia.

To identify species in this genus correctly, a microscope and good monograph are essential.

---

p.49 *Melastiza chateri* (W.G. Smith) Boudier
Otideaceae, Pezizales

Description: *Apothecia small, orange-red cup fungus, with a brownish margin.*
Apothecia up to 1.5 cm (0.6 in) diameter, cup-shaped or discoid, sometimes deformed where they are squashed together, orange-red, the margin brownish scurfy. *Microscopic features* Sp 17–20 × 9–10 μm, fusiform, hyaline, with a fairly fine reticulate ornamentation and often having a pointed appendage at each end. Asci 250–300 × 12–15 μm, no reaction with iodine. Paraphyses septate, the tip clavate, up to 10 μm, with orange granules, turning green in iodine. Excipulum hairs cylindric-wavy, 60–200 × 12–18 μm, brown, thick-walled, septate, blunt-tipped, in tufts.

Season: September–April.
Habitat: Rather damp sandy soil.
Distribution: Widespread. Sometimes quite rare.

Quite distinct (habitat and brownish scurfy margin).

---

p.48 **Orange Peel Fungus**
Aleuria aurantia (Pers.:Fr.) Fuckel
Otideaceae, Pezizales

Description: *Cup medium or rather large, bright orange thin-fleshed.*
Apothecia up to 10 cm (4 in) diameter, sessile, cup-shaped then spreading wide or lobed, bright orange, margin thin, wavy then slightly split, brittle. Excipulum paler, sometimes whitish. *Microscopic features* Sp 15–25 × 9–12 µm, elliptic, hyaline, having a very prominent wide, angularly meshed reticulum, and a pointed appendage at each end. Asci 180–220 × 10–13 µm, no reaction with iodine. Paraphyses septate, a little inflated at the tip, with orange granular contents, green in iodine. Margin and excipulum glabrous.
Season: Almost all year round.
Habitat: Bare ground in open places, ruts, banks, woodland edges, etc.
Distribution: Widespread. Common.

Very easy to recognize; this species can be quite large.

---

p.43 *Aleuria bicucullata* Boudier
Otideaceae, Pezizales

Description: *Apothecia small, egg-yellow, cup-shaped with a smooth margin.*
Apothecia up to 8 mm (0.3 in) diameter, cup-shaped then a slightly depressed disc, circular or deformed where many grow together, quite bright yellow, sometimes egg-yellow (slight orange tones). Margin and excipulum smooth. *Microscopic features* Sp 15–22 × 8–11 µm, elliptic, hyaline, ornamented with prominent, irregularly anastomosing ribs (including two diametrically opposite rings, when the spores are still in the ascus), and spectacular spines at the apex. Asci 170–210 × 8–12 µm,

blunt-tipped, no reaction with iodine. Paraphyses septate, slightly inflated and sometimes curved at the tip, up to 7 μm, with orange granular or refractive contents, greening in iodine. Margin and excipulum smooth, with globular, sometimes elongated cells just under the surface.

Season: June–October.
Habitat: Bare ground, woody debris or excrement.
Distribution: Recorded in Denmark, France, Norway and former Czechoslovakia.

This species, unlike the preceding one, cannot be determined without a microscope. There are quite a number of more or less yellow cup fungi occurring on the ground.

---

p.47 *Tazzetta catinus* (Holmskj.:Fr.) Korf & Rogers *Otideaceae*, Pezizales

Description: *Apothecia urn-shaped, buff, with a short stalk, the margin scurfy.*
Apothecia up to 4 cm (1.6 in) diameter, stalked (stalk up to 8 mm (0.3 in)), subglobular, then a hemispherical bowl, sometimes more open, ochraceous cream, the margin toothed, roughish, the outside paler and sometimes set with a brownish granular scurf. *Microscopic features* Sp 20–25 × 11–13 μm, elliptic, hyaline. Asci 280–350 × 15–20 μm, blunt, no reaction with iodine. Paraphyses septate, branched, the tip lobed, diverticulate, curved. Excipulum with large globular cells.

Season: June–November.
Habitat: On the ground, in woods, along paths, etc.
Distribution: Widespread. Common.

Easy to identify to generic level; the species are hard to separate.

---

p.50 **Hare's Ear**
Otidea onotica (Pers.:Fr.) Fuckel
Otideaceae, Pezizales

Description: *Cup large, inrolled like an ear, erect, orange-yellow.*

Apothecia up to 10 cm (4 in) tall, asymmetrical, with a broad cleft down one side and inrolled like a hare's ear, orange-yellow, sometimes with pinkish tones, finally often with rusty spots. Outer surface paler and matt. *Microscopic features* Sp 10–13 × 5–6 µm, elliptic, hyaline, with two oil drops. Asci 200–250 × 8–12 µm, blunt, no reaction with iodine. Paraphyses septate, slightly forked, the tip a little inflated and markedly hooked.

Season: September–November.
Habitat: Bare ground and humus, in woods, especially under oak.
Distribution: Widespread but sometimes quite rare.

The genus is quite easy to determine in the field, on account of the asymmetrical, split apothecia, inrolled on one side. The species are more difficult to identify without a microscope.

---

p.50 *Otidea bufonia* (Pers.:Fr.) Boudier
Otideaceae, Pezizales

Description: *Cup medium to large, dark brown, split and inrolled on one side.*
Apothecia up to 6 cm (2.4 in) tall and 4 cm (1.6 in) diameter, split down one side and inrolled, slightly asymmetrically cup-shaped, dark-brown to reddish-brown. Outer surface concolorous or slightly paler. *Microscopic features* Sp 13–15 × 6.5–7 µm, elliptic, hyaline, with two oil drops. Asci 160–200 × 10–12 µm, no reaction with iodine. Paraphyses multiseptate, forked, strongly curved at the tip.

Season: August–October.
Habitat: On the ground, in woods, sometimes in scrub.
Distribution: Appears to be widespread throughout Europe. Rarer than the preceding species.

This second *Otidea* is an example of the group of dark species (microscopic details are essential to separate them). There are some taxa whose apothecia are intermediate in colour.

p.52 *Peziza badia* Pers.:Fr.
Pezizaceae, Pezizales

Description: *Apothecia medium-sized, cup-shaped, dark brown with olivaceous tones.*
Apothecia up to 10 cm (4 in) diameter, quite deep cup-shaped then spreading, flesh rather thin, dark brown with olive tones, especially in the centre at maturity, the outer surface reddish brown, slightly scurfy at the margin. *Microscopic features* Sp 17–20 × 9–12 μm, elliptic, hyaline, with two oil drops, with a rumpled appearance. Asci 300–330 × 12–15 μm, blunt, blueing in iodine at the apex. Paraphyses septate, weakly clavate.
Season: June–November.
Habitat: On the ground, usually on sandy soils. Open places, tracks, woods.
Distribution: Widespread. Common.

This species is one of the commonest cup fungi, especially in temperate lowland districts, but there are many others with which it can be confused. The olivaceous colouring near the centre of the hymenium is, however, fairly characteristic.

---

p.52 *Peziza phyllogena* Cooke
Pezizaceae, Pezizales

Description: *Apothecia medium-sized, cup-shaped, dark reddish-brown, sometimes with olivaceous tones.*
Apothecia up to 10 cm (4 in) diameter, deeply cup-shaped then a little more expanded, the flesh rather thin, dark red-brown then reddish brown, sometimes with olivaceous tones near the centre, the outer surface slightly scurfy. *Microscopic features* Sp 17–21 × 8–10 μm, rather narrowly elliptic, hyaline, with two oil drops, ornamented with fine, sparsely distributed and very distinct, sometimes long, warts. Asci 280–320 × 10–13 μm, the wall coloured blue at the apex in iodine. Paraphyses septate, weakly inflated at the tip.
Season: July–November.
Habitat: On the ground, in shady broad-leaved woodland.

Distribution: Rarer than the preceding species. Also appears to be very widespread.

Close to *P. badia* and has been called *P. badioconfusa* Korf. There are other very similar brown cup fungi.

---

p.51 *Peziza petersii* Berk. & Curt.
Pezizaceae, Pezizales

Description: *Apothecia quite large, cup-shaped, violaceous brown.*
Apothecia up to 5 cm (2 in) diameter, cup-shaped then expanded, violaceous brown then pale ochraceous lilac to brownish, the flesh quite thin, the excipulum greyish to buff. *Microscopic features* Sp 10–12 × 5–6.5 µm, elliptic, with two oil drops, finely warted or punctate. Asci 160–200 × 8–10 µm, coloured blue at the apex in iodine. Paraphyses septate, slightly inflated and often curved above, up to 7 µm, with brownish granular contents.
Season: June–November.
Habitat: Burnt ground and bonfire sites.
Distribution: Widespread. Never very common.

Cup fungi with violet tones must be looked at under the microscope.

---

p.51 *Peziza gerardii* Cooke
Pezizaceae, Pezizales

Description: *Apothecia small, cup-shaped or discoid, violaceous.*
Apothecia up to 1.5 cm (0.6 in) diameter, shallow and soon flat, sometimes top-shaped, wholly lilac-purple, the margin slightly rough or finely toothed, the excipulum scurfy. *Microscopic features* Sp 25–30 × 8–10 µm, elliptic or fusiform, often with many oil drops, hyaline to somewhat pale crimson, in one or two rows in the ascus. Asci 250–300 × 15–20 µm, the apex coloured blue in iodine. Paraphyses septate, thickened slightly at the tip, up to 9 µm.

288 Discomycetes

Season: July–October.
Habitat: On the ground, in woods.
Distribution: Very rare.

---

p.51 *Peziza ampelina* Quélet
Pezizaceae, Pezizales

Description: *Apothecia rather medium-sized, cup-shaped, dark violaceous, the outer surface paler.*
Apothecia up to 5 cm (2 in) diameter, deeply cup-shaped then irregular, with a dark purplish hymenium and pale buff, scurfy excipulum. *Microscopic features* Sp 18–22 × 9–11 µm, elliptic, hyaline, with two main oil drops. Asci 250–320 × 12–15 µm, blunt, the wall coloured blue at the apex in iodine. Paraphyses septate, slightly clavate at the tip where there are also violaceous brown contents.
Season: March–June.
Habitat: On the ground, in woods and urban or artificial sites and woods, sometimes on bonfire sites.
Distribution: Rather rare.

Quite easy to recognize by its colour and occurrence in spring.

---

p.51 *Peziza cerea* Sow.:Fr.
Pezizaceae, Pezizales

Description: *Cup yellowish buff with a brittle, whitish, excipulum.*
Apothecia up to 10 cm (4 in) diameter, soon expanding and taking on the contours of the substrate, brittle, the hymenium light yellowish-buff, browner in the centre with age. Margin and excipulum very scurfy, whitish. Sometimes shortly stalked.
Microscopic features Sp 14–18 × 8–10 µm, elliptic, hyaline, smooth, lacking oil drops. Asci 320–350 × 12–16 µm, cylindric, blunt, the apex coloured blue in iodine. Paraphyses septate with no distinctive features.
Season: All year round.
Habitat: Places disturbed by Man, even inside houses.

Distribution: Widespread.

This taxon occurs in the same specialized habitat as *P. varia* (Hedw.:Fr.) Fr., light-buff to greyish-buff with a whitish, scurfy excipulum and brittle, stratified flesh. *P. vesiculosa* Bull.:Fr., a large, similarly coloured species, occurs on dung. It has urn-shaped apothecia when young and is distinctly scurfy, even rough at the margin.

---

p.50 *Peziza fimeti* (Fuckel) Seaver
Pezizaceae, Pezizales

Description: *Cup small, brownish buff, on dung.*
Apothecia up to 2 cm (0.8 in) diameter, cup-shaped then almost flat, brownish-buff to reddish. Margin thin or slightly rough. Excipulum slightly scurfy. *Microscopic features* Sp 18–22 × 9–12 µm, elliptic, hyaline. Asci 220–280 × 12–15 µm, blunt, the apex coloured blue in iodine. Paraphyses septate, sometimes a little constricted at the septa, the tip occasionally slightly inflated.
Season: September–November.
Habitat: Cow dung.
Distribution: Not fully understood.

There are many cup fungi occurring on dung.

---

p.44 *Peziza merdae* Donadini
Pezizaceae, Pezizales

Description: *Cup small, rather rufous brown.*
Apothecia up to 5 cm (2 in) diameter, weakly cup-shaped then spreading, adhering to the substrate, the hymenium light rufous brown. Margin thin and excipulum slightly scurfy. *Microscopic features* Sp 15–18 × 7–8.5 µm, elliptic or slightly fusiform, hyaline, lacking oil drops, very finely warted. Asci 210–240 × 10–13 µm, coloured blue in iodine. Paraphyses septate, the terminal cells inflated and hooked, up to 12 µm.

Season: January–February (winter).
Habitat: Human excrement.
Distribution: Recorded only in France.

---

p.51 *Peziza succosa* Berk.
Pezizaceae, Pezizales

Description: *Cup small to medium-sized, buff, exuding a juice which turns yellow.*
Apothecia up to 5 cm (2 in) diameter, cup-shaped then expanded, brownish buff to yellowish or hazel to olivaceous, the blunt margin slightly inrolled when young. Excipulum almost smooth, concolorous. The flesh exudes a clear juice, which yellows rapidly in air. *Microscopic features* Sp $17–22 \times 9–12$ μm, elliptic, hyaline, with two oil drops and very large warts, sometimes slightly elongated or anastomosing. Asci $300–350 \times 13–18$ μm, blunt, coloured blue in iodine. Paraphyses cylindric, septate, slightly inflated.
Season: June–November.
Habitat: On the ground, in woods.
Distribution: Widespread. Common.

The margin of unbroken apothecia may show some yellow coloration.

---

p.51 *Peziza ammophila* Durieu & Montagne
Pezizaceae, Pezizales

Description: *Cup deep with a long pseudorhiza covered in sand.*
Apothecia up to 5 cm (2 in) diameter, initially a hollow sphere, then opening by a pore on top, and finally forming a more or less spreading star, attached by a pseudorhiza enveloping grains of sand. Flesh very brittle. *Microscopic features* Sp $14–16 \times 9–10$ μm, elliptic, hyaline. Asci blunt, coloured blue in iodine. Paraphyses septate, slightly thickened at the tip up to 8 μm.
Season: September–December.
Habitat: Dunes, among marram and various other plants.

Distribution: Widespread, from the North Sea to the Atlantic and in a few sandy sites inland.

Avoid confusing with *P. pseudoammophila* Bon & Donadini, which has larger spores (16–18 × 9–10.5 µm) and is cup-shaped with heterogeneous flesh (fine darker median line).

---

p.48 **Scarlet Elf Cap**
Sarcoscypha coccinea (Scop.:Fr.) Lambotte
Sarcoscyphaceae, Pezizales

Description: *Apothecia stalked, bright scarlet, cup-shaped.* Apothecia up to 4 cm (1.6 in) diameter, deeply cup-shaped, scarlet, sometimes dull when old, some forms more orange or yellow. Excipulum slightly paler, pink, scurfy. *Stipe* pale, rather short. *Microscopic features* Sp 25–45 × 11–14 µm, long elliptic with blunt or flattened apex, the oil drops concentrated at each end. Asci 400–450 × 16–18 µm, long and narrow, wavy at the base, no iodine reaction. Paraphyses septate, clavate at the tip, readily anastomosing, with orange-red granular contents.
Season: December–April.
Habitat: On moss-covered wood of beech, elm, members of the *Rosaceae* (hawthorn, blackthorn, etc.).
Distribution: Widespread. Lowlands to submontane.

S. austriaca (Beck ex Sacc.) Boudier, on grey alder, willow, sycamore, false acacia, etc., has wider spores (400–450 × 12–16 µm) with flattened apex, from which elliptic conidia bud off; usually montane. *S. jurana* (Boudier) Baral, recorded on lime trees, has shorter spores (less than 35 µm) embedded in a mucilage forming a halo and is found mostly central and southern Europe. Also described is *S. macaronesia* Baral & Korf, with smaller apothecia (up to 2 cm (0.8 in)) and spores (22–35 × 9–12 µm), in the Canaries and perhaps in the Mediterranean on plants associated with those regions.

p.52 *Rhizina undulata* Fr.:Fr. (= *R. inflata*)
Helvellaceae, Pezizales

Description: *Apothecia large red-brown with a white margin. Rhizoids present.*
Apothecia up to 15 cm (6 in) diameter, sometimes confluent, discoid or rippled, often convex, reddish brown to blackish, the margin pale or white. The underside is attached to debris on the substrate by rhizoids. Leathery when old. *Microscopic features* Sp 25–40 × 7–11 µm, fusiform, hyaline, finely roughened or punctate, as if scribbled on, with a hood-like appendage at both ends, in one or two irregular rows in the ascus. Asci 350–400 × 15–20 µm, blunt, no reaction with iodine. Paraphyses septate, weakly inflated at the tip, encrusted in brown matter. Stiff, dark-brown, thick-walled setae, reaching just below the surface of the hymenium, have also been observed.
Season: May–November.
Habitat: Burnt ground where conifers have been cleared.
Distribution: Widespread.

Most carbonicolous species have very specific ecological preferences, namely the high temperatures required for spore germination and the pH of the substrate, which is generally very high on new ash.

---

p.54 **False Morel**
Gyromitra esculenta (Pers.:Fr.) Fr.
Helvellaceae, Pezizales

Description: *Frb large with a brown brain-like head and stout white stipe.*
Frb up to 15 cm (6 in) tall. *Stipe* very pale, stout, grooved and chambered, brittle. The fertile portion brain-like, rounded, yellow-brown to dark-brown, sometimes with reddish tones, united with the stipe. *Microscopic features* Sp 17–22 × 8–11 µm, elliptic, hyaline, with two

yellowish oil drops. Asci 300–350 × 16–20 µm, blunt, no reaction with iodine. Paraphyses septate, sometimes branched, slightly thickened.

Season: April–July.
Habitat: Mostly montane, on acid soils, under conifers (cleared areas, sawdust, etc.).
Distribution: Occasional in continental and northern Europe.

Although the name means 'edible', this fungus may be deadly under certain conditions. It contains hydrazinic derivatives (including mono-methyl-hydrazine), which attack the liver and are said to be carcinogenic. Poisoning may be immediate or take effect later.

---

p.53 *Helvella crispa* (Scop.:Fr.) Fr.
Helvellaceae, Pezizales

Description: *Frb tall, white or very pale with a lobed cap and grooved stipe.*
Frb up to 12 cm (4.7 in) tall. *Stipe* long, fairly robust, deeply grooved and chambered, white or very pale. *Cap* (apothecia) saddle-shaped, with two or three irregular whitish lobes, the margin and underside sometimes more buff. *Microscopic features* Sp 17–21 × 10–13 µm, elliptic, hyaline, smooth when ripe, with a wide central oil drop. Asci 250– 370 × 15–18 µm, blunt, no reaction with iodine. Paraphyses cylindric, septate, the terminal cell clavate up to 10 µm.
Season: July–November.
Habitat: Open, especially broad-leaved, woodland, sides of paths, banks, etc.
Distribution: Widespread.

This pale *Helvella* is relatively common but the following are rare: var. *pithyophila* Boudier (more cream to ochraceous, under conifers) and *H. lactea* Boudier (damper sites, river margins, milk-white then yellowish-buff or brownish in age and when dry).

p.53 *Helvella sulcata* Afz.:Fr.
Helvellaceae, Pezizales

Description: *Frb tall, grey to black with a lobed cap and grooved stipe.*
Frb up to 10 cm (4 in) tall. *Stipe* tall, fairly robust, very grooved and ribbed, dark grey to blackish. *Cap* (apothecia) saddle-shaped with two or three lobes, fairly regular, dark-grey to black. *Microscopic features* Sp 15–20 × 10–13 µm, elliptic, hyaline, smooth at maturity, with a wide central oil drop. Asci 280–400 × 12–18 µm, no reaction with iodine. Paraphyses cylindric, septate, the terminal cell long and clavate, up to 8 µm, brownish to brown.
Season: July–October.
Habitat: Often on bare ground, in woods.
Distribution: Widespread. Common.

Some authors combine this species with *H. lacunosa* Afz.:Fr. which has a contorted to a somewhat brain-like cap.

---

p.53 *Helvella elastica* Bull.:Fr.
Helvellaceae, Pezizales

Description: *Frb tall, slender with a pale buff head and whitish cylindric stalk.*
Frb up to 12 cm (4.7 in) tall, complex. *Stipe* slender, cylindric, not grooved or rarely slightly pinched in. *Cap* (apothecia) regularly saddle-shaped then lobed (often bilobed), cream, sometimes pale greyish buff. *Microscopic features* Sp 18–25 × 11–15 µm, elliptic, hyaline, smooth when ripe, with a wide central oil drop. Asci 300–350 × 13–20 µm, no reaction with iodine. Paraphyses cylindric, septate, the terminal cell long and clavate up to 12 µm.
Season: July–November.
Habitat: Woods, often on the edges of tracks.
Distribution: Widespread. Common.

In subgenus *Leptopodia* (smooth stalk and saddle-shaped head), *H. atra* Oeder:Fr., almost black, is smaller.

p.53 *Helvella macropus* (Pers.:Fr.) P. Karsten
Helvellaceae, Pezizales

Description: *Frb small, slender, with a cup-shaped cap and cylindric, grey stipe.*
Frb up to 7 cm (2.8 in) tall. *Stipe* long, cylindric, slender, grey, felty. *Cap* cup-shaped up to 3 cm (1.9 in) diameter, grey, pubescent beneath. *Microscopic features* Sp 20–30 × 10–12 µm, broadly fusiform, hyaline, with a wide central oil drop. Asci 250–350 × 15–20 µm, blunt, no reaction with iodine. Paraphyses septate, the terminal cell long and clavate, up to 15 µm.
Season: July–October.
Habitat: Shady, humus-rich woodland.
Distribution: Widespread. Common.

Species in the subgenus *Cyathipodia* (smooth stalk and cup-shaped head) are all rather similar. The most distinctive is *H. corium* (Web.) Massce, a small black species, under willow.

p.53 *Helvella unicolor* (Boud) Dissing
Helvellaceae, Pezizales

Description: *Cup ochraceous brown, flaring, with a short, deeply veined stalk.*
Apothecia up to 5 cm (2 in) diameter, quite deeply cup-shaped then expanded, yellowish or light ochraceous brown, with a short, thick, deeply veined stalk. *Microscopic features* Sp 18–21 × 11.5–14 µm, elliptic, hyaline, with a large central oil drop. Asci 250–350 × 12–18 µm, blunt, no reaction with iodine. Paraphyses cylindric, septate, the terminal element long and clavate, up to 9 µm.
Season: March–June.
Habitat: On the ground, in open woodland.
Distribution: Not yet assessed. Not fully understood.

The subgenus *Paxina* groups together the vernal species with an apothecia supported by a stout stalk decorated with pronounced forked veins. *H. acetabulum* (L.:Fr.) Quélet has an apothecia up to 7 cm (2.8 in) diameter, deeply cup-shaped, brown to dark-brown. The excipulum is pubescent, whitish grey at the base to brown-grey near the margin. On rather rich calcareous soil. The next species is a further example.

---

p.53 *Helvella leucomelaena* (Pers.) Nannfeldt
Helvellaceae, Pezizales

Description: *Cup large, deep, blackish grey-brown, pale outside.*
Apothecia up to 10 cm (4 in) diameter, partly buried then emergent, cup-shaped, the inside grey-brown to blackish and the outside brownish grey, the base almost white. Base wrinkled and tapering. *Microscopic features* Sp 20–25 × 10–13.5 µm, elliptic, hyaline. Asci 270–350 × 13–18 µm, blunt, no reaction with iodine. Paraphyses septate, the terminal cell long and clavate, up to 7.5 µm, brownish.
Season: December–May.
Habitat: Calcareous soil, on conifer debris.
Distribution: Widespread, mostly southern.

---

p.52 *Disciotis venosa* (Pers.:Fr.) Arnault
Morchellaceae, Pezizales

Description: *Cup large, with a light buff, very veined disc. Smell of bleach. Spring.*
Apothecia up to 15–20 cm (6–8 in) diameter, shallow then expanded and sinuate or lobed, brownish-buff to yellow-buff, with pronounced veins and folds in the centre, paler on the outer surface, becoming whitish at the base, sometimes more or less stalked. Flesh whitish, with a strong smell of bleach. *Microscopic features* Sp 19–25 × 12–15 µm, elliptic, hyaline, lacking oil drops but with a few granulations on the outside at the apex. Asci 320–440 × 17–23 µm, blunt, no reaction with iodine. Paraphyses cylindric, septate, sometimes forked, the terminal cell clavate to distinctly capitate, up to 10 µm, with brownish contents.

Season: March–May.
Habitat: On the ground in shady woodland, coppice woods, hedges, sometimes in meadows. Prefers moist conditions and neutral-to-calcareous soils.
Distribution: Appears to be widespread.

An edible species whose unpleasant smell disappears when cooked. The external granules near the apex of the spores and the absence of internal oil drops are characteristic of the *Morchellaceae*. These features separate them from the *Helvellaceae* which generally has spores with oil drops.

---

p.54 *Mitrophora semilibera* (DC:Fr.) Léveillé
Morchellaceae, Pezizales

Description: *Frb with a conical alveolate head, borne on a pale, cylindric stipe.*
Frb up to 20 cm (8 in) tall (often less), complex. *Cap* rather short, ochraceous brown to blackish brown, conical, with longitudinal and transverse ribs separating the irregular alveoli. *Stipe* wrinkled and scurfy, white to buff, often very long in relation to the head. Very deep fold between the edge of the cap and the top of the stipe. *Microscopic features* Sp $20–27 \times 12–16$ μm, elliptic, hyaline, lacking oil drops. Asci $270–310 \times 20–28$ μm, no reaction with iodine. Paraphyses cylindric, septate, sometimes forked, the terminal cell often inflated capitate, up to 15 μm.
Season: April–May.
Habitat: Gardens, damp sites in scrub, especially on calcareous soils.
Distribution: Widespread. Less abundant in the north of Scandinavia.

The head, which is almost free from the stipe, is reminiscent of *Verpa*, whose cap looks like an upside down thimble, borne at the top of the stipe. *Verpa conica* (Müll.: Fr.) Sw. has a bell-shaped cap, almost smooth or slightly wrinkled while *Ptychoverpa bohemica* (Krombh.) Boud. has a rounded, almost morchelloid cap with contorted and anastomosing wrinkles. This species has two-spored asci, which is very unusual among the Ascomycetes.

p.54 Morel
Morchella vulgaris (Pers.:Fr.) Boudier
Morchellaceae, Pezizales

Description: *Cap conical then rounded, blackish with whitish ribs. Stipe white.*
Frb up to 13 cm (5 in) tall (often less), complex. *Cap* bluntly conical, becoming rounded as it matures, with black then dark-brown alveoli, irregular to rippled within, separated by ribs that are white at first, then rather rufous or spotted with orange. *Stipe* short and stout, white then pale buff, scurfy. Furrow between cap and stipe almost absent to absent. *Microscopic features* Sp 19–22 × 10–13 µm, elliptic, hyaline, lacking oil drops but with small external granules at the apex. Asci 220–280 × 14–20 µm, blunt, no reaction with iodine. Paraphyses cylindric, septate, sometimes forked, the terminal cell long and clavate up to 10 µm, with brownish contents.

Season: March–May.
Habitat: Mycorrhizal with ash, in wet shady woodland on calcareous soils, woods carpeted with ivy, etc.
Distribution: Widespread.

M. rotunda (Pers.:Fr.) Boudier, another species with a broadly rounded cap, is much lighter in colour, brownish buff to orange yellow. Morels must not be eaten raw because they have haemolytic toxins; however, these toxins are destroyed by cooking.

p.54 Morel
Morchella elata Fr.:Fr.
Morchellaceae, Pezizales

Description: *Medium-sized morel, the cap a tall, brown cone. Stipe pale.*

Frb up to 18 cm (7 in) tall (often less), complex. *Cap* conical, the alveoli compound, rather light-brown, separated by almost parallel dark to blackish ribs. *Stipe* fairly short, pale then buff, scurfy. Furrow between cap and stipe faint but not absent. *Microscopic features* Sp 18–25 × 11–15 μm, elliptic, hyaline, no oil drops. Asci 250–300 × 15–20 μm, no reaction with iodine. Paraphyses cylindric, septate, sometimes forked, The terminal cell long and clavate, up to 15 μm, pale brown.

Season: April–May.
Habitat: A species of woodland or waste places. In orchards, rubble, disturbed ground, etc.
Distribution: Widespread over most of Europe.

M. costata (Vent.) Pers. is characterized by its distinctly parallel ribs. The head of var. *acuminata* Kickx, a smaller species, is tall like an imp's cap.

---

p.244 **Truffle**
Tuber uncinatum Chatin
Tuberaceae, Pezizales

Description: *Rather large black truffle, with pyramidal scales. Flesh brown, with white veins.*
Frb up to 10 cm (4 in) diameter, spherical to irregularly lobed, soon black, with pyramidal furrowed warts. Flesh light brown, marbled with white veins. Smell strong, rather pleasant. *Microscopic features* Sp 25–39 × 12–25 μm, elliptic, brownish yellow, with a wide-meshed network × 5–12 μm, the ridges measuring up to 5 μm high. Asci 1–6 spored, elliptic or globular, 80–100 × 55–85 μm.

Season: August–December.
Habitat: Hypogeous, especially under oak, sometimes hazel.
Distribution: Throughout Europe, as far north as Scandinavia.

This species belongs to the summer truffle group. It has very tall ridges on the spores and occurs in the far north, while a southern species (*T. aestivum* Vitt.) has lower ridges (up to 2 µm). The Perigord truffle also belongs to the black truffle group; it has black spores, with very slender and very dense spines.

---

p.244 *Balsamia vulgaris* Vittadini
Balsamiaceae, Pezizales

Description: *Small irregular, rufous truffle. Flesh white, with labyrinthine cavities.*
Frb tuber-like, up to 3 cm (1.2 in) diameter, with hollows or depressions, hazel-beige then somewhat rufous brown in patches, smooth, not warted. Flesh very pale, with numerous close folds, forming small sinuous or labyrinthine cavities, the largest being the same colour as the outside. *Microscopic features* Sp 25–32 × 10–18 µm, cylindric, the apex very blunt, hyaline to yellowish, generally with three oil drops. Asci 50–80 × 20–40 µm, elliptic to fusoid, stalked.
Season: December–March.
Habitat: Grows underground, in woods or parks, under various broad-leaved trees.
Distribution: Mostly in warm places over much of Europe, but almost absent in the north.

The asci, which are reminiscent of those in the Pezizaceae and closely related families, the pale spores, and the cavities which have the same colour as the excipulum (evidence of folds in a former apothecia), together indicate that this species may be less advanced in tuberoid evolution than the true truffles.

BASIDIOMYCETES (SUBDIVISION OF THE BASIDIOMYCOTINA)

Like the Ascomycetes, the Basidiomycetes are a very large group. However, they have fewer recorded species and less diversity of shape and structure. Generally speaking, the sporophores or fruit bodies are very easy to see. This subdivision includes well known fungi such as boletes, fly agarics, button mushrooms and bracket fungi. They have a considerable impact on our lives economically (as phytopathogens and decomposers of timber, as an important food source and in commercial culture), in medicine (occasional fatal poisoning) and in ecology (the fundamental role of the ectomycorrhiza in the balance of nature). The Basidiomycotina are divided into three main classes:

- The **Ustomycetes** (rusts and smuts), which are parasitic on higher plants, do not develop sporophores but are detected by the deformations caused to plants and by their more or less conspicuous reproductive bodies. Economically, they are very important as pathogens of many cultivated plants (cereals, etc.). They are not addressed in this book.
- The **Phragmobasidiomycetes** include species with septate basidia (completely or incompletely) and spores capable of producing secondary spores (*see* below).
- The **Homobasidiomycetes** group together all species with nonseptate basidia and spores not capable of forming secondary spores. They include most of the fungi known to the general public and nearly all the species illustrated in this book (*see* p.306).

I. *Phragmobasidiomycetes 'Jelly Fungi'*

Classification of this group is based on the type of septation in the basidia: transverse (Auriculariales), longitudinal (Tremellales), or incomplete (Dacrymycetales).

p.59 Jew's Ear
Auricularia auricula-judae (Bull.:Fr.) Wettstein (= *Hirneola a.*)
Auriculariaceae, Auriculariales

Description: *Gelatinous, ear-shaped. Scurfy to shortly pubescent, the inner surface smooth or veined, red-brown.*
Frb laterally attached, cup-shaped then more or less regularly ear-shaped, 2–12 × 1–7 cm (0.8–4.7 × 0.4–2.8 in). Outer surface shortly pubescent to scurfy, matt, greyish red-brown. Texture gelatinous, very hard when dry. Hymenium smooth or veined, glossy, red-brown or crimson-brown. *Microscopic features* Sp 13–20 × 5–8 μm, curved. Bas transversely septate.
Season: All year round, persisting in the dry state then regaining their usual form.
Habitat: Elder, rarely willow, maples, etc.
Distribution: Widespread. Common, strangely absent from certain areas.

Edible, although with little flavour (black fungus of Asian cookery).

p.59 Tripe Fungus
Auricularia mesenterica (Dicks.:Fr.) Pers.
Auriculariaceae, Auriculariales

Description: *Frb semi-resupinate, elastic, strigose. Hymenium crimson-brown, wrinkled.*
Frb resupinate then making small lateral brackets, 1–3 cm (0.4–1.2 in) wide, confluent. Surface strigose, more or less zoned, whitish to brownish grey. Hymenium radially wrinkled, crimson-brown to brownish. Flesh gelatinous, cartilaginous when dry, capable of regaining its usual form. *Microscopic features* Sp 15–18 × 5–7 μm, curved. Bas transversely septate.
Season: Especially in spring.
Habitat: On dead wood of various broad-leaved trees.
Distribution: Much less common than the preceding species but widely distributed.

These two taxa are the only Auriculariales (transversely septate basidia) easy to see and determine.

---

p.60 **Yellow Brain Fungus**
Tremella mesenterica Retz.:Fr.
Tremellaceae, Tremellales

Description: *Brain-like, gelatinous and wobbling, orange-yellow to yellowish or even white.*
Frb sessile, 1–5 cm (0.4–2 in) across, irregular then brain-like or leaf-like, bright orange-yellow then paling to yellow, yellowish or even whitish to hyaline, glossy, rather smooth. Hymenium covering the entire surface. Consistency gelatinous and wobbling. Often more highly coloured and cartilaginous in dry weather. *Microscopic features* Sp 8–15 × 6–10 µm, broadly elliptic to ovoid. Bas longitudinally septate, with four long thick sterigmata.
Season: All year round but especially in autumn.
Habitat: Dead wood of broad-leaved trees.
Distribution: Widespread. Common almost everywhere.

Some Asiatic *Tremella* spp. are cultivated and eaten.

---

p.60 *Tremella foliacea* Pers.:Fr.
Tremellaceae, Tremellales

Description: *Leaf-like lobes, brown, gelatinous and wobbling.*
Frb sessile, 3–12 cm (1.2–4.7 in) across, brain-like then with very distinct leaf-like lobes, red-brown to brown or brownish. Surface slightly shiny. Hymenium covering the entire surface. Consistency gelatinous and wobbling, shrivelled and cartilaginous when dry. *Microscopic features* Sp 8–13 × 6–9 µm, subglobose. Bas like *T. mesenterica*.
Season: Almost all year round.
Habitat: On broad-leaved trees, rarer on conifers.
Distribution: Fairly common.

T. encephala Pers.:Fr. is quite rare; it is beige to ochraceous brown or pinkish brown, more brain-like, as the name suggests, 3–4 cm (1.2–1.6 in) diameter and the lobes less separated.

p.60 Witch's Butter
Exidia glandulosa (Bull.:Fr.) Fr.
Tremellaceae, Tremellales

Description: *Misshapen or almost brain-like, brown to black, gelatinous.*
Frb sessile, 1–5 cm (0.4–2 in) diameter, 0.5–2 cm (0.2–0.8 in) thick, irregular then brain-like or lobed, almost leafy when old, often quite tightly lobed, black-brown to black. Surface shiny, irregular, with small soft pimples. Consistency gelatinous and wobbling, cartilaginous when the weather is dry but almost deliquescent in very damp conditions or when old. *Microscopic features* Sp 10–16 × 4–6 µm, cylindric and curved. Bas 4-spored, longitudinally septate, with four long sterigmata.
Season: Practically all year round.
Habitat: On dead wood of broad-leaved trees, especially oak.
Distribution: Widespread.

The main difference between *Tremella* and *Exidia* spp. is in the shape of the spore, which is subglobular in the former and curved in the latter.

p.63 *Pseudohydnum gelatinosum* (Scop.:Fr.) P Karsten (= *Tremellodon g.*)
Tremellaceae, Tremellales

Description: *Hydnoid, shortly pubescent, translucent, white to grey or brownish. Stipe eccentric or lateral.*
Cap 2–10 cm (0.8–4 in) diameter, domed or convex, spathulate or fan-shaped. Margin translucent. Upper surface shortly pubescent. White or greyish sometimes buff to brownish. Hymenophore with white or translucent, rather close, 1–4 mm (0.02–0.12 in) spines. *Stipe* eccentric or lateral, often tapering, short and stout, the entire fruit body up to 12 cm (4.7 in) long,

concolorous with the upper surface, shortly pubescent to strigose, quite firm.
Flesh gelatinous, translucent. *Microscopic features* Sp 5–8.5 × 4.5–7.5 µm, subglobular or slightly irregular. Bas small, with four rather short sterigmata.

Season: July–December.
Habitat: Rotten wood, especially of conifers.
Distribution: Mostly continental.

---

p.60 *Tremiscus helvelloides* (D.C.:Fr.) Donk
Tremellaceae, Tremellales

Description: *Petaloid, split, orange-pink, soft.*
Cap up to 15 cm (6 in) tall, shaped like a trumpet or horn then asymmetrical and like an ear or a thick petal, salmon pink becoming orange-pink or pinkish brown when old. Margin more or less recurved or brittle. Surface finely pruinose. Hymenium on the upper part of the outside, smooth, slightly more brightly coloured. *Stipe* not well differentiated, whitish at the base. *Flesh* cartilaginous or soft, concolorous. *Microscopic features* Sp 9–12 × 4–6 µm, cylindro-elliptic or asymmetrical, less curved on one side. Bas with two or four long sterigmata.

Season: August–November.
Habitat: On the ground or in contact with pieces of buried rotten wood, rotten stumps.
Distribution: Tends to be continental or montane.

Easy to determine.

---

p.74 *Calocera cornea* (Batsch:Fr.) Fr.
Caloceraceae, Calocerales

Description: *Small, horn-like, orange to orange-yellow.*
Frb 4–15 mm (0.2–0.6 in) tall, like a slender, erect horn (1–2 mm (0.04–0.08 in)), yellow, sometimes bright orange, especially when dry. Apex tapering and pointed, simple or rarely forked. *Stipe* not differentiated. Consistency gelatinous or slightly cartilaginous, rather soft. *Microscopic*

features Sp 7–11 × 2.5–4.5 μm, cylindro-elliptic to curved, often (once) septate when mature. Bas like a tuning fork, two-spored. Hyphae without clamps.

Season: Almost all year round.
Habitat: Slightly rotten dead wood, on broad-leaved trees.
Distribution: Widespread. Fr.equency variable.

A distinctive genus on account of its clavarioid stature (*see also C. viscosa*). It could be confused with *Typhula*, a slender, but not gelatinous, genus in the Clavariales.

---

p.75 *Calocera viscosa* (Pers.:Fr.) Fr.
Caloceraceae, Calocerales

Description: *Club-like, branched, viscid or gelatinous, bright orange.*
Frb 2–10 cm (0.8–4 in), like a small dichotomously branched shrublet. Bright orange, sometimes orange-red when dry. Apex pointed, often forked. *Stipe* deeply rooting. Consistency slightly cartilaginous, rather soft. *Microscopic features* Sp 7–10 × 3.5–4.5 μm, like those in *C. cornea*; basidia and hyphae also similar.
Season: Practically all year round.
Habitat: Rotten conifer wood.
Distribution: Widespread. Common.

The beginner may confuse this species with a club fungus until it is collected. No club fungus feels this greasy and viscid.

II. Homobasidiomycetes

The Homobasidiomycetes have nonseptate basidia and their basidiospores are not normally able to form secondary spores.

They are considerably varied in shape and placed in three morphological groups which do not correspond to three phylogenetic lines nor to three homogenous groups. However, they are retained for reasons of convenience. They are:

- **Aphyllophoromycetideae** (typically without gills, *see* below);
- **Agaricomycetideae** (typically with gills, *see* p.363);
- **Gasteromycetideae** (typically fertile on an internal structure, *see* p.865).

A. APHYLLOPHOROMYCETIDEAE

This subclass is composed of several independent lines and, in a complex and indirect way, links the Heterobasidiomycetes to the gilled fungi. The visual key (*see* p.34) introduces the main morphological types. These are presented in the following order:

- **Corticioid** (hymenophore smooth or practically smooth to alveolate and fruit body more or less completely resupinate);
- **Stereoid** (hymenophore smooth but fruit body with a pileus);
- **Merulioid** (as above but hymenophore ribbed alveolate);
- **Hydnoid** (hymenophore with small spines);
- **Poroid** (hymenophore tubular-poroid; the structure leathery to woody);
- **Thelephoroid** (fruit body variable; in this book, those that are more or less erect with an almost smooth hymenophore but very distinctive spores);
- **Clavarioid** (fruit body erect, simple to branched and tree-like, the hymenium growing all round the fruit body);
- **Cantharelloid** (hymenophore smooth to ribbed, on the underside of a fruit body differentiated into cap and stipe).

In the microscopic descriptions, the following characters, which are considered constant, are not repeated: spores hyaline, smooth, non-amyloid; basidia (Bas) with clamps; cystidia (Cyst) absent (or if present, hyaline, not encrusted, thin-walled with clamps).

Abbreviations are used for *features* of the anatomical structure of the flesh which can, in fact, be formed from one to three

distinct types of hyphae. A monomitic structure involves only one type of hypha, a dimitic structure two and a trimitic structure three different types. These hyphal types are as follows:
- Generative hyphae (GH), typically thin-walled, branched, septate (with or without clamp connections);
- Skeletal hyphae (SH), typically thick to very thick-walled, not branched, not septate;
- Binding hyphae (BH), typically thick-walled, more or less branched, not septate.

---

p.56 *Lyomyces sambuci* (Pers.:Fr.) P Karsten
Corticiaceae, Corticiales

Description: *Crust-like, very thin, whitish, tightly attached.*
Frb resupinate, thin, white then pale greyish, smooth then finely cracked. Margin attached, more or less delimited. *Microscopic features* Sp 5–7 × 3.5–4 µm, elliptic to cylindro-elliptic. Bas 20–25 × 3.5–4.5 µm, cylindro-clavate. Cyst 30–50 × 4–5 µm, cylindric or slightly irregular, capitate, hyaline. Monomitic. Hyphae 2–5 µm diameter, more or less encrusted or finely granular.
Season: All year round.
Habitat: On the bark of fallen or standing branches of elder, rarer on other species.
Distribution: Widespread. Common.

The specific host is the best field character for this species. Most corticioid species, however, require examination under the microscope.

---

p.57 *Peniophora aurantiaca* (Bres.) v H. & Litsch.
Peniophoraceae, Corticiales

Description: *Crust-like, tightly attached, confluent, smooth to tubercular, orange.*
Frb resupinate, making small circular cushions, confluent into a continuous crust, orange to orange red. Margin thin, whitish, slightly fibrillose at first, attached or slightly

raised. Surface tuberculate then indented and remaining uneven. Spore print yellowish to pale orange. *Microscopic features* Sp 14–17 × 7.5–10 μm, ellipsoid to cylindro-elliptic. Bas 55–90 × 8–15 μm, cylindro-clavate, sinuous. Cyst of two kinds: metuloid (lamprocystidia) cylindro-fusiform, the apex encrusted, 30–50 × 6–12 μm, thick-walled, and sulfocystidia, (staining in sulfo-aldehyde reagents) cylindric to fusoid tapering above or irregular, up to 150 × 18 μm, hyaline, smooth. Monomitic. Hyphae 2–5 μm diameter.

Season: July–October.
Habitat: Bark of green alder.
Distribution: Restricted to alpine areas.

The habitat is useful in identifying this species in the field. *P. eriksonii* Boidin is rather similar, but without clamps and has spores measuring 15–20 × 10–13 μm; it occurs on other alders. There are other orange *Peniophora*. *P. laeta* (Fr.:Fr.) Donk is frequent, in lowland areas, on broad-leaved trees (especially hornbeam). It grows under the bark, forming large cylindrical outgrowths. *P. incarnata* (Pers.:Fr.) P Karsten does not push off the bark.

---

p.56 *Bulbillomyces farinosus* (Bres.) Jülich
Corticiaceae, Corticiales

Description: *Crustose, velvety, grey patches, the margin surrounded by small white bulbils.*
Frb resupinate, thin, grey, smooth to finely velvety or pubescent, slightly verrucose or irregular. Margin attached, quite well delimited. *Microscopic features* Sp 6–10 × 4.5–6.5 μm, elliptic, slightly thick-walled. Bas 25–35 × 5–7 μm, subcylindric to urn-shaped. Cyst metuloid (lamprocystidia), numerous, 50–100 × 4–15 μm, inflated at the base but with a long cylindric or tapering upper part, the walls thin then distinctly thickened, strongly encrusted. Monomitic. Hyphae 3–5 μm diameter.

Season: September–January.
Habitat: Very damp wood of broad-leaved trees.

Distribution:	Uneven. Rare in the perfect stage (*see* below).

Very easy to identify on account of the imperfect stage which is always present and which indeed often occurs alone. It is composed of small, pure white bulbils (hence the generic name), measuring 0.1–0.4 mm (0.004–0.016 in) diameter and often very gregarious.

---

p.56 *Hyphoderma roseocremeum* (Bres.) Donk
Corticiaceae, Corticiales

Description: *Crustose patches, ochraceous cream with pinkish areas.*
Frb resupinate, thin, ochraceous cream to pale yellowish beige, often with pinkish patches or zones. Margin thin, whitish, slightly fibrillose when young, attached. *Microscopic features* Sp 9–12 × 3–4 µm, cylindro-elliptic or slightly curved. Bas 25–35 × 6–8 µm, cylindric constricted. Cyst cylindric, with a few slight constrictions, or sinuous, 50–130 × 6–8 µm. Hyphidia more or less dendroid. Monomitic. Hyphae 2–4 µm diameter.
Season: August–November.
Habitat: Decorticated and rotting wood of broad-leaved trees, rarely on conifers.
Distribution: Uneven. Rather rare.

There are a few other, more or less pinkish corticioid species.

---

p.62 *Hyphoderma radulum* (Fr.:Fr.) Donk
Corticiaceae, Corticiales

Description: *Crustose patches, circular, ochraceous yellow, odontoid.*
Frb resupinate, circular or confluent, whitish then cream to yellowish buff. Margin thin, whitish, fimbriate, attached. Surface with erect, irregular teeth up to 5 mm (0.2 in) long. *Microscopic features* Sp 9–11 × 3–3.5 µm, narrowly allantoid. Bas 20–25 × 4–6 µm, often slightly constricted. Cyst quite

rare, 50–70 × 5–8 µm, cylindric but often with many constrictions. Monomitic. Hyphae 3–4 µm diameter.

Season: July–December.
Habitat: On dead wood of broad-leaved trees; rarer on conifers.
Distribution: Widespread. Fairly common.

When on bark, this corticioid forms more circular patches than it does on decorticated wood.

---

p.57 *Hyphodontia aspera* (Fr.) J. Eriksson
(= *Grandinia granulosa*)
Corticiaceae, Corticiales

Description: *Crustose patches, finely granulose spiny, ochraceous cream.*
Frb resupinate, whitish then cream to brownish buff. Margin whitish, ill-defined, irregular pruinose, tightly attached. Hymenophore smooth then finely tuberculate to granulose spiny or papillate, the outgrowths up to 1 mm (0.04 in) long, with smooth areas between. *Microscopic features* Sp 5–6.5 × 3.5–5 µm, elliptic. Bas 18–29 × 4–5 µm, cylindro-clavate, often slightly constricted. At the top of the spines on the hymenium are bundles of protruding hyphae, with tapering or capitate apex, sometimes crowned with a small labile crystalline lobe. Monomitic. Hyphae 2–3 µm diameter.

Season: August–November.
Habitat: On wood of conifers and broad-leaved trees.
Distribution: Quite widespread. Relatively common.

This kind of hymenial surface (called grandinioid) is frequent among corticioids. The following species, from a different group, is a further example.

---

p.56 *Trechispora farinacea* (Pers.:Fr.) Liberta
Corticiaceae, Corticiales

Description: *Resupinate, whitish to greyish cream, grandinioid.*

Frb resupinate, whitish then cream to greyish beige. Margin thin, whitish, fimbriate or felty, attached. Hymenophore initially smooth, arachnoid and very thin then finely tuberculate or even distinctly grandinioid as rather dense, irregular spines develop. *Microscopic features* Sp very small, 3–4.5 × 2.5–4 µm, shortly ovoid elliptic to subglobose, warted, slightly thick-walled. Bas 10–15 × 4.5–6 µm, cylindric, often slightly constricted. Monomitic. Hyphae 2–5 µm diameter, often inflated at the septa.

Season: Practically all year round.
Habitat: Rotten broad-leaved and conifer wood.
Distribution: Widespread. Common.

This rather characteristic species (colour, hymenophore) belongs to a very difficult genus.

---

p.57 *Cerocorticium confluens* (Fr.:Fr.) Jül. & Stalp. *Corticiaceae*, Corticiales

Description: *A waxy corticioid species, ochraceous cream with bluish tones.*
Frb resupinate, circular then confluent forming long whitish then cream to buff patches with blue tones, at least towards the margin in moist specimens. Margin thin, whitish, shortly fimbriate, attached or slightly less tightly attached at the edge. Surface smooth then finely tuberculate or irregularly warted. *Microscopic features* Sp 7.5–12 × 5–9 µm, cylindro-elliptic to subglobose, the wall very slightly thickened. Bas 30–60 × 6–10 µm, clavate, with many oil drops. Monomitic. Hyphae 1–3 µm diameter.

Season: May–December.
Habitat: On bark and rotten wood of broad-leaved trees, sometimes on conifers; damp places.
Distribution: Widespread. Fairly common.

When dry, the beautiful bluish tones of this species are lost and it may even become chalk-white.

p.57 *Pulcherricium caeruleum* (Schr.:Fr.) Parm.
Corticiaceae, Corticiales

Description: *Crustose patches, dark blue to bright blue, more dull when old.*
Frb resupinate, orbicular then confluent, slightly membranous, deep blue when fresh, paling when old and when dry, blue-grey to dingy bluish-brown. Margin thin, becoming slightly detached when old. Hymenophore smooth then more or less warty, finally cracked when dry. *Microscopic features* Sp 7.5–10 × 4–6.5 µm, elliptic. Bas 30–50 × 5–7 µm, quite narrowly cylindro-clavate. Cyst 25–40 × 4–7 µm, of basidial origin, with rather short contorted branching. Monomitic. Hyphae 2–5 µm diameter, sometimes with secondary septa.
Season: July–November.
Habitat: On broad-leaved trees. Requires warm conditions.
Distribution: Tends to be southern.

When fresh, this bright, deep-blue species is really spectacular.

---

p.57 *Hymenochaete cruenta* (Pers.:Fr.) Donk
(= *H. mougeotii*)
Hymenochaetaceae, Hymenochaetales

Description: *Crustose species, bright red then rust-coloured, the edges reflexed.*
Frb resupinate, circular or irregular, tightly attached then with slightly reflexed edges, bright red then rust-coloured in patches. Margin well delimited. Surface smooth then slightly irregular and warty or finely cracking when old, finely velvety or with erect hairs (strong lens). *Microscopic features* Sp 5–8 × 2–3 µm, cylindro-elliptic or cylindric, sometimes faintly curved. Bas 20–30 × 3–4.5 µm, cylindro-clavate, slightly contorted. Hymenial setae brown, 40–90 × 4–8 µm, more or less projecting, tapering to a point, thick-walled. Monomitic. Hyphae 2–5 µm diameter, slightly thick-walled and brown near the host. No clamps.

Season: July–December.
Habitat: On fir.
Distribution: Widespread. Continental or montane.

The rare *Cytidia salicina* (Fr.) Burt, which is even more red (as it has no brown setae), another continental species, occurs on willow in damp places.

---

p.50 *Aleurodiscus amorphus* (Pers.:Fr.) Schroet.
Aleurodiscaceae, Corticiales

Description: *Crustose species, discoid, orange to pinkish-beige.* Frb circular, 1–7 mm (0.04–0.28 in) diameter, somewhat orange to fairly bright pinkish-beige, sometimes more dull when old, slightly membranous or leathery, the margin slightly reflexed, fibrillose or pubescent. Surface pruinose. *Microscopic features* Sp 20–30 × 20–25 µm, subglobose, with very fine blunt spines, amyloid. Bas 80–180 × 20–30 µm, cylindro-clavate, without clamps. Cyst 100–200 × 8–15 µm, cylindric or strongly constricted. Monomitic. Hyphae in the context 2–4 µm diameter. No clamps.
Season: Especially in autumn.
Habitat: Mainly on fir and spruce.
Distribution: Uneven. A continental species.

The pruina results from the huge spores that can be seen in fours, under a dissecting microscope. Most species of *Aleurodiscus* are discoid.

---

p.59 *Stereum hirsutum* (Willd.:Fr.) S.F. Gray
Stereaceae, Corticiales

Description: *Crustose species often forming a strigose, buff, zoned bracket. Orange-yellow underneath.* Frb resupinate then one edge recurving to form a bracket, up to 3 cm (1.2 in). Upper surface zoned, with strigose or erect hairs, greyish to ochraceous beige. Margin thin, fimbriate or entire. Hymenophore smooth, sometimes faintly zoned, yellow to orange-buff or brownish. Elastic, cartilaginous,

hard when dry. Fruit bodies confluent.
Microscopic features Sp 5–8 × 2–4 µm,
cylindro-elliptic to cylindric, amyloid. Bas
25–60 × 3–5 µm, cylindro-clavate,
without clamps. Cyst of two kinds:
pseudo-cystidia arising from deep trama
and reaching the base of the hymenium,
5–10 µm, bluntly cylindric, thick-walled
except towards the apex, with oily
contents; hymenial cyst, fusoid tapering,
level with or protruding beyond the
basidia, 20–30 × 2–4 µm. Monomitic.
Hyphae in the context 4–8 µm, thin- or
thick-walled. No clamps.

Season: All year round
Habitat: On wood of broad-leaved trees; rarer on conifers.
Distribution: Very common.

S. ochraceoflavum (Schw.) Ellis (= *S. rameale*)
is smaller and paler, rounded resupinate
then confluent, adult fruit bodies with
circular marks where they joined.

---

p.59 *Stereum insignitum* Quélet
Stereaceae, Corticiales

Description: *Crustose bracket-forming species, the upper surface strongly and rather brightly zoned.*
Frb forming lateral brackets, up to 3–6 cm
(1.2–2.4 in) wide, sometimes tapering
towards the host with a short pseudostipe. Upper surface zoned, alternatively
glabrous and finely velvety, brownish to
red-brown, the margin orange-red, orange
or bright yellow. Margin thin.
Hymenophore smooth, cream to pale
ochraceous grey. Cartilaginous when fresh,
hard when dry. *Microscopic features* Sp
4.5–6.5 × 2–3 µm, cylindric, amyloid. Bas
20–40 × 3–5 µm, cylindro-clavate,
without clamps. Cyst of two types:
pseudocystidia, extensions of the
conducting system, 5–10 µm, cylindric
obtuse; pseudoacanthophyses 2–3 µm,
slightly diverticulate at the apex. Dimitic:
SH 3–6 µm; GH 2–4 µm. No clamps.

Season: Especially in autumn.

Habitat: On wood of broad-leaved trees, usually beech.
Distribution: Mainly southern.

The brightly coloured margin is often very conspicuous. *S. subtomentosum* Pouzar, less brightly coloured and more damp-loving, is frequent on willow.

---

p.58 **Silver-leaf Fungus**
Chondrostereum purpureum (Pers.:Fr.) Pouz.
(= *Stereum p.*)
Stereaceae, Corticiales

Description: *Crustose species, confluent, purple then violaceous pink, forming small brackets.*
Frb with the edge soon reflexed forming a bracket, sometimes 4 cm (1.6 in) wide. Upper surface faintly zoned, velvety to strigose, greyish to ochraceous beige. Margin paler, wavy, lilac to purple. Hymenophore smooth, purple to lilac pink then purple-brown or brownish. Fruit bodies confluent. *Flesh* distinctly multi-layered. *Microscopic features* Sp 5–10 × 2.5–4 µm, cylindro-elliptic to cylindric, non-amyloid. Bas 25–50 × 3–5 µm, cylindro-clavate. Cyst fusoid or cylindric obtuse, 50–80 × 5–8 µm, exserted, sometimes with a few crystals towards the apex. Monomitic. Deeply originating hyphae 3–5 µm diameter, thin- or thick-walled. Clamps abundant.
Season: Practically all year round.
Habitat: On broad-leaved trees.
Distribution: Widespread. Very common to common.

Saprotrophic or parasitic (causes silver-leaf disease on fruit trees). *Trichaptum abietinum* (Dicks.:Fr.) Ryv., the same colour but with laciniate poroid hymenophore, occurs on conifers.

---

p.58 *Hymenochaete tabacina* (Sow.:Fr.) Léveillé
Hymenochaetaceae, Hymenochaetales

Description: *Crustose species, confluent, forming a zoned brown bracket.*

Frb resupinate then forming brackets, up to 1.5 cm (0.6 in), very confluent. Margin thin, yellow, wavy to lobed. Upper surface more or less zoned, orange-brown to brown. Hymenophore tobacco brown, irregular, finely velvety or with bristling (strong lens). Membranous. *Microscopic features* Sp 5–7 × 1.5–2 µm, cylindric to curved. Bas 15–25 × 3–4.5 µm, cylindro-clavate. Hymenial setae brown, 50–100 × 6–12 µm, more or less projecting, tapering to a point, thick-walled. Monomitic. Hyphae 2–8 µm diameter, thin- or slightly thick-walled and brown towards the host. No clamps.

Season: Especially in autumn.
Habitat: Damp-loving, on willow and hazel.
Distribution: Widespread. Fairly common.

H. rubiginosa (Dicks.:Fr.) Léveillé, which makes a larger bracket, occurs mainly on oak stumps.

---

p.58 *Meruliopsis corium* (Fr.:Fr.) Ginns
(= *Merulius papyrinus*)
Corticiaceae, Corticiales

Description: *Crustose species, smooth or with low pits, white to yellowish buff. Brackets narrow.*
Frb resupinate, circular then very confluent. Margin thin, wavy, forming small brackets, up to 1 cm (0.4 in) (often less). Faintly zoned above, velvety to strigose, whitish to greyish beige or ochraceous beige. Hymenophore smooth then with a network of irregular pits, white then yellowish-cream to buff. Consistency papery, leathery when dry. *Microscopic features* Sp 5–7 × 2.5–3.5 µm, cylindro-elliptic. Bas 25–35 × 4–6 µm, cylindro-clavate, without clamps. Monomitic. Hyphae 2–5 µm, thin-walled or the lower layers slightly thick-walled and there sometimes finely encrusted. Clamps absent.

Season: All year round.
Habitat: On the bark and wood of various broad-leaved trees.
Distribution: Widespread. Very common.

p.58 *Merulius tremellosus* Schrad.:Fr.
Meruliaceae, Corticiales

Description: *Soft crustose, bracket-forming with strigose hairs, pale. Hymenophore merulioid, orange-buff.*
Frb orbicular then forming soft or elastic, confluent brackets, projecting up to 5 cm (2 in). Margin thin, wavy or undulating. Pileus faintly zoned, velvety to strigose, whitish to greyish beige. Hymenophore soon more or less radially wrinkled and pitted, somewhat orange to ochraceous pinkish. Consistency hard when dry. *Microscopic features* Sp 3.5–4.5 × 1–1.5 μm, very small, cylindro-allantoid. Bas 15–25 × 3–5 μm, cylindro-clavate. Cyst quite rare, cylindric or subcapitate, 30–60 × 3–8 μm, smooth or slightly encrusted. Monomitic. Hyphae 2–5 μm diameter.

Season: August–November.
Habitat: Rotten wood, especially of broad-leaved trees.
Distribution: Widespread. Common.

Because of its soft consistency and reticulate pitted hymenophore, this species resembles dry rot, *Serpula lacrymans* (Wulf.:Fr.) Schroeter, a feared and destructive species (in another family), which colonizes worked wood and timber in buildings.

---

p.84 *Plicaturopsis crispa* (Pers.:Fr.) Reid
Meruliaceae, Corticiales

Description: *Bracket rufous brown. Hymenophore radially ribbed, white.*
Frb circular then making a small bracket, up to 2.5 cm (1 in), mounded against the host or very markedly tapered into a very short stipe. Margin inrolled, slightly lobed to undulating or crenate. Pileus faintly zoned, velvety, whitish then brownish, yellow-brown to rufous brown, the edge remaining white. Hymenophore almost smooth, soon developing more or less anastomosing radial wrinkles, white. Quite leathery, stiffer when dry. Gregarious. *Microscopic features* Sp 3–4.5 × 1–1.5 μm,

narrow, curved. Bas 12–20 × 3–5 µm, cylindro-clavate. Monomitic. Hyphae 3–6 µm diameter, thin-walled or slightly thick-walled in the cuticle.

Season: Practically all year round.
Habitat: Especially on hazel, beech, alder, etc.
Distribution: Uneven, mainly continental.

This species may be expanding towards the Atlantic in certain districts. In others, it appears to be declining.

---

p.62 *Mycoacia uda* (Fr.) Donk
Corticiaceae, Corticiales

Description: *A crustose hydnoid species, fairly bright or pale yellow. Slender pointed spines.*
Frb resupinate, bright yellow to pale yellow, sometimes more yellowish buff. Margin felty or fimbriate, tightly attached. Hymenophore with dense, slender, pointed spines, 1–2 mm (0.04–0.08 in) long, shorter near the margin. *Chemical test* reaction with KOH: brownish red or crimson red. *Microscopic features* Sp 4.5–6 × 2–3 µm, long elliptic. Bas 15–20 × 4–5 µm, cylindro-clavate. Cyst inconspicuous, fusoid, 20–35 × 3–5 µm. Monomitic. Hyphae 2–3 µm diameter. Trama with numerous crystals at maturity.

Season: Practically all year round.
Habitat: Rotten wood of various broad-leaved trees.
Distribution: Widespread. Common or fairly common.

The reaction with KOH is an important (macroscopic) character since *M. aurea* (Fr.) J. Erikss. & Ryv., which is very similar, has no reaction. *Sarcodontia setosa* (Pers.) Donk, which is also quite similar in colour and hymenophore, has no reaction with KOH; however, it has longer spines, a strong smell and occurs especially on old apple trees.

---

p.62 *Mycoacia fuscoatra* (Fr.:Fr.) Donk
Corticiaceae, Corticiales

Description: *A crustose hydnoid species, yellow-buff to black-brown. Thick spines.*

Frb resupinate. Margin fimbriate, tightly attached, pale. Hymenophore with conical spines, 1–3 mm (0.04–0.12 in) long, dense, shorter near the margin. Yellow-buff, soon browning, the mature spines blackish brown or black at the base and more or less dark buff at the apex. *Microscopic features* Sp 4.5–6 × 2–2.5 µm, elliptic to cylindro-elliptic. Bas 15–25 × 4–5 µm, cylindro-clavate. Cyst fusoid, 20–45 × 3–5 µm. Spine apex strongly encrusted with protruding hyphae. Monomitic. Hyphae 2–4 µm diameter, thin- or slightly thick-walled.

Season: July–December.
Habitat: Woods, on rotten wood of broad-leaved trees.
Distribution: Widespread. Less frequent than the previous species.

This taxon colonizes the lower surface of fallen branches.

---

p.62 *Steccherinum ochraceum* (Pers.:Fr.) S.F. Gray
Steccherinaceae, Polyporales

Description: *A crustose hydnoid species with more or less reflexed edges. Rather short, salmon-buff spines.* Frb resupinate then reflexed, bright salmon-buff or with orange tones. Margin whitish or pale. Hymenophore with conical or cylindric spines, 0.5–1 mm (0.02–0.04 in) long, dense, shorter near the margin. *Microscopic features* Sp 3–4 × 2–2.5 µm, elliptic. Bas 15–20 × 4–5 µm, cylindro-clavate. Cyst especially in the spines, thick-walled, cylindric and encrusted at the apex, 60–120 × 5–10 µm. Dimitic: SH 2–8 µm; GH 2–3.5 µm, with clamps.

Season: Practically all year round.
Habitat: On broad-leaved trees, rarer on conifers.
Distribution: Widespread. Fairly common.

This species bears a striking resemblance to *Junghuhnia nitida* (p.326) but the hymenophore is very different (poroid in the latter).

p.63 *Hydnellum concrescens* (Pers.) Banker
Bankeraceae, Thelephorales

Description: *Frb hydnoid, zoned, pinkish buff to dark brown. Spines rather crimson.*
Frb stipitate. *Cap* often confluent, 3–7 cm (1.2–2.8 in), soon depressed, zoned, radially wrinkled or ribbed, pinkish but soon becoming brown, vinaceous brown or orange-brown. Spines 1–3 mm (0.04–0.12 in), whitish then crimson-brown. *Stipe* 1–5 × 0.3–1 cm (0.4–2 × 0.12–0.4 in), clavate, velvety, concolorous or darker. *Flesh* thin. Smell mealy.
Microscopic features Sp 5.5–6 × 3.5–4.5 µm, elliptic, with warts and lobes, light brown. Bas 25–35 × 5–7 µm. Monomitic. Hyphae 1.5–4 µm diameter, the wall sometimes brownish. No clamps.

Season: August–October.
Habitat: Broad-leaved and coniferous woods.
Distribution: Uneven. Sometimes rare.

Many hydnoid species are continental or submontane. Most, being very sensitive to pollution, are in decline or threatened with extinction.

---

p.63 *Hydnellum peckii* Banker
Bankeraceae, Thelephorales

Description: *Frb hydnoid, white then red-brown, with large red droplets when young.*
Frb stipitate. *Cap* often confluent, 3–7 cm (1.2–2.8 in), slowly becoming depressed, uniformly coloured or zoned, radially fibrillose or scaly, white, with bright red droplets when young, then pinkish to vinaceous brown. Spines 1–5 mm (0.02–0.2 in), white then crimson-brown. *Stipe* 1–6 × 0.5–2 cm (0.4–2.4 × 0.2–0.8 in), fusoid, velvety or felty, concolorous. *Flesh* thick, sour smell, acrid taste.
Microscopic features as *H. concrescens* but Sp 5–5.5 × 3.5–4 µm, Bas 25–30 × 5–7 µm, hyphae 2.5–6.5 µm diameter.

Season: August–October.
Habitat: Under conifers, especially pine and spruce.

Distribution: Uneven. Globally rare.

H. ferrugineum (Fr.:Fr.) P. Karsten also has red drops but the smell is mealy and the taste mild.

---

p.64 *Phellodon niger* (Fr.:Fr.) P. Karsten
Bankeraceae, Thelephorales

Description: *Frb hydnoid, black with a white margin. Spines grey-black.*
Frb stipitate. *Cap* often confluent, 2–7 cm (0.8–2.8 in), soon flat or depressed, wrinkled to tomentose or zoned at the edge, blackish purple-blue to jet black, the margin white. Spines 1–3 mm (0.04–0.12 in), white or dark bluish-grey. *Stipe* 1–5 × 0.5–2 cm (0.4–2 × 0.2–0.8 in), cylindric or bulbous, velvety or felty, black. *Flesh* quite thin, forming two distinct layers. *Microscopic features* Sp 3.5–4.5 × 2.5–3.5 µm, ovoid or elliptic, finely spiny. Bas 20–30 × 4–7 µm, cylindro-clavate, without clamps. Monomitic. Hyphae 2–5.5 µm diameter, thin- or slightly thick-walled, sometimes brownish. No clamps.
Season: August–December.
Habitat: Coniferous and broad-leaved woodland.
Distribution: Uneven. Fairly frequent but absent from certain regions.

---

p.64 *Sarcodon imbricatum* (L.:Fr.) P. Karsten
Bankeraceae, Thelephorales

Description: *Frb hydnoid, dark brown, scaly. Spines brown.*
Frb stipitate. *Cap* solitary, 5–20 cm (2–8 in), progressively depressed, coarsely scaly, brown, darker to blackish at the scale tips. Spines 4–10 mm (0.16–0.39 in), whitish then dark brown. *Stipe* 5–10 × 1–5 cm (2–4 × 0.4–2 in), cylindro-clavate, scurfy or smooth, brownish. *Flesh* thick, weak smell, mild taste or becoming bitter. *Microscopic features* Sp 6.5–8.5 × 5–5.5 µm, ovoid or rounded, with large unequal warts, brown. Bas 30–45 × 7–8.5 µm, cylindro-clavate. Monomitic. Hyphae 2–17 µm

diameter, cylindric or sometimes inflated in the flesh.
Season: July–November.
Habitat: Coniferous woods.
Distribution: Tends to be continental or submontane. Declining.

One of the more robust and spectacular hydnoid fungi.

---

p.64 **Ear-pick Fungus**
Auriscalpium vulgare S.F. Gray
Auriscalpiaceae, Hericiales

Description: *Frb hydnoid, slender, brown, hirsute or strigose. Stipe lateral, dark.*
Frb stipitate. *Cap* 0.5–1 cm (0.2–0.4 in), often kidney-shaped or spathulate, felty to hispid, brown. Spines 1–3 mm (0.04–0.12 in), pinkish beige then dark brown-grey. *Stipe* lateral, perpendicular to the cap, 1–15 × 0.1–0.4 cm (0.4–6 × 0.04–0.16 in), hispid, brownish to dark brown. *Flesh* thin. *Microscopic features* Sp 4.5–5.5 × 3.5–4.5 µm, elliptic, finely spiny, amyloid. Bas 2- or 4-spored, 10–25 × 5–6 µm, cylindro-clavate. Cyst 20–45 × 3–6 µm, fusoid-clavate, with refractive granular contents. Dimitic: SH 2.5–4 µm, walls brownish, often irregular; GH 1.5–3.5 µm, thin- or slightly thick-walled and brown in the cuticle.
Season: July–December.
Habitat: Conifer cones, especially pine cones.
Distribution: Widespread. Fairly common.

Unusual on account of the lateral cap and the habitat.

---

p.63 *Hericium clathroides* (Pall.:Fr.) Pers.
(= *H. ramosum*)
Hericiaceae, Hericiales

Description: *Frb hydnoid, lignicolous, compound, white or very pale. Spines up to 1 cm (0.4 in).*
Frb complex, 10–25 cm (4–10 in) diameter, branching, with a thick stalk,

dividing into sometimes confluent branches, white to yellowish cream when old. Spines up to 1 cm (0.4 in), white, hanging, in linear rows under the branches. *Microscopic features* Sp 3.5–4.5 × 3–3.5 µm, elliptic, smooth, amyloid. Bas 15–20 × 3.5–4.5 µm, cylindro-clavate. Exserted hymenial hyphae with refractive contents. Monomitic. Hyphae 5–20 µm diameter, thin-walled or the lower layers slightly thick-walled, sometimes inflated.

Season: September–November.
Habitat: On beech; rare on other broad-leaved trees.
Distribution: Uneven. Rare to very rare.

A magnificent species, strongly threatened in most countries. Very close to *H. coralloides* (Scop.:Fr.) Pers. (a coral-like hydnoid species), with irregularly arranged spines, which occurs on conifers (especially firs).

---

p.62 *Hericium erinaceum* (Bull.:Fr.) Pers.
Hericiaceae, Hericiales

Description: *Frb hydnoid, rounded, lignicolous, with long hanging white spines.*
Frb complex, 10–35 cm (4–14 in) diameter, arising from a thick lateral stipe, simple or with short, thick branches. White to brownish yellow when old. Spines up to 3 cm (1.2 in), hanging and often slightly wavy, like hair, white. *Microscopic features* Sp 5–6 × 4–5 µm, elliptic, finely spiny (difficult to see), amyloid. Bas 20–35 × 5–8 µm, cylindro-clavate. Exserted hymenial hyphae with refractive contents, 5–15 µm. Monomitic. Hyphae 3–20 µm diameter, as in *H. clathroides*.

Season: August–December.
Habitat: Especially oak and beech, more rarely on other broad-leaved trees.
Distribution: Uneven. Rare to very rare.

In the same group, *Creolophus cirrhatus* (Pers.:Fr.) P. Karsten has shorter spines, under more developed yellowish to buff caps (rare, especially on willow and birch).

p.64 **Hedgehog Fungus**
Hydnum repandum L.:Fr.
Hydnaceae, Cantharellales

Description: *Frb hydnoid, fleshy, irregular, white to buff or rufous. Spines brittle.*
Frb stipitate, terrestrial, whitish to ochraceous cream, sometimes more rufous or reddish. *Cap* up to 15 cm (6 in), often irregularly lobed, matt or felted, furrowed or cracked. Spines up to 3 mm (0.12 in), brittle, white then pinkish beige. *Stipe* 2–7 × 1–4 cm (0.8–2.8 × 0.4–1.6 in), cylindric or irregular, often furrowed or tapered, firm and fleshy, slightly brittle. *Flesh* white or very pale, yellowing or browning somewhat at the stipe base, brittle. *Microscopic features* Sp 6–9 × 5–7 µm, elliptic. Bas 30–50 × 5–8 µm, cylindro-clavate. Monomitic. Hyphae in the flesh 2–15 µm, thin-walled, often thick-walled near the septa.
Season: June–November.
Habitat: Broad-leaved and coniferous woodland.
Distribution: Widespread. Fairly common.

A good edible species. Specimens collected under conifers have an unpleasantly strong taste.

---

p.64 *Hydnum rufescens* Sch.:Fr.
Hydnaceae, Cantharellales

Description: *Frb hydnoid, the cap often with a hole, fairly regular, rufous. Stipe central.*
Frb stipitate, terrestrial, rufous or orange to reddish brown. *Cap* up to 6 cm (2.4 in), usually circular and sometimes with a hole in the centre. Surface matt, soon glabrous and smooth. Spines up to 5 mm (0.2 in), brittle, buff then ochraceous beige. *Stipe* 2–7 × 0.3–1 cm (0.8–2.8 × 0.12–0.4 in), cylindric or irregular, brittle and often hollow when old. *Flesh* white or turning yellow to brown, brittle. *Microscopic features* as *H. repandum* (*see* above).
Season: August–December.

Habitat: Broad-leaved and coniferous woodland.
Distribution: Common; like the previous species.

H. rufescens is less robust than *H. repandum* (*see* p.325) and more often tastes bitter.

---

p.66 *Junghuhnia nitida* (Pers.:Fr.) Ryvarden
Steccherinaceae, Polyporales

Description: *Resupinate polypore, orange-buff, the pores very small, regular.*
Frb more or less attached, up to 12 cm (4.7 in) long and 3 mm (0.12 in) thick, orange-buff to a pretty, quite bright pinkish buff colour, sometimes paler. Margin delimited, whitish to cream. Pores fairly regular, 5–7 per mm. *Microscopic features* Sp $3.5–6 \times 2–3$ μm, elliptic. Bas $10–16 \times 4–5$ μm, clavate. Cyst abundant, $40–200 \times 5–12$ μm, arising from the hymenium and distinctly exserted, cylindric, thick-walled, the apex strongly encrusted. Dimitic: SH 2–4 μm; GH 2–4 μm, with clamps.
Season: July–November.
Habitat: On wood of broad-leaved trees; much more rarely on conifers.
Distribution: Widespread. Common.

In colour and appearance similar to *Steccherinum ochraceum* (p.320), which has a spinose hymenium.

---

p.66 *Physisporinus vitraeus* (Pers.:Fr.) P. Karst
Rigidoporaceae, Polyporales

Description: *Resupinate polypore, whitish or pale, glassy.*
Frb fully resupinate, sometimes covering large areas, up to 5 mm (0.2 in) thick, white or sometimes with bluish tones or pale buff. Margin fibrillose. Appearance and consistency glassy to waxy. Pores quite regular, 3–6 per mm (0.04 in), sometimes unequal or partly torn. *Microscopic features* Sp $4.5–6 \times 4–5.5$ μm, subglobose. Bas $15–20 \times 5–8$ μm, clavate, without clamps. Hyphae exserted, the apex sometimes

encrusted. Monomitic. Hyphae 2–6 μm diameter, thin- or slightly thick-walled. No clamps.

Season: Practically all year round.
Habitat: On the ground or rotten wood.
Distribution: Widespread. Common.

Follows the contours of the host.

---

p.66 *Schizopora paradoxa* (Schrad.:Fr.) Donk
Schizoporaceae, Polyporales

Description: *Resupinate species with regular to laciniate pores; white or ochraceous cream.*
Frb more or less attached, sometimes covering large areas, up to 5 mm (0.2 in) thick, white to ochraceous cream. Margin fibrillose, pale. Pores soon torn (1–3 per mm (0.04 in)), sometimes like large teeth or unequal plates, often appearing labyrinthine. *Microscopic features* Sp 4.5–6.5 × 3–4.5 μm, elliptic to ovoid. Bas 12–20 × 4–5 μm, cylindro-clavate or slightly constricted. Cyst small, 15–30 × 3–6 μm, often capitate. Dimitic: SH deep and sometimes rare, 3–4 μm, encrusted with numerous small crystals; GH 2–3 μm.

Season: All year round.
Habitat: Dead wood; rarer on conifers.
Distribution: Widespread.

The pores appear to be less laciniate in *S. flavipora* (Cke) Ryv., which has smaller pores (3.5–5 × 2.5–3.5 μm) and tapering encrusted cystidia and *S. radula* (Pers.:Fr.) Hall, which has spores 4–5.5 × 3–4 μm and fusoid to subcapitate cystidia.

---

p.72 *Coltricia perennis* (L:Fr.) SF Gray
Coltriciaceae, Hymenochaetales

Description: *Stipitate polypore, cap funnel-shaped, zoned, rust-coloured. Pores irregular.*
Cap 2–10 cm (0.8–4 in), fairly regular, soon funnel-shaped. Margin splitting. Surface zoned, alternately velvety and

glabrous, matt or glancing and silky; rusty brown to cinnamon grey-brown. *Stipe* 1–4 × 0.5–1 cm (0.4–1.6 × 0.2–0.4 in), central or slightly excentric, thin, dark brown. *Flesh* leathery. Pores fairly regular, 2–4 per mm (0.04 in), then wider or slightly laciniate, especially near the stipe, yellow brown, shine varies with the light. *Microscopic features* Sp 6–10 × 3.5–5.5 μm, elliptic to cylindro-elliptic, yellowish brown. Bas 15–25 × 5–7.5 μm, clavate, without clamps. Monomitic. Hyphae 2–8 μm, thin-walled or the lower layers slightly thick-walled, hyaline or brownish. No clamps. Setae absent.

Season: August–November.
Habitat: On the ground, especially under conifers.
Distribution: Uneven. Locally threatened.

There are other similar *Coltricia*.

---

p.68 *Inonotus hispidus* (Bull.:Fr.) P. Karsten
Coltriciaceae, Hymenochaetales

Description: *Bracket with pores, thick, strigose, rufous brown. Pores glancing.*
Cap fairly thick, up to 8 cm (3.2 in), semicircular or wider than long, up to 30 × 15 cm (11.8 × 6 in), with erect strigose hairs, reddish orange then rust-coloured, dark red-brown to blackish. *Flesh* fibrous, dense then leathery. Pores (1–3 per mm (0.04 in)) yellow then orange-brown to dark rusty-brown, shine varies with the light. *Microscopic features* Sp 7–11 × 6–8 μm, ovoid to subglobose, brown, slightly thick-walled. Bas 20–35 × 8–11 μm, clavate, without clamps. Setae (sometimes absent), 15–30 × 6–10 μm, fusiform, pointed, thick-walled, brown. Monomitic. Hyphae 3–6 μm diameter. No clamps.

Season: July–September.
Habitat: On standing trunks of various broad-leaved trees.
Distribution: Widespread. Quite rare.

Appears to be in quite rapid decline in Scandinavia.

p.68 *Inonotus dryadeus* (Pers.:Fr.) Murrill
Coltriciaceae, Hymenochaetales

Description: *Polypore, thick, felted, orange brown. Margin yellow with amber droplets. Pores glancing. Cap* fairly thick, up to 15 cm (6 in), semicircular or wider than long, up to 35 × 25 cm (13.8 × 10 in), felted or velvety, pale chamois then dark orange-brown. Margin thick, yellow or paler, with large amber-brown or red-brown droplets which leave spots. *Flesh* fibrous, dense then leathery. Pores (4–6 per mm (0.04 in)) white then yellow-grey to rather rusty brown, very glancing. *Microscopic features* Sp 6–9 × 5–7.5 µm, elliptic to subglobose, thick-walled. Bas 10–15 × 5–9 µm, elliptic, without clamps. Setae (fairly numerous) 20–40 × 10–15 µm, fusiform, pointed, often curved, thick-walled, brown. Monomitic. Hyphae 5–15 µm diameter. No clamps.

Season: August–November.
Habitat: On oak. Very occasionally on conifers.
Distribution: Quite widespread.

I. radiatus (Sow.:Fr.) P. Karsten, is rather small and gregarious, radially grooved and occurs on alder.

---

p.69 *Phellinus hippophaecola* Jahn
Phellinaceae, Hymenochaetales

Description: *Polypore, nodulose, dark grey to rusty-brown.* Polypore, nodulose or hoof-like, up to 10 × 7 cm (4 × 2.7 in), thick, woody, smooth then cracked, concentrically ridged, very hard, dark grey-brown or sometimes blackish but often green with microscopic algae. Pores 5–7 per mm (0.04 in), pale grey then brown-grey to dark brown. *Microscopic features* Sp 6–8 × 5.5–7 µm, subglobose, slightly thick-walled. Bas 10–15 × 8–10 µm, elliptic, without clamps. Cyst and setae absent but occasional presence, between the basidia, of bottle-shaped cells 15–30 × 2–7 µm. Dimitic: SH 2–8 µm; GH 2–4 µm. No clamps.

Season: Practically all year round.
Habitat: On sea buckthorn.
Distribution: Dunes (North Sea and English Channel) and beds of some mountain streams.

Close to *P. robustus* (Karst.) Bourd. & Galz., which is much larger, with brown pointed setae.

---

p.70 **Artist's Fungus**
Ganoderma lipsiense (Batsch) Atkinson
(= *G. applanatum*)
Ganodermataceae, Ganodermatales

Description: *Polypore making thin, zoned, grey-brown brackets. Pores brownish, bitter.*
Brackets semicircular or wider than long, 10–40 × 5–20 cm (4–16 × 2–8 in), often thin, smooth, zoned to tuberculate, the crust very hard, varnished, brown grey but covered with brown spores. *Flesh* woody, pale then becoming brown. Pores 5–6 per mm (0.04 in), white to medium brown, browning when handled, bitter and often deformed by conspicuous galls. *Microscopic features* Sp 6–8.5 × 4.5–6 µm, brown, ovoid, the apex truncate, GP hyaline, the surface punctate. Bas 10–25 × 5–10 µm, clavate. Trimitic: SH 3–7 µm, brown; GH 2–5 µm, with clamps; BH rare, more slender.
Season: Practically all year round.
Habitat: Trunks and stumps of broad-leaved trees.
Distribution: Widespread. Common.

Sometimes up to 1 m (39 in) across. Easy to recognize by the galls deforming the hymenium, caused by the larva of a fly (*Agathomyia wankowiczi*). *G. australe* (Fr.) Patouillard is thicker and usually occurs in well-lit, open places.

---

p.69 *Ganoderma lucidum* (Curt.:Fr.) P. Karsten
Ganodermataceae, Ganodermatales

Description: *Polypore, with a lateral stipe, lacquered, red-brown. Pores white then brown-grey.*

Bracket spathulate or wider than long, 4–10 × 3–8 cm (1.6–4 × 1.2–3.2 in), smooth, zoned, the superficial crust hard, shiny as if lacquered, bright red-brown to orange-brown or yellow at the margin, covered with brown spores. *Flesh* woody, pale then becoming brown. Pores 4–5 per mm (0.04 in), white then brownish cream, browning when handled. *Stipe* 5–20 × 1–3 cm (2–8 × 0.4–1.2 in), lateral, perpendicular to the bracket, cylindric to irregularly tuberculate, shiny as if lacquered. *Microscopic features* Sp 7–11 × 6–8 µm, as in *G. lipsiense*. Bas 10–25 × 5–10 µm, clavate. Dimitic: SH 2–7 µm, yellow-brown, very thick-walled; GH 2–3 µm, with clamps.

Season: June–November.
Habitat: On the ground (buried wood) or stumps of broad-leaved trees; rare on conifers.
Distribution: Quite widespread. Favours warm sites.

G. carnosum Pat., similar but softer, usually occurs under or on conifers.

---

p.68 **Beefsteak Fungus**
Fistulina hepatica (Sch.:Fr.) With.
Fistulinaceae, Polyporales

Description: *Polypore, fleshy, red to red-brown. Tubes not joined together.*
Bracket tuberculate then expanded, semicircular or spathulate, 5–20 × 5–15 cm (2–8 × 2–6 in), smooth to radially streaky, soft when young, bright red then red brown. Tubes separate from each other (4–6 per mm (0.04 in)), sometimes quite long, up to 1 cm (0.4 in). Pores whitish then pinkish to red, finally concolorous. *Microscopic features* Sp 3.5–5 × 2.5–4 µm, ovoid. Bas 15–25 × 5–7 µm, cylindro-clavate. Cyst sometimes present at the mouth of the tubes, cylindric, 30–75 × 5–7 µm. Monomitic. Hyphae 4–20 µm diameter.

Season: July–October.
Habitat: At the base of trunks and stumps of broad-leaved trees; almost always on oak.

Distribution: Widespread. Quite frequent.

The only polypore which is really fleshy when young. Edible but the taste is not memorable.

---

p.70 *Grifola frondosa* (Dicks.:Fr.) S.F. Gray
Grifolaceae, Polyporales

Description: *Polypore, branched with brown caps. Pores pale. Consistency brittle.*
Cap numerous, 3–8 cm (1.2–3.2 in), semicircular or spathulate, indistinctly laterally stipitate, grey, sometimes with violaceous tones then brownish to brown. Frb up to 50 cm (19.7 in) tall and across, fibrous then brittle. Smell sour. Pores white, 2–4 per mm (0.04 in). *Stipe* confluent into a thick, branching collective stem. *Microscopic features* Sp 5–7 × 3.5–5 µm, elliptic. Bas 20–35 × 6–8 µm, cylindro-clavate. Monomitic (or dimitic according to some authors, with SH 2–6 µm). Hyphae 2–5 µm diameter.
Season: August–October.
Habitat: Stumps of broad-leaved trees, especially oak.
Distribution: Widespread. Threatened in some countries.

Flesh fairly tender when young. At this stage, quite a good edible species.

---

p.70 *Dendropolyporus umbellatus* (Pers.:Fr.) Jül.
Polyporaceae, Polyporales

Description: *Polypore, branched, with circular caps, grey to brownish. Pores white.*
Cap very numerous, 1–4 cm (0.4–1.6 in), circular or orbicular to depressed, with a central or slightly eccentric stipe, glabrous or subsquamulose, greyish white then beige to brownish. Frb up to 50 cm (19.7 in) tall and often less across, fleshy then slightly brittle. Pores white, 1–3 per mm (0.04 in), slightly decurrent. *Stipe* confluent into a white or cream stem. *Microscopic features* Sp 7.5–10 × 2.5–4 µm, cylindric. Basidia 2- or

4-spored, 30–45 × 6–10 µm, cylindro-clavate. Dimitic: SH 8–15 µm, few, branched in a tree-like manner; GH 2–12 µm.

Season: August–October.
Habitat: Stumps and roots of various broad-leaved trees.
Distribution: Widespread but declining.

Edible when young. It merits protection on account of its rarity.

---

p.70 *Meripilus giganteus* (Pers.:Fr.) P. Karsten
Grifolaceae, Polyporales

Description: *Very large thin caps, in layers, ochraceous brown. Pores blackening.*
Clump-forming polypore sometimes more than 1 m (3.3 ft) across. *Cap* numerous, up to 30 cm (12 in) diameter, semicircular or spathulate to broadly shell-shaped, tapered near the base, radially fibrillose to squamulose, zoned, yellowish buff then brown, in concentric zones. Short and massive collective stem, brownish or pale. Pores white or pale brownish cream, very small (3–5 per mm (0.04 in)). Blackening everywhere when handled and when rotting. Fleshy then fibrous to brittle. *Microscopic features* Sp 6–7 × 4.5–6 µm, elliptic or subglobose. Bas 20–40 × 7–10 µm, cylindro-clavate, without clamps. Sometimes with cylindric or fusoid hymenial cells, 15–40 × 5–8 µm. Monomitic. Hyphae 3–15 µm diameter, thin- or thick-walled. Clamps rare.

Season: August–November.
Habitat: Stumps of broad-leaved trees, especially oak and beech; very rare on conifers.
Distribution: Uneven. Appears to be rare in some countries, in Scandinavia particularly.

One of the largest polypores in our geographic area, *Bondarzewia mesenterica* (p.346), may look slightly similar but does not blacken. The microscopic characters are also entirely different.

p.67 *Abortiporus biennis* (Bull.:Fr.) Singer
Grifolaceae, Polyporales

Description: *Polypore, pink or reddish brown, often terrestrial. Pores laciniate.*
Cap sessile or laterally and coarsely stalked, circular, funnel-shaped or very irregular, felted to subsquamulose, more or less zoned. Margin often laciniate. White then pinkish to reddish brown or dingy brown. Pores concolorous, 1–3 per mm (0.04 in), decurrent, sinuous then torn into teeth or uneven labyrinthine plates. Frb up to 15 cm (6 in) tall, soft then leathery.
Microscopic features Sp 4–6.5 × 3.5–5 µm, elliptic to subglobose. Bas 20–35 × 4–7 µm, cylindro-clavate. Cyst refractive (gloeocystidia), sometimes rare, 30–100 × 5–10 µm, cylindro-clavate, sometimes constricted. Monomitic. Hyphae 2–5 µm diameter, thin- or slightly thick-walled.
Season: June–November.
Habitat: Often on buried woody débris of various broad-leaved trees; very rare on conifers.
Distribution: Widespread. Locally rare.

The irregular shape, labyrinthine appearance of the hymenophore and pink staining when young make this an easy fungus to recognize. Not edible.

---

p.67 *Hapalopilus rutilans* (Pers.:Fr.) P. Karst
Phaeolaceae, Polyporales

Description: *Polypore, thick, spongy, rufous to cinnamon-coloured. Purple with KOH.*
Bracket, 2–10 cm (0.8–4 in), semicircular or wider than long, subtomentose to felty, looking like a sponge when old, light rufous then cinnamon-rufous. Pores concolorous or somewhat paler, 2–4 per mm (0.04 in), slightly irregular. Consistency spongy then full of holes and brittle when old and dry. *Chemical test* striking purple reaction with strong bases (sodium or potassium hydroxide).
Microscopic features Sp 3.5–5.5 × 2–3 µm, cylindro-elliptic. Bas 15–25 × 5–7 µm,

cylindro-clavate. Hyaline, fusoid, cystidioles 15–25 × 4–6 µm possibly present. Monomitic. Hyphae 2–10 µm diameter, thin- or slightly thick-walled, with extracellular brownish crystals.

Season: June–December.
Habitat: On broad-leaved trees; very rare on conifers.
Distribution: Widespread. Common.

Consistency as in *Oligoporus*.

---

p.71 *Daedalea quercina* (L.:Fr.) Fr.
Deadaleaceae, Polyporales

Description: *Polypore, thick, leathery, greyish beige. Pores labyrinthine or gill-like.*
Bracket 5–20 cm (2–8 in), thick, glabrous, more or less zoned or tuberculate, greyish beige to rufous brown. Pores labyrinthine or almost gill-like, at least near the host, with very thick (up to 3 mm (0.12 in)) concolorous or paler pore walls. Woody.
Microscopic features Sp 5–7 × 2.5–3.5 µm, cylindro-allantoid. Basidia very difficult to see, 20–30 × 5–8 µm, clavate. The hymenial tips of SH erect like a palissade, obtuse or pointed. Trimitic: SH very numerous, 3–6 µm, very thick-walled, light yellowish brown; GH 1.5–4 µm, with clamps; BH 2–6 µm.

Season: All year round.
Habitat: On broad-leaved trees, usually oaks. Parasitic but also saprotrophic, on fence posts, for example.
Distribution: Widespread. Common or fairly common.

---

p.68 **Hoof Fungus, Tinder Fungus**
Fomes fomentarius (L.:Fr.) Fr.
Fomitopsidaceae, Polyporales

Description: *A brownish grey, thick, woody polypore. Pores brownish cream.*
Broadly laterally attached, 10–30 cm (4–12 in), like a hoof (taller than wide) or a rather thin, glabrous, concentrically ridged bracket. Greyish to brownish. Pores circular (4–5 per mm (0.04 in)), concolorous to

brownish grey. Very hard. *Flesh* dark rufous brown. *Microscopic features* Sp 12–20 × 3.5–7.5 µm, cylindric. Bas 20–25 × 5–9 µm, urn-shaped. Cystidioles fusoid, frequent, 20–40 × 3–8 µm. Trimitic: SH 3–7 µm, very thick-walled, light yellowish brown; GH 2–3 µm, with clamps; BH 3–5 µm, strongly branched.

Season: All year round.
Habitat: Various broad-leaved trees; very rare on conifers.
Distribution: Widespread. Common.

The dense fibrous flesh, which burns very slowly, may have been used by prehistoric man to keep the fire alive.

---

p.68 *Fomitopsis pinicola* (Sw.:Fr.) P. Karsten
Fomitopsidaceae, Polyporales

Description: *Polypore, brownish grey, very leathery with a red-brown margin.*
Similar to *F. fomentarius* but the growing edge often red-brown to orange. Pores circular (4–5 per mm (0.04 in)), whitish to pale brownish cream. Very hard. *Microscopic features* Sp 6–9 × 3.5–4.5 µm, cylindro-allantoid. Bas 15–25 × 7–9 µm, clavate. Trimitic: SH 3–6 µm, very thick-walled; GH 2–5 µm, with clamps; BH 1.5–4 µm.
Season: All year round.
Habitat: On conifers; rare on broad-leaved trees.
Distribution: Widespread. Fairly common.

Heterobasidion annosum (Fr.:Fr.) Brefeld, with a distinctly bicoloured fruit body, is also the cause of very serious rot in conifer plantations (appears very low on the trunk, as a reddish brown bracket with a white margin).

---

p.71 *Lenzites betulinus* (L.:Fr.) Fr.
Coriolaceae, Polyporales

Description: *Polypore, flat, zoned, hirsute, greyish to buff, the hymenophore sublamellate.*
Bracket with a broad lateral attachment, 3–12 cm (1.2–4.7 in), fairly thin (up to

1.5 cm (0.6 in) thick), velvety to hirsute, concentrically zoned, pale, whitish to greyish, sometimes green showing to microscopic algae. Hymenophore gilled or almost gilled, the dissepiments often forked or anastomosing near the host, brownish, often with pinkish tones. Leathery, flexible.
Microscopic features Sp 4.5–6.5 × 2–3 μm, cylindric. Bas 20–25 × 5–7 μm, cylindro-clavate. Trimitic: SH 3–7 μm, very thick-walled; GH 2–5 μm, thin- or thick-walled, with clamps; BH 3–10 μm.

Season: Especially in autumn.
Habitat: Usually on birch.
Distribution: Widespread. Common or fairly common.

L. warnieri Dur. & Mont., a more southern species, is larger, slightly thicker and glabrous.

---

p.70 **Blushing Bracket**
Daedaleopsis confragosa (Bolt.:Fr.) Schroet.
Daedaleaceae, Polyporales

Description: *Polypore, flat, zoned, brownish. Pores becoming reddish brown on handling.*
Polypore, resupinate, then forming a bracket 3–15 cm (1.2–6 in), quite thin (up to 2 cm (0.8 in)), sometimes on both sides of the host, semicircular or wider than long, matt, concentrically zoned and ridged, buff, often pale or white at the edge. Pores often slightly radially elongated near the host, white then brownish, spotted purplish pink or red-brown if handled when fresh. Woody.
Microscopic features Sp 7.5–11 × 2–2.5 μm, curved. Bas 20–40 × 3–5 μm, cylindro-clavate. Numerous branched hyaline hyphidia, × 2–3 μm. Trimitic: SH 3–7 μm, very thick-walled; GH 2–6 μm, thin- or thick-walled, with clamps; BH 2–5 μm.

Season: Practically all year round.
Habitat: Broad-leaved trees, especially alder and birch.
Distribution: Widespread. Very common.

There are two other species in the genus with a lamellate hymenophore (*see* below).

p.71 *Daedaleopsis tricolor* (Pers.) Bond. & Sing.
Daedaleaceae, Polyporales

Description: *Polypore, lamellate, flat, dark red-brown.*
Similar to the preceding species but brown-grey then bright red to dark red-brown. Hymenophore lamellate, the 'gills' often forked, thin, brownish then dark brown to grey, the edge paler. Leathery, quite flexible, rigid when dry. *Microscopic features* like *D. confragosa* but hyphidia slightly thick-walled and brownish.

Season: August–November.
Habitat: On various broad-leaved trees.
Distribution: Fairly common in certain areas (although much rarer than the preceding species) but absent in others, particularly in Scandinavia.

In Northern Europe, it is replaced by *D. septentrionalis* (P. Karsten) Niemelä, which remains whitish to brownish; almost exclusively on birch.

---

p.71 *Gloeophyllum saepiarium* (Wulf.:Fr.) Karst.
Gloeophyllaceae, Polyporales

Description: *Polypore, lamellate, flat, dark brown with a yellow margin.*
Polypore resupinate then making a medium-sized bracket, 3–12 cm (1.2–4.7 in), fairly thin (up to 1 cm (0.4 in) thick), wider than long, velvety at the edge then glabrous or radially wrinkled, irregularly concentrically zoned and tuberculate, dark red-brown to black near the host but the margin whitish to bright yellow. Hymenophore lamellate or almost so, the 'gills' distant, often parallel rather than radiating, anastomosing near the host, brown. Leathery, flexible. *Microscopic features* Sp 8.5–11.5 × 3–5 µm, cylindric to allantoid. Bas 20–50 × 4–7 µm, cylindric. Cyst cylindro-fusoid, rather rounded and acute at the apex, tapered, 30–80 × 3–8 µm. Di- to trimitic: SH 3–6 µm, very thick-walled, golden

brownish; GH 2–5 µm, thin- or slightly thick-walled, with clamps; BH rare, 2–5 µm.

Season: Mainly in summer and autumn.
Habitat: Dead conifer wood, especially spruce.
Distribution: Continental to submontane or Nordic.

Quite frequent on wood piles or fences. *G. abietinum* (Bull.:Fr.) P. Karsten, has similarly distant 'gills' which do not, as a rule, anastomose towards the host.

---

p.67 *Pycnoporus cinnabarinus* (Jacq.:Fr.) Karst.
Coriolaceae, Polyporales

Description: *Polypore, cinnabar red. Pores very small.*
Bracket 3–13 cm (1.2–5.1 in), quite thin or thick (up to 4 cm (1.6 in)), semicircular or wider than long (2–7 cm (0.8–2.8 in)), irregularly tuberculate. Margin thick. Orange, orange-red or ochraceous reddish, becoming paler or darkening with age. Pores (2–4 per mm (0.04 in)), concolorous or a darker red. Consistency fibrous, soon dry, not very dense. *Microscopic features* Sp 4–6.5 × 2–3 µm, cylindro-allantoid. Bas 10–20 × 4–7 µm, clavate. Trimitic: SH 2.5–10 µm, very thick-walled; GH 2–5 µm, thin-walled, with clamps; BH 2–5 µm.

Season: August–December.
Habitat: Various broad-leaved trees; very rare on conifers.
Distribution: Continental and submontane but undoubtedly spreading into lowland and Atlantic areas.

No other European polypore is as conspicuous. Other species in the same genus occur in the tropics.

---

p.69 **Chicken of the Woods**
Laetiporus sulfureus (Bull.:Fr.) Murrill
Polyporaceae, Polyporales

Description: *Polypore, orange-yellow then fairly bright yellow.*

Polypore, broadly attached, gregarious and forming more or less compact layers. Frb 10–50 cm (4–20 in), thick then fairly thin (up to 5 cm (2 in)), a thick tubercle, then making fairly thin, semicircular or spathulate brackets, smooth or warted to radially coarsely wrinkled. Margin blunt, undulating, thinner when old. Orange-yellow, bright yellow to yellowish when dry. Pores all but invisible in young material then angular (2–4 per mm (0.04 in)), sulphur-yellow then yellowish to whitish. Slightly fleshy, soon woody or brittle with age. *Microscopic features* Sp 5–8 × 3–5 µm, elliptic. Bas 15–25 × 5–9 µm, clavate, without clamps. Dimitic: GH 5–12 µm, scarcely branched; BH 3–20 µm. Clamps absent.

Season: March–November.
Habitat: Broad-leaved trees, usually oak; exceptionally on conifers.
Distribution: Widespread. Very common.

The bright colour is diagnostic.

---

p.69 **Birch Polypore**
Piptoporus betulinus (Bull.:Fr.) P. Karsten
Polyporaceae, Polyporales

Description: *Polypore, brownish beige, smooth. Margin blunt. Pores very small, white.*
Polypore, attached laterally or on the underside, a globular tubercle, then forming a bracket, 3–26 cm (1.2–10.2 in) wide, thick (up to 6 cm (2.4 in)), often quite regular, very smooth, with a polished then papery, separable, outer crust sometimes cracked when old. Margin thick often undulating. White then beige to brownish chamois. Pores very small and almost invisible to the naked eye, becoming rounded (3–6 per mm (0.04 in)), white. Soon leathery to woody. *Microscopic features* Sp 5–6.5 × 1.5–2 µm, allantoid. Bas 15–25 × 5–6 µm, cylindro-clavate. Dimitic or trimitic: SH 3–12 µm, sometimes branched and acting as BH; GH 2–5 µm, with clamps.

Season: Mainly in summer and autumn.
Habitat: On living or dead birch.
Distribution: Widespread. Very common.

At the end of the season, the hymenophores of old rotten fruitbodies (sometimes fallen) support a parasite which forms white then bright yellow spotted cushions of an Ascomycete, *Hypocrea pulvinata* Fuckel.

---

p.67 **Many-zoned Polypore**
Trametes versicolor (L.:Fr.) Lloyd
Coriolaceae, Polyporales

Description: *Polypore in close-packed tiers, zoned, multicoloured, thin. Pores white.*
Polypore, resupinate on the underside of the host then (sometimes on both sides of the host branch) forming a bracket, 1–10 cm (0.4–4 in), very thin (up to 0.5 cm (0.2 in) at the point of attachment), with alternating, shiny, finely velvety and glabrous zones. Margin sharp, often undulating. Colour variable, with alternating zones of white, grey, bluish, buff, yellow, reddish, brown, violaceous, etc. Margin often white, at least when young and fresh. Pores circular then slightly radially elongated (4–5 per mm (0.04 in)), white to pale brownish. Leathery. *Microscopic features* Sp 5–7 × 1.5–2.5 µm, cylindric to allantoid. Bas 2- or 4-spored, 15–20 × 4–6 µm, cylindro-clavate. Cystidioles fusoid, rare 10–20 × 4–5 µm. Trimitic: SH 4–8 µm; GH 2–3 µm, with clamps; BH 2–5 µm, very branched.
Season: All year round.
Habitat: Broad-leaved trees; rare on conifers.
Distribution: Very common everywhere in Europe.

T. multicolor (Sch.) Jülich (= *T. zonatella*), rather similar but lacking blue or purple colours, is thicker at the point of attachment and triangular in section (edges parallel in *T. versicolor*). The other *Trametes* are less brightly coloured.

p.66 *Trametes pubescens* (Schum.:Fr.) Pilat
Coriolaceae, Polyporales

Description: *Polypore, rather thin, whitish to greyish, very pubescent.*
Like *T. versicolor*, but thick, up to 0.8 cm (0.32 in), less distinctly zoned, all one colour, very pubescent to velvety, becoming glabrous. White to ochraceous beige. Pores (2–5 per mm (0.04 in)) white then buff. Leathery to woody. *Microscopic features* Like *T. versicolor*.
Season: May–November.
Habitat: Dead wood of various broad-leaved trees.
Distribution: Widespread but of varying frequency.

T. hirsuta (Wulf.:Fr.) Pil is hispid and slightly more distinctly zoned; *T. gibbosa* (Pers.:Fr.) Fr. is often larger, thicker (up to 5 cm (2 in) at the point of attachment), the pileus concentrically zoned and tuberculate, whitish, very often green with algae.

p.66 *Bjerkandera adusta* (Willd.:Fr.) P. Karst.
Bjerkanderaceae, Polyporales

Description: *Polypore, resupinate then forming a bracket, blackening. Pores dark grey.*
Bracket 1–4 cm (0.2–1.6 in) (rarely more), thin (up to 1 cm (0.4 in) at the point of attachment), semicircular or fan-shaped, radially wrinkled, more or less zoned in well developed specimens. Margin thin. Buff to brown, blackening. Pores circular then slightly radially elongated (4–6 per mm (0.04 in)), ash-grey to dark grey. Leathery. *Microscopic features* Sp 4.5–6 × 2–3.5 µm, elliptic to subcylindric. Bas 10–15 × 4–5 µm, cylindro-clavate. Monomitic, hyphae 2–6 µm diameter, thin- or thick-walled.
Season: Practically all year round.
Habitat: Broad-leaved trees; rarer on conifers.
Distribution: Widespread. Common.

The grey pores are the best field character.

p.67 *Oligoporus caesius* (Schr.:Fr.) Gilbn & Ryv.
Bjerkanderaceae, Polyporales

Description: *Polypore, spongy, white then blue.*
Polypore, resupinate then forming a bracket 1–6 cm (0.4–2.4 in) (rarely more), thin (but up to 1.5 cm (0.6 in) at the point of attachment), with bristling hairs then pubescent, often reticulate, at least when dry. Margin quite thin. White then with bluish tints, sometimes dark blue when old. Pores frequently angular or irregular (3–5 per mm (0.04 in)), white then bluish to blue. Spongy, more rigid when old and dry. *Microscopic features* Sp 4.5–6 × 1.5–2 μm, subcylindric to curved, weakly amyloid. Bas 10–25 × 5–8 μm, cylindro-clavate. Monomitic, with hyphae 2–6 μm diameter, thin- to very thick-walled, branched.

Season: August–December.
Habitat: Conifers; very rare on broad-leaved trees.
Distribution: Widespread. Common.

O. subcaesius (David) Donger, less intensely bluish, is often thicker, with narrower spores (up to 1.2 μm) and usually occurs on broad-leaved trees.

---

p.71 *Hexagonia nitida* Dur.& Mont. (= *Scenidium nitidum*)
Polyporaceae, Polyporales

Description: *Polypore, brownish grey. Pores hexagonal.*
Bracket 2–10 cm (0.8–4 in) (rarely more), often rather thick (up to 4 cm (1.6 in) at the point of attachment), like a small hoof or semicircular (projecting up to 7 cm (2.8 in)), glabrous, concentrically zoned and tuberculate, brown, sometimes blackish when old, the margin often paler. Pores remarkably regular, in a hexagonal, honeycomb pattern (1–3 mm (0.04–0.12 in) diameter), brownish. Leathery, woody or corky. *Microscopic features* Sp 9–15 × 3.5–5 μm, elliptic to subcylindric. Bas 25–40 × 5–8 μm, clavate. Trimitic: SH 4–6 μm, brown; GH 2–5 μm, with clamps; BH 2–5 μm, yellow-brown.

Season: Practically all year round.
Habitat: Oaks of Mediterranean or southern regions.
Distribution: Southern Europe and around the Mediterranean, reaching the Atlantic coast in Fr.ance. Recorded from Fr.ance, Greece, Italy, Portugal, Spain, Turkey and former Yugoslavia.

The only European representative of the (mainly tropical) genus.

---

p.69 **Dryad's Saddle**
Polyporus squamosus (Huds.:Fr.) Fr.
Polyporaceae, Polyporales

Description: *Polypore with a very short stipe, black at the base. Pileus pale yellow to brownish, with dark scales. Pores white, large.*
Frb up to 40 cm (15.7 in) diameter, often thin (up to 5 cm (2 in)), spathulate then semicircular, pale yellow to yellowish buff then brownish, with warm brown to dark brown often rather regular triangular scales. *Stipe* lateral, short, the base black. Pores angular or slightly elongated radially, fairly large (0.5–2 mm (0.02–0.08 in) diameter), white then yellowish, regular then split. Fleshy then leathery. *Microscopic features* Sp 12–17 × 4–6 µm, narrowly elliptic to cylindric. Bas 40–70 × 6–12 µm, cylindro-clavate. Cystidioles fusoid, rare 20–35 × 5–8 µm. Dimitic: SH 2–8 µm, with tapering apex; GH 3–5 µm, with clamps.
Season: July–November.
Habitat: Trunks and stumps of various broad-leaved trees.
Distribution: Widespread. Common.

P. tuberaster (Jacq.:Fr.) Fr., rather similar, usually occurs on the ground, in contact with woody debris and has a large black hypogeous sclerotium. *P. lentus* Berk. (= *P. forquignonii*), with a more or less anise-like smell, is very close but has no sclerotium and occurs on fallen branches. *P. mori* (Poll.:Fr.) Fr. (= *Favolus europaeus*) has very large and radially elongate pores (up to 5 × 3 mm (0.2 × 0.12 in)).

p.71 *Polyporus durus* (Timm) Kreisel
(= *P. badius* = *P. picipes*)
Polyporaceae, Polyporales

Description: *Polypore, very thin, dark red-brown, shiny, with a lateral stipe. Tubes very short and pores very small. Cap* generally medium-sized (up to 20 cm (8 in) diameter), often thin (up to 1 cm (0.4 in)), spathulate then semicircular or lobed, smooth or with fine radial wrinkles or streaks then shiny, pale greenish grey, finally brownish to dark red-brown, sometimes even black near the point of attachment. *Stipe* more or less lateral, the base black. Tubes very short and pores extremely small (5–8 per mm (0.04 in)), white. Flexible, becoming leathery. *Microscopic features* Sp 6–9 × 3–5 µm, elliptic to cylindric. Bas 15–30 × 6–10 µm, clavate, without clamps. Cystidioles fusoid, rare, 15–20 × 4–7 µm. Dimitic (sometimes trimitic): SH 2–7 µm, with tapering apex; GH 3–5 µm, with clamps, rare in adult specimens.

Season: Usually in summer and autumn.
Habitat: Various broad-leaved trees.
Distribution: Widespread. Common.

P. leptocephalus (Jacq.:Fr.) Fr. (= *P. varius*) is similar but thicker (up to 10 cm (4 in)), buff to brownish yellow; the stipe is eccentric, often better defined and with a black base.

---

p.72 *Polyporus lepideus* Fr.:Fr.
Polyporaceae, Polyporales

Description: *Polypore with a mottled, central stipe. Pileus brownish with a ciliate margin. Pores white, small.*
Frb up to 10 cm (4 in) diameter, thin (up to 0.5 cm (0.2 in)), circular, fibrillo-squamulose then almost smooth. Margin ciliate or fibrillose. Dull brownish with fibrils or darker squamules. Tubes short and pores circular (5–7 per mm (0.04 in)), white then cream, regular. *Stipe* central 2–6 × 0.2–0.8 cm (0.8–2.4 × 0.08–0.95 in), mottled with dull brown on a paler background. Leathery when old.

Microscopic features Sp 5–7 × 1.5–2.5 µm, cylindric or curved. Bas 10–20 × 3.5–6 µm, clavate. Cystidioles fusoid, rare 10–22 × 3–6 µm. Dimitic: SH 2–8 µm, tapering; GH 2–10 µm, with clamps.

Season: Mainly in spring, then in autumn.
Habitat: Buried or fallen wood of broad-leaved trees.
Distribution: Widespread. Common.

P. brumalis (Pers.:Fr.) Fr., a similar species, typically occurs in winter; it has a smooth central stipe and a glabrous margin.

---

p.72 *Albatrellus pes-caprae* (Pers.:Fr.) Pouzar
Scutigeraceae, Cantharellales

Description: *Stipitate polypore. Cap dark brown, scaly. Pores wide, pale.*
Cap up to 10 cm (4 in) diameter, rather thick (up to 2 cm (0.8 in)), spathulate then reniform or irregularly lobed, fibrillo-squamulose to squamulose, dry, brown to blackish when old. Pores large (0.8–1.5 mm (0.01–0.06 in) diameter), angular, white then yellowish. *Stipe* brownish yellow. Fleshy then more leathery to brittle.
Microscopic features Sp 7–11 × 5–8 µm, elliptic. Bas 25–50 × 8–12 µm, cylindro-clavate. Monomitic. Hyphae 4–15 µm diameter, thin- or thick-walled.

Season: August–October.
Habitat: On the ground. Conifer woodland; rare under broad-leaved trees in montane districts.
Distribution: Tends to be continental or submontane. Rare everywhere.

A. cristatus (Sch.:Fr.) Kotl. & Pouz. is more greenish yellow to olivaceous. *A. confluens* (Fr.:Fr.) Kotl. & Pouz. is orange chamois and reddens when dry.

---

p.72 *Bondarzewia mesenterica* (Sch.) Kreisel
(= *B. montana*)
Bondarzewiaceae, Russulales (?)

Description: *Polypore, coarsely stipitate with small fibrillosely scaly brownish caps. Pores pale.*

Polypore, laterally stipitate, with caps (up to 15 cm (6 in) thick), usually spathulate, often confluent. Surface radially fibrillosely scaly to subsquamulose. Buff to ochraceous yellow or brown to crimson-brown. Pores large (1–2 per mm (0.04 in)), angular, white or greyish then yellowish buff. Leathery to brittle when old. *Microscopic features* Sp 6–8 × 5–7 μm, elliptic to subglobose, with long amyloid warts, 1 μm long. Bas 30–55 × 8–12 μm, cylindro-clavate, without clamps. Dimitic: SH 3–8 μm, scarcely branched; GH 2–8 μm. Clamps absent.

Season: August–November.
Habitat: On conifer wood.
Distribution: Quite widespread in western and central Europe. Globally rare.

Similar to *Meripilus giganteus* (p.333) but does not blacken or attain the same huge size. Placed in Russulales on account of the nature of the spores which closely resemble the type found in this order.

---

p.72 *Boletopsis leucomelaena* (Pers.:Fr.) Fayod
Boletopsidaceae, Thelephorales

Description: *Stipitate polypore. Cap black. Pores white then brown.*
Cap up to 10 cm (4 in) diameter, rather thick (up to 4 cm (1.6 in)), circular or slightly lobed, almost smooth to fibrillose-scaly or wrinkled, grey-black to black. Pores small (1–3 per mm (0.04 in)), angular, white then olivaceous grey to grey-brown. *Stipe* thick. Fleshy, brittle then hard. *Microscopic features* Sp 4.5–6.5 × 4–5 μm, broadly elliptic but with irregular angular warts, pale brown. Bas 20–35 × 5–10 μm, cylindro-clavate. Monomitic. Hyphae 3–20 μm diameter.

Season: August–November.
Habitat: On the ground under conifers.
Distribution: Widespread and uneven. Rare everywhere.

B. grisea (Peck) Bond. & Sing. which is more robust and more brownish grey, occurs on conifer-clad heathland and on sandy soil; it is very rare but may lead to confusion.

---

p.78 *Cotylidia pannosa* (Sow.:Fr.) Reid
Corticiaceae, Corticiales

Description: *Frb in rosette-like clusters. Hymenophore smooth. Pale.*
Frb composed of erect, rather thick and undulating plates, arranged more or less in a rosette or in circular groups on a common base (up to 10 cm (4 in) tall and diameter). Sterile surface fibrillose or with small bristling fibrillose scales. Hymenophore smooth, pale, with alternating zones of whitish then buff to brownish, sometimes reddish buff when old. *Microscopic features* Sp 6–9 × 3.5–5.5 µm, elliptic. Bas 50–60 × 5–7 µm, cylindro-clavate, without clamps. Cyst cylindric, 100–150 × 7–13 µm, blunt, slightly thick-walled. Monomitic. Hyphae 2–6 µm diameter, thin- or slightly thick-walled. No clamps.
Season: August–November.
Habitat: Broad-leaved and conifer woodland, sometimes grassy places. On the ground (buried wood).
Distribution: Uneven. Rather rare to rare.

Formerly placed with the *Thelephora* spp., but has the wrong spores for that group.

---

p.78 **Earth Fan**
Thelephora terrestris Ehr.:Fr.
Thelephoraceae, Thelephorales

Description: *Frb petaloid, rosette-like, or festooning the host, fibrillose to fibrillosely scaly. Hymenium brown.*
Frb sessile or erect, thin or fairly thick, up to 10 cm (4 in) tall (often less). Margin thin, fimbriate. Upper surface sterile, fibrillose to fibrillosely scaly, sometimes with bristling when old, brownish grey to brown. Hymenophore smooth to warted, purplish chocolate-brown when mature. No smell. *Microscopic features* Sp 8–12 × 6–9 µm, crimson-brown, with angular

tubercles and warts on the tubercles 0.5 µm long. Bas 40–90 × 8–13 µm, cylindro-clavate. Monomitic. Hyphae 3–10 µm diameter, brownish, slightly thick-walled.

Season: All year round.
Habitat: On the ground or on fallen woody débris. Mycorrhizal, especially with conifers.
Distribution: Widespread. Very common.

---

p.78 *Thelephora palmata* Fr.:Fr.
Thelephoraceae, Thelephorales

Description: *Frb club-like with flattened branches, dark grey brown. Smell fetid.*
Frb often stipitate, like a minute tree, up to 10 cm (4 in) tall and 5 cm (2 in) diameter. Tips flattened, entire. Surface smooth to grooved, grey-brown to dark chocolate-brown. Tips white then paler. Smell strongly fetid. *Microscopic features* Sp 8–11 × 6–9 µm, lobed, brown, with spiny warts on the lobes 0.5–1.5 µm. Bas 60–100 × 9–12 µm, cylindro-clavate. Monomitic. Hyphae 2.5–7 µm diameter, brownish, coloured blue in KOH.

Season: August–November.
Habitat: Woods and heaths, under conifers.
Distribution: Uneven. Fairly common to rare.

The fetid smell is a good diagnostic character.

---

p.78 *Thelephora penicillata* Pers.:Fr.
Thelephoraceae, Thelephorales

Description: *Frb thelephoroid to clavarioid, branched, brush-like, brownish.*
Frb sessile to substipitate, with irregular branches, like a shaggy paint brush, up to 5 cm (2 in) tall and 15 cm (6 in) across. Surface smooth. Buff to brownish, but white to yellowish at the tip. No smell. *Microscopic features* Sp 7–10 × 5–7 µm, irregularly lobed, brown, with spines on the lobes 0.5–1.5 µm. Bas 30–75 × 7–11 µm, cylindro-clavate. Monomitic. Hyphae 3–10 µm diameter, brownish.

Season: Practically all year round.
Habitat: Broad-leaved and coniferous woodland, coppice woodland, clearings, wood edges, etc.
Distribution: Widespread. Fairly common.

---

p.78 *Artomyces pyxidatus* (Pers.:Fr.) Jülich
Clavicoronaceae, Hericiales

Description: *Club-like, the branches tipped with small trumpets. Lignicolous.*
Frb club-like, up to 10 cm (4 in), elegant, with whorled layers of branches. White or pale yellow then rather pale ochraceous yellow. Tips trumpet-shaped, with short appendages. *Microscopic features* Sp 4–5.5 × 2–3 µm, elliptic or slightly tear-shaped, with very fine spines, amyloid. Bas 20–30 × 3.5–4.5 µm, cylindric. Cyst 15–45 × 3–7 µm, cylindro-fusoid. Monomitic. Hyphae 3–12 µm diameter, with oily or cloudy and refractive contents, exserted.
Season: September–December.
Habitat: Very rotten wood of broad-leaved trees; rarely on conifers.
Distribution: Uneven. Rare to very rare, tends to be continental or submontane.

Very characteristic appearance.

---

p.77 *Pterula multifida* (Chev.) Fr.
Pterulaceae, Clavariales

Description: *Club-like with very fine branches, violaceous cream then brownish, fetid.*
Frb up to 6 cm (2.4 in), in hair-like tufts, whitish then lilac-grey to ochraceous yellow or rather pale brown. Tips very slender, pale. Smell fetid. *Microscopic features* Sp 5.5–8 × 2.5–3.5 µm, elliptic or clavate. Bas 2- or 4-spored, 25–35 × 5–8 µm, cylindric. Dimitic: SH 3–6 µm; GH 3–5 µm, with clamps.
Season: August–December.
Habitat: Conifers; very rare under broad-leaved trees.
Distribution: Widespread but uneven. Rare.

p.77 **Cauliflower or Brain Fungus**
Sparassis crispa (Wulf.:Fr.) Fr.
Sparassidaceae, Clavariales

Description: *A voluminous club fungus with flattened, wavy branches. White or pale ochraceous beige.*
Frb up to 40 cm (15.7 in) tall and across, with numerous flattened, wavy branches, looking more or less like a cauliflower, white or pale ochraceous yellow to pale yellowish-brown. *Microscopic features* Sp 4.5–7 × 3–4.5 µm, elliptic or subovoid. Bas 40–60 × 5–8 µm, cylindric. Hyphae 2–40 µm diameter, thick or inflated in the flesh.
Season: August–November.
Habitat: Conifer roots, especially pine.
Distribution: Widespread. Rather rare.

Edible but should be respected on account of its rarity. *S. laminosa* Fr. is fairly frequent and differs in its longer and more flattened, more flexible, even drooping branches. Restricted to broad-leaved trees.

---

p.76 *Clavulina cristata* (Holmskj.:Fr.) Schroet.
Clavulinaceae, Clavariales

Description: *Club-like with crested tips. White or pale.*
Frb up to 12 cm (4.7 in) (often less), irregularly branched, white or pale, sometimes with grey, purplish, beige or buff tones. Tips delicate, finely crested, sometimes brush-like. Relatively flexible or brittle. *Microscopic features* Sp 7–11 × 6–10 µm, subglobose. Bas 2-spored, 40–60 × 5–8 µm, cylindro-clavate, with large, broadly curving sterigmata. Hyphae 3–15 µm diameter.
Season: June–December.
Habitat: Undergrowth, sometimes clearings, wood edges, etc.
Distribution: Very widespread. Common.

Very variable. *C. cinerea* (Bull.:Fr.) Schroet., which is less crested, is grey. *C. rugosa* (Bull.:Fr.) Schroet., which is less branched (sometimes simple) and with more or less misshapen or wrinkled fruit bodies, is often white.

p.74 *Clavaria falcata* Pers.:Fr. (= *C. acuta*)
Clavariaceae, Clavariales

Description: *Simple club fungus. Short sterile stipe. White.*
Frb simple, up to 8 cm (3.2 in) tall and 3 mm (0.12 in) diameter, the sterile base (stipe) short and more slender (5–20 × 1–1.5 mm (0.08–0.8 × 0.04–0.06 in)), often slightly translucent, white. Rather brittle. *Microscopic features* Sp 7–10 × 5–9 µm, elliptic to subglobose. Basidia 2- or 4-spored, 25–55 × 8–10 µm, cylindric, with a medallion clamp at the base. Hyphae in the flesh 4–30 µm, often inflated, without clamps.
Season: July–December.
Habitat: Woods or short turf and meadows; in grass.
Distribution: Widespread. Common or fairly common.

Appears to be declining. The white species with simple fruit bodies are a complex group.

---

p.76 *Clavaria zollingeri* Léveillé
Clavariaceae, Clavariales

Description: *Branched club fungus, very fragile, violaceous to lilac.*
Frb up to 8 cm (3.2 in), elegant, branching irregularly or sometimes simple, purple, becoming paler when old. Tips blunt or sometimes more densely branched. *Microscopic features* Sp 4–7 × 3–5 µm, elliptic to subglobose. Bas 40–60 × 5–9 µm, cylindro-clavate, without clamps. Hyphae 4–25 µm diameter, often inflated.
Season: August–November.
Habitat: Undergrowth, often on bare ground, especially on calcareous clays.
Distribution: Uneven. Becoming very rare in several countries.

---

p.75 *Clavulinopsis corniculata* (Sch.:Fr.) Corner
Clavariaceae, Clavariales

Description: *Club-like, very branched. Orange-yellow.*

Frb up to 8 cm (3.2 in), dichotomously branched, orange-yellow to rather pale ochraceous yellow or brownish from the base up. Tips slender and often rather pointed and divergent. *Microscopic features* Sp 4.5–7 µm, spherical or irregularly subglobose, hyaline or with internal yellow pigments. Bas 35–60 × 7–9 µm, cylindro-clavate. Hyphae 2–12 µm diameter, often inflated.

Season: September–January.
Habitat: Short, unimproved grassland and meadows, also in scrub.
Distribution: Widespread. Rather rare.

---

p.75 *Clavulinopsis fusiformis* (Sow.:Fr.) Corner
Clavariaceae, Clavariales

Description: *Simple club fungus, fusiform. Bright yellow.*
Frb up to 15 cm (6 in), simple but in tufts, fusoid, bright yellow, tapering, often rufous above. *Microscopic features* Sp 5–9 × 4.5–8.5 µm, subglobose to elliptic, hyaline or with internal yellow pigment. Bas 40–65 × 6–9 µm, cylindro-clavate. Lower layers of hyphae 2–10 µm, often inflated.
Season: September–December.
Habitat: Woods or open places.
Distribution: Uneven. Rather rare, declining.

---

p.74 *Typhula quisquilaris* (Fr.:Fr.) Corner
Typhulaceae, Clavariales

Description: *Minute white club fungus with a clavate head and finely pubescent stipe.*
Frb up to 5 mm (0.2 in) tall, simple, clavate, the head more or less differentiated, blunt, the stipe more or less distinct, slightly pubescent. White, the stipe sometimes more or less translucent. *Microscopic features* Sp 9–11.5 × 4.5–5 µm, elliptic, subcylindric or slightly almond-shaped. Bas 40–70 × 6–9 µm, cylindro-clavate. Hyphae 2–8 µm diameter.
Cyst (stipe) inflated at the base with a long cylindric blunt neck, thick-walled, 50–120 × 2–8 µm.

Season: September–November.
Habitat: On dead bracken stalks.
Distribution: Uneven. Overlooked.

Typhula species such as this are typically very hard to see. The (many) species are often associated with very specific substrates.

---

p.75 *Macrotyphula fistulosa* (Vahl.:Fr.) Petersen
Clavariadelphaceae, Clavariales

Description: *Club-like, cylindric-clavate, hollow, blunt. Brownish.*
Frb up to 25 cm (9.8 in), simple or very rarely forked, long and cylindric to faintly club-shaped, often longitudinally grooved, blunt, hollow, buff to brownish.
Microscopic features Sp 10–18 × 5–8 µm, elliptic to fusiform. Bas 50–80 × 6–12 µm, cylindro-clavate. Hyphae 2–20 µm diameter.
Season: August–December.
Habitat: On the ground or on wood, in woods or wood edges.
Distribution: Widespread. Common.

Var. *contorta* (Holmskj.:Fr.) Courtecuisse, is smaller, often misshapen and contorted; it occurs more often in tufts and is always lignicolous.

---

p.75 *Clavariadelphus pistillaris* (L.:Fr.) Donk
Clavariadelphaceae, Clavariales

Description: *Robust club fungus. Yellow-buff.*
Frb up to 30 cm (11.8 in) tall, simple, rather broadly clavate (up to 2 cm (0.8 in) diameter), the tip rounded, yellow-buff, browner when old and staining violaceous brown when handled. *Flesh* pale, slightly bitter. *Chemical test* dark green in $FeSO_4$. *Microscopic features* Sp 11–15 × 6–10 µm, elliptic. Bas 50–75 × 9–12 µm, cylindro-clavate. Hyphae 5–16 µm diameter.
Season: August–December.

Habitat:	Broad-leaved trees, often under beech on rather calcareous soils.
Distribution:	Uneven. Rather rare.

Thought to be threatened in several countries. *C. truncatus* (Q.) Donk is robust, truncated or flattened at the tip.

---

p.76 *Ramariopsis kunzei* (Fr.:Fr.) Donk
Clavariaceae, Clavariales

Description: *Club-like, very pale, finely branched, the basal stem rather slender, small.*
Frb up to 10 cm (4 in) (often less), elegant, finely and rather densely branched, the base rather slender, white to pale cream, rarely pale brownish. Tips often pointed.
Microscopic features Sp 3–5.5 × 2–4.5 µm, elliptic to subglobose, finely spiny. Bas 20–45 × 5–7 µm, cylindro-clavate. Hyphae 3–12 µm diameter.
Season: August–December.
Habitat: Woods or wood edges.
Distribution: Uneven. Rather rare or very rare.

---

p.76 *Ramaria gracilis* (Pers.:Fr.) Quélet
Ramariaceae, Clavariales

Description: *Club-like, very branched, slender, pale, anise-like smell.*
Frb up to 8 cm (3.2 in) tall and 4 cm (1.6 in) diameter, rather densely branched, the base rather thin, with forked and sometimes entangled branches, white to pinkish beige or pale brownish. Smell of anise. *Microscopic features* Sp 5–7 × 3–4.5 µm, elliptic, finely warted or almost spiny. Bas 25–45 × 5–7 µm, cylindro-clavate. Hyphae 2–10 µm diameter.
Season: August–December.
Habitat: Conifers.
Distribution: Uneven. Sometimes very rare.

The smell of anise distinguishes this species from *Ramariopsis kunzei* (*see* above) and *Clavulina cristata* (p.351).

p.76 *Ramaria ochraceovirens* (Jungh.) Donk
Ramariaceae, Clavariales

Description: *Medium-sized club fungus, branched, mustard yellow then dark green.*
Frb up to 6 cm (2.4 in) tall and 4 cm (1.6 in) diameter, quite densely branched, base of medium thickness, with irregular, forked branches, buff to mustard yellow, quickly olive or dark green with handling or when old. *Microscopic features* Sp 5.5–9.5 × 3–5 µm, elliptic or slightly tear-shaped, hyaline to pale brownish, finely warted or roughened. Bas 30–50 × 5–8 µm, cylindro-clavate. Hyphae 2–10 µm diameter.
Season: August–December.
Habitat: Conifer woods.
Distribution: Uneven. Common or rather rare.

R. abietina (Pers.:Fr.) Quélet is very similar and found in same habitat; does not green.

---

p.76 *Ramaria stricta* (Pers.:Fr.) Quélet
Ramariaceae, Clavariales

Description: *Club-like with long branches. Yellow buff, becoming brown.*
Frb up to 10 cm (4 in) tall, often narrow, but up to 8 cm (3.2 in) diameter, the base medium-sized, with very strict, rather dense branches, yellow buff to brownish buff. Tips more yellow, spotting reddish brown when handled. *Microscopic features* Sp 7–10 × 4–5 µm, elliptic, brownish, finely warted. Bas 25–40 × 7–10 µm, cylindro-clavate. Hyphae 3–10 µm diameter.
Season: August–November.
Habitat: Rotten wood of broad-leaved trees.
Distribution: Widespread. Fairly common or quite rare.

One of the commonest *Ramaria* spp.

---

p.77 *Ramaria formosa* (Fr.:Fr.) Quélet
Ramariaceae, Clavariales

Description: *Club-like, robust, pinkish, with yellow tips.*
Frb up to 20 cm (8 in) tall and 15 cm (6 in)

diameter, robust, very branched, the base thick, ochraceous pink, beige, salmon, orange. Tips yellow. *Flesh* rather thick, white, flushing slightly pink or pinkish brown when cut. *Microscopic features* Sp 8–15 × 4–6 μm, cylindro-elliptic, hyaline to yellowish, warted, the warts more or less in lines and confluent. Bas 40–60 × 7–10 μm, cylindro-clavate. Hyphae 3–12 μm diameter.

Season: August–November.
Habitat: Broad-leaved trees, usually beech.
Distribution: Uneven. Continental or southern, sometimes rare to very rare.

Threatened in several countries. Poisonous: causes a gastro-intestinal condition and violent diarrhoea. Other *Ramaria* spp. cause the same adverse reactions but a few species are edible. The genus is difficult to identify so discretion is essential.

---

p.77 *Ramaria botrytis* (Pers.:Fr.) Ricken
Ramariaceae, Clavariales

Description: *Club-like, robust, pale cream, the tips vinaceous pink, flesh white.*
Frb up to 15 cm (6 in) tall and 20 cm (8 in) diameter, stout, very branched, the base very thick, whitish cream to very pale beige. Tips pink or purplish pink. *Flesh* thick, white, unchanging. *Microscopic features* Sp 11–17 × 4–8 μm, elliptic to subfusoid, hyaline to yellowish, faintly longitudinally striate. Bas 50–70 × 7–12 μm, cylindro-clavate. Hyphae 3–12 μm diameter.

Season: August–November.
Habitat: Broad-leaved trees and sometimes conifers.
Distribution: Uneven: continental or in hilly districts.

Threatened in many areas. An example of an edible *Ramaria* (*see R. formosa*).

---

p.77 *Ramaria largentii* Marr & Stuntz
Ramariaceae, Clavariales

Description: *Club-like, robust, orange or orange yellow, flesh white.*

Frb up to 15 cm (6 in) tall and 12 cm (4.7 in) diameter, stout, very branched, the base very thick, orange-yellow to bright orange, paler when old. Tips darker or more brightly coloured. *Flesh* white, unchanging. *Microscopic features* Sp 10–16 × 4–6 μm, cylindro-elliptic, hyaline to pale yellowish, with irregularly arranged warts. Bas 60–100 × 8–15 μm, cylindro-clavate. Hyphae 3–8 μm diameter.

Season: July–November.
Habitat: Conifers.
Distribution: Mountain areas of western Europe.

Perhaps the commonest club fungus in some areas of the Alps. In lowland areas, it is replaced by other bright yellow species such as *R. aurea* (Sch.) Q., which is less coralloid and occurs under broad-leaved trees.

---

p.80 **Horn of Plenty**
Craterellus cornucopioides (L.:Fr.) Pers.
Craterellaceae, Cantharellales

Description: *Frb trumpet-shaped, hollow, black. Hymenium smooth.*
Frb up to 10 cm (4 in) tall and 7 cm (2.8 in) diameter. Margin thin, wavy, often split. Inner surface fibrillose to finely scaly, black or dark grey. Fertile surface smooth to finely wrinkled, paler (pale grey, bluish grey, etc.). *Stipe* not clearly differentiated, hollow right to the base. *Flesh* very thin, brittle. *Microscopic features* Sp 10–17 × 6–11 μm, elliptic. Bas 2-spored, 60–100 × 7–10 μm, cylindro-clavate. Hyphae 3–15 μm diameter, not very septate. No clamps.

Season: June–November.
Habitat: Broad-leaved woodland on calcareous soils, damp places.
Distribution: Widespread, frequency uneven.

Despite its widespread distribution, this species appears to be threatened in some countries. It is excellent to eat. *C. konradii* Bourdot & Maire is smaller and usually rather bright yellow.

p.80 *Pseudocraterellus undulatus* (Pers.:Fr.)
S. Rauschert (= *P. sinuosus*)
Cantharellaceae, Cantharellales

Description: *Frb stalked, funnel-like, lobed, brownish buff. Hymenophore smooth or nearly so, grey.*
Frb more or less hollow, up to 5 cm (2 in) tall and 4 cm (1.6 in) diameter, wavy or frilly. Surface smooth to fibrillose–finely scaly. Brown to yellowish buff or grey to dark brown. Fertile surface smooth or with indistinct branching folds, lighter in colour, grey. *Stipe* solid, short, sometimes bearing several caps. *Flesh* rather thin. Smell faint. *Microscopic features* Sp 9–13 × 7–9 µm, elliptic. Bas 2- or more often 4-spored, 60–120 × 6–12 µm, cylindro-clavate. Hyphae 3–15 µm diameter, the septa close together. No clamps.
Season: June–November.
Habitat: Broad-leaved woodland, sometimes in tufts.
Distribution: Widespread. Common or fairly common.

Var. *crispus* (Sow.) Courtecuisse, which is paler and occurs in damp places, has an even more tightly crisped margin.

p.82 **Chanterelle**
Cantharellus cibarius Fr.:Fr.
Cantharellaceae, Cantharellales

Description: *Frb funnel-shaped, fleshy, orange-yellow. Forked folds. Stipe solid.*
Frb top-shaped, more or less hollow, up to 15 cm (6 in) tall and 10 cm (4 in) diameter (often less). Typically orange-yellow. Fertile surface with rather dense forked folds, concolorous or paler. *Stipe* solid, fairly short, concolorous. *Flesh* rather thick. *Microscopic features* Sp 8–11 × 5–6.5 µm, elliptic. Bas 50–110 × 6–10 µm, long and cylindro-clavate. Hyphae 2–5 µm diameter.
Season: June–December.
Habitat: In woods, under broad-leaved trees and conifers.
Distribution: Widespread. Appears to be seriously declining in some countries.

This excellent edible species has a number of colour varieties, including var. *amethysteus* Q which has small lilac scales on the cap and var. *alborufescens* Mal, a southern species, which is very pale but soon spotted rusty. *C. friesii* Q, which is a much brighter orange, and often slender, smells strongly of apricot. See also Hygrophoropsis aurantiaca (p.680).

---

p.81 *Cantharellus tubiformis* Fr.:Fr.
Cantharellaceae, Cantharellales

Description: *Frb funnel-shaped, not very fleshy, grey-brown. Folds buff to yellow.*
Frb more or less hollow when mature, up to 8 cm (3.2 in) tall and 4 cm (1.6 in) diameter. Margin undulating, thin. Surface smooth to fibrillose-finely scaly, grey-brown with a yellowish buff edge, sometimes more uniformly yellowish. Folds forked, yellowish grey to fairly bright yellow. *Stipe* solid or slightly hollow, tall, often grooved, yellowish grey or fairly bright yellow. *Flesh* rather thin. Smell weak. *Microscopic features* Sp 8–11 × 5.5–8 µm, broadly elliptic. Bas 60–90 × 7–11 µm, cylindro-clavate. Hyphae 3–15 µm diameter.

Season: June–December.
Habitat: Conifer woods and also under broad-leaved trees.
Distribution: Widespread, frequency uneven.

A very good edible species. Brighter yellow forms are sometimes confused with *C. lutescens* (see below) which differs in having a smooth hymenophore.

---

p.80 *Cantharellus lutescens* (Pers.:Fr.) Fr.
Cantharellaceae, Cantharellales

Description: *Similar to* C. tubiformis *but the hymenophore smooth and more brightly coloured.*
Frb warm brown with a cinnamon or yellow margin. Fertile surface smooth or nearly so, lighter in colour, yellowish cream to fairly bright yellow. *Stipe* solid or hollow, bright yellow to a luminous orange-yellow. *Flesh*

rather thin. Smell faint. *Microscopic features* Sp 9–12 × 7–9 µm. Other characters as for *C. tubiformis*.

Season: June–December.
Distribution: Declining.

Despite the smooth or almost smooth hymenophore, this is a true *Cantharellus*.

---

p.82 *Cantharellus cinereus* (Pers.:Fr.) Fr.
Cantharellaceae, Cantharellales

Description: *Frb funnel-shaped, very dark. Folds light grey. Smell fruity.*
Frb more or less hollow, up to 10 cm (4 in) tall and 6 cm (2.4 in) diameter. Margin wavy. Surface fibrillose to very finely scaly. Grey-brown to blackish or black. Folds prominent, forking, light grey or almost white. *Stipe* solid or hollow. *Flesh* rather thin. Smell strongly fruity. *Microscopic features* Sp 7–12 × 5–9 µm, elliptic or ovoid. Bas 4-spored (sometimes with five or six sterigmata), 50–90 × 6–10 µm, cylindro-clavate. Hyphae 3–12 µm diameter.
Season: July–November.
Habitat: Broad-leaved woodland or clearings.
Distribution: Uneven. Fairly common or rare.

---

p.81 *Cantharellus melanoxeros* Desm.
Cantharellaceae, Cantharellales

Description: *Frb funnel-shaped. Folds violaceous grey. Cap yellow. Blackening.*
Frb more or less hollow, up to 10 cm (4 in) tall and 7 cm (2.8 in) diameter. Margin flexuous or undulating. Surface smooth to fibrillose or finely scaly. Yellowish buff to dingy yellow, sometimes with purplish colours at the edge. Blackening from the margin. Folds forking and anastomosing, violaceous grey. *Stipe* solid, thick, egg-yellow, blackening. *Flesh* rather thin, fragile, cream to yellowish, blackening. *Microscopic features* Sp 9–12 × 6–8 µm, elliptic. Bas 60–120 × 8–12 µm, cylindro-clavate. Hyphae 5–15 µm diameter.

Season: August–November.
Habitat: Broad-leaved coppice woodland on damp calcareous soils.
Distribution: Uneven and not fully understood.

Rare, often confused with *C. ianthinoxanthus* (Mre) Kühn, which is also rare, but is often smaller, stout, the cap darker, the folds pinkish lilac, not very distinct, the flesh white and not blackening. It occurs in the same habitats.

---

p.81 *Gomphus clavatus* (Pers.:Fr.) S.F. Gray
Gomphaceae, Cantharellales

Description: *Frb purple, funnel-shaped and fleshy. Folds thick and anastomosing.*
Frb fleshy, not very hollow, up to 10 cm (4 in) tall and 6 cm (2.4 in) diameter. Surface smooth or tuberculate. Purple to purplish brown, becoming paler when old. Folds forking and partially anastomosing, concolorous or more yellowish. *Stipe* thick and short. *Flesh* thick, pale. Smell faint. *Microscopic features* Sp 10–16 × 4–7 µm, elliptic to cylindro-elliptic, yellowish, warted. Bas 50–90 × 8–11 µm, cylindro-clavate. Hyphae 3–12 µm diameter.
Season: August–November.
Distribution: Mountain districts (very rare elsewhere).

Related to the club fungi, especially *Ramaria*, by virtue of the ornamented spores.

---

p.84 **Split-gill**
Schizophyllum commune Fr.:Fr.
Schizophyllaceae, Polyporales

Description: *Pleurotoid, strigose, grey. Gills split.*
Frb pleurotoid. *Cap* diameter 1–3 cm (0.4–1.2 in), semicircular or reniform, domed then flat. Margin inrolled then expanded, sometimes lobed and scalloped. Surface strigose. White to more or less dark grey. Gills radiating, converging towards the point of attachment, almost concolorous or paler, sometimes buff or

slightly violaceous, divided into two little flanges which recurve in opposite directions. *Flesh* elastic becoming leathery when dry. *Microscopic features* Sp 5.5–7 × 2–2.5 µm, cylindro-elliptic or slightly allantoid. Bas 15–25 × 3–6 µm, clavate. Monomitic. Hyphae 3–5 µm diameter.

Season: All year round.
Habitat: On wood. Ubiquitous.
Distribution: Cosmopolitan.

Despite the gilled hymenophore, this species is closer to some corticioid fungi than to the pleurotoid group.

B. AGARICOMYCETIDEAE

This second subclass of the Homobasidiomycetes groups together all the typically gilled forms. Some, however, are poroid (in particular, the boletes). The systematic arrangement of the principal orders and families in this subclass is as follows:

- **Order Tricholomatales**: spores white or pale; gills decurrent to adnate-notched or ascending to almost free; partial veil forming a cortina or sometimes a ring, or absent; universal veil absent.

    Pleurotaceae (*Pleurotus* and related genera, *see* pp.683–8).
    Hygrophoraceae (*Hygrocybe* and related genera, *see* pp.562–91).
    Tricholomataceae (*Tricholoma* and related genera, *see* pp.805–65).
    Marasmiaceae (*Marasmius* and related genera, *see* pp.620–72).
    Dermolomataceae (*Dermoloma* and related genera, *see* pp.625, 672–9).

- **Order Agaricales**: spores white or brown to black (exceptionally red or green); gills adnate to notched or free; partial veil absent or present, floccose to ring-like; universal veil absent or very reduced.

    Lepiotaceae (*Lepiota* and related genera, *see* pp.591–620).
    Agaricaceae (the genus *Agaricus*, *see* pp.365–77).
    Coprinaceae (*Coprinus* and related genera, *see* pp.434–54).

- **Order Amanitales:** spores white; gills free; partial veil present, ring-like, sometimes reduced; universal veil present.
    Amanitaceae (*Amanita* and related genera, see pp.377–96).
- **Order Pluteales:** spores pink; gills free; partial veil absent; universal veil absent or present.
    Pluteaceae (*Pluteus* and related genera, see pp.689–98).
- **Order Entolomatales:** spores pink; gills decurrent, adnate or notched to almost free; partial veil absent; universal veil absent.
    Entolomataceae (*Entoloma* and related genera, see pp.535–59).
    Macrocystidiaceae (*Macrocystidia*, see p.560) and **Rhodotaceae** (*Rhodotus*, see p. 559) will be included here.
- **Order Cortinariales:** spores brown to black or violaceous; gills often adnate, notched, sometimes almost free; partial veil absent or present, forming a cortina or a ring; universal veil absent or very reduced.
    Cortinariaceae (*Cortinarius* and related genera, see pp.455–521).
    Crepidotaceae (*Crepidotus* and related genera, see pp.521–34).
    Bolbitiaceae (*Bolbitius* and related genera, see pp.396–408).
    Strophariaceae (*Stropharia* and related genera, see pp.788–805).
- **Order Russulales:** symbiotic (mycorrhizal) species; spores pale (white, cream, buff or yellow); gills decurrent or notched; partial veil absent; universal veil absent (traces exceptionally); flesh with sphaerocysts (hence the brittle–ranular texture); spores with amyloid ornamentation.
    Russulaceae (*Russula* and related genera, see pp.698–788).
- **Order Boletales:** saprotrophic or symbiotic (mycorrhizal) species; spores white to brown or black; hymenophore gilled or tubular-poroid, decurrent, adnate or ascending and almost free; partial veil absent or present (ring-like);

universal veil normally absent.
Hygrophoropsidaceae (*Hygrophoropsis* and related genera, *see* pp.680–1).
Omphalotaceae (*Omphalotus* and related genera, *see* pp.681–3).
Paxillaceae (*Paxillus* and related genera, *see* p.679).
Gomphidiaceae (*Gomphidius* and related genera, *see* pp.560–2).
Boletaceae (*Boletus* and related genera, *see* pp.409–10, 412–34).
Gyrodontaceae (*Gyroporus*, *see* pp.410–12).
Strobilomycetaceae (*Strobilomyces*, *see* pp.408–9).

In this work, the main families are listed in alphabetical order: *Agaricaceae, Amanitaceae, Bolbitiaceae, Boletaceae* and allies, *Coprinaceae, Cortinariaceae, Crepidotaceae, Entolomataceae* and allies, *Gomphidiaceae, Hygrophoraceae, Lepiotaceae, Marasmiaceae* and allies, *Paxillaceae* and allies, *Pleurotaceae, Pluteaceae, Russulaceae, Strophariaceae,* and *Tricholomataceae*. A definition of each of these families is given before the description of the species illustrated.

Agaricaceae: the 'mushrooms'
The *Agaricaceae* comprises the genus *Agaricus*, formerly *Psalliota*, which includes species such as the field mushroom and the cultivated mushroom. This family is defined thus: fleshy species with a fibrous texture, free gills, dark brown to black spores, partial veil present, more or less complex.

Classification of the genus *Agaricus* is based on two main criteria: the nature and complexity of the veil and the colour of the flesh when cut or handled. Another significant character is the presence or absence of cheilocystidia. In the following descriptions, the characters that are common to all are not detailed. These are: gills close, free, ventricose; smell like the cultivated mushroom; spores dark brown; basidia 4-spored.

p.176 Field Mushroom
Agaricus campestris L.:Fr.
Agaricaceae, Agaricales

Description: *Cap pale. Gills bright pink then black. Stipe fusiform with a fleeting ring.*
Cap 4–10 cm (1.6–4 in), convex, smooth, silky, occasionally fibrillose-scaly (variety). White, occasionally pale beige or darker (varieties). *Gills* bright pink then rather pinkish brown to dark brown. Gill edge concolorous.
Stipe 4–6 × 0.5–1.5 cm (1.6–2.4 × 0.2–0.6 in), tapering toward the base. Veil mixed but fleeting, often cortinate or a simple ring. *Flesh* white, becoming pink. *Microscopic features* Sp 7–9 × 4–5.5 µm, ovoid to elliptic. Cheilo absent.
Season: July–November.
Habitat: Meadows and grassy places.
Distribution: Widespread. Appears to be declining.

The field mushroom is a widely collected edible species. Variable cap colour and scaliness.

---

p.180 *Agaricus cupreobrunneus* (J. Schaeffer & Steer ex Moeller) Pilát
Agaricaceae, Agaricales

Description: *Cap fibrillose, golden brown. Gills pink then brown. Stipe fusiform with a fragile ring.*
Cap 4–8 cm (1.6–3.2 in), the top occasionally flattened or slightly depressed, streaky, fibrillose or with appressed scales, sometimes shining, vinaceous brown, with golden lights. *Gills* pink then dark brown. Gill edge thin, not frosted, concolorous.
Stipe 3–5 × 1–2 cm (1.8–2 × 0.4–0.8 in), tapering below, fibrillose-floccose to mottled below the ring. Ring mixed, fragile and rather transient. *Flesh* white or becoming dirty pink. *Microscopic features* Sp 7–10 × 4–5.5 µm, ovoid or elliptic. Cheilo absent.
Season: September–December.
Habitat: Meadows and short turf, heathland, dunes or coastal meadows.

Distribution: Uneven. Rare or very rare.

A. porphyrocephalus Moeller is vinaceous brown but not golden, the cuticle is more fibrillose-woolly than lustrous and streaky, and it has smaller spores (5–7 × 3–4.5 µm).

---

p.176 *Agaricus bisporus* (Lange) Imbach
Agaricaceae, Agaricales

Description: *Cap brownish. Gills pinkish grey to brown. Ring ascending or double, quite large.*
Cap 5–10 cm (2–4 in), convex or flat, smooth, fibrillose or fibrillose-scaly, white to dull, somewhat rufous brown.
Gills greyish pink then dull grey-brown. Gill edge whitish to white when mature.
Stipe 3–6 × 1–2 cm (1.2–2.4 × 0.4–0.8 in), firm, cylindric, smooth to fibrillose-fleecy. Veil making a double ring, the ascending one dominant, quite large but thin and fragile. *Flesh* white or dirty pink, flushing pink. *Microscopic features* Sp 5.5–8.5 × 4–6 µm, elliptic to subglobose. Bas 2-spored. Cheilo numerous, clavate or occasionally irregular, 20–30 × 5–15 µm.
Season: August–December.
Habitat: Wood edges, road sides, gardens, parks.
Distribution: Widespread, frequency uneven.

The well-known 'shop' mushroom is the cultivated form of this species.

---

p.180 *Agaricus bohusii* M. Bon
Agaricaceae, Agaricales

Description: *Clump-forming. Cap brown, scaly. Stipe fusiform.*
Cap 6–15 cm (2.4–6 in), finely scaly to scaly, the scales occasionally recurved, more or less concentric. Dark brown to red-brown on a rufous to brownish background.
Gills arched then slightly ventricose, dingy pink then rather pinkish brown to dark brown. Gill edge frosted, paler.
Stipe 6–20 × 1–2.5 cm (2.4–8 × 0.4–1 in), fusiform, tapering (a caespitose species),

more or less garlanded with scales below. Ring complex, double, the upper element dominant. *Flesh* sepia-beige, reddening then rather brownish or dark. Smell faint but rather unpleasant. *Microscopic features* Sp 6–7.5 × 4.5–6 µm, elliptic to subglobose. Cheilo 20–40 × 7–15 µm, cylindro-clavate.

Season: August–November.
Habitat: Disturbed sites, parks, scrub.
Distribution: Uneven. Rather rare.

This species is easily recognized by its caespitose habit, the fusiform tapering stipe and the pronounced scales concentrically arranged on the cap.

---

p.179 *Agaricus subperonatus* (Lange) Singer
Agaricaceae, Agaricales

Description: *Cap with fibrillose scales, brown. Gills dingy. Stipe with mottled bands.*
Cap 5–12 cm (2–4.7 in), with small fibrillose scales or subsquamulose, rufous brown to yellowish rufous but whitish at the margin. *Gills* pinkish grey then pinkish brown. Gill edge frosted, greyish. *Stipe* 4–8 × 1–2.5 cm (1.6–3.2 × 0.4–1 in), faintly clavate below, white or very pale. Veil forming a double ring, the upper part dominant, the lower ascending part forming two or three bands mottled with white to brownish colours.
Flesh dingy beige-grey, becoming slightly pink. Smell fungal, more or less fruity. *Microscopic features* Sp 6.5–8.5 × 4.5–5.5 µm, elliptic to ovoid. Bas 4-spored. Cheilo numerous, clavate, 30–50 × 8–12 µm.

Season: August–November.
Habitat: Disturbed or enriched places, gardens, etc. Often in groups, even in tufts.
Distribution: Widespread but frequency uneven.

A. vaporarius (Pers.) Cappelli is more robust, the mottled veil less conspicuous, the smell soon strong and unpleasant and it has less abundant cheilocystidia.

p.178 *Agaricus bitorquis* (Quélet) Saccardo
Agaricaceae, Agaricales

Description: *Very hard and compact. Cap white. Double ring, the ascending part like a volva.*
Cap 5–10 cm (2–4 in), smooth or subfibrillose, white or very pale.
Gills rather dull pale pinkish then brownish pink to dingy dark-brown. Gill edge pale. *Stipe* 5–12 × 1–3 cm (2–4.7 × 0.3–1.2 in), cylindric, short and squat, occasionally slightly conical below, whitish. Veil double, the upper part forming a rather transient, hanging ring, the lower part like a volva, membranous and slightly flaring or appressed below the ring. *Flesh* hard, dingy cream, flushing slightly pink or becoming dingy. Smell fungal, strong.
Microscopic features Sp 5.5–7 × 4.5–6 µm, ovoid or shortly elliptic. Cheilo abundant, 12–35 × 7–15 µm, clavate to sphaeropedunculate.

Season: May–November.
Habitat: Trampled areas, urban lawns, roadsides, etc. Can push up through tarmac.
Distribution: Widespread. Fairly common.

Well-known for its ability to come up through tarmac, occasionally in towns. This species, which is considered edible, can accumulate heavy metals and other pollutants if the site is contaminated.

---

p.176 *Agaricus devoniensis* Orton
Agaricaceae, Agaricales

Description: *Cap fibrillose-silky, white. Gills pink initially. Ring ascending, fragile.*
Cap 3–6 cm (1.3–2.4 in), soon expanded, smooth to fibrillose-silky, white or very pale. Margin strongly appendiculate.
Gills tender pink then brown. Gill edge paler. *Stipe* 4–7 × 0.7–1.2 cm (1.6–2.8 × 0.28–0.47 in), rather fragile, slightly clavate at the base, white or very pale, floccose or glabrous. Veil double, forming a fragile, ascending ring, with an upper layer of mottled, whitish, occasionally transient,

fleecy scales. *Flesh* thin, white or pinkish to dirty beige, becoming pink above the gills. Smell faint. *Microscopic features* Sp 5.5–7 × 4–6 µm, ovoid-elliptic to subglobose. Cheilo fairly abundant, 25–40 × 10–15 µm, clavate.

Season: September–December.
Habitat: Dunes, open scrub, edges of pine woods.
Distribution: In most countries that have a Red List, this species is known to be threatened.

A. littoralis (Wak. & Pears.) Konr. & Maubl. is more delicate, has less white, with almost unchanging flesh, and has narrower cheilocystidia (up to 8 µm).

---

p.177 *Agaricus bernardii* (Quélet) Saccardo
Agaricaceae, Agaricales

Description: *Robust. Cap pale, cracked. Ring ascending, thick. Flesh becoming pink.*
Cap 6–15 cm (2.4–6 in), with small fibrillose scales then cracking irregularly, white to rather pale buff, sometimes finally brownish on the central scales. *Gills* pinkish grey then brownish to dark crimson-brown. Gill edge frosted and paler. *Stipe* 5–10 × 1–4 cm (2–4 × 0.4–1.6 in), firm, quite short, the base shortly tapering, white or ochraceous cream. Ring ascending, fairly wide but rather fragile, occasionally with faint mottled zones beneath. *Flesh* firm, white but slowly dirty vinaceous pink when cut, rather more yellowing below. Smell unpleasant. *Microscopic features* Sp 6.5–9.5 × 5–6.5 µm, elliptic to ovoid. Cheilo numerous, 25–50 × 8–14 µm, clavate to cylindro-fusoid or bottle-shaped, occasionally irregular.

Season: June–November.
Habitat: Halophilic, at the edge of coastal saline meadows, more rarely inland.
Distribution: Rather uneven. Rare.

A. maleolens Moeller, which has a matted layer over the cap giving it a cottony feel, is smooth or only fibrillose-scaly.

p.179 *Agaricus spissicaulis* Moeller
Agaricaceae, Agaricales

Description: *Cap brownish, smooth to scaly. Stipe swollen/fusoid. Ring hanging.*
Cap 5–8 cm (2–3.2 in), fleshy, smooth to finely fibrillosely scaly, beige then rather rusty brown. *Gills* narrow, arched, dirty pink then reddish brown to dark brown. Gill edge thin. *Stipe* 3–5 × 1–2.5 cm (1.2–2 × 0.4–1 in), squat, fusoid or swollen, tapering at the base where there are a few mycelial threads. Whitish to more or less rufous. Ring hanging, fragile, striate above. *Flesh* quite firm, white then turning dirty yellow and dull pink. Smell fungal, occasionally complex or mixed. *Microscopic features* Sp 5–7 × 4–5.5 μm, elliptic to subglobose. Cheilo abundant, 20–40 × 7–10 μm, cylindro-clavate.
Season: August–November.
Habitat: Rather saline or disturbed meadows, especially near the sea where it may occur high up on the beach.
Distribution: Rather rare or very rare, often locally abundant.

Because of the inconclusive colour change in the flesh when cut, this species is difficult to classify.

---

p.178 *Agaricus romagnesii* Wasser
Agaricaceae, Agaricales

Description: *Cap with small brown scales. Stipe with rhizoids. Ring hanging.*
Cap 4–7 cm (1.6–2.8 in), trapezoidal, fibrillose-scaly or with distinct appressed scales, occasionally forming a star in the centre. Brown to dull brownish on a paler dingy cream to whitish ground. *Gills* not very ventricose, pink then violaceous blackish brown. Gill edge paler.
Stipe 5–12 × 1–2 cm (2–4.7 × 0.4–0.8 in), whitish then browning slightly, clavate with white, branching rhizoids, conspicuous if the fungus is gathered with care. Ring hanging, fragile, white.

Flesh white then pinkish to pinkish grey, yellowing somewhat below. Smell faint, often unpleasant. *Microscopic features* Sp 6.5–8 × 4.5–5.5 μm, ovoid to elliptic. Cheilo often few, 25–40 × 8–15 μm, cylindro-clavate, occasionally septate toward the base.

Season: June–November.
Habitat: Short turf, grassy places, occasionally in rather trampled or disturbed areas. Particularly frequent on urban lawns.
Distribution: Rather uneven. Appears to be absent in Scandinavia.

Has a certain level of toxicity. It is frequent in towns and is responsible for quite a large number of adverse reactions.

---

p.179 *Agaricus koelerionensis* (M. Bon) M. Bon
Agaricaceae, Agaricales

Description: *Cap with fine, lilac brown scales. Smell sour.* *Cap* 5–15 cm (2–6 in), with thin, regular scales, more fibrillose toward the margin. Lilac brownish on a whitish, pinkish beige or beige ground. *Gills* often distinctly free from the stipe, fairly bright pink, then brown. Gill edge often paler. *Stipe* 5–12 × 0.8–1.5 cm (2–4.7 × 0.3–0.6 in), tall and slender, the base often clavate to almost bulbous, whitish to pinkish beige or rather lilac below. Ring hanging, quite full, fragile. *Flesh* white then pinkish to pale pinkish grey. Smell sour, like a *Lepiota*. *Microscopic features* Sp 6–8 × 3–4 μm, elliptic or rather trapezoidal. Cheilo abundant, 20–35 × 10–18 μm, broadly clavate to sphaeropedunculate, occasionally septate.
Season: September–December.
Habitat: Coastal dunes and sheltered scrub on fixed dunes.
Distribution: Not well known and rarely recorded.

Close to *A. variegans* Moeller which also has the sour *Lepiota* smell, but even stronger. It has duller or darker colours and occurs in woods or wasteland.

p.180 *Agaricus silvaticus* Sch.:Fr.
Agaricaceae, Agaricales

Description: *Cap scaly, brown. Ring fragile. Reddening.*
Cap 4–10 cm (1.6–4 in), fibrillose-scaly, brownish to tawny brown on a paler, sepia or dingy cream ground.
Gills greyish pink then soon brownish grey to dark brown. *Stipe* 5–12 × 1–1.5 cm (2–4.7 × 0.4– 0.6 in), the base clavate, white to greyish, finally brownish grey from the base up. Ring hanging, white, fairly fragile and soon torn or breaking up. *Flesh* dingy cream, reddening, especially above the gills and at the stipe apex. Smell fungal, sour.
Microscopic features Sp 4.5–6.5 × 3.5–4 µm, elliptic or ovoid. Cheilo fairly abundant, 20–50 × 7–15 µm, cylindro-clavate or more or less broadly clavate.

Season: August–November.
Habitat: Conifers; rarer under broad-leaved trees.
Distribution: Widespread. More or less common.

A. haemorrhoidarius Schulzer, reddening very strongly and with a slightly thicker ring, under broad-leaved trees, has spores the same size or slightly longer. *A. langei* (Moeller) Moeller, with whitish flesh, reddening quickly then becoming vinaceous brown, under conifers, has larger spores (up to 9.5 × 5 µm).

---

p.177 *Agaricus arvensis* Sch.:Fr.
Agaricaceae, Agaricales

Description: *Cap white, yellowing. Ring with a cog-wheel beneath. Smell of aniseed.*
Cap 6–20 cm (2.4–8 in), often like a truncated cone, white then yellowish to rather rufous yellow. Margin occasionally with fleecy white scales. *Gills* pale greyish pink then rather pinkish brown to dark brown. Gill edge whitish.
Stipe 6–12 × 1.5– 3 cm (2.4–4.7 × 0.6–1.2 in), often clavate, white or yellowish to rather rufous yellow below. Ring hanging, full and with a cog-wheel

underneath. *Flesh* rather thick, white then yellowing. Smells strongly of aniseed. *Microscopic features* Sp 7–9 × 4.5–6 µm, rather long elliptic. Cheilo numerous, 15–30 × 8–15 µm, clavate to sphaeropedunculate, occasionally septate at the base.

Season: June–November.
Habitat: Meadows and open places, parks and gardens. Often in rings.
Distribution: Widespread but frequency uneven.

A. nivescens (Moeller) Moeller, with typically grey gills when young, stays pure white for quite a long time and smells faintly of bitter almonds.

---

p.177 *Agaricus essettei* M. Bon
Agaricaceae, Agaricales

Description: *Cap white, yellowing. Stipe bulbous.*
Cap 4–10 cm (1.6–4 in), smooth to finely woolly-floccose, white then rather rufous yellow to dirty orange, in patches.
Gills distant from the stipe, pinkish grey then dull dark-brown. Gill edge paler.
Stipe 5–10 × 1–2.5 cm (2–4 × 0.4–1 in), the base abruptly bulbous, white then yellowish to rather rufous yellow. Ring hanging, full, with a fleecy, but rather fragile cog-wheel beneath. *Flesh* yellowing. Smells rather strongly of aniseed, occasionally mixed with bitter almonds. *Microscopic features* Sp 6.5–8.5 × 4–5 µm, rather long elliptic. Cheilo numerous, 15–30 × 10–20 µm, elliptic to very clavate.

Season: September–November.
Habitat: Woods dominated by conifers; rarely under pure stands of broad-leaved trees.
Distribution: Probably widespread but needs confirmation.

A. silvicola (Vitt.) Sacc. differs in the gills, which are less distant from the stipe, a more orange flesh change and its simply clavate stipe. It favours broad-leaved trees.

p.179 *Agaricus porphyrrhizon* Orton
Agaricaceae, Agaricales

Description: *Cap with small fibrillose scales, crimson. Yellowing. Smell of bitter almonds.*
Cap 4–9 cm (1.6–3.5 in), with small fibrillose scales to radially fibrillose, crimson to vinaceous on an ochraceous to lilac-beige ground. *Gills* pinkish grey to quite dark brown. Gill edge paler. *Stipe* 4–8 × 1–2 cm (1.6–3.2 × 0.4–0.8 in), the base clearly rather clavate, fibrillose or with slightly floccose patterning, whitish or pinkish grey but bright yellow from the base up. Ring hanging, thin and fragile. *Flesh* fairly fragile, whitish then yellow. Smell of bitter almonds. *Microscopic features* Sp 4–6 × 3–4 µm rather shortly elliptic to ovoid. Cheilo numerous, 20–35 × 8–15 µm, broadly clavate to elliptic.
Season: September–December.
Habitat: Broad-leaved coppice and thickets on sandy soils, usually in warm sites.
Distribution: Rather rare or very rare.

The largest in the Minores section, most species of which are more slender or sometimes even very small, distinctly yellowing, with a smell of bitter almonds.

p.177 *Agaricus xanthoderma* Génevier
Agaricaceae, Agaricales

Description: *Cap a flattened cone, white, yellowing strongly. Stipe bulbous. Smell of iodoform.*
Cap 5–10 cm (2–4 in), a flattened cone, smooth or slightly floccose-scaly, white but yellowing strongly, at least when handled. *Gills* greyish pink then pinkish brown to brown. Gill edge paler. *Stipe* 6–12 × 1–2 cm (2.4–4.7 × 0.4–0.8 in), the base distinctly bulbous, occasionally marginate, white then yellowing strongly. Ring hanging, with a rather fragile cog-wheel underneath. *Flesh* yellowing strongly. Smells very strongly and unpleasantly of ink or

iodoform when handled. *Microscopic features* Sp 5–6 × 3–4 µm, elliptic or ovoid. Cheilo fairly numerous, 15–25 × 8–16 µm, fairly broadly clavate or elliptic and bladder-like.

Season: August–November.
Habitat: Open places, wood edges, meadows, broad-leaved woodland, often preferring disturbed sites.
Distribution: Common or fairly common.

Slightly toxic. The unpleasant smell should deter people from eating it. More or less grey or scaly varieties occur.

---

p.178 *Agaricus praeclaresquamosus* Freeman
(= *Psalliota meleagris*)
Agaricaceae, Agaricales

Description: *Cap with small fibrillose scales, brownish. Yellowing. Smell of ink.*
Cap 5–15 cm (2–6 in), finely or coarsely fibrillose-scaly, the centre sometimes remaining smooth and forming a star-shaped pattern, dull, brownish to brown on a lighter, greyish beige to buff ground. *Gills* pinkish grey then rather pinkish brown to dark brown. Gill edge paler. *Stipe* 5–12 × 1–3 cm (2–4.7 × 0.4–1.2 in), more or less bulbous, occasionally marginate. White or greyish then yellowing strongly, finally dingy and dull. Ring hanging, with a few fleecy scales beneath, fairly fragile. *Flesh* greyish to palish brown, yellowing strongly. Smells fairly strongly of ink or iodine. *Microscopic features* Sp 4.5–6 × 3.5–4.5 µm, elliptic or ovoid. Cheilo numerous, 15–30 × 10–18 µm, cylindro-clavate to broadly clavate or globose.
Season: August–November.
Habitat: Disturbed coppice or damp broad-leaved woodland, occasionally on fixed dunes or in peat bogs.
Distribution: Not fully understood. Rather rare or very rare.

Toxic, like the preceding species.

p.178 *Agaricus menieri* M. Bon (= *Psalliota ammophila*)
Agaricaceae, Agaricales

Description: *Cap white, yellowing strongly. Smell of iodine or ink.*
Cap 5–12 cm (2–4.7 in), smooth or rather fibrillose-silky, white or very pale greyish cream, yellowing strongly. *Gills* fairly bright pink then pinkish grey to rather pinkish brown or brown when mature. *Stipe* 5–12 × 2–5 cm (2–4.7 × 0.8–2 in), robust, slightly clavate, not bulbous, white to pale beige, yellowing strongly. Ring hanging, fairly thick but fragile.
Flesh white, yellowing strongly. Smells strongly of iodine or phenol when handled and when old. *Microscopic features* Sp 7–10 × 5–7 µm, elliptic. Cheilo abundant, 15–30 × 8–12 µm, clavate or cylindro-clavate, occasionally septate.
Season: September–January.
Habitat: Coastal dunes, bare sand, marram grass, occasionally coastal grassland or edges of some scrub. Initially completely covered in sand.
Distribution: Rare, restricted to coastal areas. Recorded in Fr.ance and Italy.

Belongs to the same group as the two preceding species (toxic).

Amanitaceae
The *Amanitaceae*, is composed of two genera, *Amanita* and *Limacella*. This family is well known, both in the field of gastronomy, because of the excellent edible species known as Caesar's mushroom, but also in the field of toxicology since *Amanita* spp. in the group including *A. phalloides* are responsible for 95% of cases of fatal poisoning in Europe. The genus *Amanita* is therefore of great interest, for better or for worse, to all those who enjoy eating fungi.

The family has the following characters: fruit bodies fleshy with fibrous texture, flesh thick and sometimes brittle, gills free (generally), spores white, universal veil present (structure protecting the primordium like an eggshell) and bilateral gill trama. Furthermore, the species are mycorrhizal. A quick guide to identification is the fact that in *Amanita* the veil is membranous to floccose, while in *Limacella* it is slimy.

Classification of the genus *Amanita* is based on several criteria: presence or absence of a ring; striation of the cap margin; spore reaction to Melzer's iodine reagent (amyloid, i.e. reacts like starch, becoming blue-black with iodine, or has no reaction); type of universal veil (bag-like, etc., *see* below); spore shape, etc.

Universal veil: four types of volva can be distinguished:

- Bag-like volva. The veil ruptures at the top of the primordium, which pushes cleanly through the opening, the cap not carrying up fragments of veil (e.g. *A. phalloides* (left), ring present, margin lacking striations or *A. vaginata*, (right), ring absent, margin striate).
- Circumsessile volva. The veil ruptures 'at the equator' of the primordium, the cap carrying up the upper part (which breaks up further into patches as the cap diameter increases), the lower part remaining at the stipe base as half a sphere (e.g. *A. citrina*).
- Fleecy volva. The veil 'explodes', breaking up into tufts which are dispersed over the cap surface and remain close together on the bulb at the stipe base (e.g. *A. muscaria*).
- Volva almost absent. The veil ruptures at the base of the primordium and all of it is carried up by the cap as the stipe lengthens. It subsequently shatters into patches and spots as the cap diameter increases (e.g. *A. rubescens*).

More complex instances occur and it is not always obvious into which group to place a species. Species lacking a ring belong to subgenus *Amanitopsis*, those with a ring are placed in subgenus *Amanita*.

In the descriptions of this family, I+ and I– indicate, respectively, a positive (amyloid i.e. blue-black) or negative spore reaction to iodine. UV introduces a description of the structure of the universal veil.

---

p.190 *Limacella guttata* (Pers.:Fr.) Konr. & Maubl.
Amanitaceae, Amanitales

Description: *Cap pale tan- or flesh-coloured, rather slimy. Stipe with ring. Smell of cucumber.*
Cap 8–15 cm (3.2–6 in), somewhat slimy, smooth or slightly cracked, revealing the white, flesh-beige or pale tan. *Gills* close, white, edge entire. *Stipe* 10–15 × 1.5–3 cm (4–6 × 0.6–1.2 in), swollen to bulbous at the base, whitish or cream, somewhat slimy below in damp weather but soon dry. Ring membranous and large. *Flesh* pale, smell mealy, often strong, occasionally even like cucumber. *Microscopic features* Sp 4.5–6.5 × 4–5 µm, subglobose, almost smooth, I–. Clamps present.
Season: August–October.
Habitat: Broad-leaved and coniferous woodland, on calcareous soils.
Distribution: All Europe. Not very common.

The only *Limacella* with a good ring.

---

p.191 *Limacella illinita* (Fr.:Fr.) Murrill
Amanitaceae, Amanitales

Description: *Cap medium-sized, umbonate, slimy to dripping with slime, pale. Smell faint.*
Cap 4–8 cm (1.6–3.2 in), umbonate, dripping with slime, slimy even in quite dry weather, whitish to pinkish cream. *Gills* not very close, white.
Stipe 7–10 × 0.5–0.8 cm (2.8–4 × 0.2–0.3 in), slimy beneath a high, appressed, thick rounded ring zone, pruinose at the apex, almost concolorous with the cap. *Flesh* with a faint mealy smell. *Microscopic features* Sp 4.5–6 × 4–5 µm, subglobose, rough to finely or sparsely spiny. Clamps present.

Season: August–November.
Habitat: Mixed open woodland, preferring conifers in the north and broad-leaved trees in the south.
Distribution: Widespread, seems rarer towards Scandinavia.

There are not many *Limacella* spp. with a very slimy stipe. *L. ochraceolutea* Orton has a pale yellow or rather rufous cap, yellowish gills and a rather strong mealy smell. It is a very rare species which also occurs in warm sites.

---

p.185 *Amanita ceciliae* (Berk. & Br.) Boud. ex Bas
(= *A. strangulata*; = *A. inaurata*)
Amanitaceae, Amanitales

Description: *Cap olivaceous fawn, striate. Stipe fleecy, dark grey. Volva fragile. Ring absent.*
Cap 8–15 cm (3.2–6 in), olivaceous fawn to brownish, striate, with short-lived, grey to blackish patches or spots. *Gills* white to pale grey, edge occasionally greyish.
Stipe 13–20 × 1.5–3 cm (5.1–7.9 × 0.6–1.2 in), thicker below, with a zigzag to mottled pattern or sometimes fleecy, quite dark mouse-grey. Ring absent. Volva thick, cottony then fragile and very crumbly, ash grey. *Flesh* fairly fragile, whitish to grey.
Microscopic features Sp 9.5–12.5(14) µm, globose, I–. Cheilo sphaeropedunculate to clavate. UV in particular with sphaerocytes 30–70 µm diameter.
Season: July–October.
Habitat: Broad-leaved woodland, especially under hornbeam on calcareous clays. Occasionally under conifers.
Distribution: Rare.

---

p.184 *Amanita submembranacea* (M. Bon) Gröger
Amanitaceae, Amanitales

Description: *Cap olivaceous brown, striate. Volva fairly fragile, brownish. Ring absent.*
Cap 6–8 cm (2.4–3.2 in), umbonate, striate, smooth, more or less dark olivaceous brown, occasionally bearing the

remains of the greyish or pale brownish universal veil. *Gills* white or very pale. *Stipe* 6–10 × 1–2 cm (2.4–4 × 0.4–0.8 in), hollow, thicker below, whitish, but mottled olivaceous sepia below. Ring absent. Volva grey to brownish, fairly crumbly.
Flesh white or pale. *Microscopic features* Sp 10–13 µm, subglobose, I–. Cheilo broadly clavate, occasionally septate, 30 × 20 µm. UV with quite abundant sphaerocytes 20–70 µm diameter, mixed in more or less equal quantity with filamentous hyphae.

Season: August–October.
Habitat: Conifer woodland in continental or submontane districts; broad-leaved woodland in lowland areas. Prefers acid soils.
Distribution: Rather rare.

A. groendlandica Bas ex Knudsen & Borgen, which is more grey or grey-brown and less striate, is a northern-alpine species, under birch and willow. *A. beckeri* Huijsman, with a hazel cap, a zigzag patterned stipe and white veil, is a species of warmer calcareous sites. *A. battarrae* (Boudier) M. Bon has a veil that remains more intact and an olivaceous brown cap, with a darker band adjacent to the marginal striations (usually found in lowlands); *A. fuscoolivacea* (Kühner ex Contu) Romagnesi is similar but more squat and is montane.

p.184 *Amanita pseudofriabilis* Courtecuisse
Amanitaceae, Amanitales

Description: *Very slender, the cap hazel, striate. Ring absent. Cap* 2–5 cm (0.8–2 in), umbonate, smooth, often pale hazel-beige, occasionally with olivaceous or brownish tones, always with a large whitish patch of the universal veil. *Gills* distant from the stipe, not very close, white. *Stipe* 6–12 × 0.4–1.2 cm (2.4–4.8 × 0.2–0.5 in), very fragile, slender, white, slightly mottled. Ring absent. Volva sheathing, fragile and soon lost or flaking into broad transient patches. *Flesh* very thin and fragile, white. *Microscopic features* Sp 9–12 µm, globose or occasionally subglobose, I–. Cheilo in occasional clusters, septate, the terminal cell clavate, 20–25 × 15–18 µm. UV sphaerocytes dominating, up to 75 × 50 µm diameter.

Season: August–October.
Habitat: Broad-leaved woodland on average to damp soils.
Distribution: Not fully understood.

This species has been confused with small forms of *A. submembranacea* or *A. battarae*. The name is a reference to *A. friabilis* (P. Karsten) Bas, a slender but very rare species, confined strictly to alder, which has a grey, friable, and even fleetingly floccose veil.

---

p.184 *Amanita malleata* (Piane ex M. Bon) Contu
Amanitaceae, Amanitales

Description: *Cap grey-brown, dented as if hammered, striate. Stipe mottled. Ring absent.*
Cap 5–10 cm (2–4 in), striate, marked with flat or depressed patches (less obvious on some specimens), grey to greyish cream or greyish brown, with unequal, occasionally somewhat pyramidal patches. Veil whitish to greyish or rather rufous. *Gills* whitish. Gill edge occasionally greyish. *Stipe* 10–15 × 1.5–2.5 cm (4–6 × 0.6–1 in), whitish to greyish, lacking a ring but with zigzags that occasionally form bands which may be disrupted below. Volva white to rather rufous grey, fairly friable. Flesh white or very pale. *Microscopic features* Sp 11–13 µm, globose or subglobose, I–. Cheilo sphaeropedunculate or clavate, occasionally pyriform, 40–55 × 15–20 µm. UV comprising mainly sphaerocytes 30–70 µm diameter, mixed with slender hyphae.

Season: August–November.
Habitat: Grassland in parks or in woods. Appears to prefer poplars; usually on calcareous soils.
Distribution: Not well known as this is a controversial species.

p.183 **Grisette**
Amanita vaginata (Bull.:Fr.) Vittadini
Amanitaceae, Amanitales

Description: *Cap umbonate, striate. Stipe lacking a ring. Volva membranous.*
Cap 4–12(–15) cm (1.6–4.6(–6) in), ovoid then campanulate, blunt, striate, smooth, usually without velar remains, grey, but occasionally more brownish. *Gills* white. *Stipe* 10–20 × 0.5–1.2 cm (4–8 × 0.2–0.5 in), fragile, thicker below, hollow, whitish or very pale, pruinose to slightly mottled. Ring absent. Volva bag-like, membranous and remaining intact, whitish or occasionally slightly coloured.
Flesh white, fragile. *Microscopic features* Sp (9–)10–12(–14) µm, globose, I–. Cheilo sphaeropedunculate, clavate or deformed bottle-shaped, 20–40 × 10–25 µm. UV primarily filamentous.

Season: July–October.
Habitat: Woodland, under various species.
Distribution: All Europe.

Among common related species is *A. fulva* (Sch.:Fr.) Fr. (more slender, cap orange rufous to fawn, volva with rust-coloured tones).

p.184 *Amanita crocea* (Quélet) Singer
Amanitaceae, Amanitales

Description: *Cap orange yellow, umbonate, striate. Ring absent.*
Cap 6–15 cm (2.4–6 in), umbonate, striate, smooth, velar remains absent (apart from exceptional cases), saffron to orange-yellow. *Gills* white to pale yellowish. *Stipe* 15–20 × 1.5–2.5 cm (6–8 × 0.6–1 in), thickening downwards, with rather rufous orange-yellow, zigzag or mottled marking. Ring absent. Volva usually remaining intact, orange-cream inside. *Flesh* whitish except just below the surface. *Microscopic features* Sp 9–13 µm, globose to subglobose, I–. Cheilo as in *A. vaginata*. Volva generally with a mainly filamentous structure.

Season: July–October.

Habitat: Under broad-leaved trees in particular, also conifers; distinctly preferring acid soils.
Distribution: Rather rare.

The only species similar in size and general characters is *A. badia* (Sch.) M. Bon & Contu, which has a chestnut to dark bay-brown cap and occurs in damp or marshy places in the mountains.

---

p.183 *Amanita mairei* Foley
Amanitaceae, Amanitales

Description: *Cap silvery grey, striate. Ring absent. Volva large.*
Cap 7–15 cm (2.8–6 in), not umbonate, striate, without velar remains, silvery grey. *Gills* white or whitish. *Stipe* 15–20 × 1–1.5(–2) cm (6–8 × 0.4–0.6(–0.8) in), thickening from top to bottom, whitish to greyish, not mottled. Ring absent. Volva a large bag, whitish. *Flesh* white. *Microscopic features* Sp 10–15 × 8–12 µm, elliptic, occasionally subglobose or cylindric, I–. Cheilo clavate, ordinary. Volva with a dominant filamentous structure.
Season: April–December.
Habitat: Open broad-leaved and coniferous woodland, open places, occasionally on sandy soils.
Distribution: Tends to have a Mediterranean–Atlantic distribution. Rather rare.

A. lividopallescens (Gill.) Gilb. & Kühn., found under broad-leaved trees, has a pale buff to hazel, sometimes almost umbonate cap, a fibrillose-fleecy stipe and is cream inside the volva.

---

p.185 *Amanita nivalis* Greville
Amanitaceae, Amanitales

Description: *Small and squat, pale, ring absent.*
Cap 3–8 cm (1.2–3.2 in), shortly striate, white to whitish then dingy greyish or brownish, velar patches occasionally confluent, white to dirty beige. *Gills*

white or slightly dingy. *Stipe* 5–10 × 1.5–2.5 cm (2–4 × 0.6–1 in), squat, very pale. Ring absent. Volva bag-like, remaining intact to papery when old, white or spotted pinkish to brownish. *Flesh* white. *Microscopic features* Sp 9–13 µm, globose to subglobose, I–. Cheilo clavate.

Season: July–September.
Habitat: Alpine; with mats of dwarf willow (*Salix herbacea* but also *S. retusa* and *S. reticulata*), also occasionally birch (*Betula nana*). Prefers acid soils.
Distribution: Typically northern and alpine. Occurs at sea level in Iceland, but only in high mountain sites in continental Europe.

A. arctica Bas, Knudsen & Borgen, which also occurs in northern or alpine zones, has thicker flesh, a pruinose to fleecy-scaly stipe, and a volva with a low ring zone where it has ruptured. The spores are less round. *A. lactea* Mal., Romagn. & Reid, a Mediterranean species, is more fleshy and white, but with a large volva and a stipe which has a low, fragile, often broken ring.

---

p.186 **Caesar's Mushroom**
Amanita caesarea (Scop.:Fr.) Pers.
Amanitaceae, Amanitales

Description: *Cap orange, striate. Gills and stipe yellow. Volva bag-like, white.*
Cap 10–20 cm (4–8 in), striate, orange to reddish orange, without spots, but occasionally with a large white patch of veil. *Gills* golden yellow. *Stipe* 10–15 × 1.5–3 cm (4–6 × 0.6–1.2 in), pale yellow to golden yellow. Ring yellowish, pendant. Volva bag-like, membranous, white. *Flesh* thick, yellowish just below the surface. *Microscopic features* Sp 10–13 × 6–8 µm, more or less long elliptic, I–. Cheilo rare, clavate. Volva filamentous.
Season: August–October.
Habitat: Woods, coppice and wood pasture, in warm regions or sites.
Distribution: Southern but can occur in quite northern areas; it appears very sporadically.

An excellent edible species. An inattentive collector could gather old material of *A. muscaria* by mistake for this species (*see* below).

---

p.185 **Fly Agaric** (poisonous)
Amanita muscaria (L.:Fr.) Pers.
Amanitaceae, Amanitales

Description: *Cap bright red with white fleecy spots, striate. Gills and stipe white. Volva fleecy-scaly.*
Cap 8–20(–25) cm (3.2–8(–10) in), striate, scarlet, fading quite commonly to orange or orange-yellow, with rather regular white or yellowish fleecy spots but occasionally bare when very old or following rain. *Gills* white.
Stipe 10–25 × 1–4(–6) cm (4–8 × 0.4–1.6 (–2.4) in), bulbous, white, smooth to finely fleecy. Ring large, white but with fleecy, sometimes yellowish, transient scales beneath. Volva crumbly, forming large, fairly regular, fleecy scales on the bulbous base. *Flesh* with orange-yellow tones under the cap cuticle. *Microscopic features* Sp 7.5–12 × 6–8.5 µm, elliptic, I–. Cheilo clavate to bladder-like.
Season: July–November.
Habitat: Under birch and spruce on acid soils.
Distribution: Common.

Var. *formosa* (Gonn. & Rabenh.) Fr., which discolours a little more, has a yellow veil (fleecy spots); f. *aureola* (Kalchbr.) Vesely, which prefers damp sites, is very slender, the cap orange-yellow, fading strongly from the margin and the universal veil in tatters above the bulb, but with few spots on the cap. Toxicity complex, of the pantherin type.

---

p.185 *Amanita muscaria* ssp. *americana* (J.E. Lange) Singer
Amanitaceae, Amanitales

Description: *Like A. muscaria, but the cap yellow.*
Characteristics as for the typical fly agaric but cap colour yellow. Known primarily

from North America where it is frequent. Exceptionally occurs in Europe and the precise identity of these collections is so far not certain. Differs from discoloured individuals of *A. muscaria* var. *formosa* by its white veil. The spots on the cap are perhaps more tightly attached.

A. gemmata (Paulet) Bertillon, is smaller, bright or pale yellow, has a fragile ring and a volva that commonly forms bracelets, occasionally in spirals, on the basal bulb.

---

p.186 **Panther Cap** (poisonous)
Amanita pantherina (D.C.:Fr.) Krombh.
Amanitaceae, Amanitales

Description: *Cap brown with small white fleecy spots. Margin striate. Volva forming spiralling bands.*
Cap 5–10 cm (2–4 in), striate, occasionally faintly so, under the spots at the extreme edge where they are more mealy or confluent. More or less light brown or dark, with small, regular, pure white fleecy spots. *Gills* white. *Stipe* 7–10 × 1–1.5 cm (2.8–4 × 0.4–0.6 in), bulbous, white. Ring membranous. Volva friable, with spiralling bands above a thick ring at the top of the basal bulb. *Flesh* rather thin. *Microscopic features* Sp 7–12.5 × 6.5–9 µm, elliptic, I–. Cheilo clavate to sphaeropedunculate. UV has a mixture of filamentous hyphae and sphaerocytes, the latter more numerous in the spots on the cap, ovoid or elliptic, up to 50 × 35 µm diameter.
Season: July–October.
Habitat: Broad-leaved trees, but also conifers.
Distribution: No longer occurs north of latitude 60°N.

Dangerously poisonous. Beware of the similarity to *A. spissa* (*see* p.388), which is occasionally eaten, although of only average quality. The pure white fleecy spots, the striate margin and the thick rimmed volva with spiral bands are identifying characters.

p.189 *Amanita spissa* (Fr.) Kummer
Amanitaceae, Amanitales

Description: *Cap brown with flat confluent greyish spots, striate. Volva absent.*
Cap 5–15 cm (2–6 in), not striate, except when very old (striations then short and fine), brown, occasionally light sepia or dark brown, with flat confluent dirty greyish or beige-grey veil remains. *Gills* white or greyish. *Stipe* 8–15 × 1–3 cm (3.2–6 × 0.4–1.2 in), often clavate to almost bulbous, pale, whitish to patterned with brownish zigzags. Ring like a skirt, striate above, white or grey. Volva absent. *Flesh* with a smell of oil-seed rape. *Microscopic features* Sp 8–10 × 7–8 µm, ovoid to subelliptic, I+. Cheilo clavate. UV dominated by sphaerocytes 30–60 µm diameter.
Season: August–October.
Habitat: Broad-leaved trees and conifers.
Distribution: Common everywhere.

Do not confuse this edible species with *A. pantherina* (see p.387), which is dangerously poisonous. *A. spissa* is very variable: var. *excelsa* (Fr.:Fr.) Dörfelt & Roth, is found mainly under broad-leaved trees (paler, the stipe rooting, no smell); var. *valida* (Fr.) Gilb. is found mainly under conifers (discolouring brown, with greyish spots).

---

p.188 **The Blusher**
Amanita rubescens Pers.:Fr.
Amanitaceae, Amanitales

Description: *Cap brownish, with confluent flat patches and fleecy spots. Volva absent. Flesh reddening.*
Cap 3–15 cm (1.3–6 in), not striate, whitish to brownish buff, with confluent and irregular, whitish to dingy ochraceous cream fleecy spots and flat patches of veil. *Gills* reddening from the edge inwards. *Stipe* 6–15 × 1.5–4 cm (2.4–6 × 0.6–1.6 in), almost bulbous, whitish to brownish, often mottled, reddening where damaged, especially from the basal bulb up. Ring pendant, white to reddish. Volva

absent. *Flesh* white, reddening (damage).
Microscopic features Sp 8–10 × 6–7.5 μm, ovoid to elliptic, I+. Cheilo clavate to sphaeropedunculate. UV dominated by sphaerocytes.

Season: August–November.
Habitat: Broad-leaved trees and conifers.
Distribution: Common everywhere.

A good edible species when thoroughly cooked (contains haemolytic toxins that are destroyed by heat); f. *annulosulfurea* Gillet is often more slender and has a yellow ring.

---

p.188 *Amanita franchetii* (Boudier) Fayod
(= *A. aspera*)
Amanitaceae, Amanitales

Description: *Cap greenish yellow to olive. Flesh not reddening. Veil yellow.*
Cap 5–15 cm (2–6 in), not striate, brownish buff, often rather dull, with confluent fleecy spots and flat patches of veil, bright yellow, yellow-buff, yellowish brown. *Gills* ordinary. *Stipe* 8–12 × 1.5–3 cm (3.2–4.7 × 0.6–1.2 in), almost bulbous, whitish to pale beige, pruinose to mottled yellowish. Ring white with a yellow or yellowish edge. Volva limited to a few yellow or yellowish traces on the edge of the basal bulb. *Flesh* white, unchanging.
Microscopic features as for *A. rubescens*.

Season: August–October.
Habitat: Woodland, especially broad-leaved trees.
Distribution: Widespread.

It is very similar to the Blusher, but does not redden and has a yellow veil. A form has been described, f. *lactella* (Gilb. ex Bert.) M. Bon & Contu, which has an almost white cap, is rare and prefers warm sites.

---

p.188 *Amanita strobiliformis* (Paulet) Bertillon
Amanitaceae, Amanitales

Description: *Cap with short-lived polyhedral veil remains. Ring like cream. Volva almost absent.*

Cap 10–20 cm (4–8 in), the margin thick, not striate, often appendiculate, the veil remains with a texture like cream. White then dingy pale brownish grey, with short-lived, irregular, polyhedral patches and spots. *Gills* thick, white to cream.
Stipe 18–25 × 2.5–4 cm (7–10 × 1–1.6 in), occasionally deeply buried, white or very pale. Ring set very high, texture like cream, often sticking to the gills or appendiculate at the margin. Basal bulb ovoid, with a few bands. Volva absent. *Microscopic features* Sp (9–)10–13(–15) × (6–)7–8.5(–9.5) μm, elliptic to cylindric, I+. Cheilo clavate or septate, 20–30 × 15–20 μm. UV with broadly elliptic sphaerocytes 50–70 μm diameter. Clamps absent.

Season: July–October, occasionally later.
Habitat: Broad-leaved woods, especially beech–hornbeam or even coniferous woodland, always on calcareous soils. Prefers warm sites, and tends to be southern.
Distribution: Increasingly rare northwards.

The largest *Amanita* in the European flora (with *A. ovoidea*).

---

p.187 *Amanita echinocephala* (Vitt.) Quélet
Amanitaceae, Amanitales

Description: *Cap whitish, with acute pyramidal, fleecy spots. Basal bulb with acute pointed scales.*
Cap 10–15 cm (2–6 in), not striate, pale, white to whitish, with short-lived, regular and fairly dense, pointed, pyramidal warts. *Gills* cream, with greenish tones.
Stipe 12–18 × 1.5–3 cm (4.7–7 × 0.6–1.2 in), white or very pale, the basal bulb like a truncated cone with rows of pointed pyramidal warts. Ring not like cream, hanging, white. *Flesh* tends to turn slightly green. *Microscopic features* Sp (8–)9–13(–15) × 6–9(–10) μm, ovoid-elliptic to cylindric, I+. Cheilo often clavate to irregular, 45–70 × 10–13 μm. UV dominated by sphaerocytes, 50–75 μm diameter or with large elliptic hyphae and ordinary hyphae.

Season: August–October.
Habitat: Broad-leaved trees and conifers; calcareous soils, usually in warm sites.
Distribution: Widespread.

A. strobiliformis and *A. echinocephala* are 'leaders' of a group of southern *Amanita* spp.

---

p.189 *Amanita beillei* (Beaus.) Mesp. ex Bon & Contu
Amanitaceae, Amanitales

Description: *Cap white, with pyramidal fleecy spots. Gills pink.*
Cap 4–8 cm (1.6–3.2 in), white to cream or buff, with more or less pyramidal, rather small, white fleecy velar spots. *Gills* pink from the outset, salmon or even reddish, duller when old (spore print salmon pink). *Stipe* 5–12 × 1–3 cm (2–4.7 × 0.4–1.2 in), rooting bulbous, white to ochraceous cream, scaly below the fragile, white or cream ring. Volva soon broken up into short-lived fragments, pink to cream or pinkish beige. *Flesh* ochraceous beige to rather rufous. *Microscopic features* Sp 11–14 × 6–8 μm, elliptic to cylindric, I+. Cheilo clavate, 35–50 × 10–18 μm. UV mostly composed of sphaerocytes in the remains on the cap, mostly filamentous in the volva.
Season: April–May.
Habitat: Sandy shores along the Atlantic and the Mediterranean.
Distribution: Extremely rare, recorded in Fr.ance and Italy.

A. boudieri Barla is a more continental or montane-Mediterranean species; it is more squat, the gills whitish initially, the stipe less scaly and the flesh and spore print are white.

---

p.188 *Amanita singeri* Bas
Amanitaceae, Amanitales

Description: *Small species. Gills salmon pink. Stipe not bulbous.*
Cap 4–8 cm (1.6–3.2 in), the margin slightly appendiculate, not striate, whitish

to greyish cream, with irregular, brownish, rather persistent velar remains.
Gills yellowish cream then salmon yellow.
Stipe 4–8 × 0.8–1.5 cm (1.6–3.2 × 0.3–0.6 in), fairly squat, tapering below, not bulbous, white to brownish. Ring replaced by rows of scaly and irregular, fleeting belts. Volva almost absent. *Flesh* white or very pale. *Microscopic features* Sp 8.5–10(–11) × 6.5–8 µm, subglobose to elliptic, I+. Cheilo rare. UV with ellipsoid to fusiform cells.

Season: August–November.
Habitat: Especially cypress, e.g. *Chamaecyparis lawsoniana*, and town parks or meadows.
Distribution: Fr.ance and Sardinia. A South American species, spreading from western Fr.ance.

Probably introduced. This species is globally rare, although locally abundant.

---

p.186 *Amanita ovoidea* (Bull.:Fr.) Link
Amanitaceae, Amanitales

Description: *Large white to pale cream species. Margin and ring like cream. Volva bag-like, ochraceous cream.*
Cap 15–30(40) cm (6–12(16) in), the margin like cream, not striate, with a few white to cream or pale ivory veil fragments.
Gills white. Gill edge like cream.
Stipe 12–15 × 2.5–5 cm (4.7–6 × 1–2 in), thick, bulbous and somewhat rooting under the basal bulb, white, floccose. Ring like cream, soon sticking to the stipe then disappearing. Volva bag-like, large, membranous to cream-like, pale cream-buff. *Flesh* thick, softening with age. Smell slightly nauseous. *Microscopic features* Sp 9–12 × 6–7.5 µm, ovoid to subcylindric, I+. Cheilo abundant, clavate to pyriform, 45–60 × 10–20 µm. UV as *A. beillei*.
Season: August–October.
Habitat: Woods and wood pasture; calcareous, usually warm sites.
Distribution: Very rare in more northern parts of Europe but fairly common in the south.

A. proxima Dumée. is poisonous while *A. ovoidea* is edible. The former is less tall, more slender, has a ring less like cream and a rufous volva.

---

p.187 **Death Cap** (deadly poisonous)
Amanita phalloides (Vaill.:Fr.) Link
Amanitaceae, Amanitales

Description: *Cap yellow-green, with innate radial fibrils. Margin not striate. Stipe mottled. Volva bag-like. Smell of withered roses when old.*
Cap 4–15 cm (1.6–6 in), not striate, lacking velar remains, typically greenish yellow to bronze, with darker, innate, radial fibrils. *Stipe* 5–20 × 1–3(–5) cm (2–8 × 0.4–1.2(2) in), bulbous, white but with a zigzag mottling concolorous with the cap, occasionally more brownish grey. Ring pendant, white, fairly fragile. Volva bag-like, large, membranous, white but often greenish inside. *Flesh* white. Smell of old roses as it ages or dries. *Microscopic features* Sp 8–10 × 7–9 µm, shortly elliptic to subglobose, I+. Cheilo rare, inflated pyriform or clavate, occasionally septate, 20–45 × 8.5–20 µm. Volva filamentous. Clamps absent.

Season: July–October.
Habitat: Prefers acid soils, especially under oak.
Distribution: Throughout Europe.

Cap colour rather variable. There are white (f. *alba* (Vitt. ex Gilb.) Vesely), yellow, green or brown forms, sometimes causing confusion but all equally poisonous. Other variants have colonized particular habitats (*see below*).

---

p.187 *Amanita dunensis* (Heim) ex Bon & Andary
Amanitaceae, Amanitales

Description: *Slender, cap with fine striations. Stipe white, slender. Volva bag-like. Deeply buried in the sand.*
Cap 3–8 cm (1.2–3.2 in), the margin finely striate or striate, faintly radially streaky, whitish to olivaceous buff.

Gills white. *Stipe* 12–15 × 1–2 cm
(4.7–6 × 0.4–0.8 in), tall and slender,
deeply buried in sand. Ring fragile and
soon disintegrating. Volva bag-like, white.
Flesh thin, white. *Microscopic features*
Sp 8–11 × 6–7.5 µm, ovoid to elliptic, I+.
Cheilo like *A. phalloides*.
Volva filamentous.

Season: October–November.
Habitat: Sand, near pine and holm oak.
Distribution: Atlantic coast of France.

The sand, which invariably coats this
species, should be a deterrent to the
imprudent.

---

p.186 **Destroying Angel** (deadly poisonous)
Amanita virosa (Lamarck) Bertillon
Amanitaceae, Amanitales

Description: *Cap often ovoid, asymmetrical. Stipe markedly fleecy.*
Cap 3–10 cm (1.2–4 in), ovoid or
parabolic then convex, seemingly not set
squarely on the stipe, the margin not
striate, without velar remains. White or
ivory, rarely with beige tones when old.
Gills white. *Stipe* 8–15 × 0.5–1.5(–3) cm
(3.2–6 × 0.2–0.6(1.2) in), fairly fragile,
bulbous, white to pale cream, with
fibrillose-woolly wisps below the ring,
fragile. Volva bag-like, large, white.
Flesh not very thick. *Microscopic features*
Sp 9.5–12 µm diameter, globose, I+.
Cheilo absent, or a few marginal cystidia as
in *A. phalloides*. Clamps absent.

Season: July–October.
Habitat: Woodland on acid soils, often under
conifers, also under birch, especially in peat
bogs.
Distribution: Throughout Europe (mainly in the north).

A. verna (Bull.:Fr.) Lamark, which has a
larger and more regular white to pale cream
cap, a more robust white, smooth to silky
stipe with a white membranous ring and a
bag-like volva, is mainly a species of warm
sites. It is equally deadly.

p.189 **False Death Cap**
Amanita citrina (Sch.) Pers.
Amanitaceae, Amanitales

Description: *Cap yellow or whitish, with patches of veil. Volva half a sphere.*
Cap 5–10 cm (2–4 in), not striate, more or less bright lemon yellow, occasionally yellowish, with irregular and occasionally confluent, whitish or dingy cream patches. *Gills* white or with lemon tones. *Stipe* 8–15 × 1.5–2(–4) cm (3.2–6 × 0.6–0.8 (1.6) in), bulbous, white or yellowish. Ring large, white or yellow. Volva circumsessile, edged with a sharp rim. Smells strongly of raw potato. *Microscopic features* Sp 8–10 × 7–9 µm, subglobose to elliptic, I+. Cheilo rare, clavate. UV filamentous in the volva, sphaerocytes more numerous on the cap.
Season: July–November.
Habitat: Woods.
Distribution: All Europe.

There is a white form (f. *alba* (Price) Q. & Bat.), just like the type except for the colour. *A. asteropus* Sabo ex Romagnesi is much rarer (cap cream then spotted rufous and with patches of brown veil, stipe fleecy below the ring, discolouring slightly brown when handled, volva yellowish above a very well-developed bulb and opening like a star) and is usually found with conifers.

p.189 *Amanita porphyria* Alb. & Schw.:Fr.
Amanitaceae, Amanitales

Description: *Like A. citrina, but violaceous grey.*
Cap 5–10 cm (2–4 in), not striate, radially fibrillose, porphyrian grey with violaceous or crimson tones, velar patches brownish grey. *Gills* white or pale greyish. *Stipe* 8–12 × 1.5–2 cm (3.2–4.7 × 0.6–0.8 in), bulbous marginate (*see A. citrina*), whitish or flushed violaceous grey, discolouring brown when handled and when old. Ring grey, soon sticking to the stipe. Volva like *A. citrina*. Smells strongly of raw potato. *Microscopic features* Same characters as *A. citrina*.

Season: August–October.
Habitat: Usually broad-leaved trees; tends to occur on acid soils.
Distribution: Widespread. Never very common and even rare in some areas.

BOLBITIACEAE

The *Bolbitiaceae* is a very homogeneous family. It is defined as follows: fleshy or slender, usually saprotrophic (or occasionally parasitic); spores tobacco-buff to brown or dark brown to black; spore usually with a germ pore (GP); cap cuticle (CC) much evolved, often a hymeniderm. It includes one group with tobacco-buff to brown spores (tribe Bolbitieae) and another with darker brown to black spores (tribe Panaeoleae).

Constant characters in *Bolbitiaceae* are: gills ventricose, ascending, rather close; gill edge finely frosted or paler; stipe often hollow when mature; smell insignificant; spores smooth, slightly thick-walled.

---

p.148 *Agrocybe aegerita* (Brig.) Fayod
(= *A. cylindracea*)
Bolbitiaceae, Cortinariales

Description: *Cap cracking, beige to brownish. Ring white.*
Cap 2–15 cm (0.8–6 in), convex or wavy, occasionally slightly lubricious, soon wrinkled or even cracking, white to warm dark brown, occasionally with mixed colours. *Gills* adnate to decurrent, sinuate or arched, soon sepia beige to tobacco-brown. *Stipe* 5–15 × 1–2.5 cm (1–6 × 0.4–1 in), often curved, white or very pale. Ring large, white, fairly fragile. In clusters. *Flesh* fairly thick, white. Smell slightly fruity or mealy to radish-like.

Microscopic features Sp 7.5–11.5 × 4.5–6.5 μm, elliptic to cylindric-elliptic, occasionally with a flatter ventral face, GP almost absent or very narrow. Bas (occasionally 2-spored) 30–40 × 5–10 μm. Cheilo numerous, 20–45 × 10–15 μm, clavate, cylindro-clavate or fusiform-clavate. Pleuro identical or more lageniform, occasionally papillate to mucronate (rarer). CC with pyriform or clavate hyphae, 20–40 × 8–20 µm. Clamps present.

Season: June–November.
Habitat: On stumps and trunks of broad-leaved trees, often on poplar.
Distribution: Widespread.

A good edible species.

---

p.143 *Agrocybe praecox* (Pers.:Fr.) Fayod
Bolbitiaceae, Cortinariales

Description: *Cap whitish to buff, cracking. Ring white occasionally appendiculate.*
Cap 2–8 cm (0.8–3.2 in), the margin often appendiculate with fleeting, white velar remains. Soon wrinkled or cracking, white to ochraceous beige, rarely brownish and occasionally with pinkish tones. *Gills* adnate-notched, pale beige then sepia to quite dark tobacco-brown. *Stipe* 4–10 × 0.5–1 cm (1.6–4 × 0.2–0.4 in), often with white rhizoids, the apex as if powdered, white or almost concolorous. Ring large but fragile, white. *Flesh* white or pale. Taste slightly or slowly bitter to mealy. *Microscopic features* Sp 8.5–11 × 5–6.5 µm, elliptic, occasionally with a slightly flatter ventral face, GP 1–1.5 µm diameter. Bas 25–35 × 6–9 µm. Cheilo 25–65 × 14–20 µm, clavate to sphaeropedunculate. Pleuro identical or fusiform to lageniform, rarer. CC with pyriform or clavate hyphae, 30–50 × 12–25 µm. Clamps present.
Season: April–October.
Habitat: Undergrowth, edges of wood, paths, occasionally meadows or coppices.
Distribution: Common or fairly rare.

Occurs mainly in spring, but occasionally in autumn. *A. molesta* (Lasch) Singer (= *A. dura*) grows up to 12 cm (4.7 in) diameter, is more fleshy, distinctly cracking when old, and the gills are grey with a faint touch of mauve then brownish; it occurs in fields, meadows and open places.

p.143 *Agrocybe pediades* (Fr.:Fr.) Fayod
Bolbitiaceae, Cortinariales

Description: *Small. Cap pale yellow-buff. Gills ventricose. Stipe without a ring.*
Cap 1–4 cm (0.4–1.6 in), hemispherical then convex, occasionally with fine wrinkles or cracks, more or less pale yellow-buff. *Gills* notched to almost free, ventricose, soon brownish, and quite dark. *Stipe* 2–7 × 0.2–0.5 cm (0.8–2.8 × 0.08–0.2 in), cylindric, pale to almost concolorous below. *Flesh* pale. Smell and taste slightly mealy. *Microscopic features* Sp 10.5–14 × 6.5–8.5 µm, broadly elliptic, symmetrical, GP 1.5–2 µm diameter. Bas 30–40 × 7–15 µm. Cheilo 25–55 × 7–15 µm, lageniform, the base slightly inflated and the neck subcapitate. Pleuro almost absent. Clamps present.

Season: May–November.
Habitat: Lawns, fallow ground and woodland tracks.
Distribution: Not very well known.

A. semiorbicularis (Bull) Fayod, another common species in this group, has 2-spored basidia.

---

p.142 *Agrocybe vervacti* (Fr.:Fr.) Singer
Bolbitiaceae, Cortinariales

Description: *Cap finely wrinkled, yellow to yellowish buff. Gills brownish.*
Cap 1–4 cm (0.4–1.6 in), convex and often dented to wrinkled, quite bright ochraceous yellow, occasionally with orange or rufous tones, fading to pale yellowish buff and even occasionally whitish.
Gills adnate to subdecurrent, pale beige then brownish buff. *Stipe* 1.5–4 × 0.2–0.5 cm (0.6–1.6 × 0.08–0.2 in), cylindric, white or pale buff. *Flesh* white or pale. *Microscopic features* Sp 8–9.5 × 5–6 µm, elliptic to shortly ovoid, GP quite narrow. Bas 30–35 × 6.5–8 µm. Cheilo 35–50 × 8–10 µm, lageniform with a rather long, blunt or subcapitate neck. CC with clavate hyphae, 35–65 × 8–20 µm, the

pileocystidia similar to the cheilocystidia or up to 70 × 12 µm. Clamps present.

Season: August–November.
Habitat: Lawns and open grassy places.
Distribution: Uneven. Rare to very rare.

---

p.128 *Bolbitius vitellinus* (Pers.:Fr.) Fr.
Bolbitiaceae, Cortinariales

Description: *Slender. Cap very striate, viscid, bright yellow. Gills rust-coloured.*
Cap 1–6 cm (0.4–2.4 in), campanulate then almost flat, with long striations, almost deliquescent. Surface viscid, bright egg-yellow, occasionally paler, also paler when old. *Gills* almost free, narrow, straw-yellow then brownish and finally rust-coloured, almost deliquescent. *Stipe* 4–12 × 0.1–0.5 cm (1.6–4.7 × 0.04–0.2 in), very fragile, pruinose, white or pale yellow. *Flesh* very thin, fragile. *Microscopic features* Sp 11–16 × 6–8.5 µm, elliptic, the ventral face slightly more flat, GP wide. Bas 15–30 × 10–18 µm, broadly clavate to sphaeropedunculate, with pseudoparaphyses in between. Cheilo 30–50 × 14–20 µm, lageniform to broadly utriform. Pleuro rare, similar. CC gelatinized, with clavate or sphaeropedunculate hyphae 25–50 × 10–15 µm. Clamps absent.

Season: Practically all year round.
Habitat: Dung, rotting plant remains; often in open places.
Distribution: Widespread but frequency uneven.

Var. *titubans* (Bull.:Fr.) M. Bon & Courtec., smaller and slender, has a translucent cap with very long striations.

---

p.113 *Bolbitius tener* Berk. & Br
Bolbitiaceae, Cortinariales

Description: *Slender. Cap campanulate, white or pale. Gills rust-coloured. Stipe white.*
Cap 1–3 cm (0.4–1.2 in), like the finger of a glove, white to cream or beige, rarely pale

buff when old. Margin striate, soft.
Gills almost free, narrow, soon rust-coloured, almost deliquescent.
Stipe 4–8 × 0.1–0.3 cm (1.6–3.2 × 0.04–0.1 in), very fragile, cylindric or slightly bulbous, pruinose or striate, white.
Flesh very thin, fragile. *Microscopic features* Sp 10–15 × 7–10 µm, elliptic, flatter on the ventral face, GP wide (up to 3.5 µm). Bas 15–25 × 10–18 µm, broadly clavate to sphaeropedunculate, with pseudoparaphyses in between (20–30 × 15–25 µm). Cheilo 15–25 × 7–12 × 3–5 µm, like a skittle. CC slightly gelatinized, with clavate or sphaeropedunculate hyphae, 20–80 × 15–40 µm, and a few skittle-shaped pileocystidia, the neck occasionally elongated. Clamps rare.

Season: June–December.
Habitat: Lawns.
Distribution: Uneven. Fairly common to very rare.

Occasionally very abundant on urban lawns but may be totally absent in some areas.

---

p.128 *Conocybe tenera* (Sch.:Fr.) Kühner
Bolbitiaceae, Cortinariales

Description: *Slender. Cap rusty fawn, finely striate. Gills rust-coloured. Stipe pruinose.*
Cap 1–2 cm (0.4–0.8 in), campanulate, hygrophanous, striate, brownish fawn to cinnamon, buff when dry. *Gills* almost free, pale buff to rust-coloured. *Stipe* 3–8 × 0.1–0.4 cm (1.2–3.2 × 0.04–0.2 in), fragile, slightly bulbous, pruinose, pale cream above, more buff or rust-coloured toward the base which may be dark. *Flesh* thin, fragile.

Microscopic features Sp 9–14 × 5.5–7.5 µm, elliptic, the ventral face flatter, GP fairly wide (about 2 µm). Bas 25–30 × 8–11 µm. Cheilo 15–25 × 7–10 µm × 4–5.5 µm, skittle-shaped. CC with clavate or sphaeropedunculate hyphae 40–50 × 15–20 µm, with skittle-shaped pileocystidia (25–40 × 7–9 µm). Clamps present. Needle-shaped crystals develop in ammonia.

Season: June–November.
Habitat: Lawns, tracks, grassy places.
Distribution: Widespread. Common.

A genus with very many species, almost impossible to identify without a microscope.

---

p.127 *Conocybe dunensis* Wallace
Bolbitiaceae, Cortinariales

Description: *Slender. Cap reddish brown to fawn. Gills rusty buff. Stipe very pale.*
Cap 1–2.5 cm (0.4–1 in), bell-shaped, hygrophanous, buff to reddish brown or date-brown, fading when dry. *Gills* like *C. tenera* or yellowish cream when young. *Stipe* 3–8 × 0.2–0.4 cm (1.2–3.2 × 0.08–0.2 in), fragile, slightly bulbous, pruinose or finely striate, white then cream or buff from the base up. *Flesh* thin, fragile. *Microscopic features* Sp 11–15 × 7–9 µm, elliptic, the ventral face flatter, GP fairly wide (about 2.5 µm). Bas 25–30 × 8–12 µm. Cheilo 15–30 × 7–12 µm × 3–4.5 µm, skittle-shaped. CC with clavate or sphaeropedunculate hyphae, 40–50 × 15–25 µm. Clamps absent.
Season: August–November.
Habitat: Dunes, grassy, inland side of a dune system or short grassland on fixed dunes.
Distribution: Uneven. Rare.

---

p.128 *Conocybe inocybeoides* Watling
Bolbitiaceae, Cortinariales

Description: *Cap reddish beige. Gills rust-coloured. Stipe bulbous, whitish to rather rufous.*
Cap 1–4 cm (0.4–1.6 in), bell-shaped then almost flat, hygrophanous, not very striate, brownish beige or greyish beige, the centre more reddish or date-brown, paler when dry, yellowish buff. *Gills* almost free, white or buff then rust-coloured. *Stipe* 2–8 × 0.1–0.6 cm (0.8–3.2 × 0.04–0.3 in), fairly fragile, bulbous (up to 1.2 cm (0.5 in)), pruinose to floccose at the apex, pale cream, honey-coloured or rather rufous in the

middle. *Flesh* thin. Smell sour when cut. *Microscopic features* Sp 13–19(–25) × 6–9 µm, elliptic or almond-shaped, the ventral face flatter, GP medium-sized. Bas 2-spored, 15–28 × 8–10 µm. Cheilo 15–30 × 7–10 × 3–4 µm, skittle-shaped. CC with clavate or sphaeropedunculate hyphae, 30–45 × 15–25 µm, the pileocystidia filiform or cylindric (3–7 µm diameter). Clamps absent.

Season: August–November.
Habitat: Lawns, coppice, wood edges, etc.
Distribution: Uneven. Rare.

Bulb very pronounced, characteristic.

---

p.129 *Pholiotina arrhenii* (Fr.) Singer
Bolbitiaceae, Cortinariales

Description: *Cap rufous buff, finely striate. Stipe white with a striate ring.*
Cap 1–3 cm (0.4–1.2 in), campanulate then expanded, hygrophanous, striate, often finely wrinkled, rufous brown to tawny buff, the margin often more yellow, ochraceous yellow when dry. *Gills* pale buff to bright rusty buff. *Stipe* 2–5 × 0.1–0.4 cm (0.8–2 × 0.04–0.16 in), fairly fragile, slightly bulbous, pale then brownish to brown from the base up. Ring more or less median, like a short skirt, striate or even grooved above, white, very fragile. *Flesh* thin, fragile. *Microscopic features* Sp 6.5–9 × 4–5.5 µm, elliptic or almond-shaped, GP very narrow. Bas 25–30 × 8–10 µm. Cheilo 25–55 × 3–10 µm, bottle-shaped or cylindric. CC with clavate or sphaeropedunculate hyphae, 30–45 × 15–20 µm. Clamps present.

Season: August–November.
Habitat: Damp road sides, wood edges, lawns, undergrowth on clay soils.
Distribution: Widespread. Common or fairly rare.

P. hadrocystis (K.vW.) Courtec. is similar, although often slightly less tall and with a smell of geranium. *P. aporos* (K.v.W) Clémençon is a spring look-alike, often more dull.

p.129 *Pholiotina vestita* (Fr.) Singer
Bolbitiaceae, Cortinariales

Description: *Slender. Cap fawn, the margin appendiculate with white velar remains.*
Cap 1–3 cm (0.4–1.2 in), bell-shaped then expanded, hygrophanous, finely striate when fresh, ochraceous fawn to brownish rufous, paler when dry, ochraceous cream, appendiculate with transient, large, triangular, white veil remains.
Gills strictly adnate, ochraceous yellow then bright rusty buff. *Stipe* 2–5 × 0.2–0.3 cm (0.8–2 × 0.08–0.12 in), fairly fragile, slightly bulbous, often fibrillose toward the base, glabrescent, white but occasionally dark brown to brown-black from the base up. *Flesh* thin, fragile.
Microscopic features Sp 6–9 × 4.5–5.5 μm, elliptic, GP very narrow. Bas 20–25 × 6–8 μm. Cheilo 30–50 × 5–10 μm, bottle-shaped or cylindric. CC with clavate or sphaeropedunculate hyphae, 20–40 × 10–15 μm. Clamps present.
Season: August–November.
Habitat: Undergrowth in cool, damp broad-leaved woodland.
Distribution: Not very well known. Rather rare.

Other taxa have smaller veil remains, sometimes forming regular, white teeth, when young.

p.121 *Panaeolus rickenii* Hora
Bolbitiaceae, Cortinariales

Description: *Tall. Cap dark brown. Gills black, mottled. Stipe red-brown.*
Cap 0.5–2 cm (0.2–0.8 in), campanulate, hygrophanous, finely wrinkled or occasionally cracking, dark brown, with blackish or violaceous tones, fading from the centre which becomes red-brown then rather pinkish brown. *Gills* strictly adnate, grey then black in mottled patches. Gill edge fimbriate, weeping fine clear droplets when young. *Stipe* 4–10 × 0.1–0.3 cm (1.6–4 × 0.04–0.12 in), fairly

fragile, slightly bulbous, with whitish pruina on a fairly bright red-brown, striate ground. *Flesh* fragile, brownish to vinaceous brown, very dark at the stipe base. *Microscopic features* Sp 11–16 × 8–11 × 6–8 µm, flattened, lemon-shaped or almost hexagonal in side view, GP axial, fairly wide, very dark. Bas 15–25 × 8–10 µm, clavate. Cheilo 25–40 × 3–10 µm, bottle-shaped, with a cylindric neck, more or less wavy. CC with clavate hyphae, 10–15 × 8–15 µm, and occasional pileocystidia similar to the cheilocystidia. Clamps present.

Season: May–November.
Habitat: Lawns and open grassy places.
Distribution: Widespread but frequency uneven.

The cap of *P. acuminatus* (Sch.) Quélet is conical, with an almost pointed umbo.

---

p.135 *Panaeolus sphinctrinus* (Fr.) Quélet
Bolbitiaceae, Cortinariales

Description: *Cap brownish grey. Margin toothed.*
Cap 1–4 cm (0.4–1.6 in), conico-campanulate, hygrophanous, the margin with appendiculate teeth, micaceous, often finely wrinkled or cracking, slightly brownish or dark olivaceous grey, fading. *Gills* strictly adnate, grey then black in mottled patches. *Stipe* 5–12 × 0.2–0.5 cm (2–4.7 × 0.08–0.2 in), fairly stiff, greyish pruinose on a dark brown-grey ground, blackish from the base up. *Flesh* fairly thin, very dark at the base. *Microscopic features* Sp 14–18 × 9–12 × 7–9 µm, flattened, lemon-shaped or hexagonal in face view, GP axial, very wide, very dark. Bas 20–30 × 12–15 µm, clavate. Cheilo 25–80 × 2–7 µm, cylindric to almost lageniform, wavy. CC with clavate hyphae, 13–30 × 10–15 µm.

Season: July–December.
Habitat: Lawns and manured meadows, even on dung.
Distribution: Widespread but frequency uneven.

The toothed margin, often accentuated by velar remains or white fleece, are characteristic.

---

p.135 *Panaeolus ater* (Lange) Kühn. & Romagn. ex M. Bon
Bolbitiaceae, Cortinariales

Description: *Small. Cap brown. Gills black, mottled. Stipe dirty brown.*
Cap 1–5 cm (0.4–2 in), hemispherical then expanded, hygrophanous, finely wrinkled to wrinkled, occasionally cracking, brown, sometimes blackish in the centre, fading. *Gills* adnate, grey then black in mottled patches. *Stipe* 2–5 × 0.2–0.4 cm (0.8–2 × 0.08–0.16 in), fairly fragile, pruinose, finely striate, more or less dark, dirty brown from the base up. *Flesh* thin, brownish to dark brown. *Microscopic features* Sp 10–13 × 6.5–8 µm, slightly lemon-shaped, GP oblique, fairly wide, displaced dorsally, very dark. Bas 15–25 × 5–10 µm, broadly clavate. Cheilo 15–45 × 6–12 µm, lageniform to cylindro-clavate, occasionally with a wavy or subcapitate neck. Pleuro occasional, 15–25 × 5–8 µm, clavate or fusoid, contents staining bright yellow in ammonia (chrysocystidia). CC with clavate hyphae, 15–25 × 10–25 µm. Clamps present.
Season: August–December.
Habitat: Lawns and meadows, occasionally wood edges.
Distribution: Uneven. Rather rare to rare.

Characterized in particular, under the microscope, by the presence of chrysocystidia. In the field, it can be identified by its dark colour.

---

p.135 *Panaeolus obliquoporus* M. Bon
Bolbitiaceae, Cortinariales

Description: *Cap dark with a reddish margin. Gills distant. Stipe short.*
Cap 2–6 cm (0.8–2.4 in), convex, hygrophanous, finely wrinkled to wrinkled,

occasionally cracking, more or less dark reddish brown but the margin often reddish to red, fading slightly when dry.
Gills fairly deep and not very close, grey then black in mottled patches. *Stipe* 1–3 × 0.2–0.7 cm (0.4–1.2 × 0.08–0.28 in), cylindric, pruinose at the apex, almost concolorous with the cap or paler.
Flesh thin, fairly fragile, brownish.
Microscopic features Sp 10–15 × 7–8 × 5–7 μm, slightly flattened, elliptic to ovoid, GP oblique, displaced toward the dorsal face, very dark. Bas 25–35 × 10–12 μm, clavate. Cheilo 25–45 × 5–10 μm, fusi-lageniform, the neck the same or slightly thickened. Pleuro occasional, 30–60 × 15–30 μm, broadly fusoid, occasionally mucronate, the contents bright yellow in ammonia (chrysocystidia). CC with clavate hyphae, 15–25 × 10–25 μm. Clamps present.

Season: September–December.
Habitat: Short turf on dunes.
Distribution: A rare species only recorded in Fr.ance.

---

p.134 *Panaeolus dunensis* M. Bon & Courtecuisse
Bolbitiaceae, Cortinariales

Description: *Cap brown to brownish beige, very dented. Gills black.*
Cap 0.5–4 cm (0.2–1.6 in), convex, dented or occasionally cracking, hygrophanous, brown, sometimes dark at the centre, fading when dry. *Gills* notched, not very close, grey then black in mottled patches. *Stipe* 2–4 × 0.2–0.6 cm (0.8–1.6 × 0.08– 0.24 in), fairly fragile, pruinose at the apex, paler than the cap and occasionally with violaceous tones, especially toward the base. *Flesh* thin, fairly fragile, brownish to pinkish brown, dark at the stipe base. *Microscopic features* Sp 10–15 × 7–10 × 6–7 μm, slightly flattened, elliptic to angular in face view, GP distinctly displaced toward the dorsal side, fairly wide. Bas 25–35 × 8–12 μm, clavate. Cheilo 25–45 × 6–10 × 3–5 μm, bottle-shaped, the neck not capitate. Clamps present.

Season: April–December.

Habitat: Short turf on dunes.
Distribution: Uneven. Rare or not well known.

P. fimicola (Fr.) Gill., which is more regular, with a more hygrophanous cap, and has less ventricose gills and a taller stipe, has an axial GP and capitate cheilocystidia. It occurs on short grass, even inland.

---

p.120 *Panaeolina foenisecii* (Pers.:Fr.) Maire
Bolbitiaceae, Cortinariales

Description: *Cap pinkish brown. Gills reddish buff, mottled.*
Cap 0.5–2.5 cm (0.2–1 in), convex, hygrophanous, often finely wrinkled, occasionally cracking, date-brown to reddish brown, drying rather pinkish brown or pinkish buff. *Gills* not very close, pinkish buff to pinkish brown in mottled patches. *Stipe* 3–7 × 0.1–0.3 cm (1.2–2.8 × 0.04–0.12 in), fragile, slightly bulbous, pruinose. *Flesh* thin, fairly fragile, rather pinkish brown. *Microscopic features* Sp 12–17 × 7–9 µm, elliptic or almond-shaped, strongly warted, GP axial, wide, reddish brown when mature. Bas 25–35 × 10–15 µm, broadly clavate. Cheilo 35–55 × 6–12 µm, bottle-shaped, the neck fairly thick, occasionally wavy, blunt. CC with clavate hyphae, 20–50 × 10–25 µm. Clamps present.
Season: Practically all year round.
Habitat: Lawns and grassy places.
Distribution: Widespread but frequency uneven.

On lawns in towns in particular. It is considered poisonous.

---

p.113 *Anellaria semiovata* (Sow.:Fr.) Pearson & Dennis (= *Panaeolus s.*)
Bolbitiaceae, Cortinariales

Description: *Cap ovoid, pale, viscid. Gills black. Stipe tall, with a ring.*
Cap 2–6 cm (0.8–2.4 in), ovoid, viscid then shining, occasionally cracking, white or greyish to dirty brownish when old.

Gills strictly adnate, grey then mottled black. *Stipe* 5–15 × 0.3–0.5 cm (2–6 × 0.12–0.2 in), fragile, finely striate, concolorous or discolouring dingy from the base up, the ring rather high, ascending, thin and soon sticking to the stipe.
Flesh fairly thin, fragile, brownish.
Microscopic features Sp 18–22 × 10–13 µm, elliptic, GP often displaced toward the dorsal face, fairly wide. Bas 25–40 × 15–20 µm, clavate. Cheilo 30–50 × 8–15 µm, inflated bottle-shaped, the neck long, occasionally wavy. Pleuro occasional, 45–75 × 15–20 µm, clavate or broadly fusoid, often mucronate or rostrate, the contents bright yellow in ammonia (chrysocystidia). CC with clavate hyphae, 15–25 × 8–15 µm. Clamps present.

Season: August–October.
Habitat: Gregarious. On dung, in meadows.
Distribution: Fairly widespread but frequency uneven.

Boletaceae: Boletes and Related Genera

This morphological family comprises the poroid Boletales: Boletaceae, Strobilomycetaceae and Gyrodontaceae (an order which also includes lamellate families).

Boletes and their allies have the following characters: fleshy, with a fibrous, sometimes brittle texture; tubular-poroid hymenophore; spores usually brown to olive-brown (rarely different), very often fusiform with a plage. They are generally ectomycorrhizal on trees or woody plants.

The main constant characters are: cap convex, fleshy, tubes adnate, horizontal, pores small, concolorous with the tubes, stipe solid, firm, flesh thick, smell and taste insignificant; spores fusiform with a plage, brown, smooth, without a germ pore, dextrinoid (red-brown in Melzer's iodine reagent); basidia 4-spored, clavate; clamps absent.

---

p.228 **Old Man of the Woods**
Strobilomyces strobilaceus (Scop.:Fr.) Berk.
Strobilomycetaceae, Boletales

Description: *Cap scaly, dark grey. Stipe woolly, with a ring.*

Cap 5–12 cm (2–4.7 in), appendiculate, coarsely woolly, rather pale grey then brown-grey, with large pyramidal scales, blackish at the tips. Tubes notched to subdecurrent. Pores about 0.5 mm (0.02 in) diameter, sooty-black to blackish, reddening slightly where handled. Spore print crimson-black. *Stipe* 6–15 × 1–2 cm (2.4–5 × 0.4–0.8 in), woolly to fleecy under a fibrillose-woolly, fairly large, but quickly disappearing ring, concolorous with the cap. *Flesh* pale, flushing or reddening when cut then blackish. *Microscopic features* Sp 9–12 × 8–11 μm, subglobose, with fairly prominent network of ridges in a honeycomb pattern. Cyst 60–80 × 15–35 μm, fusiform and broadly ventricose, sometimes clavate or bottle-shaped. CC a trichodermium, hyphae 5–15 μm diameter. Pigment intracellular.

Season: August–November.
Habitat: Under beech especially, rarer under conifers.
Distribution: Uneven, mainly southern.

Easy to identify, on account of the very distinctive cap cuticle.

p.228 *Porphyrellus porphyrosporus* (Fr.) Gilbert
Boletaceae, Boletales

Description: *Cap velvety, olivaceous grey-brown. Stipe concolorous.*
Cap 5–12 cm (2–4.7 in), velvety, indented to finely wrinkled, matt, brownish grey to blackish, often with purplish or crimson tones. Tubes adnate, sometimes notched. Pores rounded (0.5–1 mm (0.02–0.04 in)), sooty violaceous then crimson-brown to blackish. Spore print reddish brown. *Stipe* 6–12 × 1–3 cm (2.4–4.7 × 0.4–1.2 in), often clavate, finely velvety, matt, concolorous with the cap. *Flesh* white, reddening or violaceous to blackish grey when cut. *Microscopic features* Sp 12–20 × 5.5–8 μm. Cyst 50–135 × 10–20 μm (marginal cyst largest), cylindro-clavate or fusoid, pigment intracellular, brown, cloudy. CC a trichoderm, with

hyphae 7–12 μm diameter. Pigment intracellular.
Season: August–November.
Habitat: Conifers, rarely broad-leaved trees (in lowland areas).
Distribution: Uneven. Rather continental, quite rare to rare.

---

p.228 *Gyroporus castaneus* (Bull.:Fr.) Quélet
Gyrodontaceae, Boletales

Description: *Cap very dry, rufous chestnut. Pores pale. Stipe brittle.*
Cap 3–10 cm (1.2–4 in), wavy or dented, dry, often deeply cracked when old, quite warm rufous chestnut, sometimes paler or marbled yellowish buff. Tubes adnate or notched. Pores rounded, fairly small or medium-sized, white then slightly yellowish or spotted yellowish beige. Spore print pale lemon. *Stipe* 4–8 × 1–3 cm (1.6–3.2 × 0.4–1.2 in), brittle, with spaces or cavities, clavate or dented below, scurfy to almost smooth, concolorous.
Flesh unchanging, very brittle. *Microscopic features* Sp 8–12 × 4.5–6.5 μm, elliptic to almost bean-shaped, hyaline. Cheilo 25–40 × 10–12 μm, broadly fusiform ventricose. CC a trichoderm, with hyphae 8–15 μm diameter. Pigment intracellular. Clamps present.
Season: July–December.
Habitat: Broad-leaved trees and conifers on acid soils, especially on sand.
Distribution: Uneven. Mainly in warm sites.

The very dry cuticle and brittle stipe, with large cavities when mature, are characteristic of the genus.

---

p.228 *Gyroporus cyanescens* (Bull.:Fr.) Quélet
Gyrodontaceae, Boletales

Description: *Cap dry, yellowish white. Stipe brittle, with cavities. Blueing strongly.*
Cap 4–10 cm (1.6–4 in), often dented, very dry, sometimes cracked, whitish to

yellowish, sometimes mottled pale yellowish buff, blueing strongly (intense purple-blue) where handled. Tubes notched. Pores rounded, white then slightly yellowish or spotted yellow, blueing. Spore print pale yellow. *Stipe* 4–10 × 1–2.5 cm (1.6–4 × 0.4–1 in), with cavities when mature, brittle, often dented, matt, pruinose to almost smooth, whitish to yellowish, blueing. *Flesh* blueing vividly. *Microscopic features* Sp 7–11 × 3.5–6 µm, elliptic to cylindric-elliptic, the ventral face sometimes slightly depressed, hyaline. Cheilo 20–55 × 8–15 µm, broadly fusiform or cylindro-clavate, sometimes septate. CC a cutis with trichodermial cells, hyphae 7–15 µm diameter. Pigment mainly encrusted intraparietal. Clamps present.

Season: July–December.

Habitat: Broad-leaved or coniferous trees, sometimes in sunny situations, especially on sandy, acid soils.

Distribution: Widespread. Rare to very rare.

The extraordinary blueing, contrasting strongly with its pale colour, is a very reliable character.

---

p.228 *Gyrodon lividus* (Bull.:Fr.) P. Karsten
Gyrodontaceae, Boletales

Description: *Cap putty-coloured or olivaceous. Tubes short, decurrent, blueing.*
Cap 5–10 cm (2–4 in), slightly convex, lubricious or velvety, whitish to yellowish olivaceous or with dingy buff marbling, blueing slightly. Tubes very short, decurrent. Pores rounded or radially elongated to labyrinthine, yellowish, becoming blue-green where handled. Spore print dingy yellow-buff to olivaceous brownish. *Stipe* 3–8 × 0.5–1.5 cm (1.2–3.1 × 0.2–0.6 in), fusiform, tapering, sometimes almost rooting, yellowish buff then dingy red-brown from the base. *Flesh* spongy, whitish to dingy yellowish buff, red-brown at the stipe base, becoming blue-green when cut. *Microscopic features* Sp

5–7 × 3.5–4.5 µm, broadly elliptic, pale. Cheilo rare, 35–50 × 5–10 µm, cylindric-fusoid, sometimes septate. CC a cutis, hyphae 2.5–5 µm diameter. Pigment intracellular. Clamps present.

Season: August–November.
Habitat: Damp places, under alder.
Distribution: Uneven. Rare to very rare.

The very short, decurrent tubes and the habitat are good diagnostic characters.

---

p.229 *Boletinus cavipes* (Klotzsch ex Fr.) Kalchbr.
Gyrodontaceae, Boletales

Description: *Cap umbonate, felty, rufous brown. Tubes decurrent. Pores large. Ring cottony.*
Cap 5–12 cm (2–4.7 in), convex with a rather pronounced umbo, dry, fibrillose to finely felty scaly, rusty rufous brown. Margin appendiculate. Tubes short, decurrent. Pores large, radially elongated (up to 3 mm (0.12 in)), compound, dingy or brighter yellow to olivaceous. Spore print dingy yellow to olivaceous yellow. *Stipe* 4–10 × 0.5–1.5 cm (1.6–4 × 0.2–0.6 in), hollow, tapering, yellowish to reddish brown, the ring fibrillose-cottony, pale, fleeting. *Flesh* pale, unchanging. *Microscopic features* Sp 8–11 × 3–4 µm, pale. Cyst 50–80 × 7–10 µm, cylindric or clavate-fusoid. CC a trichoderm, the hyphae 10–15 µm diameter. Pigment intracellular.

Season: July–October.
Habitat: Only under larch.
Distribution: Continental. Rare.

---

p.229 **Slippery Jack**
Suillus luteus (L.:Fr.) Roussel
Boletaceae, Boletales

Description: *Cap lubricious, crimson chocolate brown. Ring violaceous.*
Cap 5–12 cm (2–4.7 in), viscid or greasy, slightly fibrillose-innately fibrillose, dark chocolate to rufous brown, sometimes

mottled or blotched with lighter areas. Margin appendiculate. Tubes adnate. Pores small, pale yellow to pale olivaceous buff. Spore print tawny brown. *Stipe* 5–10 × 1.5–2.5 cm (2–4 × 0.6–1 in), yellowish under the tubes, brownish to rust-coloured towards the base, with a large, membraneous, transient, often violaceous ring. *Flesh* firm, white, unchanging. *Microscopic features* Sp 7–10 × 3–4.5 µm, quite pale. Cyst 35–60 × 5–8 µm, cylindric or clavate-fusoid. CC an ixocutis, the hyphae 2–8 µm diameter. Pigment primarily epimembranary.

Season: July–November.
Habitat: Pine (especially Scots pine).
Distribution: Widespread. Fairly common.

Edible despite the slightly slimy and repellent texture after cooking. It can cause gastric upsets (remove the slimy skin on the cap before cooking!).

---

p.229 **Larch Bolete**
Suillus grevillei (Klotzsch:Fr.) Singer
(= *B. elegans* = *B. flavus*)
Boletaceae, Boletales

Description: *Cap viscid, orange-yellow. Pores yellow. Ring whitish.*
Cap 4–12 cm (1.6–4.7 in), viscid or lubricious, smooth or with slight bright yellow to orange-yellow dots or granulations, sometimes mottled. Margin appendiculate. Tubes adnate. Pores small, yellow, sometimes with rusty spots. Spore print yellow-brown. *Stipe* 4–10 × 1–2 cm (1.6–4 × 0.4–0.8 in), orange rufous towards the base, with a membranous fairly broad, white or whitish ring. *Flesh* whitish to yellow, sometimes rusty spotted. *Microscopic features* Sp 7–11 × 3–4.5 µm, quite pale. Cyst 30–50 × 4.5–7 µm, like *S. luteus*. CC an ixocutis or ixotrichodermium, the hyphae 2–6 µm diameter.

Season: July–November.
Habitat: Only under larch.

Distribution: Widespread. Follows the host.

Also found under larch: *S. tridentinus* (Bres.) Singer has a viscid then almost dry, fibrillose-finely scaly cap, reddish brown on a paler ground; *S. viscidus* (L.) Roussel, has a viscid cap, ochraceous beige to putty-coloured, marbled with brown-grey or olivaceous greenish colours, the pores flesh-beige then dingy olivaceous grey.

---

p.229 *Suillus collinitus* (Fr.) O. Kuntze (= *S. fluryi*) Boletaceae, Boletales

Description: *Cap lubricious, mottled with chocolate and somewhat rufous fawn colours. Stipe base lilaceous pink.*
Cap 3–10 cm (1.2–4 in), viscid or lubricious, smooth or innately fibrillose. Colours often mixed, chocolate brown with rufous or yellow-buff. Tubes adnate. Pores small, rather pale yellow to yellow or olivaceous, with milky or clear droplets when young. Spore print ochraceous brown. *Stipe* 3–9 × 0.7–1.5 cm (1.2–3.5 × 0.3–0.6 in), pale yellow, granular at the tip, brownish or somewhat rufous below, the extreme base purple or lilac to pink. *Flesh* quite firm, whitish to yellowish. *Microscopic features* Sp 8–12 × 3.5–5 µm, quite pale. Cyst 30–50 × 5–10 µm, clavate or cylindric-clavate. CC an ixocutis, the hyphae 2–7 µm diameter. Pigment membranal and intracellular.
Season: August–December.
Habitat: Two-needled pines, on calcareous soils.
Distribution: Fairly widespread, mainly southern.

S. granulatus (L.:Fr.) Roussel has a more uniformly coloured cap, yellowish beige to reddish brown and has no pink at the stipe base. *S. bellinii* (Inzenga) Watling has white marbling on the cap and rather red granulations on the stipe; it is a warmth-loving species of the Mediterranean–Atlantic regions.

p.230 *Suillus bovinus* (L.:Fr.) Roussel
Boletaceae, Boletales

Description: *Cap sticky, cinnamon buff. Pores olivaceous.*
Cap 2–10 cm (0.8–4 in), lubricious then greasy, quite bright ochraceous yellow or more brownish, sometimes cinnamon or orange. Tubes adnate, fairly short. Pores large, compound, angular, olivaceous mustard. Spore print olive-brown. *Stipe* 2–10 × 0.5–1.5 cm (0.8–4 × 0.2–0.6 in), dingy yellow to rather fawn-buff, often pinkish below. *Flesh* soon soft, whitish to reddish buff, sometimes only slightly blueing. *Microscopic features* Sp 7–11 × 3–4 µm, olivaceous. Cyst 30–60 × 4–8 µm, clavate or cylindric-clavate. CC an ixocutis, the hyphae 2–6 µm diameter. Pigment mainly intracellular.

Season: July–November.
Habitat: Two-needled pines; usually on damp soils.
Distribution: Widespread, frequency uneven.

Chalciporus piperatus (Bull:Fr.) Bataille is the same size; the fruit body often just one colour, yellow-brown to cinnamon, but the mycelium is bright yellow (stipe base and flesh in the base) and flesh very peppery.

p.230 *Suillus variegatus* (Sw.:Fr.) Rich. & Roze
Boletaceae, Boletales

Description: *Cap speckled, brownish buff. Pores olivaceous mustard.*
Cap 5–15 cm (2–6 in), soon dry, with dull ochraceous brown speckles, fibrils or fine scales on a yellowish buff to olivaceous buff, sometimes more rufous ground. Tubes adnate, fairly short. Pores like *S. bovinus* (*see* above) or larger. Spore print olivaceous brown. *Stipe* 4–10 × 0.8–2.5 cm (1.6–4 × 0.3–1 in), rather rufous buff to rather tawny. *Flesh* more or less blueing. *Microscopic features* Sp 8–12 × 3–4.5 µm, olivaceous. Cyst 30–60 × 5–8 µm, cylindric-clavate. CC a (ixo) trichoderm, the hyphae 2–15 µm diameter. Pigment intracellular.

Season: August–November.

Habitat: Two-needled pines; acid soils.
Distribution: Widespread, frequency uneven.

---

p.231 *Xerocomus subtomentosus* (L.:Fr.) Quélet
Boletaceae, Boletales

Description: *Cap olivaceous brown, dry. Pores bright yellow. Stipe cylindric.*
Cap 3–10 cm (1.2–4 in), dry, velvety, brownish to buff, often olivaceous, sometimes darker, more rufous or paler. Tubes notched. Pores large, bright yellow, sometimes blueing. Spore print olive-brown. *Stipe* 3–10 × 0.5–2 cm (1.2–4 × 0.2–0.8 in), variable, pale yellow, bright yellow, sometimes red-brown from the base, sometimes wrinkled near the apex. *Flesh* pale, almost unchanging. *Microscopic features* Sp 10–16 × 4–5.5 µm, brownish. Cyst 30–65 × 6–12 µm, cylindric-fusoid or clavate. CC a trichoderm with hyphae 5–15 µm diameter and a long terminal element.
Season: July–November.
Habitat: Broad-leaved and coniferous trees; rather ubiquitous.
Distribution: Widespread, frequency uneven.

Very variable.

---

p.231 *Xerocomus armeniacus* (Quélet) Quélet
Boletaceae, Boletales

Description: *Cap orange-pink to apricot. Stipe tapering, orange below.*
Cap 2–10 cm (0.8–4 in), velvety, reddish pink sometimes with violaceous tones, then orange to apricot, more dull with age. Tubes subdecurrent, fairly short. Pores yellow then greenish yellow, blueing. Spore print olive-brown. *Stipe* 3–12 × 0.5–1.5 cm (1.2–4.7 × 0.2–0.6 in), long tapering, yellow at the tip, more orange or cinnamon towards the base. *Flesh* yellowish to cinnamon-orange below. *Microscopic features* Sp 10–15 × 4–6 µm. Cyst 30–60 × 6–10 µm, cylindric-fusoid to fusiform. CC like *X. subtomentosus*.

Season: August–October.
Habitat: Open grassy undergrowth in broad-leaved woodland, wood edges, clearings, parks, etc.
Distribution: Rare.

Similar to *X. rubellus* (p.418), but the structure of the cap cuticle places it nearer to the *X. subtomentosus* group.

---

p.231 *Xerocomus chrysenteron* (Bull.) Quélet
Boletaceae, Boletales

Description: *Cap often cracked, the underlying flesh pink. Stipe streaked with red.*
Cap 2–8 cm (0.8–2.4 in), dry, often cracked, olivaceous brownish, sometimes quite pale or mottled. Flesh under the cuticle violaceous pink, visible in the cracks or when scratched. Tubes adnate. Pores yellow or greenish yellow, blueing weakly. Spore print olive-brown.
Stipe 2–10 × 0.4–2 cm (0.8–4 × 0.2–0.8 in), often slender, yellow, frequently with red fibrils or irregular striations from the base upwards, sometimes entirely red. *Flesh* soft, whitish to yellow, reddish under the cap, sometimes blueing or reddening.
Microscopic features Sp 11–16 × 4–6 µm, brownish. Cyst 50–80 × 8–15 µm, fusoid. CC almost a hymeniderm, the terminal cells short, clavate or pyriform, 8–15 µm diameter.
Season: May–December.
Habitat: Ubiquitous.
Distribution: Very widespread. Common.

X. pruinatus (Fr.) Quélet, with a dark umber brown cap, is very pruinose and often wrinkled and brain-like; it has a bright yellow stipe, bruises blue when handled and has bright yellow flesh. *X. porosporus* Imler is easy to identify under the microscope on account of the almost poroid spores and often has a very fissured, olivaceous-putty cap; it has white underlying flesh and hardly any red on the stipe, blackening below.

p.230 *Xerocomus rubellus* Quélet (= *X. versicolor*)
Boletaceae, Boletales

Description: *Cap cinnabar to raspberry.*
Characters as for *X. chrysenteron* (see p.417), but blood red, bright red, cinnabar to raspberry, sometimes discolouring or orange-pink with age. Pores bright yellow. *Stipe* bright yellow to bright red.
Season: August–November.
Habitat: Open, grassy scrub, wood edges.
Distribution: Uneven. Fairly common or rare.

---

p.230 *Xerocomus badius* (Fr.:Fr.) Gilbert
Boletaceae, Boletales

Description: *Cap warm brown, sometimes lubricious. Pores yellow, blueing. Stipe fairly robust.*
Cap 4–12 cm (1.6–4.7 in), viscid in wet weather, drying out and then velvety, warm brown. Tubes adnate, horizontal. Pores yellow or greenish yellow, blueing strongly when pressed or rubbed. Spore print olive-brown. *Stipe* 3–10 × 1–3 cm (1.2–4 × 0.4–1.2 in), relatively robust, yellow above, more rufous or brownish orange from the base. *Flesh* fairly thick, firm, whitish to yellowish, blueing slightly when cut.
Microscopic features Sp 10–16 × 4–6.5 µm, olivaceous. Cyst 35–70 × 8–15 µm, fusoid to fusoid-clavate. CC a trichoderm, with hyphae 3–10 µm diameter.
Season: July–November.
Habitat: Broad-leaved and coniferous woods; fairly ubiquitous.
Distribution: Widespread, frequency uneven.

Edible (the other *Xerocomus* are very mediocre) when young, but this species accumulates certain radioactive elements.

---

p.236 *Xerocomus parasiticus* (Bull.:Fr.) Quélet
Boletaceae, Boletales

Description: *Cap olivaceous brownish. Tubes decurrent.*
Cap 1–6 cm (0.4–2.4 in), convex, velvety, sometimes cracked, olivaceous beige to dirty

brownish. Tubes decurrent. Pores fairly large, irregular near the stipe, ochraceous yellow or reddish. Spore print red-brown. *Stipe* 2–6 × 0.4–1 cm (0.8–2.6 × 0.2–0.4 in), tapering and curved, concolorous or darker below. Flesh average, inclined to be soft, yellowish to brownish in the base, almost unchanging. *Microscopic features* Sp 12–17 × 4–6 µm, yellow-brown. Cyst 30–60 × 5–12 µm, fusi-lageniform. CC a trichoderm, the hyphae 3–8 µm diameter. Pigment intracellular.

Season: September–November.
Habitat: On Earthballs (*see* p.866).
Distribution: Uneven. Quite rare.

If collected with the host, this bolete is easy to identify. In an exhibition, the usual problem of the fungus being shown without the Earth ball can easily be overcome by observing the tapering stipe, curved at the base, and the shortly decurrent tubes.

---

p.169 *Phylloporus rhodoxanthus* (Schw.:Fr.) Bresadola (= *P. pelletieri*)
Boletaceae, Boletales

Description: *General appearance of a* Xerocomus *but hymenophore with anastomosing gills, bright yellow-buff.*
Cap 2–8 cm (0.8–3.2 in) dry, velvety, dark brown to reddish brown, sometimes marbled. Hymenophore bright yellow to orange-yellow, with adnate to subdecurrent, thick, distant, gills that anastomose to form a large almost poroid pattern at the base of the gills. Spore print rusty brown.
Stipe 2–6 × 0.6–1.5 cm (0.8–2.4 × 0.24–0.6 in), tapering, reddish brown to brownish, often bright yellow at each end.
Flesh rather soft, yellowish to reddish, slightly violaceous when cut. *Microscopic features* Sp 10–14 × 4–5.5 µm, rather cylindric, quite pale. Cyst 60–130 × 10–20 µm, cylindric-clavate. Trama bilateral. CC a trichoderm, with hyphae 8–20 µm diameter. Pigment intracellular.
Season: August–November.

Habitat: Broad-leaved and coniferous woods.
Distribution: Uneven. Rare to very rare.

A true bolete, related to the *Xerocomus* spp.

---

p.229 *Boletus pulverulentus* Opatowski
Boletaceae, Boletales

Description: *Cap velvety, polychrome. Pores yellow. Bruising blue-black.*
Cap 3–9 cm (1.2–3.5 in) lubricious then velvety, copper, vinaceous or sometimes olivaceous buff, always with some reddish brown, bruising blue-black. Tubes horizontal. Pores small, yellow, blue-black where touched. Spore print olivaceous brown. *Stipe* 4–8 × 1–2 cm (1.6–3.2 × 0.4–0.8 in), cylindric, pruinose or finely wrinkled, bright yellow above, rather red below, darkening below to red-brown and staining dark blue at the slightest touch. *Flesh* yellow but blueing instantly and intensely when cut. *Microscopic features* Sp 11–15 × 4–5 µm, yellowish. Cyst 30–60 × 6–10 µm, fusoid-lageniform. CC a trichoderm, with hyphae 3–6 µm diameter.
Season: August–November.
Habitat: Broad-leaved and coniferous woods.
Distribution: Uneven. Quite rare to rare.

The most intensely blueing of all the boletes.

---

p.233 *Boletus radicans* Pers.:Fr. (= *B. albidus*)
Boletaceae, Boletales

Description: *Robust. Cap whitish to grey or greenish. Pores yellow. Stipe swollen, reticulate. Blueing. Bitter.*
Cap 6–20(30) cm (2.4–8(12) in), velvety, whitish to dingy ochraceous cream, sometimes with greenish or brownish spots. Tubes notched, horizontal. Pores small, yellow then dirty beige or light olivaceous, blueing. Spore print olive-brown.
Stipe 4–15 × 3–7 cm (1.6–6 × 1.2–2.8 in), firm, swollen, obese with a rooting base, reticulate, whitish to quite bright yellow, blueing. Gregarious, even caespitose.

Flesh thick, white to yellowish, blueing when cut then fading. Smell rather unpleasant. Taste bitter. *Microscopic features* Sp 10–15 × 4–6 μm, brownish. Cyst 25–55 × 7–10 μm, fusoid-lageniform. CC a trichoderm, with hyphae 5–10 μm diameter.

Season: August–October.
Habitat: Broad-leaved trees; usually on calcareous soils.
Distribution: Becoming rarer northwards.

---

p.233 *Boletus calopus* Pers.:Fr.
Boletaceae, Boletales

Description: *Fleshy. Cap greyish beige. Pores yellow. Stipe reticulate, bright red below. Blueing. Bitter.*
Cap 5–15 cm (2–6 in), velvety, whitish to pale ochraceous beige. Pores pale yellow then brighter, sometimes spotted olivaceous grey, blueing. Spore print olive-brown. *Stipe* 5–12 × 1–4 cm (2–4.7 × 0.4–1.6 in), distinctly clavate, reticulate, with a white network, quite bright yellow at the apex and intensely red towards the base, blueing. *Flesh* white to yellow in the cap, blueing when cut. Smell weak or slightly vinegary. Taste bitter. *Microscopic features* Sp 10–16 × 4–5.5 μm, brownish. Cyst 25–75 × 6–12 μm, fusiform. CC a trichoderm, with hyphae 3–9 μm diameter.

Season: July–October.
Habitat: Broad-leaved and coniferous trees.
Distribution: Warm sites, rare in the north.

---

p.232 *Boletus appendiculatus* Sch.
Boletaceae, Boletales

Description: *Robust. Cap tawny brown. Pores yellow. Stipe reticulate, yellow. Scarcely blueing.*
Cap 5–18 cm (2–7 in), velvety, smooth or slightly disrupted, warm brown to tawny brown, sometimes with reddish or yellow tones. Pores pale yellow, sometimes with rusty spots. Spore print olive-brown.
Stipe 5–15 × 1.5–5 cm (2–6 × 0.6–2 in), cylindric to clavate, with a yellow reticulation over a yellow ground.

Flesh white to yellow, almost unchanging or blueing slightly. Taste mild. *Microscopic features* Sp 11–16 × 3.5–6 μm, yellowish. Cyst 30–65 × 5–15 μm, fusiform. CC a trichoderm, with hyphae 3–7 μm diameter.

Season: July–October.
Habitat: Broad-leaved trees, usually beech and oak.
Distribution: Uneven. Mainly in lowland areas.

A good edible species. *B. subappendiculatus* Derm., Laz. & Ves., found in the mountains and under conifers, has a more reddish cap and more yellow and unchanging flesh.

---

p.234 *Boletus regius* Krombholz
Boletaceae, Boletales

Description: *Robust. Cap pinkish or raspberry red. Pores yellow. Stipe reticulate, pale yellow.*
Cap 5–20 cm (4–8 in), slightly lubricious then velvety, subtomentose to dented, pinkish to raspberry red, sometimes with diffuse patches of ochraceous yellow. Pores yellow. Spore print olive-brown.
Stipe 5–12 × 2–6 cm (2–4.7 × 0.8–2.4 in), cylindric-clavate, rarely obese, with a yellow network over a yellow ground.
Flesh pale yellow. Taste mild. *Microscopic features* Sp 11–16 × 4–5 μm, yellowish. Cyst 20–75 × 5–15 μm, fusiform. CC a trichoderm, with hyphae 2–7 μm diameter.

Season: July–October.
Habitat: Broad-leaved trees, on calcareous soils.
Distribution: Mainly southern.

B. pseudoregius (Hubert) ex Estades is duller (old rose to reddish fawn) with the stipe reddish below and reticulate only at the apex; the flesh and tubes blue when handled.

---

p.232 **Penny Bun, Cep**
Boletus edulis Bull.:Fr.
Boletaceae, Boletales

Description: *Fleshy. Cap hazel with a narrow white edge. Pores pale. Stipe reticulate.*
Cap 5–25 cm (2–10 in), lubricious then

dry, often dented, hazel, sometimes pale or marbled with lighter areas, the edge white. Pores white then yellowish or flushed olivaceous, unchanging. Spore print olive-brown. *Stipe* 5–20 × 2–8 cm (2–8 × 0.8–3.2 in), obese then cylindric or clavate, with a white reticulum in the upper half, white to pale buff. *Flesh* white, unchanging, mild. *Microscopic features* Sp 14–20 × 3.5–6 µm, greenish yellow. Cyst 35–60 × 4.5–12 µm, fusiform or lageniform-fusiform. CC a trichoderm, with hyphae 5–10 µm diameter.

Season: July–October.
Habitat: Broad-leaved trees, conifers, sometimes wood edges.
Distribution: Widespread. Rarer to the north.

There are many colour variants: *B. persoonii* M. Bon is entirely white or very pale, while *B. venturii* M. Bon is wholly yellow. *B. aestivalis* (Paulet) Fr. (= *B. reticulatus*), which is very close to the Penny Bun (mid-brown, sometimes slightly reddish, greyish or yellowish cap, all one colour right to the edge, stipe reticulate to the base), occurs under broad-leaved trees, or in mixed woods, in spring or early summer. This species and the rest of this group (*see* below) are excellent to eat (especially when young). Only these taxa (excluding others on the market) may be called cep.

---

p.232 *Boletus pinophilus* Pilát & Dermek
(= *P. pinicola*)
Boletaceae, Boletales

Description: Like B. edulis *but the cap is dark red-brown.* *Cap* dark red-brown right to the edge, often rather coarsely roughened or dented. *Stipe* reddish brown, paler than the cap, with a rather well-developed network.
Season: July–October.
Habitat: Conifers, especially Scots pine.
Distribution: Less widespread than *B. edulis*.

B. mamorensis Redeuilh is similar but is a warm rufous chestnut colour, the stipe is fusoid rooting, white then rufous buff in the middle, and the network concolorous and often limited to the apex. It is found under cork oak in the Mediterranean (Morocco) and is sold commercially in France (December–January).

---

p.232 *Boletus aereus* Bull.:Fr.
Boletaceae, Boletales

Description: *Like B. edulis but the cap blackish brown. Cap* blackish brown, sometimes marbled with dark olive-brown. *Stipe* ochraceous beige to brownish grey.
Season: July–October.
Habitat: Broad-leaved trees, especially oak.
Distribution: Less frequent than *B. edulis*.

The darkest *Boletus* in this group.

---

p.232 *Boletus depilatus* Redeuilh
Boletaceae, Boletales

Description: *Cap dented, glabrous, brownish to chamois. Stipe almost rooting, not straight.*
Cap 5–12 cm (2–4.7 in), dry, glabrous, battered or with brain-like indentations, rufous tawny brown to pinkish chamois, sometimes more dull or greyish beige, often marbled. Pores yellow to yellow olivaceous, unchanging. Spore print olive-brown.
Stipe 7–15 × 1–4 cm (2.8–6 × 0.4–1.6 in), rooting, wavy or curved, not reticulate, whitish then yellowish, flushed somewhat rufous to brownish and sometimes with a rather vinaceous rufous band near the apex.
Flesh pale lemon or white, sometimes vinaceous rufous where damaged. Taste mild. Smell rather strong, of iodine.
Microscopic features Sp 11–15 × 5–6 µm, greenish yellow. Cyst 30–60 × 6–10 µm, fusiform. CC a trichoderm, the terminal elements sometimes spherical or inflated, 10–30 µm diameter, so that it appears to be an hymeniderm, even an epithelium.
Season: July–October.
Habitat: Hornbeam especially, on calcareous soils.

Distribution: Not yet fully understood.

Fairly recently separated from *B. impolitus* Fr. which can be distinguished by the velvety to filamentous felty, beige-grey cap with greenish, pinkish or brown tones and the straighter, robust stipe.

---

p.234 *Boletus erythropus* Pers.
Boletaceae, Boletales

Description: *Cap dark brown. Pores orange-red. Stipe dotted red. Blueing strongly.*
Cap 5–20 cm (2–8 in), tacky then velvety, often finely wrinkled to dented, dark brown, sometimes with crimson or olivaceous tones, spotting dark blue where handled. Pores bright orange-red then orange-yellow from the margin, blueing strongly. Spore print olive-brown.
Stipe 5–15 × 2–4 cm (2–6 × 0.8–1.6 in), clavate to cylindric, punctate, with bright red to orange-red dots on a rather yellow ground which is dirty brown below, blueing strongly. *Flesh* bright yellow, blueing intensely and rapidly. Taste mild.
Microscopic features Sp 12–18 × 4.5–6.5 µm, yellowish. Cyst 20–55 × 6–12 µm, fusiform or lageniform-fusiform. CC a trichodermial palisade, with hyphae 3–6 µm diameter.
Season: July–November.
Habitat: Rather ubiquitous.
Distribution: Widespread.

An excellent edible species when cooked (when raw, it is toxic) – a perfect illustration of the fact that blueing of the flesh is not synonymous with toxicity.

---

p.235 *Boletus queletii* Schulzer
Boletaceae, Boletales

Description: *Cap red-brown. Pores orange. Flesh dark red at the base. Blueing.*
Cap 5–15 cm (2–6 in), slightly velvety, brick red to red-brown, sometimes with orange or

flesh-coloured tones, even with dark olive patches. Pores olivaceous yellow then orange, finally fading, blueing. Spore print olive-brown. *Stipe* 5–13 × 1.5–3 cm (2–5.2 × 0.6–1.2 in), clavate or tapering, smooth or very finely longitudinally wrinkled, not reticulate, yellow above then flushed red to dirty red-brown below. *Flesh* whitish to yellow but beetroot red at the base, blueing rather faintly. Taste mild. *Microscopic features* Sp 10–18 × 4.5–7.5 µm, olive-yellow. Cyst 30–60 × 6–14 µm, fusiform, sometimes with brownish intracellular pigment. CC a more or less palisadic trichoderm, with hyphae 4–7 µm diameter.

Season: August–October.
Habitat: Broad-leaved or coniferous trees, grassy places.
Distribution: Uneven.

The cap is rather variable in colour and different varieties have been described. Var. *rubicundus* Maire (very rare) has a brick red or garnet cap while var. *discolor* (Quélet) Alessio has a rather orange to orange-yellow cap, pores that discolour more and a pale stipe with reddish dots on a yellow ground.

---

p.234 *Boletus dupainii* Boudier
Boletaceae, Boletales

Description: *Cap scarlet, viscid. Pores orange-red. Stipe reticulate above.*
Cap 4–12 cm (1.6–4.7 in), viscid to lubricious, shiny, blood-red to scarlet, fading slightly with age. Pores orange or red, orange-yellow from the margin, blueing. Spore print olive-brown.
Stipe 4–12 × 1.5–4 cm (1.6–4.7 × 0.6–1.6 in), broadly clavate or tall, with a red network at the apex and dots elsewhere, on a yellow to reddish ground, or a red ground at the stipe base. *Flesh* yellowish to yellow, blueing. *Microscopic features* Sp 10–18 × 3.5–8 µm, olive-yellow. Cyst 35–60 × 6–10 µm, fusiform. CC an ixotrichoderm, with hyphae 3–5 µm diameter.
Season: July–November.

Habitat: Beech and oak in particular, on calcareous soils. Nearly always on warm sites.
Distribution: Rare to very rare, almost nonexistent northwards.

---

p.234 *Boletus luridus* Sch.:Fr.
Boletaceae, Boletales

Description: *Cap olive-brown or brick red. Pores reddish orange, blueing. Flesh red under the tubes. Stipe with a red network. Blueing.*
Cap 5–20 cm (2–8 in), dry, matt to felty, yellowish ochraceous brown, often with olive or sometimes brick red to red tones, spotting blue-black where handled. Pores yellow then orange to reddish, fading, blueing. Spore print warm olive-brown. *Stipe* 5–15 × 1–4.5 cm (2–6 × 0.4–1.8 in), obese then more tall and slender, with a fine then large red network, the ground colour yellow to red-brown or vinaceous below. *Flesh* yellowish, sometimes vinaceous in the stipe base, red above the tubes, blueing quickly when cut. Taste mild. *Microscopic features* Sp 10–17 × 5–9 µm, yellowish. Cyst 25–75 × 5–12 µm, lageniform-fusiform. CC subtrichodermial, with hyphae 4–10 µm diameter.
Season: August–October.
Habitat: Broad-leaved and coniferous trees, on calcareous soils.
Distribution: Widespread, frequency uneven.

Species in this group are toxic when raw (edible if well cooked).

---

p.233 *Boletus luteocupreus* Bertéa & Estades
Boletaceae, Boletales

Description: *Short, thickset. Cap orange-yellow to copper. Pores orange-red. Stipe obese with a rather fine network. Blueing strongly.*
Cap 5–15 cm (2–6 in), slightly felty, chrome yellow then marbled with orange-yellow to coppery red, with dark blue spots when handled. Pores bright red, then orange or orange-yellow from the margin,

blueing strongly. Spore print olive-brown.
Stipe 6–10 × 2–5 cm (2.4–4 × 0.8–2 in),
obese then clavate, with a fine bright red
network, the ground colour bright yellow,
more orange-yellow or vinaceous red below,
blueing. *Flesh* bright yellow, blueing
strongly then light blue and finally reddish.
Taste sharp. *Microscopic features* Sp 12–17 ×
4.5–7 µm, yellowish Cyst 25–55 ×
5–8 µm, fusiform. CC a subtrichodermial
cutis, with hyphae 3–8 µm diameter.

Season: August–October.
Habitat: Usually under oak; warm sites.
Distribution: Southern but occurring as far north as the Parisian basin.

One of the very difficult to identify boletes of the *Purpurei* group occurring in the south or in warm sites, with an extraordinary range of colours.

---

p.233 **Devil's Bolete**
Boletus satanas Lenz
Boletaceae, Boletales

Description: *Robust. Cap whitish to dirty olivaceous beige. Pores orange-red. Stipe obese, with a red network. Only slightly blueing.*
Cap 5–20 cm (2–8 in), slightly viscid then velvety, whitish then marbled greyish, olivaceous, beige to brownish. Pores blood red, fading to orange or yellowish near the margin, scarcely blueing. Spore print olive-brown. *Stipe* 6–12 × 5–10 cm (2.4–4.7 × 2–4 in), rounded to obese, with a fine red network, ground colour yellowish above, reddish in the middle and dingy below. *Flesh* thick, whitish to yellowish, blueing slightly. Smell nauseous, more so in adult specimens. *Microscopic features* Sp 11–16 × 4.5–7 µm, yellowish. Cyst 35–60 × 6–9 µm, fusiform, more or less ventricose. CC a subtrichodermial cutis, with hyphae 4–9 µm diameter.

Season: July–October.
Habitat: Under broad-leaved trees and in coppice woodland, on calcareous soils; warm sites.
Distribution: Mainly southern.

Considered poisonous or strongly indigestible, causing severe reactions. Numerous related species should be avoided. They have pale caps, more or less red pores and more or less reticulate stipes.

---

p.231 *Tylopilus felleus* (Bull.:Fr.) P. Karsten
Boletaceae, Boletales

Description: *Cap brownish beige. Pores pink. Network coarse, brownish. Bitter.*
Cap 3–12 cm (1.2–4.7 in), velvety, brownish beige, sometimes olivaceous brown to reddish brown. Pores fairly large, white then pink in mature specimens. Spore print dingy pink. *Stipe* 4–10 × 1–4 cm (1.6–4 × 0.4–1.6 in), clavate, with a conspicuous, brownish or brown network, the ground colour buff or brownish beige. *Flesh* white, unchanging. Very bitter. *Microscopic features* Sp 11–17 × 3.5–5 µm, hyaline. Cyst 20–60 × 5–10 µm, fusiform. CC a subtrichodermial cutis with hyphae 3–6 µm diameter.
Season: August–November.
Habitat: Broad-leaved and coniferous trees; on acid soils.
Distribution: Widespread, frequency uneven.

Superficially resembles *B. edulis* (p.422), but lacks the white margin, has pink pores, a different network on the stipe and tastes bitter. They are so bitter that just one in a dish of ceps is enough to ruin the dish.

---

p.236 *Leccinum aurantiacum* (Bull.) S.F. Gray
Boletaceae, Boletales

Description: *Cap orange. Stipe white, with small browning then blackish fibrillose scales.*
Cap 5–20 cm (2–8 in), slightly felty, reddish orange, sometimes more yellow or more red. Pores white then greyish to brownish. Spore print olive-brown. *Stipe* 6–15 × 1.5–4 cm (2.4–6 × 0.6–1.6 in), tall, cylindric, white, sometimes bluish green below, with white, browning

then blackening squamules. *Flesh* white, violaceous grey to blackish when cut.
Microscopic features Sp 13–17 × 3.5–5.5 µm, brownish. Cyst 35–50 × 8–15 µm, fusoid-lageniform, sometimes clavate. CC a trichoderm with cylindric hyphae 10–20 µm diameter. Pigment mixed, mostly intracellular.

Season: August–November.
Habitat: Poplars, especially aspen.
Distribution: Widespread, frequency uneven.

Among the scaly stiped boletes with an orange cap is *L. quercinum* (Pilát) ex Pilát (cap orange-brown to red-brown, stipe often blue-green below, squamules initially rufous then blackish, dense below) which occurs mainly under oak. *L. vulpinum* Watling is rather similar and occurs under pine, while *L. piceinum* Pilát & Dermek, with flesh that blackens or flushes purple more rapidly, occurs under fir and spruce. In the same group, *L. versipelle* (Fr.) Snell, with a more felty, dull orange buff cap that has a markedly overhanging margin and a stipe covered with rapidly blackening squamules, is restricted to birch. All *Leccinum* are edible.

---

p.235 *Leccinum duriusculum* (Schulzer) Singer
Boletaceae, Boletales

Description: *Cap greyish brown to somewhat rufous, wrinkled-cracked. Stipe clavate almost rooting, pale, with small brownish fibrillose scales. Flesh flushing violaceous pink.*
Cap 5–15 cm (2–6 in), felty to slightly fibrillosely scaly, often wrinkled to cracked, quite hard, greyish brown. Pores pale cream, greyish with pinkish then chocolate brown tones. Spore print olive-brown.
Stipe 8–20 × 2–5 cm (3.2–8 × 0.8–2 in), rather short and squat, swollen then elongating, clavate, almost rooting, whitish then greyish to brownish, with small dense, white then brownish fibrillose scales.
Flesh white, violaceous pink when cut.
Microscopic features Sp 12–18 × 4–5.5 µm,

brownish. Cyst 30–50 × 6–12 µm, fusoid-clavate mucronate, sometimes sublageniform. CC a trichoderm, the filamentous hyphae 4–12 µm diameter. Pigment intracellular.

Season: August–October.
Habitat: Under poplar.
Distribution: Widespread, frequency uneven.

The photograph illustrates f. *robustum* Lannoy & Estades, which is more squat.

p.235 **Birch Bolete**
Leccinum scabrum (Bull.:Fr.) S.F. Gray
Boletaceae, Boletales

Description: *Cap fawn-brown to rufous-brown. Stipe with small brownish to black fibrillose scales. Almost unchanging.*
Cap 5–20 cm (2–8 in), velvety, brownish, buff, fawn-brown, sometimes marbled buff or rufous. Pores pale cream then ochraceous beige to brownish. Spore print olive-brown. *Stipe* 8–20 × 1.5–3 cm (3.2–8 × 0.6–1.2 in), cylindric-clavate, pale or brownish grey, no green at the base, with rather fine, dense, grey then dark brown to black squamules. *Flesh* rather soft when mature, white, almost unchanging or becoming slightly rufous pink. *Chemical tests* $FeSO_4$, blue-grey; formalin, pink. *Microscopic features* Sp 15–20 × 4.5–6 µm, brownish. Cyst 20–60 × 8–14 µm, fusiform. CC a trichoderm, the filamentous hyphae 3–15 µm diameter. Pigment intracellular.

Season: September–November.
Habitat: Birch.
Distribution: Uneven.

Examples occur (*see* the photograph) in which the cap is mottled with warm colours, as in *L. pulchrum* Lannoy & Estades (cap with brown marbling on a yellow to rather bright yellowish buff ground colour, flesh more distinctly flushing pink or red and reaction in $FeSO_4$ more dingy green).

p.236 *Leccinum variicolor* Watling
Boletaceae, Boletales

Description: *Cap marbled grey-brown and blackish. Stipe with small dark fibrillose scales, blue-green below. Flesh flushing pink.*
Cap 4–12 cm (1.6–4.7 in), dry, mottled with grey-brown, blackish and grey colours. Pores whitish then greyish buff to chocolate brown. Spore print olive-brown. *Stipe* 7–15 × 1–3 cm (2.8–6 × 0.4–1.2 in), cylindric-clavate, pale or blackish grey, with a net-like arrangement towards the base of rather dense, grey to black squamules, the base often blue-green. *Flesh* white but intense blue-green below, flushing pink. *Chemical tests* $FeSO_4$, greenish; formalin, pink. *Microscopic features* Sp 13–20 × 5–7 µm, brownish. Cyst 30–70 × 8–15 µm, fusiform. CC a trichoderm, the hyphae cylindric, with fairly short cells 10–20 µm diameter. Pigment intracellular.

Season: September–November.
Habitat: Mixed woodland dominated by birch, often in damp to marshy places.
Distribution: Fairly widespread.

---

p.236 *Leccinum brunneogriseolum* Lannoy & Est.
Boletaceae, Boletales

Description: *Cap brownish. Pores pale. Stipe felty to almost fibrillosely scaly, brownish grey, blue-green below.*
Cap 5–10 cm (2–4 in), silky to micaceous or shiny, greyish beige to brown. Pores whitish then greyish beige to buff. Spore print olive-brown. *Stipe* 7–16 × 0.8–2.5 cm (2.8–6.3 × 0.3–1 in), tall, cylindric-clavate, white to beige-grey, with felty grey squamules, often blue-green at the base. *Flesh* white or blue-green below, flushing slightly pink or red. *Chemical tests* $FeSO_4$, olive-grey. Formalin, pinkish. *Microscopic features* Sp 15–22 × 5–6.5 µm, brownish. Cyst 30–60 × 6–12 µm, fusiform to clavate or mucronate. CC a trichoderm, the cylindric hyphae 10–18 µm diameter. Pigment intracellular.

Season: September–November.
Habitat: Birch, sometimes with willow or even conifers; usually in damp places.
Distribution: Widespread, frequency uneven.

One of the most widespread taxa in lowland areas and in western Europe.

---

p.236 *Leccinum carpini* (Schulzer) Moser ex Reid
Boletaceae, Boletales

Description: *Cap olivaceous brownish, dented. Stipe tall, with grey then black scales. Flesh violaceous pink then black.*
Cap 4–12 cm (1.6–4.7 in), often not very expanded, dented or wrinkled to brain-like, hard, brownish, with olivaceous, somewhat rufous, yellowish tones. Pores yellowish cream then olivaceous brownish. Spore print olive-brown. *Stipe* 5–15 × 0.5–2 cm (1–6 × 0.2–0.8 in), tall, greyish, with white squamules that become brown-black, often in lines. *Flesh* pale, violaceous pink then black. *Microscopic features* Sp 15–21 × 4.5–6.5 µm, brownish. Cyst 40–60 × 8–17 µm, fusoid-clavate or clavate. CC an hymeniderm with septate hyphae, the cells subglobose 8–30 µm diameter. Pigment intracellular.
Season: July–September.
Habitat: Hornbeam, coppice woodland and undergrowth.
Distribution: Uneven. Sometimes very frequent.

The very dented and very hard cap is recognizable to the touch.

---

p.235 *Leccinum crocipodium* (Letellier) Watling
Boletaceae, Boletales

Description: *Cap yellow-brown, cracked. Pores yellow. Stipe fusoid, with small brownish yellow fibrillose scales.*
Cap 5–15 cm (2–6 in), dented then cracked to fissured, quite hard, bright or rather warm yellow-brown, sometimes more olivaceous or brown. Pores bright

yellow then olivaceous beige to brownish in patches. Spore print olive-brown. *Stipe* 5–15 × 1.5–4 cm (2–6 × 0.6–1.6 in), cylindric, often rooting, yellowish to ochraceous yellow below, with yellow then browning squamules, often in lines. *Flesh* hard, yellowish, reddening then blackish. *Microscopic features* Sp 14–20 × 5.5–7.5 μm, brownish. Cyst 30–70 × 6–15 μm, fusoid-clavate to mucronate. CC an hymeniderm with septate hyphae, the cells subglobose 10–25 μm diameter. Pigment intracellular.

Season: August–November.
Habitat: Oak, in woods, coppice, wood edges, etc.
Distribution: Widespread, frequency uneven.

---

p.235 *Leccinum lepidum* (Bouch. ex Ess.) Quadr.
Boletaceae, Boletales

Description: *Cap yellow-brown, dented. Pores yellow. Stipe yellow or brownish. Flesh reddening then violaceous grey.*
Cap 6–12 cm (2.4–4.7 in), often dented, quite hard, dark brown, tawny brown, sometimes more yellow or darker. Pores bright yellow then spotted buff. Spore print olive-brown. *Stipe* 5–10 × 1.5–4 cm (2–4 × 0.6–1.6 in), slightly clavate, bright yellow or more rufous below, with yellow, browning squamules. *Flesh* hard, yellowish, reddening before becoming violaceous and blackish. *Microscopic features* Sp 15–22 × 5–7.5 μm, often very long, brownish. Cyst as above. CC an hymeniderm, the hyphae cylindric, septate 6–20 μm diameter. Pigment intracellular.

Season: October–January.
Habitat: Holm oak.
Distribution: Mediterranean area and extending as far as the Atlantic coast.

The flesh reddens more strongly than in *L. crocipodium*, to which this fungus is related.

COPRINACEAE
The *Coprinaceae* is comprised of *Coprinus* and *Psathyrella*. The family has the following characteristics: fleshy or fragile,

sometimes deliquescent; gills adnate to almost free; spore print dark brown to black, sometimes with blackish crimson tones; veils (partial and universal) present or absent; saprotrophic.

Theoretically, only *Coprinus* (ink caps) deliquesce when mature but some become papery. Among *Coprinus*, the presence and character of the veil are critical. Other characters, particularly microscopic ones, are important, especially in *Psathyrella*.

Typical *Coprinus* characters are: margin thin at maturity, striate, gills narrowly adnate, deliquescent, stipe fragile, flesh thin, deliquescent, smell not distinctive; spores with an axial germ pore; basidia 4-spored; pseudo-paraphyses present, between the basidia; cap cuticle hymenidermial.

Typical *Psathyrella* characters: cap hygrophanous with a thin margin, gills narrowly adnate with a paler edge, stipe hollow at maturity, flesh usually thin; smell not distinctive; spores with an axial germ pore; basidia 4-spored; marginal cells clavate to sphaeropedunculate; cap cuticle hymenidermial with clavate or sphaeropedunculate hyphae.

The abbreviation MC used here is for the marginal cells often mixed with the cheilocystidia. They are smaller and often clavate to sphaeropedunculate.

---

p.123 *Coprinus auricomus* Patouillard
Coprinaceae, Agaricales

Description: *Small, centre rufous, greyish beige elsewhere. Cap* 1–4 cm (0.4–1.6 in), ovoid then flat or umbonate, without a veil, with very small erect yellow-buff hairs (a strong lens required), rufous-brown, fading from the grooved, slightly crenate margin, which reaches almost to the centre at maturity. *Gills* arching, whitish then grey to brown-black, not very close or distant. *Stipe* 4–10 × 0.15–0.4 cm (1.6–4 × 0.06–0.16 in), very fragile, bare to faintly fibrillose or slightly tomentose below, white or pale beige-cream. *Microscopic features* Sp 10–16 × 6–8.5 µm,

elliptic or vaguely almond-shaped. Cheilo 20–70 × 10–25 µm, bladder-shaped or broader. Pleuro more cylindric-lageniform and larger, up to 130 × 40 µm. CC with awl-shaped brown setae 100–400 × 5–10 µm, thick-walled 1–2 µm diameter, more numerous in the centre. Veil absent. Clamps rare.

Season: May–October.
Habitat: Bare ground in woods, wood edges, paths.
Distribution: Widespread. Fairly common.

The rather bright colours are a help to identification in the field.

---

p.123 *Coprinus kuehneri* Uljé & Bas
Coprinaceae, Agaricales

Description: *Not very deliquescent. Cap very sulcate, the centre orange-brown.*
Cap 1–3 cm (0.4–1.3 in), bell-shaped then expanded, not umbonate, without a veil, orange-brown, then dull or fading with age to greyish beige or dingy cream. Margin sulcate striate almost to the centre. *Gills* free, distant from the stipe, narrow, distant, white then grey to black. *Stipe* 3–7 × 0.1–0.25 cm (1.2–2.8 × 0.04–0.1 in), very fragile, glabrous or faintly white fibrillose, white or dingy cream to brownish grey from the base. *Microscopic features* Sp 7.5–10.5 × 6–8 × 5–6 µm, flattened, mitriform to rhomboidal in face view, elliptic in side view, the GP slightly dorsal. Cheilo 20–55 × 15–20 µm, clavate, cylindric-clavate to lageniform. Pleuro 40–90 × 20–30 µm, cylindric-clavate. Veil absent. Clamps present.

Season: June–November.
Habitat: Ruts and muddy places, in undergrowth.
Distribution: Not known.

C. plicatilis (Sow.:Fr.) Fr., found in short grass and meadows, is very similar (cap rather tawny in the centre to greyish elsewhere; Sp 11–13 × 8–10 × 6–7 µm). In *C. leiocephalus* Orton, the centre is more distinctly eye-like, brickish tawny brown to greyish buff near the margin (Sp 9–11 × 8–10 × 5–7 µm); it generally appears in woodland.

---

p.124 **Trooping Crumble Cap**
Coprinus disseminatus (Pers.:Fr.) S.F. Gray
Coprinaceae, Agaricales

Description: *Very gregarious. Cap campanulate, striate, with fine erect hairs.*
Cap 0.5–2 cm (0.2–0.8 in), ovoid then bell-shaped, with fine erect hairs (use a strong lens in low lighting), yellow-buff then beige or grey, crimson-brown, dark brown or grey-black at maturity.
Gills grey to blackish umber. *Stipe* 1–4 × 0.1–0.3 cm (0.4–1.6 × 0.04–0.12 in), often curved, thickened or with a small bulb below, whitish then beige-grey. *Microscopic features* Sp 7–10 × 4–6 µm, elliptic or slightly almond-shaped. Cheilo 60–150 × 15–20 µm, lageniform. CC with lageniform setules 60–200 × 15–30 µm, colourless. Veil transient, with elliptic cells, 10–60 µm diameter, the walls warty, encrusted and a few rare thick-walled septate hyphae

Season: Practically all year round.
Habitat: Stumps of broad-leaved trees. Extremely gregarious.
Distribution: Widespread. Very common.

Psathyrella pygmaea (Bull.:Fr.) Singer, very similar, is less gregarious and lacks setules on the cap. Under the microscope, pleurocystidia with thick walls encrusted with refractive crystals can be seen (metuloid).

---

p.124 *Coprinus narcoticus* (Batsch:Fr.) Fr.
Coprinaceae, Agaricales

Description: *Deliquescent. Cap greyish, floccose. Smell of bitumen.*
Cap 0.5–1.5 cm (0.2–0.6 in) tall and 1–2 cm (0.4–0.8 in) diameter, ovoid or

conico-campanulate, finally almost flat, the veil abundant, floccose, forming loose then pruinose masses; greyish white to greyish, sometimes slightly brownish at the centre then blackening. Margin with long striations under the veil, splitting.
Gills narrow, not very close, white then grey to black. *Stipe* 1.5–7 × 0.1–0.3 cm (0.6–2.8 × 0.04–0.12 in), very fragile, not rooting, pubescent or almost woolly at the base, greyish. Smell strong and unpleasant, reminiscent of coal gas or bitumen.
Microscopic features Sp 10–13.5 × 5.5–6.5 µm, elliptic or almond-shaped, the membrane (perispore) loose, hyaline, with a few dots or blackish lines. Cheilo 30–80 × 15–25 µm, bladder-shaped or sphaeropedunculate. Pleuro 45–110 × 35–60 µm, broadly elliptic. Veil composed of subglobose, hyaline to brownish hyphae 35–120 µm diameter, the wall finely warted, with some filiform, inflated or diverticulate hyphae 1.5–6.5 µm diameter. Clamps present.

Season: July–November.
Habitat: Dung or manured ground.
Distribution: Uneven.

C. radicans Romagn., which has a rooting stipe and the same smell occurs on dung heaps, rotten plant remains or manured ground.

p.124 **Glistening Ink Cap**
Coprinus micaceus (Bull.:Fr.) Fr.
Coprinaceae, Agaricales

Description: *Caespitose, lignicolous. Cap tawny. Detached, white to brownish flaky granules. Stipe pruinose.*
Cap 1–5 cm (0.4–2 in), globose or ovoid then campanulate or expanded, ochraceous brownish to tawny brown, darker in the centre, with minute, often transient, white or brownish tipped scales, at least in the centre. *Gills* narrow, very close, grey-brown to fuscous black. *Stipe* 3–10 × 0.2–0.5 cm (1.2–4 × 0.08–0.2 in), white to ochraceous cream below, pruinose under the

lens. *Flesh* sometimes fairly thick.
Microscopic features Sp 6.5–10 × 4.5–6.5 × 4–5.5 µm, slightly flattened, mitriform in face view, almond-shaped in side view. Cheilo 20–120 × 20–75 µm, bladder-shaped. Pleuro 70–150 × 40–70 µm, broadly clavate. Veil of hyaline or brownish and weakly granular encrusted sphaerocytes 20–75 µm diameter with walls up to 1 µm diameter, and some cylindric hyphae. Clamps present.

Season: Practically all year round.
Habitat: Stumps of broad-leaved trees or buried wood. Often trooping.
Distribution: Widespread. Common.

Type species of a complex group.

---

p.124 *Coprinus saccharinus* Romagnesi
Coprinaceae, Agaricales

Description: *Cap bluntly conical, veil white. Stipe finely striate. Spores very dark.*
Cap 1–4 cm (0.4–1.6 in), ovoid then conico-campanulate to expanded, blunt, ochraceous brown, more or less pale, sometimes quite dark in the centre, greying with age, with minute, free, white scales. *Gills* brown-black at maturity. *Stipe* 4–10 × 0.3–0.6 cm (1.6–4 × 0.12–0.24 in), white or the base slightly buff, glabrous. *Microscopic features* Sp 7.5–10.5 × 6–8 × 4.5–6 µm, flattened, submitriform in face view, ovoid in side view, very dark under the microscope. Cheilo 70–100 × 50–60 µm, globose to elliptic. Pleuro rare or absent. Otherwise like *C. micaceus*.

Season: Practically all year round (needs confirmation).
Habitat: Stumps of broad-leaved trees. Gregarious or caespitose.
Distribution: Not fully known.

C. truncorum (Scop.) Fr. has a more truncate cap, flat in the centre, with minute brownish or brown scales in the centre, a clavate stipe and paler spores.

p.124 *Coprinus xanthothrix* Romagnesi
Coprinaceae, Agaricales

Description: *Cap bluntly conical, yellow somewhat rufous. Small pointed, floccose, rust-coloured scales.*
Cap 1–4 cm (0.4–1.6 in), conico-convex to expanded, often rounded and acute or umbonate, somewhat rufous beige to ochraceous cream, the centre sometimes rather rufous yellow-brown, fading with age, the veil abundant and relatively persistent although fragile, composed of sharp or erect floccose scales which are whitish cream at the base and rust-coloured at the tip. *Gills* dark date brown at maturity. *Stipe* 5–10 × 0.2–0.4 cm (2–4 × 0.08–0.16 in), very fragile, subbulbous, white, very finely striate and slightly pruinose when young.
Microscopic features Sp 7–10 × 4.5–6 µm, elliptic, quite pale under the microscope. Cheilo 30–50 × 5–30 µm, lageniform to clavate bladder-shaped. Pleuro 30–75 × 20–45 µm, broadly elliptic to cylindric. Veil complex, with some sphaerocytes 10–45 µm diameter, with a brownish wall and numerous chains of elliptic or cylindric cells 5–15 µm diameter, the terminal cells tapering or deformed. Cell wall of the veil 1–2 µm diameter, often coloured and sometimes encrusted. Clamps absent.
Season: August–December.
Habitat: Stumps of broad-leaved trees and buried wood.
Distribution: Not fully known.

---

p.113 *Coprinus niveus* (Pers.:Fr.) Fr.
Coprinaceae, Agaricales

Description: *Entirely white, mealy-floccose.*
Cap 1–4 cm (0.4–1.6 in), ovoid or campanulate, thickly floccose then mealy pruinose, sometimes with large micaceous granules when young. White.
Gills narrow, rather close, whitish then grey to black. *Stipe* 3–9 × 0.2–0.6 cm (1.2–3.5 × 0.08–0.24 in), mealy-floccose or simply pruinose, white. *Microscopic features* Sp 14–20 × 11–14 × 8.5–11 µm, flattened,

elliptic but slightly angular or almost lozenge-shaped in outline, the GP sometimes conspicuous. Cheilo 15–80 × 10–50 μm, inflated or bladder-shaped. Pleuro 40–150 × 25–65 μm, ovoid-elliptic. Veil with globose irregular hyphae 20–40 μm diameter, hyaline, smooth or slightly punctate, with some narrow hyphae. Clamps present.

Season: August–December.
Habitat: Horse manure and cow dung.
Distribution: Widespread. Fairly common.

---

p.125 **Shaggy Ink Cap, Lawyer's Wig**
Coprinus comatus (Müll.:Fr.) Pers.
Coprinaceae, Agaricales

Description: *Quite tall. Cap cylindric, with fibrous scales. Ring thin, detached.*
Cap 4–20 cm (1.6–8 in) tall and 2–6 cm (0.8–2.4 in) across, ovoid then bluntly cylindric, with a central starry, mainly brownish disc, covered with more or less up-turned, fibrous, pure white to brownish scales. *Gills* almost free, white then pinkish and greying, finally black and deliquescent from the cap margin. *Stipe* 10–40 × 1–4 cm (4–15.7 × 0.4–1.6 in), fibrous and brittle at maturity, thickening from top to bottom, white. Ring thin, membranous, soon detached and often falling to the ground. *Flesh* white, fairly thick, but soon deliquescent from the cap margin. *Microscopic features* Sp 10–15 × 6–8.5 μm, elliptic or slightly almond-shaped. Cheilo 20–85 × 15–40 μm, sphaeropedunculate to pyriform. Pleuro absent. CC a radially arranged cutis with hyphae 5–20 μm diameter. Veil filamentous with cylindric hyphae, the terminal cell tapering, sometimes diverticulate.

Season: Practically all year round.
Habitat: Grassy places, fallow fields, etc. Ubiquitous.
Distribution: Very common.

Varieties occur with a more ovoid cap. A good edible species when young but it should be avoided in polluted areas. This species is known to accumulate heavy metals. Really mature specimens have a very long stipe, bearing the often very reduced remains of the cap, which has almost completely broken down into a black liquid.

p.125 *Coprinus romagnesianus* Singer
Coprinaceae, Agaricales

Description: *Cap campanulate, grey to brownish, finely scaly. Stipe finely scaly beneath a low ring.*
Cap 2–7 cm (0.8–2.8 in), rounded and acute or campanulate, smooth, slightly viscid or dry, scaly and brown in the centre, brownish beige or grey to lead grey elsewhere. Margin often lobed or slightly wavy, finely striate. *Gills* almost free, very close, grey to black-brown or black at maturity. *Stipe* 4–10 × 0.5–1 cm (1.6–4 × 0.2–0.4 in), fibrous, with brown squamules beneath a very low thick ring-like zone. *Flesh* fairly thick then deliquescent. *Microscopic features* Sp 8–12 × 4.5–5.5 μm, slightly flattened, elliptic or cylindric-elliptic in face view, slightly almond-shaped in side view. Cheilo 25–120 × 12–30 μm, clavate or elliptic, sometimes with a cylindric beak. Pleuro 35–150 × 15–40 μm, cylindric to bladder-shaped or lageniform. CC a radially arranged cutis. Veil filamentous, fleeting, with hyphae 2–12 μm diameter. Pigment membranal. Clamps present.
Season: June–January.
Habitat: On the ground, on buried wood.
Distribution: Uneven.

The typical ink cap, *C. atramentarius* (Bull.:Fr.) Fr., has a rather dark grey cap, without squamules and the stipe is bare at the base. *C. acuminatus* (Romagn.) Ort., also has a smooth cap but there is a conspicuous, well-defined, often browner umbo. It is possible to eat them (well away from polluted areas), provided that no alcohol is consumed at the same time, or even in the next few days (*see* p.20, coprine poisoning).

p.125 *Coprinus ammophilae* Courtecuisse
Coprinaceae, Agaricales

Description: *Cap small, fleecy, grey. Rather slow to deliquesce. Caespitose.*
Cap 0.5–2 cm (0.2–0.8 in), conico-campanulate then almost flat, grey, the veil woolly to fleecy, disappearing. *Gills* narrow, grey to black. *Stipe* 1–3 × 0.1–0.2 cm (0.4–1.3 × 0.04–0.08 in), widening slightly below the gills, fibrillose, white. *Flesh* very thin, rather slow to deliquesce. *Microscopic features* Sp 9.5–12 × 6–7.5 µm, elliptic or slightly almond-shaped, very dark. Cheilo 30–50 × 10–25 µm, clavate or sphaeropedunculate. Pleuro 50–100 × 20–40 µm, cylindric to bladder-shaped. CC a cutis. Veil hyphae with cells 20–65 × 8–35 µm, conical at the tip.
Season: September–December.
Habitat: Coastal dunes, with marram grass.
Distribution: France and Great Britain.

C. lagopus (Fr.:Fr.) Fr. is similar but much taller, although slender to spindly and very fragile. It occurs on the ground, in ruts and cool places.

p.125 *Coprinus lagopides* P. Karsten
Coprinaceae, Agaricales

Description: *Cap dark grey under a loose, finely fibrillosely scaly, whitish veil.*
Cap 1.5–6 cm (0.6–2.4 in), conico-campanulate then gradually expanding, dark grey to brownish grey under the veil, woolly then fleecy, very loose, white. Margin finely striate, splitting. *Gills* almost free, grey then black. *Stipe* 3–11 × 0.3–1.5 cm (1.2–4.3 × 0.12–0.6 in), hollow, conical then thickening below, white, fibrillosely scaly like the cap. *Microscopic features* Sp 6–9.5 × 5–7.5 × 4.5–6.5 µm, flattened, subglobose in face view, elliptic or slightly almond-shaped in side view, the GP axial, truncate. Cheilo 40–110 × 20–45 µm, bladder-shaped or clavate, sometimes cylindric ventricose. Pleuro

70–150 × 25–55 μm, inflated. CC a cutis. Veil hyphae septate, the cells 50–200 × 10–40 μm. Clamps present.

Season: August–November.
Habitat: Burnt ground and bonfire sites.
Distribution: Widespread. Quite rare to rare.

C. cinereus (Sch.:Fr.) S.F. Gray, has a hollow stipe with a very rooting base and is found on dung. *C. macrocephalus* (Berk.) Berk is non-rooting.

---

p.125 **The Magpie**
Coprinus picaceus (Bull.:Fr.) S.F. Gray
Coprinaceae, Agaricales

Description: *Tall. Cap dark beneath whitish patches of veil.*
Cap 4–10 cm (1.6–4 in), conico-campanulate then expanded, deliquescent, ochraceous brownish to sepia brown beneath loose, irregular patches of whitish veil. *Gills* almost free, close, pinkish beige then black. *Stipe* 9–30 × 1–2 cm (3.5–11.8 × 0.4–0.8 in), fibrous, thickening downwards, white.
Flesh smelling of gas or bitumen.
Microscopic features Sp 12–20 × 10–13 μm, slightly flattened, broadly ovoid in face view, elliptic to slightly almond-shaped in side view, the GP large. Cheilo 45–80 × 15–30 μm, cylindric to bladder-shaped. Pleuro 70–200 × 25–50 μm, cylindric-fusiform. CC a cutis. Veil hyphae lying horizontally, intrawoven, branched, 4–20 μm diameter. Clamps present.
Season: August–November.
Habitat: Usually under beech; some preference for warm, calcareous sites.
Distribution: Especially in areas with a mild climate.

---

p.140 **Weeping Widow**
Psathyrella lacrymabunda (Bull.:Fr.) Moser
(= *Lacrymaria velutina*)
Coprinaceae, Agaricales

Description: *Cap fleecy, rather rufous brown. Gills mottled, black, the edge weeping. Cortina abundant.*

Cap 2–10 cm (0.8–4 in), sometimes slightly umbonate, fibrillose to fleecy, brownish with yellowish, buff tones, rather rufous or rufous in the centre. Margin appendiculate, not striate. *Gills* notched, relatively narrow, greyish then black but cloudy or dappled, edge frosted and, particularly when young, weeping opalescent droplets which sometimes roll about like beads of mercury under the cap (if it is held upside down). *Stipe* 4–12 × 0.4–1 cm (1.6–4.7 × 0.16–0.4 in), fibrous, concolorous with the cap and finely fibrillosely scaly, the cortina abundant and soon blackened by the spores. *Flesh* dingy brownish beige. *Microscopic features* Sp 8.5–11.5 × 5.5–7 μm, almond- or more or less lemon-shaped, coarsely warted, GP projecting. Cheilo numerous, 40–75 × 5–15 μm, the apex capitate, sometimes wavy. CC hyphae filamentous 10–15 μm diameter. Clamps present.

Season: Practically all year round.
Habitat: Caespitose or in groups. On grassy places, fallow land, parks, scrub, enriched often-disturbed soils.
Distribution: Widespread. Common.

Some authors retain the genus *Lacrymaria* for species with small fibrillose scales on the cap and ornamented spores.

---

p.140 *Psathyrella pyrotricha* (Holms.:Fr.) Moser
Coprinaceae, Agaricales

Description: Like P. lacrymabunda *but bright orange.*
Cap 2–10 cm (0.8–4 in), but much brighter than *P. lacrymabunda*, orange rufous or bright orange, dull with age and with finer fibrillose scales. *Stipe* brighter and cortina slightly less abundant. *Microscopic features* Sp 8–11 × 5.5–6.5 μm, as in *P. lacrymabunda*. Superficial hyphae more distinctly encrusted with a yellow pigment.
Season: August–November.
Habitat: Caespitose or gregarious; coppice woods, grassy places; usually in warm sites.
Distribution: Much rarer than the preceding species.

p.149 *Psathyrella maculata* (Parker) A.H. Smith
Coprinaceae, Agaricales

Description: *Caespitose. Cap with brown to blackish scales. Ring fragile.*
Cap 1–5 cm (0.4–2 in), not umbonate, hygrophanous, fibrillose then with coarse, broad, appressed scales; initially white or very pale, soon dingy brown to black on the scales. Margin appendiculate, fibrillose, not striate. *Gills* adnate, rather close, pinkish beige then dirty chocolate brown at maturity. *Stipe* 3–6 × 0.2–0.5 cm (1.2–2.4 × 0.08–0.2 in), often curved, fibrillose-fleecy under a rather fleeting ring or ring zone, white then with dingy brown to black fibrils at the base. *Flesh* dirty brownish to yellowish brown. *Microscopic features* Sp 4.5–5.5 × 2.5–3.5 μm, very small, elliptic, slightly flattened in side view or slightly depressed on the ventral face, GP indistinct. Bas cylindric-clavate. Pleuro 20–50 × 10–15 μm, numerous, clavate with a constricted terminal appendage. Cheilo 20–40 × 7–15 μm, numerous, similar, mixed with a few MC, 10–25 × 5–10 μm. CC a cutis, with hyphae 5–20 μm diameter. Pigment membranal, encrusted. Clamps present.
Season: September–November.
Habitat: Stumps and rotten wood.
Distribution: Uneven. Rare to very rare.

---

p.113 *Psathyrella cotonea* (Quélet) Konr. & Maubl.
Coprinaceae, Agaricales

Description: *Caespitose. Cap fibrillose-woolly, whitish to brownish. Stipe yellow at the base.*
Cap 2–7 cm (0.8–2.8 in), with a low umbo, fibrillose-woolly then with darker, whitish to brownish buff fibrils or fibrillose scales or squamules, over a pale greyish beige ground. Margin appendiculate, fibrillose, not striate. *Gills* adnate, fairly narrow, close, greyish beige then dark crimson-grey to dingy crimson-brown. *Stipe* 3–12 × 0.3–1 cm (1.2–4.7 × 0.12–0.4 in), curved or tapering, fibrillose

under a very fleeting ring zone, whitish to
dirty beige over a pale ground, yellow-buff,
sometimes bright yellow at the base.
Flesh whitish, bright yellow at the base.
Microscopic features Sp 6.5–8.5 × 3.5–
4.5 µm, narrowly elliptic, bean-shaped or
constricted in side view. Bas cylindric-
clavate. Pleuro 30–55 × 10–18 µm, fusoid
bladder- or bottle-shaped. Cheilo 30–65 ×
3–15 µm, abundant, often more slender
than the pleurocystidia, mixed with some
MC, 10–20 × 7–12 µm. CC a cutis, hyphae
3–15 µm diameter. Pigment membranal,
encrusted, fairly slight. Clamps present.

Season: September–November.

Habitat: On the ground or on rotten wood and stumps.

Distribution: Uneven. Rare to very rare.

P. maculata and *P. cotonea* belong to the *Pseudostropharia* section of *Psathyrella*, whose members are often striking but rare.

---

p.123 **Psathyrella gracilis** (Pers.:Fr.) Quélet
Coprinaceae, Agaricales

Description: *Cap brown, becoming pink as it dries out. Gill edge red. Stipe rooting.*
Cap 0.5–3 cm (0.2–1.2 in), campanulate to almost flat, micaceous when dry, warm brown to brownish buff, fading when dry and then quickly becoming vinaceous pink.
Gills not very close, brownish grey then crimson-brown, edge soon pink to red.
Stipe 3–12 × 0.1–0.3 cm (1.2–4.7 × 0.04–0.12 in), very tall and slender, tapering, rooting, faintly fibrillose or with white rhizoids at the base. *Flesh* hygrophanous, grey-brown to pinkish. *Microscopic features* Sp 10–14.5 × 5.5–7 µm, elliptic, with a narrow GP. Bas short or sphaeropedunculate. Pleuro numerous, 50–100 × 8–15 µm, fusoid-lageniform, pedicellate, sometimes wavy. Cheilo fairly numerous, 20–60 × 8–12 µm, mixed with numerous clavate or fusoid to sublageniform MC, 10–35 × 5–15 µm, slightly thick-walled and more or less coloured.

Psathyrella ammophila (Dur. & Lév.) Orton
Coprinaceae, Agaricales

p.132

Description: *Cap brownish, hygrophanous. Gills not very close. Stipe deeply buried in the sand.*
Cap 1–4 cm (0.4–1.6 in), convex or flat, not umbonate, smooth or slightly wrinkled, ochraceous beige to greyish brown, fading in radial or concentric zones. *Gills* notched, not very close, greyish beige then chocolate brown with crimson or blackish tones. Gill edge whitish. *Stipe* 4–10 × 0.2–0.5 cm (1.6–4 × 0.08–0.2 in), white or pale greyish ochraceous cream, sheathed in sand towards the base, rooting. *Flesh* fragile, dingy greyish brown. *Microscopic features* Sp 10–14 × 6.5–8 µm, elliptic, the GP 1.5–2 µm. Bas sphaeropedunculate. Pleuro 30–75 × 10–25 µm, quite rare and sometimes difficult to see, fusoid ventricose or broadly lageniform. Cheilo 25–70 × 10–25 µm, similar, rare and occasional among numerous MC, 20–40 × 10–30 µm.

Season: Practically all year round.
Habitat: Coastal dunes, especially among marram; also on sand inland.
Distribution: Widespread along the coasts. Rare on sands inland.

---

p.122 *Psathyrella conopilus* (Fr.:Fr.) Pearson & Dennis (= *P. subatrata*)
Coprinaceae, Agaricales

Description: *Tall. Cap brown, very hygrophanous. Stipe white.*
Cap 1–6 cm (0.4–2.4 in), conico-convex or ovoid, remaining bluntly conical or more or less umbonate, smooth or wrinkled, sometimes slightly micaceous when dry, dark chocolate-brown, soon hygrophanous and fading from the umbo to ochraceous

beige or yellowish beige. *Gills* narrowly adnate, close, greyish beige then dark chocolate-brown with crimson overtones. *Stipe* 8–20 × 0.2–0.5 cm (3.2–8 × 0.08–0.2 in), very tall, white or honey beige below. *Microscopic features* Sp 12.5–18 × 6.5–9 µm, elliptic, slightly compressed in side view, the GP 2–2.5 µm, often dorsally eccentric. Bas sphaeropedunculate. Cheilo 30–75 × 10–25 µm, variable, cylindric or bottle-shaped, with shorter MC. Trama yellowish brown. Pleuro absent. CC with erect setae, 100–400 × 2.5–10 µm, slightly inflated at the base, thick-walled, yellowish brown.

Season: June–December.
Habitat: Grassy places and fallow land, sometimes wood edges or scrub.
Distribution: Widespread. Common.

---

p.122 *Psathyrella hirta* Peck (= *P. coprobia*)
Coprinaceae, Agaricales

Description: *Cap red-brown, with small white fibrillose scales.*
Cap 0.5–2.5 cm (0.2–1 in), conico-campanulate then convex, with small white fibrillose scales, becoming glabrous; dark red-brown then brown, fading when dry to yellowish beige or brownish grey. *Gills* adnate, distant or not very close, brown to dark brownish grey. *Stipe* 2–5 × 0.1–0.2 cm (0.8–2 × 0.04–0.08 in), with a small bulb or swollen below, pruinose at the apex, white or very pale. *Flesh* grey-brown. *Microscopic features* Sp 9.5–13 × 4.5–7 µm, elliptic, the GP fairly distinct, 1.5–2 µm diameter. Bas sphaero-pedunculate. Trama distinctly coloured. Pleuro 25–60 × 10–18 µm, fairly numerous, fusoid ventricose to sublageniform. Cheilo 25–55 × 8–15 µm, fairly numerous, often smaller, intraspersed with MC, 10–35 × 7–25 µm. Veil filamentous.

Season: June–November.
Habitat: Dung (horse or cow) and dung heaps.
Distribution: Not fully understood. Very rare.

p.148 *Psathyrella leucotephra* (Berk. & Br.) Orton
Coprinaceae, Agaricales

Description: *Fragile. Cap brownish to yellowish. Stipe white, with a large ring. Caespitose.*
Cap 2–7 cm (0.8–2.8 in), hemispherical then convex, with fleeting, small, white scales; pale, ochraceous yellow to brownish yellow, fading to pale beige or even whitish. Margin appendiculate with white veil remains, fragile. *Gills* adnate, rather close, beige then dark greyish brown. *Stipe* 5–15 × 0.5–1 cm (2–6 × 0.2–0.4 in), very fragile, more or less mottled towards the base, white. Ring large but fragile, sometimes torn when the cap expands. *Flesh* white or pale beige. *Microscopic features* Sp 7–10 × 5–6 µm, elliptic or more or less ovoid, faintly angular, GP absent. Bas cylindric-clavate. Trama colourless. Cheilo 20–40 × 5–12 µm, lageniform or the neck slightly inflated subcapitate. Pleuro absent.
Season: August–November.
Habitat: Rotten stumps or buried wood, under broad-leaved trees.
Distribution: Uneven. Quite rare.

---

p.116 *Psathyrella candolleana* (Fr.:Fr.) Maire
Coprinaceae, Agaricales

Description: *Fragile. Cap pale. Margin with fleeting white veil remains. Gills lilac grey.*
Cap 1–7 cm (0.4–2.8 in), hemispherical then almost flat, with fibrillose white patches, but soon glabrous, fairly pale, brown, grey-brown, ochraceous brown or yellow-brown, fading when dry to yellowish cream or almost white. Margin appendiculate with fleeting white veil remains. *Gills* narrow, thin and rather close, white then greyish beige, with lilac tones when old and finally dark crimson-brown. *Stipe* 2–10 × 0.2–0.5 cm (0.8–4 × 0.08–0.2 in), with a fleeting veil. *Flesh* white or sepia cream. *Microscopic features* Sp 6.5–9.5 × 3.5–5 µm, elliptic to slightly depressed

on the ventral face, the GP 1.5–1.8 μm.
Bas clavate. Cheilo 25–70 × 10–20 μm,
cylindric ventricose or more or less
bladder-shaped, mixed with MC,
20–30 × 15–20 μm. Pleuro absent.

Season: Practically all year round.
Habitat: Lawns, wood edges and woods; ubiquitous.
Distribution: Widespread. Very common.

Very variable, and sometimes difficult to identify despite its ordinary appearance.

---

p.149 *Psathyrella piluliformis* (Bull.:Fr.) Orton
(= *P. hydrophila*)
Coprinaceae, Agaricales

Description: *Caespitose, on wood. Cap brown, hygrophanous. Veil marginal.*
Cap 1–4 cm (0.4–1.6 in), hemispherical then almost flat, dark red-brown then date brown to yellowish brown, brownish or dirty beige. Margin with whitish veil remains, which gets smaller with age but remains as small overhanging patches of veil, soon blackened by the spores. *Gills* adnate, rather narrow, close, pinkish beige to reddish brown then finally rather crimson chocolate-brown or blackish. *Stipe* 2–8 × 0.2–0.7 cm (0.8–3.2 × 0.08–0.28 in), rather fragile, often curved, pruinose at the apex, with a large, very thin white ring, which soon tears, leaving no trace except at the cap edge. *Flesh* whitish to dingy brown.
Microscopic features Sp 5–6.5 × 3–3.5 μm, elliptic in face view, the ventral face flat or slightly depressed in side view, the GP almost absent. Bas cylindric-clavate. Pleuro rather numerous, 30–55 × 8–18 μm, elliptic, clavate, fusoid or bladder-shaped. Cheilo abundant, 20–45 × 7.5–15 μm, similar and mixed with numerous MC, 8–30 × 5–15 μm.

Season: Practically all year round.
Habitat: Stumps of broad-leaved trees, rarely on conifers; often in dense tufts.
Distribution: Widespread. Very common.

p.149 *Psathyrella spadicea* (Sch.→Kummer) Singer
Coprinaceae, Agaricales

Description: *Relatively fleshy. Cap pinkish brown. Caespitose.*
Cap 2–8 cm (0.8–3.2 in), globose then convex and often lobed, finally almost flat, smooth and often matt, sometimes wrinkled or cracked with age, typically red-brown, pinkish brown or flesh beige, fading when dry from the centre.
Gills rather narrow, fairly close, flesh beige then pinkish brown, relatively pale for the genus. *Stipe* 2–8 × 0.3–1 cm (0.8–3.2 × 0.12–0.4 in), not very fragile, quite often curved, pruinose to floccose at the apex, whitish then pinkish beige to almost concolorous from the base. *Flesh* quite firm, more or less elastic then fragile, concolorous with the surfaces or paler. *Microscopic features* Sp 7–10 × 4–5.5 µm, elliptic in face view, almost bean-shaped in side view, GP absent. Bas cylindric-clavate. Gill trama yellowish brown. Pleuro 35–70 × 12–25 µm, numerous, fusiform ventricose and tapering to a fine but blunt apex, thick-walled (almost 0.5 µm diameter) at the apex, crowned with slender refractive crystals. Cheilo slightly longer, with many MC, 15–45 × 10–25 µm.

Season: August–November.
Habitat: Stumps or at the foot of standing broad-leaved trees, especially beech, in woods.
Distribution: Widespread, frequency uneven.

The relative elasticity of the flesh and the pale colour of the gills when mature can lead to difficulties in determination.

---

p.123 *Psathyrella lutensis* (Rom.) Watl. & Rich.
Coprinaceae, Agaricales

Description: *Cap with white fibrils, red-brown to brownish beige. Stipe fibrillose.*
Cap 1–3 cm (0.4–1.2 in), hemispherical then shallowly convex, with detachable small white fibrillose scales; becoming glabrous from the centre, red-brown to

brownish, fading when dry to rather pale brownish beige. *Gills* initially brown then crimson-brown to dark red-brown. *Stipe* 2–6 × 0.1–0.4 cm (0.8–2.4 × 0.04–0.16 in), with a faint zigzag fibrillose marking below, white or pinkish ochraceous beige from the base.
Flesh thin, brittle, whitish or brownish to almost concolorous near the surface.
Microscopic features Sp 8.5–11 × 4.5–5.5 µm, elliptic in face view, the ventral face almost flat in side view, the GP 1.5–2 µm. Bas cylindric-clavate. Trama yellowish brown. Pleuro occasional, 30–80 × 10–25 µm, bottle-shaped, the apex secreting a mucilage that turns green in ammonia. Cheilo numerous, 25–60 × 6–20 µm, identical or shorter, mixed with MC, 10–25 × 5–15 µm. Veil filamentous.

Season: June–December.
Habitat: Damp to muddy places.
Distribution: Widespread. Rather rare.

This species is easy to identify under the microscope by the green reaction of the mucilage secreted by the cystidia.

---

p.122 *Psathyrella artemisiae* (Pass.) Konrad & Maublanc (= *P. squamosa*)
Coprinaceae, Agaricales

Description: *Cap brown, densely covered with small whitish fibrillose scales. Stipe with cortina.*
Cap 1–3 cm (0.4–1.2 in), conico-convex then broadly convex, sometimes slightly umbonate, covered with small white fibrillose scales then becoming glabrous from the centre but always remaining fibrillose at the edge; initially dark red-brown or warm brown, fading to greyish buff when dry. *Gills* adnate, rather close, pale brownish to dark grey-brown at maturity. *Stipe* 2–6 × 1.5–4 cm (0.8–2.4 × 0.6–1.6 in), rather fragile, with small white fibrillose scales beneath a fairly abundant cortinal zone, pale. *Flesh* whitish to brownish, fading.

Microscopic features Sp 7.5–10 × 4.5–5.5 μm, elliptic ovoid or broadly cylindric-elliptic in face view, the ventral face flatter, the GP 1.5 μm. Bas cylindric-clavate. Trama brownish. Pleuro 40–80 × 6–18 μm, numerous, ventricose, with a tapering, blunt or almost sharp neck, slightly thick-walled and pale brownish near the base. Cheilo numerous, 30–65 × 5–18 μm, often with thinner walls, mixed with MC, 15–35 × 5–15 μm. Veil filamentous.

Season: August–December.
Habitat: Broad-leaved and coniferous woodland.
Distribution: Widespread. Fairly common.

---

p.123 *Psathyrella badiophylla* var. *neglecta* (Romagnesi) Kits van Waveren
Coprinaceae, Agaricales

Description: *Cap hygrophanous, red-brown to cream. Margin striate. Gills tobacco brown.*
Cap 0.5–1 cm (0.2–0.4 in), conico-campanulate then expanded to shallowly convex, sometimes subumbonate, slightly micaceous when dry, dark red-brown, fading when dry to ochraceous cream.
Gills adnate, rather close, brown to tobacco brown. *Stipe* 2–4 × 0.1–0.2 cm (0.8–1.6 × 0.04–0.08 in), brittle, slightly pruinose at the apex, white.
Flesh concolorous with the surfaces.
Microscopic features Sp 9.5–11.5 × 5.5–6.5 μm, elliptic in face view and slightly almond-shaped in side view, the GP 1.5–2 μm diameter. Bas clavate. Trama yellow-brown. Cheilo 25–40 × 10–18 μm, broadly bladder-shaped, mixed with MC. Pleuro absent. CC a loose hymeniderm.

Season: April–May (occasionally in autumn).
Habitat: On the ground, lawns and grassy places.
Distribution: A rare taxon, recorded in France and Italy.

This variety is close to *P. badiophylla* (Romagnesi) M. Bon, which differs in having more reddish gills at maturity and larger spores (11–13.5 × 6.5–7 μm), with a larger GP (2–2.5 μm diam).

Cortinariaceae

The *Cortinariaceae* is a very large family. Although there are not many genera, it is a very difficult group to identify because each genus is complex or includes a great many taxa (or both). Species in *Cortinariaceae* have the following characters: fibrous; mycorrhizal; brown spore print (various shades of brown – tobacco brown, milk coffee, rusty, etc.); partial veil present (as a cortina, possibly also sheathing, more rarely as a ring) or absent; cap cuticle not or only slightly differentiated (never a hymeniderm). The family is divided into three tribes.

The **Hebelomeae** includes *Hebeloma* and *Alnicola*. The latter are usually more delicate, have scarcely any smell and are, above all, restricted to alder and willow. Variation in *Hebeloma* is demonstrated by the presence or absence of a veil, the presence of clear droplets on the young gill edge, smell and taste.

The **Inocybeae** corresponds to the genus *Inocybe*. This is a difficult genus to identify because classification is based on microscopic characters (spore shape, nature and position of the cystidia). However, the shape of the cap, the cuticle, the abundance of the partial veil, the possible presence of pruina on the stipe and the shape of the stipe base are important field characters.

The **Cortinarieae** include *Rozites*, which have a transient whitish frosting on the cap (only one European species) and the formidable genus *Cortinarius*, which has a very large number of species. However, from an aesthetic point of view, certain groups include some very beautiful species.

Constant characters will not be repeated in the descriptions. These characters are listed below.

In *Hebeloma* and *Alnicola*: cap hemispherical or convex then expanded, shallowly convex to almost flat; gills adnate-notched, edge fimbriate or frosted, whitish; spores brown, slightly thick-walled, lacking a germ pore; basidia 4-spored, clavate, often with thick sterigmata; pleurocystidia absent; clamps abundant.

In *Inocybe*: gill edge frosted, whitish; stipe and base equal width; spores brown, slightly thick-walled, lacking a germ pore; gill edge sterile; clamps abundant. In this genus, cystidia are metuloid, thick-walled and encrusted with crystals at the apex; the often clavate, marginal cells are noted as MC.

In *Cortinarius*, it is difficult to pick out any characters common to the genus as a whole.

Chemical reagents are often used for this group, especially the strong bases such as Potassium hydroxide (KOH).

---

p.164 *Leucocortinarius bulbiger* (AS:Fr.) Sing.
Cortinariaceae, Cortinariales

Description: *Cap tan with transient veil remains. Stipe bulbous. Gills white.*
Cap 5–10 cm (2–4 in), convex, lubricious then dry, smooth or finely wrinkled, with more or less transient white veil remains; rufous brown to flesh-beige or tan-clay. Margin with some veil remains.
Gills notched, quite close, white to dingy cream. Spore print pale brownish cream.
Stipe 4–7 × 1–2.5 cm (1.6–2.8 × 0.4–1 in), bulbous or even marginate, fibrillose, white to ochraceous honey from the base. Cortina white, distinct. *Flesh* thick, white or very pale. *Microscopic features* Sp 7–8.5 × 4–5 μm, almond-shaped, very pale, smooth, slightly thick-walled. Bas 30–40 × 5–8 μm. Cyst absent. CC an (ixo)cutis with hyphae 2–5 μm diameter. Pigment intracellular.
Season: September–November.
Habitat: Conifers, especially in the mountains.
Distribution: Uneven, continental to submontane.

---

p.165 *Rozites caperatus* (Pers.:Fr.) P. Karsten
Cortinariaceae, Cortinariales

Description: *Cap brownish beige, frosted in the centre, wrinkled. Ring white, appressed.*
Cap 4–12 cm (1.6–4.7 in), subglobose then convex, with a broad umbo, white-frosted in the centre and sometimes right

to the edge, smooth then radially wrinkled, ochraceous beige to yellowish buff. *Gills* adnate, not very close, rather thick, beige then brownish to tawny brown with vertical wrinkles.
Stipe 5–12 × 1–2.5 cm (2–4.7 × 0.4–1 in), cylindro-clavate, white then almost concolorous, pruinose-fleecy at the apex, fibrillose-silky below the ring, which is quite wide, thick then appressed, white to dingy cream.
Flesh thick, yellowish cream. *Microscopic features* Sp 10–14 × 7–8 µm, almond- or lemon-shaped, brownish, with rather fine, dense, low warts. CC a cutis with hyphae 5–25 µm diameter. Pigment epimembranal, encrusting.

Season: August–November.
Habitat: Often on acid soils, continental.
Distribution: Uneven.

Very easy to recognize on account of the whitish frost-like veil and white ring.

---

p.175 *Cortinarius violaceus* (L.:Fr.) S.F. Gray
Cortinariaceae, Cortinariales

Description: *Entirely purple. Cap felted. Gills distant.*
Cap 4–12 cm (1.6–4.7 in), hemispherical then expanded to slightly umbonate, felted or with small fibrillose scales, dry, dark or bright purple, often mottled or spotted brown or bronze when old.
Gills adnate, rather distant, concolorous then rust, very dark when old.
Stipe 6–12 × 1–4 cm (2.4–4.7 × 0.4–1.6 in), cylindro-clavate, concolorous, with pure purple mycelium at the base, fibrillose-silky. Cortina soon disappearing.
Flesh purple. Smell of cedarwood.
Chemical reactions Bright red with KOH.
Microscopic features Sp 11–17 × 7–9 µm, almond-shaped, with rather low, medium-sized warts. Cyst (facial and marginal) numerous, 60–100 × 12–25 µm, lageniform, with purplish contents. CC a trichoderm with hyphae 10–15 µm diameter. Pigment intracellular.

Season: July–November.
Habitat: Broad-leaved trees.
Distribution: Widespread but uneven.

Ssp. *hercynicus* (Pers.) Brandrud which occurs under conifers, is often less robust and has wider spores. *Cortinarius violaceus* is edible but the genus is extremely difficult to identify and includes highly poisonous, even deadly species. Only experts can distinguish one species from another and on no account should any species in this genus be eaten.

---

p.142 *Cortinarius melanotus* Kalchbrenner
Cortinariaceae, Cortinariales

Description: *Cap felted, olivaceous buff then with small brown fibrillose scales. Stipe with a yellow base.*
Cap 2–7 cm (0.8–2.8 in), hemispherical then broadly umbonate, felted to finely scaly, dry, rather bright olivaceous yellow then olivaceous buff, darker on account of the browning of the central scales, finally ochraceous brown. *Gills* adnate-notched, quite close, olivaceous yellow then rusty fawn. *Stipe* 3–8 × 0.4–1.2 cm (1.2–3.2 × 0.2–0.5 in), slightly clavate, olivaceous then olivaceous yellow to rufous buff, the base more yellow, fibrillose or almost mottled. Cortina fairly abundant, in bands.
Flesh medium, pale olivaceous yellow. Smell of parsley. *Microscopic features* Sp 6.5–8.5 × 5–6.5 µm, shortly elliptic to subglobose, with medium-sized warts and larger warts at the apex. Cheilo occasional, 15–25 × 5–10 µm, cylindric. CC a trichoderm with hyphae 4–8 µm diameter.

Season: August–November.
Habitat: Broad-leaved trees and sometimes conifers (firs).
Distribution: Primarily a temperate or southern species.

C. cotoneus Fr. is similar, slightly more robust species with a more clavate stipe, a cottony to fibrillosely scaly, olivaceous yellow-brown cuticle and smell of radish.

p.138 *Cortinarius orellanus* Fr. (deadly poisonous)
Cortinariaceae, Cortinariales

Description: *Cap fibrillose-felted, orange rufous. Gills distant, broad. Stipe tapering.*
Cap 3–7 cm (1.2–2.8 in), hemispherical then convex, sometimes with a low broad umbo, fibrillose-felted or with small fibrillose scales in the centre, red brown to tawny orange rufous. Margin straight, often turned down and split. *Gills* adnate, broad, distant, brownish buff then bright rusty fawn. Edge thick, irregular.
Stipe 4–10 × 0.7–1.5 cm (1.6–4 × 0.28–0.6 in), tapering and often curved below, light golden yellow then concolorous, fibrillose. Cortina weak, yellowish. *Flesh* yellow to tawny buff. Smell of radish. *Microscopic features* Sp 8–10.5 × 5.5–6.5 µm, elliptic or slightly almond-shaped, with fine dense warts. Cheilo occasional, 15–25 × 8–10 µm, clavate and sometimes septate. CC a subtrichodermial cutis, with hyphae 3–15 µm diameter.

Season: August–November.

Habitat: Broad-leaved trees, especially oak, on rather acid soils; slight preference for warm sites.

Distribution: Mainly in Southern Europe or warm temperate zones.

The most well known of a group of deadly poisonous *Cortinarius* (see p.19, orellanine poisoning). *C. speciosissimus* Kühn. & Rom., which is more conical and with a stipe patterned with yellowish bands on a reddish brown ground, is a species of damp conifer woods.

---

p.142 *Cortinarius saniosus* (Fr.:Fr.) Fr.
Cortinariaceae, Cortinariales

Description: *Cap tawny brown. Stipe slender, with yellow bands.*
Cap 1–2.5 cm (0.4–1 in), conico-convex with a low or almost acute umbo, more or less hygrophanous, silky fibrillose or slightly fibrillosely scaly when old,

tawny brown to yellowish buff. Margin with yellow veil. *Gills* notched, not very close, brownish then rust.
Stipe 2.5–5 × 0.2–0.4 cm (1–2 × 0.08–0.16 in), fragile, yellowish brown to cinnamon, mottled or banded with yellow fibrils. Cortina yellow.
Flesh thin. Smell slightly raphanoid.
Microscopic features Sp 7–10 × 4.5–6 μm, elliptic, more or less coarsely wrinkled. Cheilo occasional, 15–20 × 5–10 μm, clavate. CC a cutis with hyphae 5–15 μm diameter.

Season: July–November.
Habitat: Damp places, under broad-leaved trees.
Distribution: Widespread but frequency uneven.

C. gentilis (Fr.:Fr.) Fr. is taller, more hygrophanous, with more distant gills, and occurs in damp conifer woods, usually in the mountains or in more northern areas.

---

p.142 *Cortinarius humicola* (Quélet) Maire
Cortinariaceae, Cortinariales

Description: *Cap conical, with rufous scales. Stipe fusoid, scaly.*
Cap 2–6 cm (0.8–2.4 in), sharply conical then conico-convex, with small rufous scales, almost erect when young, on a yellowish buff ground. *Gills* notched, not very close, yellowish beige then rust.
Stipe 4–10 × 0.4–1 cm (1.6–4 × 0.16–0.4 in), fusiform, bright rufous brown to dark cinnamon-fawn, ornamented like the cap or scales even more erect. Cortina weak. *Flesh* pale. Smell of cedar wood.
Microscopic features Sp 6.5–11 × 4.5–6.5 μm, elliptic, with rather dense warts. CC a trichoderm, with hyphae 5–20 μm.

Season: August–November.
Habitat: Broad-leaved trees, especially beech.
Distribution: Uneven, declining in the north.

The only species with an equally scaly cuticle is *C. pholideus* (Fr.:Fr.) Fr. (in a different group), which is a brown species with purplish or lilaceous gills.

p.144 *Cortinarius bolaris* (Pers.:Fr.) Fr.
Cortinariaceae, Cortinariales

Description: *Cap with red scales on a buff ground, yellowing.*
Cap 2–6 cm (0.8–2.4 in), hemispherical then almost flat, lubricious then dry, soon with bright rufous to bright red appressed scales on a yellowish buff ground. Bruises rather strongly yellow. *Gills* pale yellowish then rust. *Stipe* 3–7 × 0.5–1.5 cm (1.2–2.8 × 0.2–0.6 in), more fibrillosely mottled than scaly, bruising yellow. Cortina fairly abundant.
Flesh white but yellow when cut. *Chemical reactions* Bright yellow with strong bases.
Microscopic features Sp 6–8 × 4.5–6 µm, elliptic to subglobose, with small thick warts. Cheilo occasional, 20–30 × 5–8 µm, clavate, constricted or misshapen. CC a cutis, trichodermial in places, with hyphae 5–8 µm diameter. Pigment mixed.
Season: July–November.
Habitat: Broad-leaved trees; usually on acid soils.
Distribution: Tends to have an Atlantic distribution.

C. rubicundulus (Rea) Pearson, is a rarer species that usually has small fibrillose orange-brown scales on the cap cuticle. The flesh yellows then reddens. It has no reaction with strong bases.

p.174 *Cortinarius alboviolaceus* (Pers.:Fr.) Fr.
Cortinariaceae, Cortinariales

Description: *Cap with a low umbo, silky, bluish mauve. Stipe concolorous.*
Cap 2–8 cm (0.8–3.2 in), domed then shallowly convex, with a low broad umbo, fibrillose-silky, dry, almost whitish from the veil then pale bluish mauve.
Gills notched, light beige-grey with pale lilac tones then rust. *Stipe* 5–12 × 0.5–1.5 cm (2–4.7 × 0.2–0.6 in), clavate, concolorous, covered with white or very pale mauve fibrils. Cortina fairly abundant, white. *Flesh* whitish to pale mauve. *Microscopic features* Sp 7–10 × 5–6.5 µm, elliptic, with medium-

sized finely crested warts, CC a cutis with hyphae 3–8 µm diameter.
Season: July–November.
Habitat: Broad-leaved trees, ranging from the lowlands into the mountains.
Distribution: Widespread. Rather common.

---

p.174 *Cortinarius camphoratus* (Fr.:Fr.) Fr.
Cortinariaceae, Cortinariales

Description: *Cap lilac-blue then buff. Smell strong.*
Cap 3–10 cm (1.2–4 in), hemispherical then convex or with a broad umbo, fibrillose-silky to felted, lilaceous blue then buff to light brownish from the centre. Margin more or less wavy. *Gills* notched, quite close, purple then rust. *Stipe* 5–10 × 1–2.5 cm (2–4 × 0.4–1 in), concolorous with the cap, covered with an abundant, very pale mauve cortina. *Flesh* pale purplish mauve, cinnamon-buff below. Smell strong and unpleasant, a mixture of potato (cold, mashed) and cheese.
Microscopic features Sp 8.5–11 × 5.5–6.5 µm, elliptic or slightly almond-shaped, with fine distant warts. CC a cutis with hyphae 3–8 µm diameter.
Season: August–November.
Habitat: Conifers but also broad-leaved trees.
Distribution: Mainly montane and northern.

---

p.174 *Cortinarius traganus* (Fr.:Fr.) Fr.
Cortinariaceae, Cortinariales

Description: *Cap convex, bluish mauve then yellow-buff. Smell of pear.*
Cap 3–10 cm (1.2–4 in), convex, with or without a broad umbo, fibrillose-silky to felted or with appressed scales, mauve then becoming yellowish buff or light brownish from the centre. Margin long, turned down or almost inrolled. *Gills* notched, quite narrow and fairly close, yellow buff or rather bright yellow buff then rust.
Stipe 5–12 × 1–3 cm (2–4.7 × 0.4–1.2 in), clavate, concolorous, with pale mauve fibrillose veil. *Flesh* pale purplish but

rather saffron-cinnamon to rusty yellow in the stipe. Smell strong (sometimes heady) of pear liqueur. *Microscopic features* Sp 8–10 × 5–6 μm, elliptic, with fine warts. CC a cutis with hyphae 3–10 μm diameter.

Season: August–November.
Habitat: Broad-leaved and coniferous trees.
Distribution: Rather continental.

---

p.160 *Cortinarius anomalus* (Fr.:Fr.) Fr.
Cortinariaceae, Cortinariales

Description: *Cap shallowly convex, micaceous, rufous buff with a bluish margin.*
Cap 2–7 cm (0.8–2.8 in), hemispherical then convex or almost flat, fibrillose-silky or often micaceous, lubricious (in damp weather) then dry, buff to brownish grey, sometimes with bluish mauve tones near the margin. *Gills* notched, quite close, bluish or violaceous, soon rust.
Stipe 5–10 × 0.5–1 cm (2–4 × 0.2–0.4 in), lilac blue, almost concolorous or paler than the cap near the base, with pale yellowish buff bands. Cortina yellowish. *Flesh* violaceous to purplish blue at the stipe apex, paler and ochraceous beige elsewhere. *Microscopic features* Sp 7–9 × 6–7.5 μm, elliptic to subglobose, the warts uneven or finely crested. Cheilo occasional, 15–25 × 6–8 μm, contorted. CC a cutis with hyphae 3–10 μm diameter. Pigment mixed or intracellular.

Season: July–December.
Habitat: Broad-leaved and coniferous trees.
Distribution: Not fully known.

There are a number of closely related taxa, separated by rather subtle or variable characters. The next three fungi also belong in this group.

---

p.169 *Cortinarius caninus* (Fr.:Fr.) Fr.
Cortinariaceae, Cortinariales
Description: *Cap tawny with a lilac margin. Gills lilac. Stipe violaceous with mottled brownish zones.*

Cap 3–10 cm (1.2–4 in), subglobose then convex, fibrillose or slightly frosted micaceous, lubricious but soon dry, rufous tawny, the margin briefly pale lilac. *Gills* notched, not very close, pale purplish mauve then rust. *Stipe* 3–10 × 0.7–2 cm (1.2–4 × 0.3–0.8 in), cylindro-clavate, violaceous above, with rows of narrow, brownish bands. Cortina leaving a narrow, oblique, brown ring zone. *Flesh* violaceous grey to whitish. *Microscopic features* Sp 8–10 × 6–8 µm, broadly elliptic to subglobose, with rather regular warts. Cheilo 20–35 × 6–9 µm, cylindro-clavate. CC a cutis with hyphae 5–10 µm diameter. Pigment mixed.

Season: August–October.
Habitat: Open broad-leaved and coniferous woodland but mainly young grassy spruce forest.
Distribution: Essentially continental or submontane.

---

p.138 *Cortinarius epsomiensis* var. *alpicola* M. Bon
Cortinariaceae, Cortinariales

Description: *Cap rufous beige. Stipe pale, mauve above.*
Cap 1.5–5 cm (0.6–2 in), hemispherical then convex to umbonate, almost smooth to fibrillose then finely wrinkled or cracking from the centre, ochraceous beige to rufous tawny, not mauve. Margin long, inrolled. *Gills* adnate-notched, not very close, rarely mauve, usually yellowish beige to pale buff then rust. *Stipe* 1.5–4 × 0.2–0.8 cm (0.6–1.6 × 0.08–0.3 in), lilaceous bluish above, whitish or very pale, not mottled. Cortina weak. *Flesh* pale, violaceous at the stipe apex. *Microscopic features* Sp 7.5–10 × 7–8.5 µm, broadly elliptic, with low faintly crested warts. CC a cutis with hyphae 6–8 µm diameter. Pigment mixed or intracellular.

Season: July–September.
Habitat: Meadows and short alpine grassland on calcareous soils; sometimes with dwarf willow or *Dryas*.
Distribution: Probably throughout the Alps and perhaps in Scandinavia.

p.138 *Cortinarius pholideus* (Fr.:Fr.) Fr.
Cortinariaceae, Cortinariales

Descriptions: *Cap brown, with erect scales. Gills violaceous blue. Stipe garlanded with scales.*
Cap 3–10 cm (1.2–4 in), conico-convex then convex, often umbonate, covered with small fibrillose up-turned scales, closer in the centre, rather dark brown on a lighter ground. *Gills* notched, not very close, bluish or violaceous then rust. *Stipe* 5–12 × 0.5–1 cm (2–4.7 × 0.2–0.4 in), soon hollow, violet at the apex, and mottled with zig-zag bands of fibrillose squamules from the base up to the fairly abundant cortina. *Flesh* purple at the stipe apex, buff elsewhere.
Microscopic features Sp 6.5–8.5 × 5–6.5 µm, elliptic to subglobose, with rather coarse warts. Cheilo occasional or almost absent. CC a trichoderm, with hyphae 10–30 µm diameter. Pigment mixed.

Season: August–November.
Habitat: Birch and also conifers; usually on damp acid soils.
Distribution: Widespread. Quite rare.

p.144 *Cortinarius puniceus* (Orton) Moser
Cortinariaceae, Cortinariales

Description: *Cap, gills and stipe dark blood red. Smell none.*
Cap 1.5–4 cm (0.6–1.6 in), hemispherical then convex or flat, sometimes subumbonate, slightly hygrophanous, fibrillose-silky or glabrous, dry, blood red to brownish red from the centre.
Gills notched, rather narrow, quite close, bright blood red, then rust. *Stipe* 2–5 × 0.3–0.6 cm (0.8–2 × 0.12–0.24 in), blood-red or paler below. Cortina rather fleeting. *Flesh* concolorous. Smell none or weak.
Microscopic features Sp 6.5–8.5 × 4–5 µm, elliptic, with fine and rather distant warts. CC a cutis with hyphae 5–10 µm diameter. Pigment intracellular.

Season: August–November.
Habitat: Broad-leaved lowland woods.

Distribution: Uneven. Common or rare.

C. sanguineus (Wulf.:Fr.) S.F. Gray, which is very similar, is the equivalent under conifers, occurring in continental to submontane areas. It has a slightly darker cap and smells rather strongly of cedar wood.

---

p.138 *Cortinarius uliginosus* Berkeley
Cortinariaceae, Cortinariales

Description: *Cap umbonate, brick red to coppery. Gills yellow.*
Cap 1–5 cm (0.4–2 in), long remaining conical, then with a persistent umbo, fibrillose-silky, sometimes innately streaky, dry but often shining to lustrous, brick red then a beautiful coppery orange. Margin often wavy. *Gills* notched, not very close, saffron yellow then orange to rust.
Stipe 4–7 × 0.2–0.6 cm (1.6–2.8 × 0.08–0.24 in), yellowish at the apex and almost concolorous with the cap at the base, fibrillose. Cortina reddish, rather short-lived. *Flesh* pale yellow to brownish orange. *Microscopic features* Sp 8–12 × 5–6 µm, elliptic, with low faintly crested warts. Cheilo occasional, 25–35 × 5–6 µm, clavate or septate. CC a cutis with hyphae 5–10 µm diameter.
Season: August–December.
Habitat: Damp places, pond edges, muddy woods, especially under willow.
Distribution: Uneven. Quite rare to rare.

---

p.141 *Cortinarius cinnamomeus* (L.:Fr.) S.F. Gray
Cortinariaceae, Cortinariales

Description: *Cap reddish brown. Gills cinnamon. Stipe with yellow veil.*
Cap 2–5 cm (0.8–2 in), convex, almost depressed at maturity, with a more or less persistent umbo, fibrillose or very slightly scaly or radially innately streaky, dry, yellow brown then reddish brown. *Gills* notched, rather narrow, close, cinnamon-orange then rust. *Stipe* 3–7 ×

0.3–0.8 cm (1.2–2.8 × 0.12–0.32 in), quite light yellow, then covered with orange-rufous to red-brown fibrils from the base up. Veil yellow. *Flesh* ochraceous yellow. Smell of radish or iodine.
Microscopic features Sp 6–9 × 3.5–5.5 µm, elliptic, with rather weak warts. Cheilo 10–15 × 4–5 µm, cylindro-clavate. CC a cutis with hyphae 5–10 µm diameter.

Season: August–December.
Habitat: Coniferous but also broad-leaved woods; especially on acid soils.
Distribution: Rather northern and continental.

C. cinnamomeoluteus Orton, with a reddish brown to olivaceous brown cap and rather bright yellow gills, has an olivaceous yellow, fibrillose stipe. *C. bataillei* (Favre ex Moser) Høiland is bright orange at the stipe base and *C. malicorius* Fr. has an orange veil and dark olive flesh.

---

p.138 *Cortinarius polaris* Høiland
Cortinariaceae, Cortinariales

Description: *Cap dark brown. Gills yellowish buff. Flesh yellow.*
Cap 0.5–3 cm (0.2–1.2 in), conico-convex, with a broad domed umbo, fibrillose to innately streaky, dark brown to chestnut, sometimes fading in radial streaks. *Gills* adnate-notched, broad, distant, pale mustard yellow then yellow-brown to rust. *Stipe* 1.5–3 × 0.2–0.4 cm (0.6–1.2 × 0.08–0.16 in), fairly light yellow then yellow-brown to rust, with reddish brown fibrils. Veil yellow-brown to red-brown.
Flesh yellowish buff. Smell of iodine.
Microscopic features Sp 7.5–12.5 × 5–6.5 µm, elliptic, verrucose punctate. Cheilo occasional, 10–20 × 5–10 µm, clavate. CC a cutis with hyphae 2.5–10 µm diameter. Pigment mixed.

Season: July–September.
Habitat: Low alpine shrub community with dwarf willow.
Distribution: Alps and mountains in Norway.

p.141 *Cortinarius sphagneti* Orton
Cortinariaceae, Cortinariales

Description: *Cap yellow-brown to reddish. Gills olivaceous mustard. Stipe tall and slender.*
Cap 1–3 cm (0.4–1.2 in), conico-convex then expanded but with a more or less conspicuous umbo, fibrillose to felted, ochraceous yellow then darker, yellow-brown or reddish brown. *Gills* notched, quite narrow, not very close, olivaceous mustard yellow then olivaceous buff to rust. *Stipe* 3–12 × 0.3–0.6 cm (1.2–4.7 × 0.12–0.24 in), cylindric-wavy, pale yellowish buff, browning slightly, rather weakly fibrillose. Veil buff to brownish grey. *Flesh* pale olivaceous buff, brownish below. Smell of iodine. *Chemical reactions* Gills dark brown with KOH. *Microscopic features* Sp 7.5–12.5 × 5–7 µm, elliptic, punctate verrucose. Cheilo abundant, 10–35 × 4–8 µm, cylindro-clavate. CC a subtrichodermial cutis, with hyphae 3–15 µm diameter. Pigment mixed.
Season: July–November.
Habitat: Sphagnum on peat.
Distribution: Mainly montane or nordic.

C. olivaceofuscus Kühner occurs in temperate broad-leaved woodland (especially hornbeam), not in sphagnum. It has a fibrillose-fleecy, warm-brown cap with an olivaceous yellow margin, olivaceous yellow gills and a stipe veiled with brownish yellow fibrils.

---

p.140 *Cortinarius armillatus* (Fr.:Fr.) Fr.
Cortinariaceae, Cortinariales

Description: *Cap fibrillose-fleecy, brick-red. Stipe with red bands.*
Cap 4–10 cm (1.6–4 in), globose then convex, with a low umbo, finely fibrillose-fleecy to felted, slightly brick-red (scales) on a duller or paler ground, light brown to ochraceous yellow. Margin with reddish fibrils or fibrillose scales. *Gills* notched, quite close, pale beige then rust. *Stipe* 6–

15 × 1–3 cm (2.4–6 × 0.4–1.2 in), slightly clavate, fibrillose, whitish to ochraceous beige or greyish beige from the darkening base. Veil cinnabar, forming a series of disrupted bands under a relatively low and appressed ring. *Flesh* pale beige to brownish at the base. Smell of radish. *Microscopic features* Sp 7–12 × 5–7 μm, elliptic or almond-shaped, very finely punctate. Cheilo fairly abundant, 15–30 × 5–6 μm, cylindro-clavate. CC a trichoderm, with hyphae 4–8 μm diameter.

Season: July–November.
Habitat: Under birch, on damp, acid soils.
Distribution: Northern to submontane.

C. bulliardii (Pers.:Fr.) Fr. (under beech) has a red veil enveloping the entire base, with a rather dark chestnut cap, distant gills and a fusoid ventricose stipe.

---

p.145 *Cortinarius bicolor* Cooke
Cortinariaceae, Cortinariales

Description: *Cap conico-convex, violaceous brown. Stipe tapering, violaceous. Flesh purple.*
Cap 2–7 cm (0.8–2.8 in), conico-convex, retaining a low umbo, hygrophanous, smooth, slightly shining, dry, intense chestnut then paler, ochraceous brown, with violaceous tones at least near the margin. *Gills* deeply notched, not very close, violaceous mauve then rust brown. *Stipe* 7–10 × 0.5–1 cm (2.8–4 × 0.2–0.4 in), fragile, fusoid tapering, purple then fading to pale violaceous mauve even whitish, fibrillose-silky. Veil very faint. *Flesh* violaceous to mauve in the stipe. *Microscopic features* Sp 8–12 × 5–6.5 μm, elliptic or almond-shaped, rather coarsely warted, especially at the apex. CC a cutis, with hyphae 6–10 μm diameter.

Season: August–November.
Habitat: Broad-leaved and coniferous trees.
Distribution: Uneven. Quite rare to rare.

C. evernius (Fr.:Fr.) Fr., with a possibly darker cap, is very hygrophanous, with a raised white ring on the stipe.

Cortinarius torvus (Fr.:Fr.) Fr.
Cortinariaceae, Cortinariales

p.174

Description: *Cap brownish. Gills distant. Stipe with a sheath and flaring ring.*
Cap 3–10 cm (1.2–4 in), hemispherical then almost flat, the margin bearing wisps of buff veil, finely fibrillose-innately streaky, sometimes micaceous, dull brownish, the margin more or less violaceous. *Gills* notched, distant, rather crimson-violaceous then rust brown. *Stipe* 4–10 × 0.5–2.5 cm (1.6–4 × 0.2–1 in), fusoid ventricose, violet at the apex, with a whitish to pale buff or beige-grey sheath, up to a persisting, ascending, flaring ring. *Flesh* pale or violaceous. Smells slightly of earth, fruit or radish.
Microscopic features Sp 8–12 × 5–7.5 µm, elliptic, with medium-sized warts. CC a cutis, with hyphae 3–7 µm diameter.
Season: August–November.
Habitat: Broad-leaved trees, especially beech.
Distribution: Uneven. Rather common or rare.

Cortinarius hinnuleus Fr.
Cortinariaceae, Cortinariales

p.140

Description: *Cap conical, radially streaky. Gills distant. Stipe with a ring zone.*
Cap 2–6 cm (0.8–2.4 in), conico-convex then expanded with an obtuse umbo, hygrophanous, fibrillose-silky or glabrous, sometimes tearing radially, tawny to reddish brown, fading radially in yellow-buff streaks and with black-brown spots when old. Margin sometimes split. *Gills* notched, very distant, light buff then rust brown at maturity. *Stipe* 4–10 × 0.5–1.5 cm (1.6–4 × 0.2–0.6 in), ochraceous beige or pale yellowish, browning from the base. Veil white or very pale, fibrillose-scaly from the base, up to a more or less persistent ring zone. *Flesh* whitish to brownish or brown at the base. Smell strongly musty, earthy.
Microscopic features Sp 6.5–10 × 4.5–6.5 µm, elliptic, warted, especially at the apex. CC a cutis, with hyphae 3–5 µm diameter.

Season: August–November.
Habitat: Broad-leaved trees, especially on damp soils; much rarer under conifers.
Distribution: Widespread but frequency uneven.

C. safranopes R. Henry is also fairly common; it has no smell, the cap is paler and the flesh saffron yellow at the base of the stipe.

---

p.139 *Cortinarius fragrantior* Gaugué
Cortinariaceae, Cortinariales

Description: *Cap reddish chestnut. Gills distant. Stipe tapering, white.*
Cap 2–5 cm (0.8–2 in), conico-campanulate then expanded, umbonate, hygrophanous, fibrillose-finely wrinkled or smooth, slightly shining but soon dry, reddish chestnut, fading to ochraceous beige. *Gills* notched, distant, yellowish beige then light buff, finally rust.
Stipe 4–7 × 0.5–1 cm (1.6–2.8 × 0.2–0.4 in), rather fragile, tapering, white. Cortina fleeting. *Flesh* not very thick, whitish to brownish. Smells strongly of cedar wood. *Microscopic features* Sp 7.5–9 × 5–6 µm, elliptic to subglobose, finely warted. CC a cutis, with hyphae 3–7 µm diameter.
Season: September–November.
Habitat: Conifers, sometimes with broad-leaved trees; favours acid soils.
Distribution: Rare, possibly not fully understood.

Belongs to a very difficult group of hygrophanous, more or less brown species with a white stipe. However, the strong characteristic smell makes identification fairly easy.

---

p.139 *Cortinarius cavipes* Favre
Cortinariaceae, Cortinariales

Description: *Small. Cap dark brown with a whitish margin. Stipe violaceous blue.*
Cap 1.5–3 cm (0.6–1.2 in), conico-campanulate then umbonate or almost flat,

hygrophanous, fibrillose, dark brown then brownish tawny when dry. Margin long retaining pale grey wisps of veil.
Gills notched, not very close, ochraceous beige to rust. *Stipe* 1.5–3.5 × 0.3–0.6 cm (0.6–1.4 × 0.12–0.24 in), soon hollow, violaceous mauve then warm-brown under an abundant light beige-grey veil.
Flesh not very thick, brown to brownish. *Microscopic features* Sp 8.5–10.5 × 5–6 µm, elliptic, finely warted. CC a cutis, with hyphae 4–6 µm diameter.

Season: July–September.
Habitat: Dwarf willow, in low alpine shrub community.
Distribution: Restricted to the central European alpine zone where it appears to be very rare.

---

p.145 *Cortinarius paleifer* Svrcek
Cortinariaceae, Cortinariales

Description: *Cap fibrillose-fleecy. Gills violaceous. Stipe purple at the base. Smell of geranium.*
Cap 1.5–3 cm (0.6–1.2 in), conico-campanulate then umbonate or almost flat, hygrophanous, fibrillose, becoming glabrous from the centre, dark brown then brownish tawny when dry. *Gills* notched, violaceous, not very close, rust at maturity. *Stipe* 1.5–3.5 × 0.3–0.6 cm (0.6–1.4 × 0.12–0.24 in), hollow, lilaceous purple or warm-brown when old, under a fibrillose, light beige-grey veil. *Flesh* not very thick, brown to brownish. Smells strongly of geranium (*Pelargonium*) (sometimes with a citronella component). *Microscopic features* Sp 8.5–10.5 × 5–6 µm, elliptic, finely warted. CC a cutis, with hyphae 4–6 µm diameter.

Season: June–November.
Habitat: Damp moss in wet woodland.
Distribution: Uneven. Quite rare.

C. paleaceus (Weinm.) Fr., a very similar species, differs in its closer gills and the absence of purple tones. *C. hemitrichus* (Pers.:Fr.) Fr., which is also rather similar, has no smell and occurs particularly under birch, sometimes mixed with conifers.

p.164 *Cortinarius bivelus* (Fr.:Fr.) Fr.
Cortinariaceae, Cortinariales

Description: *Cap rufous-brown. Stipe white, with a sheathing veil forming a ring.*
Cap 3–10 cm (1.2–4 in), subglobose, domed then flat, fibrillose-innately streaky or almost smooth, rusty tawny-brown, yellowish tawny when dry, sometimes finely flecked with black-brown dots. Margin initially with appendiculate veil remains.
Gills notched, initially rusty beige then darker. *Stipe* 5–12 × 1–2 cm (2–4.7 × 0.4–0.8 in), clavate to almost bulbous, light beige at the apex, white elsewhere from the veil which sheathes the base below an almost membranous ring zone.
Flesh thick, whitish to brownish.
Microscopic features Sp 7–9 × 5–6 µm, elliptic or slightly almond-shaped, with medium-sized warts. CC a cutis, with hyphae 4–10 µm diameter.
Season: September–November.
Habitat: Birch, rarer under other broad-leaved trees or conifers; on damp, preferably acid soils.
Distribution: Continental or northern.

There are other similar taxa.

---

p.126 *Cortinarius acutus* (Pers.:Fr.) Fr.
Cortinariaceae, Cortinariales

Description: *Cap pointed, honey buff, striate. Smell of iodine.*
Cap 0.5–1.5 cm (0.2–0.6 in), very conical then with a sharp or narrow umbo, hygrophanous, with long striations, glabrous, micaceous when dry, warm buff-brown, becoming pale yellowish honey when dry. *Gills* almost free, not very close, yellowish honey then rusty buff. Edge toothed. *Stipe* 2–10 × 0.1–0.3 cm (0.8–4 × 0.04–0.12 in), fragile, wavy, yellowish honey to ochraceous beige below, with silvery white fibrils. Cortina fleeting. *Flesh* thin, whitish to pale honey. Smell often of iodine. *Microscopic features* Sp 7–10 × 4–6 µm, elliptic, with

fine, rather low warts. Cheilo abundant, 25–50 × 5–10 µm, clavate to sphaeropedunculate. CC a cutis, with hyphae 4–8 µm diameter.
Season: September–October.
Habitat: Damp conifer woods.
Distribution: Widespread. Rather common.

The *Acuti* and *Obtusi* sections are groups of small, rather mycenoid species with umbonate or pointed caps.

p.165 *Cortinarius herculeus* Malençon
Cortinariaceae, Cortinariales

Description: *Large, light rufous cap. Margin with veil remains. Stipe white with pale buff veil remains.*
Cap 6–20(25) cm (2.4–8(10) in), subglobose then expanded, convex or domed, the margin blunt, covered in patches or tufts of yellowish or rufous veil; viscid then shining, rather light rufous brown, sometimes more reddish or flecked. *Gills* notched, close, pale greyish beige then rust, edge long remaining pale. *Stipe* 5–18 × 2–5 cm (2–7 × 0.8–2 in), clavate, light honey or whitish, with a white then slightly reddening sheath, breaking up into scales or oblique bands, below a fleeting appressed ring. *Flesh* white. Smell earthy, unpleasant. *Chemical reactions* Flesh yellow with KOH. *Microscopic features* Sp 11–14.5 × 5.5–8 µm, elliptic or almond-shaped, with broad but low warts. CC an ixocutis, with hyphae 2–6 µm diameter.
Season: October–December.
Habitat: Under cedar.
Distribution: Mediterranean basin or southern districts; warm sites.

Belongs in the subgenus *Phlegmacium*, which is characterized by having a sticky cap. *C. claricolor* (Fr.) Fr., a rather similar but less gigantic species, has a yellow-buff cap that is rather rufous in the centre.

p.166 *Cortinarius triumphans* Fr.
Cortinariaceae, Cortinariales

Description: *Cap yellow-brown. Stipe with ochraceous yellow bands.*
Cap 4–12 cm (1.6–4.8 in), hemispherical then almost flat, viscid then shining, yellow-brown to yellowish rufous. Margin with a few whitish veil remnants.
Gills notched, close, pale violaceous lilac or whitish, finally rather pale rufous grey.
Stipe 5–15 × 1–2.5 cm (2–6 × 0.4–1 in), clavate fusoid or tapering, yellowish white, with three to seven finely scaly, yellowish rufous bands of veil. *Flesh* white.
Chemical reactions Flesh yellow with KOH.
Microscopic features Sp 11–13 × 6–8 µm, almond-shaped, with quite wide but low warts. CC an ixocutis, with hyphae 2–6 µm diameter.

Season: September–November.
Habitat: Under birch.
Distribution: Uneven. Rather rare.

C. olidus Lange (= *C. cephalixus*), which has a cap with a few small scales in the centre, has a duller, brownish olivaceous veil and a strong earthy smell.

p.175 *Cortinarius praestans* (Cordier) Gillet
Cortinariaceae, Cortinariales

Description: *Cap wrinkled, violaceous brown, with patches of white veil. Stipe clavate.*
Cap 6–20 cm (2.4–8 in), globose then convex, dark brown, with violaceous or chocolate-brown tones, viscid then shining, with patches of whitish or dingy yellowish veil, wrinkled. *Gills* notched, close, violaceous or bluish, then rust.
Stipe 6–20 × 1.5–5 cm (2.4–8 × 0.6–2 in), clavate, white, with a series of more or less distinct bluish mauve bands of veil. Cortina rather abundant. *Flesh* very thick, white.

Microscopic features Sp 12–17 × 8–10 µm, elliptic or almond-shaped, with large warts. Cheilo difficult to find. CC an ixocutis with hyphae 2–6 µm diameter.

476 Cortinariaceae

Season: September–November.
Habitat: Broad-leaved trees (rare under conifers) on calcareous soils; tends to occur on warm sites.
Distribution: Uneven and rather southern.

One of the few edible *Cortinarius*; however, those who are not experts are recommended not to eat *Cortinarius*, as this is a difficult genus to identify, whose species are sometimes deadly poisonous.

---

p.175 *Cortinarius caligatus* Malençon
Cortinariaceae, Cortinariales

Description: *Cap rather rufous. Gill edge bright purple. Veil in yellow-brown tiers.*
Cap 3–10 cm (1.2–4 in), hemispherical then plano-convex, lubricious then dry, fibrillose to fibrillose-squamulose in the centre, brownish beige, later more rufous. Margin with fleeting rufous veil.
Gills adnate-notched, close, pale beige but the edge bright purple. *Stipe* 5–10 × 1–3 cm (2–4 × 0.4–1.2 in), fusoid tapering, white to chestnut near the base, with a yellow-brown sheath, disrupted into a series of bands below the cortina.
Flesh thick, white. *Microscopic features* Sp 9–12 × 6–7.5 µm, elliptic or almond-shaped, with rather coarse warts. CC an (ixo)cutis, with hyphae 3–8 µm diameter.
Season: September–January.
Habitat: Oaks on warm sites.
Distribution: Southern or Mediterranean.

C. variiformis Malençon is similar: the cap is more rufous, the stipe bluish above and the gills entirely bright purple.

---

p.169 *Cortinarius subtortus* (Pers.:Fr.) Fr.
Cortinariaceae, Cortinariales

Description: *Cap fibrillose-matted, tawny olivaceous. Gills distant.*
Cap 2–7 cm (0.8–2.8 in), globose then plano-convex or wavy, lubricious then

fibrillose to almost felted, looking matted when old, yellowish brown, with olivaceous tones. *Gills* notched, rather distant and quite thick, dark grey-green or olivaceous sepia then dark rusty olivaceous brown. *Stipe* 4–9 × 0.5–1.5 cm (1.6–3.5 × 0.2–0.6 in), whitish to pale olivaceous buff. Veil not very abundant.
Flesh whitish to olivaceous buff. Smell of incense or cedar wood. *Microscopic features* Sp 7–8.5 × 5.5–6.5 µm, elliptic or subglobose, with rather small warts. Cyst 50–85 × 4–20 µm, fusiform tapering, with a sheath of crystals. CC an ixocutis, with hyphae 3–6 µm diameter.

Season: August–November.
Habitat: Conifers on marshy ground, especially with sphagnum.
Distribution: Rather continental or submontane.

The dark olive gills are characteristic of a small group, including *C. infractus* (Pers.:Fr.) Fr., a fairly common species occurring on drier ground with deciduous trees. The cap is innately and irregularly streaky, dark dingy brown, the gills olivaceous brown and the taste bitter.

---

p.175 *Cortinarius terpsichores* Melot
Cortinariaceae, Cortinariales

Description: *Cap purple-blue. Gills purple. Stipe bulbous.*
Cap 4–12 cm (1.6–4.7 in), hemispherical then convex, bright purple, becoming slightly more dull from the centre, sticky then shining, smooth or with slight innate radial streaks. *Gills* notched, narrow and close, purple then rusty beige. Edge toothed. *Stipe* 4–10 × 1–2.5 cm (1.4–4 × 0.4–1 in), with a marginate bulb, pale purple. Veil violaceous or whitish, rather fleeting. *Flesh* whitish or tinged violaceous blue to violaceous grey.
Microscopic features Sp 9.5–12.5 × 6–7 µm, elliptic, with rather fine, tall warts. Cheilo variable, numerous. CC an ixocutis, with hyphae 2–5 µm diameter.

Season: September–November.

Habitat: Broad-leaved trees, especially beech–hornbeam woods, on calcareous soils; warm sites.
Distribution: Uneven. Rather common to very rare.

An example of a spectacular, complex group with more or less intense purple colours. They can be collectively called the *C. caerulescens* (Sch.) Fr. group.

---

p.175 *Cortinarius purpurascens* (Fr.) Fr.
Cortinariaceae, Cortinariales

Description: *Cap violaceous russet. Stipe bulbous. Bruising dark crimson.*
Cap 4–12 cm (1.6–4.8 in), hemispherical then convex, sticky then shining, with innate streaks or scribbles, warm-brown with a violaceous tone, especially near the margin, bruising dark violaceous crimson. *Gills* notched, quite close, lilaceous purple then violaceous beige-grey to rust, bruising dark violaceous crimson. *Stipe* 3–8 × 1–2.5 cm (1.2–3.2 × 0.4–1 in), with a more or less marginate bulb, purple then dull, fibrillose, bruising bright or dark violaceous crimson. *Flesh* quite thick, rather dark purple then paler, bruising crimson. *Microscopic features* Sp 8–11 × 4.5–6 µm, elliptic or almond-shaped, with low, rounded warts. Cheilo occasional, 20–50 × 1.5–2 µm. CC an ixocutis, with hyphae 2–5 µm diameter.
Season: August–November.
Habitat: Conifers; usually on acid soils.
Distribution: Uneven. Rather common or rare.

The best known species in a group characterized by the colour change of all parts (violaceous crimson) when cut or rubbed.

---

p.164 *Cortinarius arcuatorum* R. Henry
Cortinariaceae, Cortinariales

Description: *Cap rosy tawny. Gills lilac. Stipe with a marginate bulb.*

Cap 5–10 cm (2–4 in), hemispherical then convex, viscid then shining, tawny to ochraceous russet, often tinged with flesh colours or pinkish beige.
Gills adnate-notched, narrow and close, rather bright lilac then paler, finally rusty beige. *Stipe* 5–8 × 1–3 cm (2–3.2 × 0.4–1.2 in), with a marginate bulb, whitish to pale buff, fibrillose. Veil quite bright lilac, especially on the bulb. Cortina rather fleeting. *Flesh* very pale, faintly lilac at the stipe apex. Smell slightly honeyish. *Chemical reactions* KOH bright pink on the cap.

Microscopic features Sp 10–13.5 × 6–7 µm, almond- or almost lemon-shaped, with slightly crested warts. Cheilo almost absent. CC an ixocutis, with hyphae 2–4 µm diameter.

Season: September–November.
Habitat: Broad-leaved coppice woodland on calcareous soils; warm sites.
Distribution: Uneven.

C. calochrous (Pers.:Fr.) S.F. Gray, with a rufous brown to yellow-buff or yellow cap, tender pinkish lilac gills and whitish to buff stipe, occurs under beech on calcareous soils.

---

p.167 *Cortinarius elegantissimus* R. Henry
Cortinariaceae, Cortinariales

Description: *Cap yellow-rufous. Gills bright yellow. Stipe with a marginate bulb.*
Cap 5–13 cm (2–5.1 in), shallowly convex then depressed, viscid then shining, smooth or with very small scales in the centre, yellow to orange-brown from the centre. *Gills* notched, close, bright yellow to lemon yellow or bluish, finally bright rust. *Stipe* 5–12 × 1–2.5 cm (2–4.7 × 0.4–1 in), with a marginate bulb, yellow sometimes lemon yellow, fibrillose. Veil yellow then rufous, colouring the bulb and the base of the stipe. Cortina abundant.
Flesh white or very pale, yellow at the edge and typically lilaceous or bluish at

the stipe apex. Smell slightly aromatic or fruity. *Chemical reactions* Flesh blood red with KOH. *Microscopic features* Sp 12–18 × 7.5–10 µm, almond- or almost lemon-shaped, with very large confluent warts. Cheilo almost absent. CC an ixocutis, with hyphae 2–6 µm diameter.

Season: September–November.
Habitat: Under beech, on calcareous soils; warm sites.
Distribution: Uneven. Quite rare to very rare.

The subgroup of *Phlegmacium* with a marginate bulb and yellow gills, gathers species that are very difficult to identify.

---

p.167 *Cortinarius splendens* R. Henry
Cortinariaceae, Cortinariales

Description: *Cap speckled with rufous on a bright yellow ground. Gills, stipe and flesh bright yellow.* *Cap* 2–6 cm (0.8–2.4 in), broadly convex then flat, viscid then shining, speckled or with small rusty fawn to rust brown appressed scales in the centre, on a bright yellow ground. *Gills* adnate-notched, close, yellow, bright rust at maturity. *Stipe* 2–7 × 0.7–3 cm (0.8–2.8 × 0.28–1.2 in), with a marginate bulb, concolorous with the cap, fibrillose. Veil yellow. Cortina rather fleeting. *Flesh* an intense yellow. *Chemical reactions* Cap olive-brown with KOH. *Microscopic features* Sp 9–12 × 6–7 µm, almond-shaped or slightly lemon-shaped, with rather low, irregular warts. Cheilo almost absent. CC an ixocutis, with hyphae 2–7 µm diameter.
Season: August–November.
Habitat: Beech woods, on calcareous soils.
Distribution: Quite widespread.

This species was thought responsible for a case of fatal poisoning but later proved not to be the cause. Var. *meinhardii* (M. Bon) Melot (= *C. vitellinus*), a conifer species, is taller, becomes more rufous to crimson, and has a peppery smell.

p.175 *Cortinarius rufoolivaceus* (Pers.:Fr.) Fr.
Cortinariaceae, Cortinariales

Description: *Cap vinaceous russet with olivaceous tones. Gills violaceous olive. Stipe greenish yellow with a mauve apex.*
Cap 3–10 cm (1.2–4 in), convex to depressed, sticky, smooth or fibrillose-innately streaky, sometimes with a few small scales, vinaceous russet in the centre, with olive, olivaceous beige, buff or lilaceous tones at the edge.
Gills adnate-notched, close, olivaceous, sometimes with violaceous tones, then olivaceous brown or rust. *Stipe* 4–8 × 0.7–3.5 cm (1.6–3.2 × 0.28–1.4 in), with a marginate bulb, pale, lilac to olivaceous at the apex, very fibrillose, lilac then wine red on the edge of the bulb. Cortina abundant. *Flesh* lilaceous to crimson, olivaceous greenish at the stipe apex, bitter. *Chemical reactions* Flesh olive then crimson with KOH. *Microscopic features* Sp 10–14 × 6–8 μm, almond- or lemon-shaped, with coarse warts. Cheilo almost absent. CC an ixocutis, with hyphae 2–7 μm diameter. Pigment epimembranal, granulate, abundant.
Season: September–November.
Habitat: Broad-leaved trees, especially beech and oak.
Distribution: Widespread but declining in northern Europe.

p.145 *Cortinarius croceocaeruleus* (Pers.:Fr.) Fr.
Cortinariaceae, Cortinariales

Description: *Cap viscid, violaceous lilac, discolouring yellow. Stipe viscid.*
Cap 2–5 cm (0.8–2 in), convex, viscid, smooth to fibrillose, sometimes finely wrinkled, violaceous lilac, discolouring yellowish buff from the centre.
Gills notched, pale lilac then saffron to rusty yellow. *Stipe* 3–10 × 0.2–1 cm (1.2–4 × 0.08–0.4 in), fragile, fusoid tapering to rooting, viscid, white to pale violaceous, yellowing from the base. Cortina fleeting. *Flesh* glassy in damp

weather, white to ochraceous yellow, very bitter. *Microscopic features* Sp 6.5–8.5 × 4–4.5 µm, elliptic or almond-shaped, rather weakly warted. CC a very well-developed ixocutis, with hyphae 1–3 µm diameter. Pigment intracellular.

Season: August–November.
Habitat: Broad-leaved trees (especially beech), on calcareous soils.
Distribution: Quite widespread.

C. vibratilis (Fr.:Fr.) Fr., with an orange-buff to golden yellow cap and white stipe, also tastes bitter if the gluten on the cap is touched with the tongue. Other bitter species are paler, even white or whitish, such as *C. ochroleucus* (Sch.:Fr.) Fr., which has a pale ochraceous white cap, white stipe and smells faintly of honey, or *C. barbatus* (Batsch:Fr.) Melot, which has a white to ivory cream cap, white to yellowish stipe, and smells of radish.

---

p.143 ***Cortinarius delibutus*** Fr.
Cortinariaceae, Cortinariales

Description: *Cap viscid, ochraceous yellow. Gills lilac. Stipe with a yellow veil.*
Cap 2–7 cm (0.8–2.8 in), hemispherical then convex, viscid, smooth, yellow or ochraceous yellow. *Gills* notched, close, pale lilac then rusty buff. *Stipe* 4–10 × 0.4–1 cm (1.6–4 × 0.16–0.4 in), cylindric or slightly clavate, viscid, violaceous or mauve at the apex, and mottled with ochraceous yellow bands from the base. Cortina fairly abundant or fleeting. *Flesh* lilaceous blue at the stipe apex. *Microscopic features* Sp 7–10 × 5–8 µm, elliptic to subglobose, with fine dense warts. Cheilo 15–45 × 6–10 µm, cylindro-clavate. CC an ixocutis, with hyphae 3–7 µm diameter. Pigment intracellular, granular.

Season: August–November.
Habitat: Broad-leaved trees (especially birch) and also conifers; usually on acid soils.
Distribution: Widespread but frequency uneven.

C. betulinus Favre (under birch, in marshy ground), is rather similar but has a dull cap, tinged violaceous at the margin. *C. salor* Fr. is distinguished by its purple or purple-blue cap, discolouring yellowish buff, and its white or lilac stipe mottled with olivaceous yellow.

---

p.168 *Cortinarius trivialis* Lange
Cortinariaceae, Cortinariales

Description: *Cap viscid, olivaceous brown. Veil viscid forming anastomosing bands.*
Cap 3–10 cm (1.2–4 in), umbonate or not, viscid, smooth, buff-brown to red-brown, sometimes with more olive tones.
Gills notched, not very close, greyish or pale lilac then rust. *Stipe* 4–10 × 0.5–2 cm (1.4–4 × 0.2–0.8 in), firm, slightly fusoid, viscid, white then browning. Sheath thick and sticky, disrupting into anastomosing bands. Cortina rather fleeting. *Flesh* white, brownish below. *Microscopic features* Sp 9–15 × 6–8 µm, almond-shaped to almost fusoid, coarsely warted. Cheilo 35–35 × 5–8 µm, cylindro-clavate. CC an ixocutis, with hyphae 2–6 µm diameter.
Season: August–November.
Habitat: Undergrowth in broad-leaved woodland.
Distribution: Widespread but frequency uneven.

C. collinitus (Sow.:Fr.) S.F. Gray, with a red-brown cap, dark in the centre, white stipe and pale purple veil often occurs in spruce forests. *C. mucosus* (Bull.) Kickx, with an orange-red brown cap, sometimes almost black in the centre, white stipe, orange-brownish below, and viscid veil, prefers pines.

---

p.139 *Cortinarius favrei* (Moser) ex Henderson
Cortinariaceae, Cortinariales

Description: *Cap viscid, red-brown. Stipe viscid with a rust brown veil.*
Cap 1–5 cm (0.4–2 in), with a low umbo, viscid, smooth, dark red-brown, sometimes more yellowish at the margin

or dark olive in the centre.
Gills notched, rather distant, greyish cream then dark rust. *Stipe* 2–4 × 0.3–1.5 cm (0.8–1.6 × 0.12–0.6 in), cylindric or tapering, viscid, white then browning. Sheath thick and sticky, rather fibrillose or mottled. Cortina fairly abundant. *Flesh* white to brownish yellow below. *Microscopic features* Sp 11–15 × 6.5–8.5 µm, almond-shaped, with roughened warts. Cheilo absent. CC an ixocutis, with hyphae 2–7 µm diameter.

Season: July–September.
Habitat: Dwarf willow in the low alpine shrub community.
Distribution: Alps and mountains in Scandinavia and the Arctic.

A very characteristic species of the alpine zone.

---

p.168 *Cortinarius stillatitius* Fr.
Cortinariaceae, Cortinariales

Description: *Cap viscid, buff brown, striate. Stipe viscid, white with a violaceous veil.*
Cap 3–8 cm (1.2–3.2 in), hemispherical then convex, viscid, smooth or slightly radially wrinkled, red-brown to yellow-brown, sometimes slightly olivaceous or darker in the centre. *Gills* notched, not very close, whitish then rusty beige. *Stipe* 5–10 × 0.8–1.5 cm (2–4 × 0.32–0.6 in), cylindric or tapering, viscid, white then slightly yellow buff. Sheath sticky, bluish mauve or light violaceous. Cortina set very high, sticky. *Flesh* white to honey buff below. Smell of honey. *Microscopic features* Sp 12–16 × 6.5–9 µm, almond- or lemon-shaped, with medium-sized, rather dense warts. Cheilo abundant, 30–40 × 15–25 µm, clavate to pyriform. CC an ixocutis, with hyphae 2–6 µm diameter.

Season: August–November.
Habitat: Mixed conifers, rare under broad-leaved trees.
Distribution: Uneven and not fully understood.

C. elatior Fr., with a radially wrinkled cap, crisping or wrinkling on the face of the gills, white, frosted gill edge and purple veil on the stipe, occurs under beech.

---

p.120 *Alnicola melinoides* (Bull.:Fr.) Kühner
(= *Naucoria escharoides*)
Cortinariaceae, Cortinariales

Description: *Cap yellow-buff, not striate. Stipe darker below.*
Cap 0.5–2.5 cm (0.2–1 in), conico-convex then almost flat, not striate, fibrillose to slightly scurfy or finely fibrillosely scaly, yellowish beige to buff, sometimes darker or reddish from the centre. *Gills* notched, not very close, yellowish cream then rufous beige. *Stipe* 2–7 × 0.1–0.5 cm (0.8–2.8 × 0.04–0.2 in), fragile, yellowish to ochraceous honey, soon brownish to dark brown from the base. *Microscopic features* Sp 9–13 × 4.5–6.5 µm, almond-shaped, coarsely warted except at the apex. Cheilo numerous, 30–55 × 6–10 × 1–2 µm, lageniform with a filiform neck, sometimes slightly capitulate. CC a subtrichodermial cutis, with septate hyphae, the cells 20–60 × 8–20 µm, the terminal cells elliptic or tapering. Pigment mixed, not roughened.
Season: July–November.
Habitat: Alders, on marshy ground.
Distribution: Widespread but frequency uneven.

Alnicola is a difficult genus.

---

p.126 *Alnicola scolecina* (Fr.) Romagnesi
(= *A. badia* = *Naucoria phaea*)
Cortinariaceae, Cortinariales

Description: *Cap reddish brown. Stipe dark brown below.*
Cap 0.5–3.5 cm (0.2–1.4 in), conico-convex then sometimes depressed, hygrophanous, striate, glabrous or with slight fibrillose scales, dark red-brown to tawny brown, fading. *Gills* notched, not very close, brown. *Stipe* 2–8 × 0.2–0.5 cm (0.8–3.2 × 0.08–0.2 in), fragile, concolorous or darker from the base.

Slightly bitter. *Microscopic features* Sp 9–12 × 6–7.5 µm, almond-shaped, with roughened warts but slightly umbonate and almost smooth at the apex. Cheilo numerous, 30–60 × 8–15 × 1–3 µm, lageniform with a filiform neck, sometimes wavy or slightly capitulate. CC a subtrichodermial cutis, with septate hyphae, the terminal elements 30–80 × 5–15 µm. Pigment mostly epimembranal, encrusting.

Season: July–November.
Habitat: Alders, on marshy ground.
Distribution: Widespread but frequency uneven.

---

p.139 *Hebeloma mesophaeum* (Pers.) Quélet
Cortinariaceae, Cortinariales

Description: *Cap brown, the margin covered with whitish cream veil. Stipe with cortina.*
Cap 1.5–6 cm (0.6–2.4 in), conico-convex then sometimes umbonate, greasy or dry, zoned, the centre more or less dark brown, from blackish brown or rufous brown to pale beige, the edge paler from the presence of the marginal veil. *Gills* notched, light greyish beige then brownish to tobacco brown. *Stipe* 2–8 × 0.2–0.7 cm (0.8–3.2 × 0.08–0.28 in), fibrillose or with fibrillose bands, pale cream then brownish to blackish brown from the base. Cortina abundant. *Flesh* concolorous, dark brown at the base, bitter. Smell of radish and chocolate. *Microscopic features* Sp 8.5–11.5 × 4.5–6.5 µm, elliptic, finely roughened to almost smooth. Cheilo numerous, 30–65 × 7–12 × 4–8 µm, lageniform with a cylindric neck. CC an ixocutis, with hyphae 2–3 µm diameter. Pigment epimembranal, encrusting.

Season: June–December.
Habitat: Undergrowth, broad-leaved coppice woodland (often birch and willow) and conifer clearings.
Distribution: Widespread. Very common or fairly rare.

Hebeloma is a difficult genus.

p.139 *Hebeloma kuehneri* Bruchet
Cortinariaceae, Cortinariales

Description: *Cap brown with a paler margin. Gills distant. Stipe browning.*
Cap 0.5–2 cm (0.2–0.8 in), conico-convex then more or less umbonate, frosted then becoming glabrous, sometimes with faint scales in the centre, dark brown, more brownish to buff towards the edge. *Gills* very notched, distant, yellow-brown then tobacco brown. *Stipe* 2–3.5 × 0.2–0.4 cm (0.8–1.4 × 0.08–0.16 in), rather fragile, fibrillose below, pruinose-fleecy at the apex, whitish then yellow-brown or dark brown from the base. Cortina fleeting. *Flesh* dark, slightly or slowly bitter. Smell of radish. *Microscopic features* Sp 11–14.5 × 6–7.5 µm, almond-shaped, roughened. Cheilo numerous, 40–90 × 7–11 × 4–6 µm, lageniform with a long neck. CC an ixocutis, with hyphae 3–5 µm diameter. Pigment epimembranal, encrusting.

Season: July–September.

Habitat: Dwarf willow in the low alpine shrub community and damp high montane areas.

Distribution: Alps and mountains in Scandinavia and the Arctic.

---

p.163 **Poison Pie**
Hebeloma crustuliniforme (Bull.) Quélet
Cortinariaceae, Cortinariales

Description: *Cap frosted, ochraceous beige. Gills with opalescent droplets. Stipe pruinose.*
Cap 3–7 cm (1.2–2.8 in), shallowly convex, umbonate, the margin straight or slightly scalloped, slightly viscid or frosted, yellowish brown to ochraceous cream, one colour throughout. *Gills* notched, close, greyish beige then mid-tobacco brown, edge with opalescent droplets in young specimens,. *Stipe* 2–8 × 0.5–1.5 cm (0.8–3.2 × 0.2–0.6 in), clavate to bulbous, pruinose-fleecy or with very small scales at the apex, concolorous or whitish. Cortina absent. *Flesh* pale. Smell of radish. *Microscopic features* Sp 10–13 × 5.5–7 µm,

almond-shaped, warted. Cheilo numerous, 40–80 × 4–12 µm, cylindro-clavate. CC an ixocutis, with hyphae 2–3 µm diameter.
Season: August–November.
Habitat: Broad-leaved trees and conifers.
Distribution: Quite widespread but frequency uneven.

In adult specimens, the weeping gills have brownish dots on the edge (the droplets dry out and are coloured by the spores).

p.164 *Hebeloma sinapizans* (Paulet) Gillet
Cortinariaceae, Cortinariales

Description: *Cap flesh beige. Stipe finely scaly, hollow.*
Cap 5–12 cm (2–4.7 in), shallowly convex umbonate, the margin straight or slightly scalloped, sometimes pitted, slightly viscid then frosted especially at the edge, flesh beige to tawny rufous, sometimes zoned. *Gills* notched, close, pale greyish beige then tobacco brown, edge not weeping. *Stipe* 5–12 × 0.8–2.5 cm (2–4.7 × 0.32–1 in), robust but soon hollow with a wick at the apex (in section), clavate, pruinose-fleecy or with small scales above, concolorous. Cortina absent. *Flesh* pale. Smell strong, of radish. Taste bitter. *Microscopic features* Sp 10–14 × 6–8 µm, almond- or lemon-shaped, warted, the apex smoother. Cheilo numerous, 40–60 × 7–12 × 5–8 µm, lageniform. CC an ixocutis, with hyphae 2–3 µm diameter.
Season: August–November.
Habitat: Broad-leaved trees, very rare under conifers.
Distribution: Quite widespread. Fairly rare to rare.

p.163 *Hebeloma edurum* Métrod ex M. Bon
Cortinariaceae, Cortinariales

Description: *Cap rufous brown. Margin slightly grooved. Stipe with small fibrillose scales, with a dark base.*
Cap 4–12 cm (1.6–4.7 in), hemispherical then broadly umbonate, viscid, then shining, rufous brown to rufous buff, sometimes more dull, the margin typically

ribbed or grooved, often paler. *Gills* adnate, flesh-buff then rather dark rufous-brown, edge not weeping. *Stipe* 5–12 × 0.5–2 cm (2–4.7 × 0.2–0.8 in), quite robust, clavate, with small fibrillose scales, browning strongly from the base which becomes cinnamon, tawny-brown to dark brown. Cortina absent. Smell strong, a mixture of chocolate and fruit. Slightly bitter. *Microscopic features* Sp 8.5–12 × 5–6.5 µm, almond-shaped, the apex slightly elongated, finely warted. Cheilo 25–50 × 4–8 µm, cylindric, sometimes septate. CC an ixocutis, with hyphae 1–3 µm diameter.

Season: August–December.

Habitat: Copses, meadows and grassy woods, on calcareous soils, usually under conifers; mainly warm sites.

Distribution: Quite widespread. Rather rare to rare.

---

p.164 *Hebeloma radicosum* (Bull.:Fr.) Ricken
Cortinariaceae, Cortinariales

Description: *Cap viscid, with small scales. Stipe rooting. Ring large.*
Cap 3–12 cm (1.2–4.7 in), globose then convex expanded, appendiculate, viscid then with small fibrillose scales or more or less coarse, appressed scales; beige to buff, sometimes more brown in the centre. *Gills* notched, close, pale beige then rufous-brown. *Stipe* 8–20 × 1–2.5 cm (3.2–8 × 0.4–1 in), clavate but deeply rooting below the swelling, finely scaly to scaly under a membraneous then torn ring. *Flesh* concolorous. Smell of bitter almonds. *Microscopic features* Sp 8–10 × 4.5–6 µm, rather pointed almond-shaped, warted. Cheilo numerous, 65–50 × 3–5 × 6–10 µm, cylindric capitate, sometimes slightly inflated at the base. CC an ixocutis (subtrichodermial), with hyphae 1–6 µm diameter.

Season: August–December.

Habitat: Broad-leaved trees, especially beech, on wood edges or open places.

Distribution: Quite widespread. Rather rare to rare.

p.155 *Inocybe heimii* M. Bon (= *I. caesariata*)
Cortinariaceae, Cortinariales

Description: *Cap felted woolly, rusty buff. Cortina forming a thick woolly band.*
Cap 3–8 cm (1.2–3.2 in), globose then convex, not umbonate, appendiculate, felted woolly or shaggy in the centre, rather dark or dull rusty buff. *Gills* decurrent with a small tooth, not very close, olivaceous yellow-brown then with rusty tones, edge concolorous or paler.
Stipe 3–7 × 0.5–1.5 cm (1.2–2.8 × 0.2–0.6 in), hollow, short, cylindric or fusoid, smooth at the apex but fibrillose under the cortina which forms a thick woolly band, often more yellow above.
Flesh yellowish beige, slightly or slowly bitter. *Microscopic features* Sp 9–15 × 4.5–7 µm, cylindro-elliptic, rather irregular and sometimes bean-shaped, with blunt apex. Cyst absent but MC 25–50 × 8–20 µm, clavate to ventricose fusiform. CC a trichoderm, with hyphae 5–8 µm diameter. Pigment mixed.

Season: August–December.
Habitat: Sandy or gravelly places.
Distribution: Quite rare although sometimes abundant.

Important characters of subgenus *Mallocybe* are the absence of cystidia, the deeply originating MC, the rather squat shape, the convex cap and relatively short stipe.

p.158 *Inocybe agardhii* (Lundell) Orton
Cortinariaceae, Cortinariales

Description: *Cap not umbonate, felted, rust brown. Cortina conspicuous.*
Cap 3–7 cm (1.2–2.8 in), hemispherical then almost flat, not umbonate, slightly

appendiculate, felted or with small fibrillose scales, rusty greyish brown, sometimes with yellowish or olivaceous tones. *Gills* broadly adnate, not very close, olivaceous buff then brown. *Stipe* 3–8 × 0.5–1 cm (1.2–3.2 × 0.2–0.4 in), often short, with a conspicuous cortina, fibrillose, concolorous or more olivaceous yellow to whitish at the apex. *Flesh* slightly or slowly bitter, yellowish beige or whitish. Smell herbaceous when bruised. *Microscopic features* Sp 8.5–11 × 4.5–6 µm, cylindro-elliptic, sometimes slightly curved, conical at the apex. Cyst absent. MC 20–45 × 7–15 µm, clavate to cylindric-fusoid, variable, sometimes septate. CC a trichoderm with hyphae 8–15 µm diameter. Pigment mixed.

Season: August–December.
Habitat: Willow.
Distribution: Widespread but uneven.

Same comment as for *I. heimii* (*see* p.490).

---

p.155 *Inocybe terrigena* (Fr.) Kühner
Cortinariaceae, Cortinariales

Description: *Cap not umbonate, yellow-buff, finely scaly. Sheath scaly.*
Cap 2–7 cm (0.8–2.8 in), hemispherical then convex, with small fibrillose or coarse scales, the scales sometimes erect in concentric bands; brownish to yellowish buff. *Gills* notched, not very close, buff then more rust brown, edge thin, concolorous or paler. *Stipe* 3–6 × 0.5–1 cm (1.2–2.4 × 0.2–0.4 in), cylindric or slightly angled or swollen below, concolorous or more yellowish. Sheath with dark scales, upturned below the cortina. *Flesh* slightly or slowly bitter, yellowish to whitish. Smell earthy. *Microscopic features* Sp 9.5–12.5 × 6–7.5 µm, elliptic or almond-shaped, with blunt apex. Cyst absent. MC clavate or with very inflated heads, 20–40 × 10–25 µm. CC a trichoderm, with hyphae 5–15 µm diameter. Pigment mixed.

Season: August–November.

Habitat: Stony places beside streams.
Distribution: Mainly in lowland continental sites.

As for *I. heimii* (see p.490).

---

p.155 *Inocybe squarrosoannulata* Kühner
Cortinariaceae, Cortinariales

Description: *Cap obtuse, rufous with small erect scales. Sheath scaly.*
Cap 1–3 cm (0.4–1.2 in), globose then almost flat, not umbonate, scaly or even squarrose on a fibrillose, rufous tawny ground, the scales browner. *Gills* notched, rather distant, brown then with chocolate-brown tones. *Stipe* 1.5–2.5 × 0.3–0.6 cm (0.6–1 × 0.12–0.24 in), hollow, the sheath concolorous, with upturned scales.
Flesh slightly or slowly bitter, pale brownish to whitish. Smell none to herbaceous. *Microscopic features* Sp 8–14 × 5–8 µm, narrowly elliptic or the ventral face straight or slightly concave, the apex slightly conical. Cyst absent. MC clavate, sometimes close packed, septate, 20–35 × 10–15 µm. CC tending to be trichodermial, with hyphae 5–12 µm diameter. Pigment mixed, more streaky in the scales.
Season: June–September.
Habitat: Dwarf willow, in high stony places in the mountains.
Distribution: Not yet fully understood.

Same comment as for *I. heimii* (see p.490).

---

p.155 *Inocybe paludosa* Kühner
Cortinariaceae, Cortinariales

Description: *Cap felted to fibrillosely scaly, yellow-brown. Stipe yellow at the base.*
Cap 1–3 cm (0.4–1.2 in), hemispherical or conico-campanulate then convex or subumbonate, finely fibrillosely scaly to woolly felted, ochraceous rufous brown, more yellow at the edge, darker when old. *Gills* notched, quite wide, not very close,

rusty buff then chocolate brown, edge thin, almost concolorous or paler. *Stipe* 10–25 × 1.5–2.5 mm (4–8 × 0.6–1 in), hollow, ochraceous beige, rather bright yellow at the base, with slightly upturned rufous scales. *Flesh* slightly or slowly bitter, concolorous or paler, fairly bright yellow at the base. Smell earthy. *Microscopic features* Sp 8–12 × 6–7.5 µm, elliptic, the ventral face often flatter or concave, variable, the apex rather blunt. Cyst absent. MC clavate, 10–30 × 8–15 µm, sometimes septate. CC a trichoderm, with septate hyphae 5–15 µm diameter. Pigment mixed.

Season: June–September.
Habitat: Dwarf willow, high damp places in the mountains.
Distribution: Alps.

Comments as for *I. heimii* (see p.490).

---

p.154 *Inocybe dulcamara* (Pers.) Kummer
Cortinariaceae, Cortinariales

Description: *Cap brownish to rather tawny, felted. Smell of honey when bruised.*
Cap 1–6(–12!) cm (0.4–2.4(4.7!) in), hemispherical to almost flat, finely scurfy to slightly scaly or felted, brownish to rather tawny. *Gills* notched, or with a small decurrent tooth, not very close. Edge almost concolorous or whitish. *Stipe* 1–5 × 0.2–1 cm (0.4–2 × 0.08–0.4 in), slightly hollow, concolorous, fibrillose, with cortina. *Flesh* slightly or slowly bitter, almost concolorous or paler. Smell of honey, at least when rubbed. *Microscopic features* Sp 7.5–11 × 4.5–7 µm, elliptic to ovoid, sometimes with a straighter ventral face, the apex blunt. Cyst absent. MC clavate, sometimes close packed, 30–50 × 6–18 µm. CC a trichoderm, with hyphae 5–12 µm diameter. Pigment mixed, not very encrusted.

Season: April–January.
Habitat: Sun-loving or in dappled shade, on sandy or gravelly ground.
Distribution: Widespread. Common.

Variable. Many varieties and forms have been described, with more or less specific habitats.

Same comment as for *I. heimii* (*see* p.490).

---

p.155 *Inocybe umbrinofusca* Kühner ex Kühner
Cortinariaceae, Cortinariales

Description: *Cap dark brown, fibrillose-woolly. Cortina rather fleeting.*
Cap 1–2.5 cm (0.4–1 in), hemispherical then flat, not umbonate, fibrillose to scurfy woolly, hygrophanous, brown with darker tones in the centre, more buff brown when dry. *Gills* notched, sometimes almost free, very deep near the cap margin, not very close, dingy brown. *Stipe* 1–3.5 × 0.2–0.5 cm (0.4–1.4 × 0.08–0.16 in), hollow, concolorous or more yellowish buff, with a rather fleeting brownish cortina. *Flesh* slightly or slowly bitter, fragile, almost concolorous or paler when dry. *Microscopic features* Sp 8–11 × 6–7.5 µm, elliptic to ovoid, sometimes slightly irregular. Cyst absent. MC clavate to broadly capitate, sometimes close packed, septate, 20–40 × 10–18 µm. CC a subtrichodermial cutis, with hyphae 5–15 µm diameter. Pigment mixed, mostly epimembranal.
Season: June–September.
Habitat: Dwarf willow communities on damp alpine sites.
Distribution: Alps.

Same comment for *I. heimii* (*see* p.490).

---

p.159 *Inocybe patouillardii* Bresadola (highly poisonous)
Cortinariaceae, Cortinariales

Description: *In spring, flushing pinkish red. Cap pale.*
Cap 4–10 cm (1.6–4 in), conico-campanulate then expanded, umbonate, radially lined or fibrous, pale, whitish to ochraceous beige, flushed or spotted pink then red where bruised and when old.

Gills notched, quite close, pale beige to greyish beige. *Stipe* 3–7 × 0.7–2 cm (1.2–2.8 × 0.28–0.8 in), fibrillose, with no cortina or pruina, concolorous with the cap and similarly reddening. *Flesh* white then reddening. Smell faintly aromatic. *Microscopic features* Sp 9–15 × 5–8 µm, elliptic or slightly bean-shaped. Cyst absent. MC cylindro-clavate to clavate, 30–60 × 8–15 µm. CC a radial cutis. Pigment mixed.

Season: May–July (August and September).

Habitat: Coppice and scrub on calcareous soils, under broad-leaved trees and also on calcareous hillsides; tends to occur on fairly warm sites.

Distribution: Quite widespread; rather southern.

Highly poisonous (severe muscarine syndrome), causing some fatalities. It should never be confused with St George's Mushroom (*Calocybe gambosa*, see p.863), a white or pale, unchanging, very mealy species. The subgenus *Inosperma* lacks metuloids but has MC which originate less deeply (from the subhymenium and not from the gill trama). The cuticle in this group is innately streaky to rimose or has small to shaggy scales.

---

p.158 *Inocybe jurana* (Patoullard) Saccardo
Cortinariaceae, Cortinariales

Description: *Cap pinkish brown, becoming vinaceous as does the entire fruit body. Stipe pale.*
Cap 4–10 cm (1.6–4 in), conico-convex then expanded, umbonate, radially lined to innately streaky, quite often with appressed scales in the centre, lilac pink to lilaceous brownish mauve, then dark vinaceous brown in patches or irregular zones. Margin slow to expand, fibrillose, often lobed. *Gills* notched to almost free, greyish beige then brownish, becoming wine-coloured. *Stipe* 5–10 × 0.7–2 cm (2–4 × 0.28–0.8 in), hollow, almost concolorous and staining vinaceous like the cap, fibrillose. Base variable. *Flesh* pale to crimson. Smell aromatic, finally rather

unpleasant. *Microscopic features* Sp 11–14 × 5–8 μm, cylindro-elliptic or slightly bean-shaped. Cyst absent. MC more or less clavate, sometimes very inflated or variable, 40–65 × 10–25 μm. CC a cutis with hyphae 5–8 μm diameter. Pigment mixed, more streaky when it originates from deeper trama.

Season: June–November.
Habitat: Broad-leaved trees (wood edges, road sides, etc.); especially on calcareous soils.
Distribution: Quite rare, Unevenly distributed.

I. rhodiola Bresadola, a pretty pink species becoming wine-red when old, is more slender and requires warm, damp ground.

---

p.157 *Inocybe fastigiata* (Sch.:Fr.) Kummer
Cortinariaceae, Cortinariales

Description: *Cap conical umbonate, innately streaky or rimose, buff to brownish.*
Cap 3–8 cm (1.2–3.2 in), rounded and acute then conical or almost flat, with a rather pronounced umbo, innately streaky to fibrous, often buff to brownish tawny. *Gills* notched, almost free, often narrow, beige with yellowish olivaceous then brownish tones. *Stipe* 5–10 × 0.4–1 cm (2–4 × 0.16–0.4 in), slightly fibrillose to fleecy at the apex, cylindric, white or very pale, slightly yellowish then buff from the base. *Flesh* white. Smell spermatic or slightly earthy. *Microscopic features* Sp 10–16 × 6–8 μm, elliptic to almond-shaped, sometimes kidney-shaped. Cyst absent. MC clavate, sometimes cylindric or fusoid, 40–75 × 10–20 μm. CC a cutis with hyphae 4–8 μm diameter. Pigment mixed, finely punctate when it originates from a deeper trama.

Season: May–December.
Habitat: Ubiquitous.
Distribution: Widespread but frequency uneven.

There are a number of similar taxa, which differ in respect of cap and gill colour, smell, spore size, habitat, etc.

p.158 *Inocybe arenicola* (Heim) M. Bon
Cortinariaceae, Cortinariales

Description: *Cap yellowish buff, with a white veil. Smell faint.*
Cap 3–6 cm (1.3–2.4 in), conico-convex then umbonate or domed, radially innately streaky to fibrillose, yellowish buff, sometimes brownish and dark when old, with fleeting whitish veil. *Gills* adnate, yellowish beige with olivaceous tones then brown. *Stipe* 5–12 × 0.5–1.5 cm (2–4.7 × 0.2–0.6 in), with a thick coating of sand, very pale. *Flesh* white or pale. Smell faint, spermatic. *Microscopic features* Sp 12–17 × 7–8.5 μm, cylindric or sometimes slightly kidney-shaped, curved. Cyst absent. MC clavate or cylindric to wavy, 50–95 × 10–18 μm. CC a cutis with hyphae 4–8 μm diameter. Pigment mixed but often very pale.
Season: October–December.
Habitat: Coastal sands.
Distribution: Not fully understood.

Once considered to be a variety of
I. fastigiata (see p.496).

---

p.158 *Inocybe maculata* Boudier
Cortinariaceae, Cortinariales

Description: *Cap conical, warm rufous brown, with patches of white veil.*
Cap 3–7 cm (1.2–2.8 in), conical then expanded with a distinct umbo, fibrillose to innately streaky, more or less bright or dark warm rufous brown to chestnut, with white veil remains, disrupted into appressed, sometimes fleeting patches. Margin often incurved then expanding, splitting. *Gills* notched, white then beige grey, slowly becoming brownish. *Stipe* 5–10 × 0.4–0.8 cm (2–4 × 0.16–0.32 in), slightly fusoid, fibrillose or slightly fleecy above, white then flushed rather rufous to rufous from the centre. *Flesh* very pale. Smells of gingerbread. *Microscopic features* Sp 8–10 × 5–6 μm, elliptic to depressed on the ventral

face. Cyst absent. MC clavate, sometimes irregular, 30–75 × 10–20 μm. CC a cutis. Pigment mixed or rather epimembranal, encrusting.

Season: July–December.
Habitat: Woodland rides, sometimes on bare damp ground in broad-leaved woodland.
Distribution: Widespread. Rather common everywhere.

Specimens which have lost their veil can be recognized by their distinctive colour, the progressive rufous flush on the stipe and the smell.

---

p.157 *Inocybe fastigiella* Atkinson
Cortinariaceae, Cortinariales

Description: *Cap conical, innately streaky, beige to buff. Stipe whitish or pale.*
Cap 2–5 cm (0.8–2 in), conical then campanulate or umbonate, fibrillose-innately streaky to rimose, buff or beige on a pale ground. *Gills* almost free, quite narrow, whitish then pale greyish beige. *Stipe* 4–8 × 0.3–0.7 cm (1.6–3.2 × 0.12–0.28 in), white then pale beige, very slightly fleecy at the apex. *Flesh* thin, white or very pale. Smell slightly spermatic. *Microscopic features* Sp 7–11 × 5–6.5 μm, elliptic or ovoid to slightly bean-shaped. Cyst absent. MC cylindric or clavate, obtuse, 25–55 × 10–15 μm. CC a cutis. Pigment mixed.

Season: August–December.
Habitat: Broad-leaved trees or scrub on damp soils.
Distribution: Not fully understood. Rather common.

Belongs to the *I. maculata* group on account of the small spores but differs in the rather pale colours, quite slender build and damp habitat.

---

p.156 *Inocybe curreyi* Berkeley
Cortinariaceae, Cortinariales

Description: *Cap umbonate, finely scaly, yellow-brown. Gills and margin flecked with yellow.*

Cap 2–5 cm (0.8–2 in), conico-convex then obtusely umbonate, streaky and disrupted into small scales around the umbo, rufous brown on a yellow-buff, sometimes slightly golden ground. Margin splitting, with bright yellow spots. *Gills* notched, yellow then yellowish beige to olivaceous brown, edge with bright yellow flecks. *Stipe* 2–5 × 0.4–1 cm (0.8–2 × 0.16–0.4 in), often short, fibrillose and slightly fleecy at the apex, whitish then yellowish beige to brownish from the base, spotted bright yellow when old. *Flesh* white or sometimes with bright patches of yellow near the surface. *Microscopic features* Sp 10–13 × 5.5–7.5 µm, cylindro-elliptic short to ovoid, sometimes slightly bean-shaped. Cyst absent. MC clavate, sometimes irregular or contracted to capitate, 30–60 × 10–20 µm. CC a cutis with a tendency to be trichodermial in the scales. Pigment mixed or rather intracellular.

Season: September–November.
Habitat: Rather damp or even muddy ground, under broad-leaved trees.
Distribution: Little known. Not very common.

I. squamata J.E. Lange differs in the less bright colours and the absence of bright yellow patches.

---

p.154 *Inocybe cookei* Bresadola
Cortinariaceae, Cortinariales

Description: *Cap umbonate, yellow-buff. Stipe bulbous, not pruinose. Smell of honey.*
Cap 2–5 cm (0.8–2 in), conico-campanulate then convex, sometimes without an umbo, almost smooth or slowly rimose, rather bright yellow-buff to rufous from the centre.
Gills notched, quite close, pale beige grey then rufous buff. *Stipe* 4–7 × 0.4–0.8 (–1.4) cm (1.6–2.8 × 0.16–0.32 (–0.55) in), bulbous submarginate, white or very pale then yellowish buff near the middle, fibrillose to fleecy at the apex.

Flesh pale. Smells strongly of honey.
Microscopic features Sp 7–10 × 4–5.5 µm, cylindro-elliptic or more or less bean-shaped. Cyst absent. MC clavate, 30–50 × 12–20 µm. CC a cutis. Pigment mixed or rather epimembranal, encrusting in deeper trama. No clamps.

Season: August–October.
Habitat: Conifers and broad-leaved trees, wood edges.
Distribution: Widespread. Common.

In this small group, which have a basal bulb and lack cystidia, *I. quietiodor* M Bon, has a brownish cap and a smell like damp laundry (like *Lactarius quietus*), and *I. kuthanii* Stangl & Veselsky has a more sharply umbonate, buff brown cap and smells mealy.

---

p.152 *Inocybe calamistrata* (Fr.:Fr.) Gillet
Cortinariaceae, Cortinariales

Description: *Cap domed, with erect scales, dark brown. Stipe dark blue-green at the base.*
Cap 2–5 cm (0.8–2 in), hemispherical then expanded, not umbonate, covered in very erect scales, especially in the centre, with smaller scales or woolly to fibrillose towards the margin, dark brown.
Gills notched, not very close, buff then dark rust-brown or chocolate brown.
Stipe 3–7 × 0.4–0.8 cm (1.2–2.8 × 0.16–0.32 in), more or less hollow, fleecy at the apex but with flat or erect scales below, concolorous, the base dark blue-green. *Flesh* brownish to dark brown, sometimes pinkish deep in the stipe and dark green below. *Microscopic features* Sp 9–13.5 × 4.5–7 µm, cylindro-elliptic or more or less bean-shaped. Cyst absent. MC clavate or cylindric to constricted, 20–50 × 10–15 µm. CC a trichoderm. Clamps few.

Season: July–November.
Habitat: Damp woods (mixed coniferous and broad-leaved trees, extending up to the upper montane zone).
Distribution: Widespread. Becoming rare.

p.159 *Inocybe cervicolor* (Pers.) Quélet
Cortinariaceae, Cortinariales

Description: *Cap convex, with small fleecy scales, brownish beige. Becoming rufous. Smell earthy.*
Cap 2.5–5 cm (1–2 in), conico-convex then expanded umbonate, with rather fine small fleecy scales, beige-grey to brownish, becoming slightly rufous or dingy when old. *Gills* almost free, quite close, beige then rusty brownish. Edge rufous when old. *Stipe* 5–10 × 0.3–1 cm (2–4 × 0.12–0.4 in), tall and slender, fleecy at the apex and slightly fibrillose-fleecy elsewhere, almost concolorous or paler, slightly rufous or becoming pink to rufous. *Flesh* pale or whitish, developing a rufous tinge. Smell not very pleasant (saucisson, musty casks, earth, etc.). *Microscopic features* Sp 11–15 × 6–8.5 µm, cylindro-elliptic to slightly bean-shaped. Cyst absent. MC clavate, 30–60 × 10–18 µm. CC a subtrichodermial cutis. Pigment mixed. Clamps few.
Season: August–November.
Habitat: Broad-leaved trees, sometimes mixed with conifers, usually on calcareous clays.
Distribution: Widespread. Rather common.

---

p.160 *Inocybe bongardii* (Weinm.) Quélet
Cortinariaceae, Cortinariales

Description: *Cap obtuse, with small brownish scales. Flushing pink-red. Smell aromatic.*
Cap 3–6 cm (1.3–2.4 in), conico-convex then obtuse, sometimes with a broad umbo and small scales, especially in the centre, more fibrillosely scaly elsewhere, greyish beige then with more brownish scales. Reddening when old. *Gills* almost free, quite close, whitish then greyish sepia, spotting dingy pink like the edge. *Stipe* 5–9 × 0.5–1.2 cm (2–3.5 × 0.2–0.47 in), barely scaly to finely fibrillosely scaly near the base, reddening slightly when handled. *Flesh* pale, flushing pinkish-red. Smell strong, aromatic, rather

difficult to describe (like friar's balsam).
Microscopic features Sp 11–14 × 6–7.5 µm, cylindro-elliptic to more or less bean-shaped. Cyst absent. MC cylindro-clavate or wavy, 30–60 × 9–20 µm. CC a trichoderm. Pigment rather epimembranal, smooth. Clamps few.

Season: August–November.
Habitat: Broad-leaved trees, sometimes with conifers, usually on clay soils.
Distribution: Widespread. Rather common everywhere.

I. pisciodora Donadini & Riousset occurs in winter and spring in the south. It is less scaly, reddens more, and has a strong smell of the sea or fish.

---

p.159 *Inocybe piriodora* (Pers.:Fr.) Kummer
Cortinariaceae, Cortinariales

Description: *Cap umbonate, rufous. Reddening. Strong pear smell.*
Cap 3–8 cm (1.2–3.2 in), campanulate then shallowly convex to broadly umbonate, fibrillose or with fine fibrillose scales, sometimes finally cracking near the margin, pale, yellowish buff then more rufous brown to reddish.
Gills almost free, quite close, greyish beige then rusty brownish.
Stipe 4–10 × 0.5–1.2 cm (1.6–4 × 0.2–0.47 in), fibrillose or very slightly fibrillosely scaly below, whitish to rather rufous beige or reddish from the base.
Flesh pale, flushing faintly pink. Smells strongly of pear liqueur. *Microscopic features* Sp 9–12 × 6–7.5 µm, almond- or lemon-shaped, the apex umbonate or elongated. Cyst 35–60 × 10–18 µm, clavate to fusoid, the wall 2–4.5 µm diameter, yellowish in ammonia. MC clavate ventricose, sometimes septate, with short cells. CC a cutis. Pigment mixed. Caulo rare and at the stipe apex.

Season: August–November.
Habitat: Broad-leaved trees, on rather heavy calcareous soils.
Distribution: Widespread. Rather common.

The subgenus *Inocybium* is characterized by smooth spores and metuloids (thick-walled, crystal encrusted cystidia), on the gill face (pleurocystidia) and edge (cheilocystidia), the latter usually accompanied by thin-walled MC, lacking any characteristic shape. *I. incarnata* Bresadola, sometimes brick-red at maturity, is very similar but reddens more obviously.

---

p.160 *Inocybe corydalina* Quélet
Cortinariaceae, Cortinariales

Description: *Cap pale, with a broad, dark green umbo. Smell aromatic.*
Cap 3–7 cm (1.2–2.8 in), conico-convex then with a broad obtuse umbo, fibrillose-innately streaky or almost smooth, whitish to ochraceous beige or brownish at the edge, the umbo tinged green under a fleeting whitish veil. *Gills* almost free, pale greyish to brownish beige. *Stipe* 5–10 × 0.5–1 cm (2–4 × 0.2–0.4 in), slightly fleecy at the apex, not pruinose, whitish or with brownish fibrils; sometimes greenish below. *Flesh* pale to almost concolorous near the surface. Smell strong, similar to *I. bongardii. Microscopic features* Sp 8–11 × 5.5–7 µm, ovoid or slightly lemon-shaped, the apex obtuse to umbonate. Cyst 40–65 × 12–20 µm, fusiform ventricose to almost urn-shaped, the wall 1–1.5 µm diameter, not reacting with ammonia. MC ovoid or clavate ventricose 20–50 × 10–15 µm. CC a cutis. Pigment mixed. Caulo almost absent.

Season: August–November.
Habitat: Broad-leaved woods and coppice, on damp soils.
Distribution: Widespread. Rather common.

I. erinaceomorpha Stangl & Veselsky, which is the same shape, has a distinctly scaly cap, the scales forming a dark brown pattern on a paler ground. The stipe is sometimes tinged greenish, and flushes more or less pink.

p.156 *Inocybe haemacta* (Berk. & Cke.) Saccardo
Cortinariaceae, Cortinariales

Description: *Cap fibrillose-scaly, brown with green tones, reddening.*
Cap 3–6 cm (1.3–2.4 in), conico-convex then broadly umbonate, fibrillose to scaly, especially in the centre where the disc may be slightly star-shaped, dull brown, darker in the centre where there are green tones, lighter and often tinged greenish and pinkish at the edge. *Gills* notched, whitish then brownish grey to brown, sometimes tinged greenish or pinkish. *Stipe* 3–6 × 0.5–1 cm (1.2–2.4 × 0.2–0.4 in), fleecy at the apex, whitish then tinged green and mottled with pink, the fibrils browning when old.
Flesh white then greenish to green (sometimes blue-green) below and flushing pink to red. Often smelling of horse droppings. *Microscopic features* Sp 8.5–11.5 × 5–6.5 µm, cylindro-elliptic or almond-shaped, the apex rounded and acute or slightly umbonate. Cyst 50–80 × 10–18 µm, fusiform ventricose, the wall 1–1.5 µm diameter, not reacting with ammonia. MC clavate ventricose 20–50 × 10–15 µm. CC a cutis. Pigment mixed or mostly intracellular. Caulo rare, at the apex.

Season: August–December.
Habitat: Damp disturbed woods and wood edges.
Distribution: Widespread and frequency uneven.

Other species have purer green tones. They are rare and usually occur on sandy soils.

---

p.152 *Inocybe hystrix* (Fr.) P. Karsten
Cortinariaceae, Cortinariales

Description: *Cap convex, brown, with pointed erect scales like the stipe. Gills pale.*
Cap 2–5 cm (0.8–2 in), hemispherical then expanded, sometimes slightly umbonate, squarrose, brown, with darker scales. Margin fibrillose or woolly appendiculate. *Gills* adnate, not very

close, white then brownish to rust brown. *Stipe* 3–6 × 0.3–0.7 cm (1.2–2.4 × 0.12–0.28 in), cylindric, fibrillose-woolly and whitish immediately under the gills, with a high ring zone, and erect pointed scales (like the cap) below. *Flesh* white or brownish cream below. Smell spermatic. *Microscopic features* Sp 9–14 × 4.5–7 μm, almond-shaped, the apex blunt or umbonate. Cyst 60–90 × 10–25 μm, fusiform to clavate, the wall 1–2 μm diameter, not reacting with ammonia. MC clavate ventricose. CC a trichoderm with septate hyphae. Caulo rare, at the apex.

Season: August–October.
Habitat: Broad-leaved trees, sometimes conifers, usually on calcareous soils; on damp soils.
Distribution: Uneven, northern.

---

p.153 *Inocybe squarrosa* Rea
Cortinariaceae, Cortinariales

Description: *Slender. Cap with pointed erect scales, brownish. Stipe fleecy.*
Cap 0.5–1.5 cm (0.2–0.6 in), hemispherical then expanded, brownish, squarrose then with coarse appressed, lighter or darker scales. Margin appendiculate with fibrous scales. *Gills* notched, distant, greyish beige later rust brown. *Stipe* 1–4 × 0.1–0.3 cm (0.4–1.6 × 0.04–0.12 in), wavy, only slightly fleecy at the apex, fibrillose or with small scales below, beige to brownish, sometimes pinkish or slightly lilac at the apex. Base sometimes with a small bulb. *Flesh* concolorous or paler. Smell herbaceous, sharp. *Microscopic features* Sp 8.5–12 × 5–6.5 μm, ovoid or elliptic almond-shaped, the apex obtuse pointed or rounded and acute. Cyst 40–70 × 10–25 μm, utriform to fusiform ventricose, the wall 0.5–1 μm diameter, not reacting with ammonia. Cyst on the gill edge more cylindric or sometimes subcapitate and MC clavate ventricose 20–50 × 10–20 μm. CC a

trichoderm with septate hyphae. Caulo rare, at the apex.
Season: July–October.
Habitat: Marshy woods, muddy tracks, pond edges.
Distribution: Uneven.

There are darker or lighter, sometimes almost white, forms.

---

p.153 *Inocybe obscura* (Pers.:Fr.) Gillet
Cortinariaceae, Cortinariales

Description: *Cap squarrose, dark brown. Gills with a brown edge. Stipe violaceous lilac at the apex.*
Cap 2–5 cm (0.8–2 in), hemispherical then expanded to almost flat, scaly in the centre, more innately streaky at the margin, brown to dark brown on a rather beige ground. *Gills* adnate, quite close, brownish then rust brown, edge dark brown, sometimes discontinuously and harder to see when old. *Stipe* 3–6 × 0.3–0.7 cm (1.2–2.4 × 0.12–0.28 in), fibrillose-fleecy or very slightly scaly near the base, not pruinose, violaceous or mauve above, at least initially.
Flesh ochraceous cream but purple or lilac at the stipe apex. Smell spermatic.
Microscopic features Sp 8–11.5 × 5–6.5 µm, almond-shaped or with an umbonate or elongated apex. Cyst 60–100 × 10–15 µm, fusiform or lageniform with a long neck, the wall 1.5–3 µm diameter, bright yellow in ammonia. Cyst on the gill edge slightly shorter or varying and MC clavate or balloon-shaped, 20–40 × 10–25 µm, brown-walled. CC a cutis or trichoderm. Pigment dark, encrusting. Caulo absent.
Season: August–November.
Habitat: Broad-leaved trees, especially on calcareous clays; on damp soils, rarely on conifer needles.
Distribution: Widespread but frequency uneven.

I. cincinnata (Fr.) Quélet is more slender. There are other species in this group characterized by violaceous stipes.

p.154 *Inocybe lacera* (Fr.:Fr.) Kummer
Cortinariaceae, Cortinariales

Description: *Cap felted to fibrillose or with small scales, brownish. Stipe dark brown below.*
Cap 1–5 cm (0.4–2 in), conico-convex then expanded with a low umbo, felted to fibrillose-woolly sometimes more fibrillose to innately streaky or more squamulose or squarrose, brownish to rufous brown. *Gills* notched, brownish beige to rather dark brown. *Stipe* 2–7 × 0.2–0.8 cm (0.8–2.8 × 0.08–0.32 in), hollow, fibrillose, with a fleeting cortina near the apex, beige to brownish. Base darker to almost black when bruised. *Flesh* pale beige to brownish, almost concolorous near the surface, dark brown to black at the stipe base. Smell slightly spermatic or sometimes herbaceous. *Microscopic features* Sp 11–17 × 4–6 µm, cylindric, the apex blunt or conical to umbonate and the base sometimes truncate, sometimes irregular or indented to contorted. Cyst 40–70 × 8–18 µm, fusiform or ventricose, sometimes cylindric, the wall 0.5–1.5 µm diameter, yellowish in ammonia. Cyst on the gill edge variable and MC clavate or ventricose, 20–50 × 10–25 µm. CC a cutis or trichoderm. Pigment usually encrusting. Caulo absent.
Season: April–December.
Habitat: Woods, heaths, etc. Ubiquitous.
Distribution: Very widespread. Common.

A very variable taxon. There are numerous forms and varieties reflecting variations in the cuticle, stature, slenderness and colour.

---

p.113 *Inocybe geophylla* (Fr.:Fr.) Kummer
Cortinariaceae, Cortinariales

Description: *Cap conical umbonate, almost smooth, entirely white. Gills clay-brown.*
Cap 1.5–3.5 cm (0.6–1.4 in), conico-convex then expanded umbonate, almost smooth or with fine radial fibrils, almost silky when dry, white. *Gills* notched, grey

then brownish, clay-brown. *Stipe* 2–4 × 0.2–0.5 cm (0.8–1.6 × 0.08–0.2 in), pruinose at the apex and smooth to slightly fibrillose in the lower half, white. *Flesh* white or greyish. Smell strong, spermatic. *Microscopic features* Sp 7.5–12.5 × 5.5–6.5 µm, almond-shaped or elliptic, the apex rather obtuse. Cyst 35–75 × 10–18 µm, lageniform, the wall 1–2.5 µm diameter, not reacting with ammonia. MC clavate. CC a cutis sometimes slightly gelatinized. Caulo at the stipe apex, often more variable than the Pleuro, 25–80 × 12–20 µm.

Season: June–December.
Habitat: Broad-leaved, sometimes mixed, woodland; prefers calcareous clays.
Distribution: Widespread but frequency uneven.

There are colour variants (*see* var. *lilacina* below).

---

p.145 *Inocybe geophylla* var. *lilacina* (Peck) Gillet
Cortinariaceae, Cortinariales

Description: *Same characters as I. geophylla but the cap and stipe violaceous lilac.*
Cap as *I. geophylla* (*see* p.507) but quite bright lilac, fading to pale violaceous, sometimes greyish when old. Umbo ochraceous yellow to yellowish. *Stipe* concolorous with the cap and similarly fading.
Habitat: Often growing with the type.

There is also a rarer var. *violacea* (Pat.) Saccardo, which is more squat and, in particular, has a bright or deep violaceous cap, lacking any trace of yellow on the umbo.

---

p.158 *Inocybe psammophila* M. Bon
Cortinariaceae, Cortinariales

Description: *Cap obtusely and broadly umbonate, with a white veil. Stipe deeply buried in sand.*
Cap 4–6 cm (1.6–2.4 in), convex then expanded with a low broad umbo, more or less covered in white veil, especially in the centre, on an innately streaky to rimose,

buff or yellowish buff ground.
Gills notched, not very close, brownish to rust brown at maturity. *Stipe* 4–8 × 0.8–1.5 cm (1.6–3.2 × 0.32–0.6 in), whitish, powdery at the apex and more fibrillose below. Base sometimes appearing swollen bulbous from the thick coating of sand. *Flesh* white, fibrous. Smell faint, spermatic. *Microscopic features* Sp 12–16 × 6–9.5 µm, very large, elliptic or depressed on the ventral face, yellow brown. Cyst 45–75 × 15–35 µm, elliptic or ovoid, obtuse, the wall 2–3.5 µm diameter, pale yellow in ammonia. MC clavate. CC a cutis. Pigment mixed, pale. Caulo at the stipe apex.

Season: October–December.
Habitat: Coastal dunes.
Distribution: Mainly northern.

---

p.160 *Inocybe pruinosa* R. Heim
Cortinariaceae, Cortinariales

Description: *Cap fibrillose-scaly, buff with a pruinose white veil. Stipe browning.*
Cap 1.5–4 cm (0.6–1.6 in), conico-convex then umbonate, with a white or pale cream veil over a ground that is fibrillose or almost scaly in the centre, ochraceous cream to yellowish beige. *Gills* almost free, not very close, yellowish beige then more rust brown. Edge whitish to yellowish. *Stipe* 2–4.5 × 0.4–0.8 cm (0.8–1.8 × 0.16–0.32 in), pruinose at the apex, smooth to fibrillose below, white then brownish beige to rufous brown from the base. *Flesh* white or pale, almost concolorous near the surface. Smell fruity rather than spermatic, faint. *Microscopic features* Sp 11–16 × 5–7 µm, cylindric, the apex slightly pointed or obtuse, sometimes irregular. Cyst 45–70 × 10–20 µm, fusiform or ventricose, the neck scarcely distinct, the wall 2–3.5 µm diameter, yellow in ammonia. MC clavate. CC a cutis. Pigment mixed, pale. Caulo at the stipe apex.

Season: August–November.
Habitat: Coastal sands, sometimes open pine woods near the sea; usually warm sites.
Distribution: Uneven. Rare or very rare.

---

p.153 *Inocybe flocculosa* (Berk.) Saccardo
Cortinariaceae, Cortinariales

Description: *Cap subumbonate, yellowish brown, felted. Stipe pale.*
Cap 2–4 cm (0.8–1.6 in), conico-convex then with a faint umbo, felted or with small woolly scales in the centre, more fibrillose at the edge, rather light yellowish brown. *Gills* notched, quite close, greyish beige then brown. *Stipe* 3–5 × 0.3–0.5 cm (1.2–2 × 0.12–0.2 in), pruinose at the apex and fibrillose elsewhere, pale yellowish buff. *Flesh* not very thick, pale or almost concolorous. Smell medium, spermatic. *Microscopic features* Sp 7–9.5 × 4.5–5.5 µm, elliptic to ovoid, the apex obtuse. Cyst 40–60 × 12–18 µm, fusiform or ventricose, the neck not very distinct, the wall 0.5–1 µm diameter, rather bright yellow in ammonia. MC clavate. CC a subtrichodermial cutis. Pigment mixed, pale. Caulo at the stipe apex.
Season: September–November.
Habitat: Especially oak, hornbeam, beech and other broad-leaved trees.
Distribution: Not yet assessed. Not fully understood.

---

p.153 *Inocybe nitidiuscula* (Britz.) Saccardo
(= *I. friesii*)
Cortinariaceae, Cortinariales

Description: *Cap brownish buff, fibrillose to slightly rimose. Stipe apex powdery, pinkish.*
Cap 2–4 cm (0.8–1.6 in), conico-convex then slightly umbonate, fibrillose then disrupting into small fibrillose scales, rather dark rufous brown, more brownish buff or greyish beige at the margin. *Gills* notched, quite close, buff to brownish. *Stipe* 2.5–5 × 0.3–0.6 cm (1–2 × 0.12–0.24 in), fibrillose, pale beige

but powdery and pinkish buff to reddish brown at the apex. *Flesh* very pale ochraceous beige to almost concolorous near the surface but distinctly reddish pink to reddish beige at the stipe apex. Smell spermatic. *Microscopic features* Sp 8–14 × 4–7 μm, cylindro-conical or almond-shaped, the apex rounded and acute or sometimes elongated. Cyst 40–80 × 12–20 μm, fusiform or lageniform, the wall 1–1.5 μm diameter, not reacting with ammonia. MC clavate. CC a cutis. Caulo at the stipe apex.

Season: June–December.
Habitat: Conifers in particular, sometimes broad-leaved trees.
Distribution: Widespread. Rather common.

An example of the complex group of *Inocybe* with pruina at the stipe apex and pink or reddish colours.

---

p.159 *Inocybe splendentoides* M. Bon
Cortinariaceae, Cortinariales

Description: *Cap red-brown to orange-chestnut, veil present. Stipe bulbous.*
Cap 5–10 cm (2–4 in), conico-convex then with a more or less domed umbo, with veil remains or frosted in the centre, fibrillose-innately streaky to rimose, pale at first but soon red-brown to orange-tawny then brownish to rather tawny towards the margin. *Gills* notched, pale beige, then rather light rust-brown. *Stipe* 5–10 × 0.5–1.5 cm (2–4 × 0.2–0.6 in), with a rounded bulb at the base, entirely powdery pruinose, white but soon becoming orange rufous to brownish from the middle. *Flesh* white to pale buff. Smell faint, slightly fruity to honeyish. *Microscopic features* Sp 8–12 × 5–7 μm, cylindro-conical or almond-shaped, the apex conical or elongated. Cyst 55–75 × 15–22 μm, fusiform or ventricose, the wall 1–2 μm diameter, yellow in ammonia. MC clavate or septate. CC a cutis. Caulo all along the stipe.

Season: October–December.
Habitat: Broad-leaved coppice woodland on sandy soils.
Distribution: Not fully assessed. Fixed coastal dunes.

The cap of *I. splendens* Heim is fibrillose-innately streaked with reddish brown or vinaceous brown over a warm-buff ground, the stipe is pure white and the gills pinkish buff. It occurs in broad-leaved woods and on grassy wood edges, in more inland sites.

---

p.153 *Inocybe vulpinella* Bruylants
Cortinariaceae, Cortinariales

Description: *Cap fibrillose or bristling, rufous brown. Stipe bulbous, entirely pruinose.*
Cap 2–4 cm (0.8–1.6 in), campanulate then convex to slightly umbonate, with erect scales in the centre, almost smooth or frosted with veil remains, with small fibrillose or woolly scales near the margin, rufous brown to rufous buff. *Gills* notched, rather distant, rufous then rather dark rufous brown. *Stipe* 2–6 × 0.4–1 cm (0.8–2.4 × 0.16–0.4 in), clavate to bulbous, not marginate, white or flushed rufous to rufous brown from the middle. *Flesh* white or rufous cream. Smell rather weak, spermatic. *Microscopic features* Sp 11–16 × 6–9 µm, elliptic to cylindro-elliptic, the apex bluntly rounded, the wall quite thick. Cyst 55–70 × 15–20 µm, fusiform or cylindro-ventricose, the wall 2–5 µm diameter thick, yellowish in ammonia. MC cylindro-clavate. CC a trichoderm. Caulo the entire length of the stipe.
Season: May–December.
Habitat: Damp sandy or gravelly places, with willow, buckthorn, etc.
Distribution: Uneven. Quite rare or rare.

---

p.117 *Inocybe kuehneri* Stangl. & Veselsky
Cortinariaceae, Cortinariales

Description: *Cap pale beige, very finely scaly. Stipe whitish, entirely pruinose.*

Cap 1.5–4 cm (0.6–1.6 in), conico-convex then obtuse or slightly umbonate, more or less disrupted, with an almost smooth, ochraceous beige central disc, scaly to finely fibrillosely scaly and whitish elsewhere. *Gills* notched, quite close, light greyish beige then brownish. *Stipe* 3–7 × 0.3–0.6 cm (1.2–2.8 × 0.12–0.24 in), often wavy, entirely pruinose, whitish to pale buff. *Flesh* whitish. Smell spermatic. *Microscopic features* Sp 7–10 × 4.5–5.5 µm, almond-shaped or ovoid, the apex obtuse. Cyst 40–80 × 8–15 µm, cylindro-fusiform, the wall 2–3 µm diameter, pale yellow in ammonia. MC clavate, numerous. CC a cutis. Pigment mixed, very pale. Caulo the entire length of the stipe.

Season: August–November.
Habitat: Broad-leaved and coniferous trees.
Distribution: Widespread but frequency uneven.

A fairly common taxon, relatively easy to identify from the pale colours frequently occurring on the central disc and pruina, extending right to the base of the stipe.

---

p.154 *Inocybe pholiotinoides* Romagnesi
Cortinariaceae, Cortinariales

Description: *Cap rufous tawny, with small fibrillose scales. Stipe with a small bulb.*
Cap 0.5–1 cm (0.2–0.4 in), conico-convex, finally expanded subumbonate, finely fibrillosely scaly to more or less squamulose around the central disc, hygrophanous, yellow-brown to rufous tawny, ochraceous cream or yellowish when dry. *Gills* notched, not very close, rather tawny then rust. *Stipe* 2–3.5 × 0.15–0.3 cm (0.8–1.4 × 0.06–0.12 in), more or less bulbous, sometimes wavy, white then honey to rufous from the middle, almost entirely pruinose. *Flesh* whitish to yellowish honey near the surface. Smell spermatic. *Microscopic features* Sp 8–11.5 × 4.5–6.5 µm, elliptic or almond-shaped, the ventral face often less domed, the apex conical. Cyst

55–95 × 5–17 µm, fusoid-lageniform, with a narrow neck, the wall 2–3.5 µm diameter, rather bright yellow in ammonia. Cyst on the gill edge more squat and MC clavate or subspherical. CC a cutis. Caulo very low on the stipe, more irregular than the Pleuro.

Season: August–November.
Habitat: Damp places, ditches, ruts, wet or clay soils under broad-leaved trees.
Distribution: Rare or overlooked.

---

p.159 *Inocybe tenebrosa* Quélet (= *I. atripes*)
Cortinariaceae, Cortinariales

Description: *Cap dark olivaceous brown in the centre. Stipe blackening from the base.*
Cap 2–5 cm (0.8–2 in), conico-convex then expanded, more or less umbonate, fibrillose, more or less squamulose around the umbo which may separate into a distinct central disc, covered with greyish veil; brown or with olivaceous tones, rufous to yellowish buff at the edge. *Gills* notched, pale yellowish grey then rust brown.
Stipe 2.5–5 × 0.3–0.8 cm (1–2 × 0.12–0.32 in), whitish at the apex, browning then blackening from the base, entirely pruinose. *Flesh* white to almost concolorous near the surface, brown at the stipe base. Smell spermatic with a strong banana component.
Microscopic features Sp 8.5–10.5 × 4.5–6 µm, ovoid or almond-shaped, the apex rounded and acute or elongated umbonate. Cyst 40–60 × 10–15 µm, short and clavate to cylindro-clavate, the wall 2–3 µm diameter, yellow in ammonia. Cyst on the gill edge sometimes more utriform and MC clavate or septate. CC a cutis. Caulo the whole length of the stipe.

Season: August–November.
Habitat: Broad-leaved trees, on rather damp, rich soils; tends to occur on warm sites.
Distribution: Uneven. Rare.

One of the more identifiable *Inocybe* spp. on account of its colours, the browning stipe and the smell of banana.

p.154 *Inocybe hirtella* Bresadola
Cortinariaceae, Cortinariales

Description: *Cap yellow buff. Stipe white, pinkish above, powdery. Smell of bitter almonds.*
Cap 1–2.5 cm (0.4–1 in), conico-convex then expanded, not very umbonate, radially fibrillose then with small fibrillose scales, yellow or yellowish buff.
Gills notched, quite close, whitish then brownish beige or slightly rust.
Stipe 2–4.5 × 0.3–0.5 cm (0.8–1.8 × 0.12–0.2 in), white then pinkish to rather bright pink at the apex, entirely pruinose. Underside of the base sometimes flat.
Flesh white or pinkish at the stipe apex. Smell strong, of bitter almonds. *Microscopic features* Sp 8–15 × 5–8 µm, almond-shaped to cylindro-fusoid, the apex rounded and acute or elongated. Bas 2- or 4-spored. Cyst 40–60 × 12–20 µm, fusoid or clavate, sometimes with a short neck, the wall 1.5–3 µm diameter, not reacting with ammonia. MC clavate. CC a cutis. Caulo the whole length of the stipe.
Season: July–December.
Habitat: Broad-leaved trees; tends to occur on damp or calcareous soils.
Distribution: Widespread but frequency uneven.

Another species with good field characters (smell, rather yellow cap and stipe apex often flushed pink).

---

p.156 *Inocybe acutella* M. Bon
Cortinariaceae, Cortinariales

Description: *Cap with a sharp umbo, dark brown. Stipe bruising brown.*
Cap 1–3 cm (0.4–1.2 in), conical then expanded but with a conspicuous rather sharp umbo, rather finely fibrillose-innately streaky, dark brown on the umbo, more chestnut towards the margin. *Gills* almost free, brownish to rust brown. *Stipe* 2–5 × 0.2–0.4 cm (0.8–2 × 0.08–0.16 in), with a faint cortina and fibrillose, pale at the apex, progressively brownish then brown at the

base, which is clavate or has a small emarginate bulb, and remains white for longer. *Flesh* pale to brownish grey. Smell almost absent. *Microscopic features* Sp 8.5–11 × 6.5–7.5 µm, rather elongated, with 8–12 low projections. Cyst 45–65 × 10–25 µm, utriform, ventricose with a short neck, the wall 0.5–1.5 µm diameter, not reacting with ammonia. Cyst on the gill edge fairly rare, shorter and squat and MC clavate. CC a cutis. Caulo absent.

Season: July–November.
Habitat: Very damp places, muddy willow carr.
Distribution: Not fully assessed. Rare or very rare.

The last subgenus, *Clypeus* has spores which are more or less numerous, sharp projections (angular spores). *I. acuta* Boudier is taller, with a sharp smell, and found under conifers in marshy places. *I. umbrina* Bresadola is smaller, with a low umbo on the cap, a stipe which is concolorous with the cap, except for the white, onion-like, bulbous base and is found on drier soils.

---

p.156 *Inocybe curvipes* P. Karsten
Cortinariaceae, Cortinariales

Description: *Cap with small fibrillose scales. Stipe fibrillose, silvery.*
Cap 2–4 cm (0.8–1.6 in), conico-convex then expanded with a more or less conspicuous umbo, fibrillose-innately streaky but with small fibrillose scales around the umbo, looking rather like a *Lepiota*, brownish beige to rufous tawny, sometimes more greyish beige.
Gills notched, whitish then brownish beige, remaining rather pale. *Stipe* 2–6 × 0.3–0.6 cm (0.8–2.4 × 0.12–0.24 in), with a cortina then fibrillose-silky, slightly silvery, bruising brown at the base and sometimes slightly violaceous at the apex. *Flesh* pale. Smell spermatic. *Microscopic features* Sp 9–12 × 6.5–8.5 µm, long, angular with rather

numerous and not very conspicuous projections. Cyst 40–70 × 15–25 μm, broadly clavate or bladder-like, with an umbo at the apex (the pleurocystidia are more fusoid-clavate but always have this umbonate apex), the wall 1–2 μm diameter, not reacting with ammonia. Cyst on the gill edge short and MC clavate. CC a cutis. Caulo absent.

Season: August–November.

Habitat: Broad-leaved trees, wood edges, open grassy clearings; often on damp soils.

Distribution: Widespread but frequency uneven.

---

p.160 *Inocybe lanuginosa* (Bull.:Fr.) Kummer
Cortinariaceae, Cortinariales

Description: *Cap obtuse, woolly to shaggy, brown. Stipe with small fibrillose scales.*
Cap 2–5 cm (0.8–2 in), hemispherical then shallowly convex or flat, with woolly to shaggy tufts or small scales, brownish or brown, often rather dull or sometimes more rufous. *Gills* narrowly adnate, whitish then brownish beige to dark brown. *Stipe* 4–8 × 0.4–0.8 cm (1.6–3.2 × 0.16–0.32 in), with a cortina, fibrillose below or with the same cuticle as the cap, concolorous. *Flesh* whitish then ochraceous beige to brownish from the base. Smell slightly spermatic to sharp. *Microscopic features* Sp 8–12 × 6–8 μm, elliptic in outline, with numerous medium-sized, blunt projections. Cyst 50–80 × 10–25 μm, cylindro-fusoid, cylindro-clavate, clavate to bladder-like ovoid, the wall about 1 μm thick, not reacting with ammonia. Cyst on the gill edge variable and MC cylindric or clavate, sometimes septate. CC a trichoderm. Caulo absent.

Season: July–November.

Habitat: Conifers or broad-leaved trees; marshy places.

Distribution: Widespread but frequency uneven.

Several varieties have been described, based in part on the shape of the cystidia.

p.152 *Inocybe calospora* Quélet
Cortinariaceae, Cortinariales

Description: *Cap with small fibrillose to bristling scales, rufous-brown. Stipe reddish.*
Cap 0.5–2 cm (0.2–0.8 in), conico-convex then shallowly convex, with fine small up-turned scales, sometimes rather fibrillose at the margin, dark rufous brown over a sometimes lighter ground. *Gills* notched, rather distant, brownish beige then rather dark rust brown. *Stipe* 1.5–3.5 × 0.1–0.3 cm (0.6–1.4 × 0.04–0.12 in), slender, wavy, reddish with orange tones, orange-buff, red, reddish brown, entirely pruinose. *Flesh* concolorous with the surfaces or paler. Smell faintly spermatic. *Microscopic features* Sp 8–15 μm, subglobose or broadly elliptic, with slender, obtuse, straight or slightly curved spines, up to 3–5 × 1 μm. Cyst 40–55 × 10–15 μm, cylindro-lageniform or clavate, the wall 2–3 μm diameter, yellowish to yellow in ammonia. Cyst on the gill edge and MC clavate. CC a trichoderm. Caulo the whole length of the stipe, more variable than the cystidia and sometimes longer.

Season: July–November.
Habitat: Broad-leaved trees usually on damp soils.
Distribution: Quite widespread. Rather rare.

---

p.120 *Inocybe petiginosa* (Fr.:Fr.) Gillet
Cortinariaceae, Cortinariales

Description: *Cap brownish with a frosted zone. Gills mustard yellow.*
Cap 0.5–1 cm (0.2–0.4 in), hemispherical then domed but not very umbonate, with whitish or greyish frosting around the disc which is smooth or very finely fibrillosely scaly, brownish to pinkish brown.
Gills almost free, yellowish then mustard yellow to rather pale rust brown.
Stipe 1–3 × 0.1–0.2 cm (0.4–1.2 × 0.04–0.08 in), slender, light reddish brown to pale orange-buff, entirely pruinose. *Flesh* almost concolorous with the surfaces. Smell spermatic, rather faint. *Microscopic features*

Sp 5.5–8 × 4.7 µm, small, rather broadly elliptic, with numerous small not very distinct projections. Cyst 40–55 × 10–15 µm, fusiform or lageniform with a more or less conical neck, the wall 2–3 µm diameter, yellow in ammonia. Cyst on the gill edge variable and MC clavate. CC a cutis under a coating (veil) of thick-walled refractive hyphae. Caulo the whole length of the stipe.

Season: August–November.
Habitat: Broad-leaved trees, often on damp, rather heavy, usually calcareous soils.
Distribution: Widespread but frequency uneven.

Easy to identify in the field from its size, concentric frosting and mustard yellow gills. Very characteristic under the microscope.

---

p.157 *Inocybe praetervisa* Quélet
Cortinariaceae, Cortinariales

Description: *Cap yellow-buff to yellow-brown. Stipe white, pruinose, bulbous.*
Cap 2–5 cm (0.8–2 in), remaining conical for a long time then expanded with a persistent central umbo, almost smooth then fibrillose-innately streaky to rimose, yellow to yellow-buff or yellow-brown. *Gills* almost free, greyish then beige with violaceous tones, finally rusty brownish. *Stipe* 4–7 × 0.5–1 cm (1.6–2.8 × 0.2–0.4 in), with a marginate bulb, white, entirely pruinose. *Flesh* white except near the surface. Smell spermatic. *Microscopic features* Sp 9.5–12.5 × 6.5–9.5 µm, rather elongated, with 10–12 conspicuous, obtuse projections. Cyst 60–80 × 15–20 µm, bottle-shaped fusoid, the base more or less ventricose, the wall about 2 µm diameter, not reacting with ammonia. Cyst on the gill edge variable and MC clavate. CC a cutis. Caulo the whole length of the stipe, often shorter than the cystidia.

Season: August–November.
Habitat: Broad-leaved trees, more rarely under conifers.
Distribution: Widespread. Rather common.

I. mixtilis (Britz) Saccardo is a much smaller version, with a cap up to 2–3 cm (0.8–1.2 in), usually found under conifers but also under broad-leaved trees.

I. margaritispora (Berk) Saccardo (found under broad-leaved trees) is similar to *I. praetervisa* but the cuticle has small appressed buff-brown scales on a yellowish buff ground and the gills do not have violaceous tones.

---

p.156 *Inocybe salicis* Kühner
Cortinariaceae, Cortinariales

Description: *Cap tawny brown. Stipe white or yellowish, bulbous, powdery.*
Cap 1–2.5 cm (0.4–1 in), conico-campanulate then expanded with a broad umbo, finely streaky rimose to fibrillosely or woolly scaly, tawny brown to yellow-brown. *Gills* almost free, pale grey then ochraceous grey to brownish. *Stipe* 1.5–3.5 × 0.4–1 cm (0.6–1.4 × 0.16–0.4 in), with a marginate bulb, white or tinged yellowish honey, entirely pruinose. *Flesh* white or almost concolorous near the surface. Smell spermatic, not distinctive. *Microscopic features* Sp 9–11 × 6–8 µm, with a rather elongated silhouette or sometimes curved, with 10–15 short blunt projections with a very characteristic semi-circular outline. Cyst 50–70 × 12–25 µm, like a swollen bottle or fusoid, the wall 2–3 µm diameter, yellowish in ammonia. Cyst on the gill edge variable and MC clavate or balloon-shaped. CC a cutis. Caulo the whole length of the stipe.
Season: August–December.
Habitat: Damp willow carr and clearings.
Distribution: Widespread. Not very common.

---

p.157 *Inocybe asterospora* Quélet
Cortinariaceae, Cortinariales

Description: *Cap very rimose, brown. Stipe orange-brown, bulbous, pruinose.*
Cap 1.5–7 cm (0.6–2.8 in), conico-convex then shallowly conical or with a persistent

umbo, brown, quite often with whitish or greyish veil in the centre, streaky rimose at the margin, forming disconnected radial stripes when mature. *Gills* notched, greyish beige then brownish grey to brown. *Stipe* 4–9 × 0.4–1.5 cm (1.4–3.5 × 0.16–0.6 in), with a strongly marginate bulb, entirely pruinose, orange rufous to warm brownish buff. *Flesh* brownish or pale. Smell strongly spermatic. *Microscopic features* Sp 8–12 × 6–10 µm, with 10–12 very conspicuous, conical star-like projections. Cyst 40–80 × 15–25 µm, bottle-shaped or fusoid, the base more or less ventricose, the wall 1–3 µm diameter, especially thick at the apex, not reacting with ammonia. Cyst on the gill edge variable and MC clavate. CC a cutis. Caulo the whole length of the stipe.

Season: July–December.
Habitat: Rather ubiquitous.
Distribution: Widespread. Common.

An example of the last section in *Inocybe*, whose species are characterized by an entirely pruinose but brightly or darkly coloured stipe with a marginate bulb. Many species have been described, varying in size, colour and habitat.

CREPIDOTACEAE

Crepidotaceae are placed in the Cortinariales. Species have the following characters: more or less fleshy or slender to very slender, stipe sometimes absent; spore print brown (yellowish buff, brownish, rust, etc.); veil present (cortina, occasionally a sheath and, more rarely, a ring, may also be present) or absent; cap cuticle not or only slightly differentiated (never a hymeniderm).

Basically, *Crepidotaceae* differ from the *Cortinariaceae* in their biology since the species are saprophytic. This is a rather heterogeneous family and only a few characters are common to all species: stipe cylindric, with parallel sides; gill edge frosted or eroded, often paler than the faces; smell not distinctive; spores lacking a germ pore; basidia 4-spored; pleurocystidia

p.85 *Crepidotus calolepis* (Fr.) P. Karsten
Crepidotaceae, Cortinariales

Description: *Pleurotoid. Cap a dingy brownish grey with rufous scales and a gelatinous layer.*
Cap 2–10 cm (0.8–4 in), domed, hygrophanous, striate at the margin, with fibrillose rufous or tawny scales, over a separable, elastic, gelatinous, brown to dull brownish grey layer. *Gills* rather close but soft and becoming more distant when old, sepia to rather dull rust brown. *Stipe* almost absent or very soon disappearing. *Flesh* rather thin, gelatinous. *Microscopic features* Sp 7–11 × 5–7.5 µm, elliptic or almond-shaped, slightly asymmetrical, smooth. Bas 20–35 × 5–10 µm, clavate. Cheilo 30–80 × 5–20 µm, cylindro-lageniform, sometimes wavy or deformed. CC an ixocutis, with trichodermial zones in the squamules, the hyphae 3–15 µm diameter. Clamps absent.
Season: July–December.
Habitat: Trunks or fallen branches of broad-leaved trees.
Distribution: Warm or Mediterranean areas.

C. mollis (Sch.:Fr.) Staude, which is very much more common everywhere, has no scales on the cap or is sometimes merely fibrillose. The gelatinous layer, which can be pinched up and completely pulled off, is the main character for this species, which is also one of the largest in the genus.

p.85 *Crepidotus variabilis* (Pers.:Fr.) Kummer
Crepidotaceae, Cortinariales

Description: *Pleurotoid. Cap white or whitish. Gills pinkish beige.*
Cap 0.5–2 cm (0.2–0.8 in), domed to almost flat, pubescent to tomentose near

the point of attachment, white or faintly buff to pale beige when old. Margin even or wavy to lobed. *Gills* rather close, pale cream then rather pale pinkish beige. *Stipe* reduced to a small whitish knob. *Microscopic features* Sp 6–7.5 × 2.5–4 µm, narrowly cylindro-elliptic, very finely spiny, pale. Bas 15–25 × 4–7 µm, cylindro-clavate. Cheilo 20–40 × 5–15 µm, irregular or variously branched at the apex. CC a trichoderm, with hyphae 2–5 µm diameter.

Season: Practically all through the year.
Habitat: Fallen or standing dead wood.
Distribution: Widespread. Common.

This species, which is frequent in mature woodland or large blocks of woodland, is often replaced, in coppice woodland for example, by the almost identical species *C. cesatii* (Rab.) Sacc., which has rather large (6–10 × 5–8 µm) very finely spiny subglobose spores.

---

p.121 *Flammulaster carpophilus* var. *autochtonoides* M. Bon
Crepidotaceae, Cortinariales

Description: *Cap yellowish to buff, granular. Gills yellowish.*
Cap 0.5–1 cm (0.2–0.4 in), hemispherical then faintly umbonate or slightly domed, finely granular or scurfy to micaceous velvety, pale, yellowish buff.
Gills notched, distant, yellowish cream then slightly more rust. *Stipe* 1–3 × 0.1–0.2 cm (0.4–1.2 × 0.04–0.8 in), wavy, concolorous or slightly paler. Veil absent or insignificant. *Flesh* thin, pale. Smells rather distinctly of *Pelargonium* (florist's geranium). *Microscopic features* Sp 6–10 × 4–6 µm, almond-shaped or ovoid, smooth, pale. Bas 15–25 × 4–7 µm, cylindro-clavate. Cheilo 40–50 × 5–7 × 1–2 µm, lageniform, the neck cylindric, sometimes wavy or slightly inflated at the apex. CC a trichoderm or almost subcellular, with chains of elliptic or subglobose cells 20–50 × 15–30 µm. Pigment almost smooth.

Season: September–December.
Habitat: Scrub on sand dunes.
Distribution: A rare or little known taxon, restricted to a few coastal habitats in France.

The type species *F. carpophilus* (Fr.) Earle, which is quite frequent, has a less specific habitat, on humus and debris, in woods.

---

p.121 *Flammulaster ferrugineus* (Maire) Watling
Crepidotaceae, Cortinariales

Description: *Cap rust-coloured, granular. Gills rust.*
Cap 0.5–3 cm (0.2–1.2 in), hemispherical then almost flat, more or less striate, granular to scurfy, rusty brown to tawny, sometimes paler towards the margin. *Gills* rather broadly adnate, not very close, tawny beige then rust. *Stipe* 1–3.5 × 0.2–0.3 cm (0.4–1.4 × 0.08–0.12 in), cylindric and curved, concolorous or slightly paler at the apex, scurfy below. Ring zone rather fleeting. *Flesh* rusty fawn. *Microscopic features* Sp 6–9.5 × 4–5.5 µm, elliptic or almond-shaped, smooth, rather dark. Bas 15–25 × 4–8 µm, cylindro-clavate. Cheilo 20–60 × 5–10 × 3–6 µm, lageniform, the neck cylindric, inflated to subcapitate. CC a trichoderm or almost subcellular, with chains of elliptic or cylindric cells 15–35 × 8–15 µm. Pigment in pronounced irregular bands.
Season: June–November.
Habitat: In woods, on bare damp ground.
Distribution: Widespread. Quite common.

---

p.127 *Galerina heterocystis* (Atk.) Sm. & Sing.
Crepidotaceae, Cortinariales

Description: *Cap hygrophanous, tawny to beige, striate. Stipe pale honey.*
Cap 1–3 cm (0.4–1.2 in), conico-convex then convex or umbonate, hygrophanous, with long striations, rather rufous tawny soon beige to pale yellowish buff when dry. *Gills* adnate, yellowish honey then

slightly more rust but remaining pale. *Stipe* 2.5–8 × 0.1–0.3 cm (1–3.2 × 0.04–0.12 in), very fragile, concolorous or paler, the base slightly fibrillose, whitish. Veil almost absent. *Microscopic features* Sp 11–16 × 5.5–9 µm, elliptic or almond-shaped, finely punctate, pale. Bas 25–35 × 8–14 µm, clavate. Cheilo 40–70 × 6–15 × 2–4 × 4–10 µm, lageniform, the neck cylindric, inflated to capitate, sometimes almost skittle-shaped. CC a cutis, with hyphae 5–15 µm diameter and with a few lageniform-lecithiform pileocystidia. Clamps absent.

Season: June–December.
Habitat: Damp moss.
Distribution: Widespread but frequency uneven.

A very difficult genus in the field.

---

p.127 *Galerina laevis* (Pers.) Singer
Crepidotaceae, Cortinariales

Description: *Cap hygrophanous, striate, honey-yellow. Gills distant. Stipe pale.*
Cap 0.5–1.5 cm (0.2–0.6 in), conico-convex then expanded and slightly domed, hygrophanous, with long striations, yellowish honey then almost whitish when dry. *Gills* ventricose, very pale yellowish honey then with slightly stronger colours. *Stipe* 2–4 × 0.1–0.2 cm (0.8–1.6 × 0.04–0.08 in), fragile, coloured or almost white at the base. *Microscopic features* Sp 7–10 × 4–5.5 µm, elliptic or almond-shaped, almost smooth, pale. Bas 25–40 × 5–8 µm, clavate. Cheilo 20–45 × 6–12 × 2–3 × 4–6 µm, lageniform, the neck cylindric, inflated to capitate. CC a cutis, with hyphae 3–5 µm diameter, pileocystidia very rare. Clamps absent.

Season: Practically all through the year.
Habitat: Lawns and grassy places.
Distribution: Widespread. Common.

The pale colours and habitat make this a fairly easy taxon to identify.

p.126 *Galerina paludosa* (Fr.) Kühner
Crepidotaceae, Cortinariales

Description: *Cap rusty buff. Stipe with abundant white veil remains.*
Cap 1–4 cm (0.4–1.6 in), conico-convex then convex or umbonate, hygrophanous, with long striations, reddish brown then rather rufous tawny or pale chestnut when dry, the margin with fibrillose veil when young. *Gills* ventricose or sinuate, not very close, almost concolorous with the cap. *Stipe* 5–15 × 0.2–0.5 cm (2–6 × 0.08–0.2 in), very tall and fragile, concolorous or more brownish to reddish buff at the apex. Garlanded with abundant fibrils of white veil, from the base up to a rather fragile almost ring-like zone. *Flesh* quite dark. Smell mealy. *Microscopic features* Sp 8–13 × 5–7 µm, almond-shaped, the apex rounded and acute, weakly verrucose or roughened to mottled, dark, the SP quite distinct. Bas 25–35 × 8–12 µm, clavate. Cheilo 25–50 × 8–15 × 4–6 × 6–10 µm, lageniform, the neck cylindric, subcapitate. CC with hyphae 10–15 µm diameter.
Season: Practically all year round.
Habitat: Sphagnum in particular, also damp moss in peat bogs.
Distribution: Uneven. Rather rare.

One of the easier *Galerina* spp. to recognize because of its habitat and the abundant veil on the stipe.

---

p.150 *Galerina marginata* (Batsch) Kühner
(deadly poisonosus)
Crepidotaceae, Cortinariales

Description: *Cap hygrophanous, rufous brown to yellowish honey. Stipe dark brown to blackish below. Ring appressed.*
Cap 1–7 cm (0.4–2.8 in), hemispherical then broadly convex or almost flat, hygrophanous, slightly striate, often shining, rufous brown to rather bright or dark rufous tawny, fading from the centre to ochraceous beige or yellowish honey. *Gills* adnate,

rather close, ochraceous honey then slightly more rust. *Stipe* 2–7 × 0.2–0.7 cm (0.8–2.8 × 0.08–0.28 in), fragile, cylindric or curved, buff at the apex, brown to dark brown or even blackish at the base, pruinose under the gills, fibrillose below. Veil ring-like, membranous, greyish or pale brownish beige, soon appressed. *Flesh* buff to concolorous. Smell and especially taste mealy (spit it out!). *Microscopic features* Sp 8–11 × 5–6.5 µm, ovoid or almond-shaped, warted, the episporium slightly detached near the smooth and very visible plage. Sp Bas 25–35 × 5–10 µm, cylindro-clavate. Cheilo and Pleuro abundant, identical, 30–80 × 10–16 × 3–8 µm, stipitate lageniform, the neck cylindric obtuse. CC a cutis, with hyphae 3–6 µm diameter.

Season: June–December.

Habitat: Gregarious (not in clusters). On stumps, trunks or fallen branches of coniferous trees, but also on broad-leaved trees.

Distribution: Widespread but frequency uneven.

Contains the same toxins as the death cap group. The danger arises from the great similarity of this species to *Kuehneromyces mutabilis* (*see* p.789) which is edible. The nearest related species is the equally deadly *G. autumnalis* (Peck) Smith & Singer, which can be distinguished by the greasy, lubricious cuticle.

---

p.121 *Galerina subfusispora* (Moeller) M. Bon
Crepidotaceae, Cortinariales

Description: *Cap brownish honey, scurfy. Margin with fleecy fibrils. Veil fibrillose, silvery.*
Cap 0.5–1.5 cm (0.2–0.6 in), hemispherical then almost flat, hygrophanous, not striate, with fleecy fibrils or small scales near the centre, pale brownish honey; centre darker and margin with rather pronounced veil remains. *Gills* adnate, distant, yellowish ochraceous cream then almost concolorous with the cap. *Stipe* 1–3 × 0.1–0.3 cm (0.4–1.2 × 0.04–0.12 in), pale honey at the apex, concolorous below or more reddish.

Veil fibrillose, whitish silvery, especially towards the base. *Microscopic features* Sp 9.5–16 × 4–6.5 μm, fusiform or long almond-shaped, sometimes with a suprahilar depression, the apex with a more or less distinct GP, smooth. Bas 25–30 × 7–10 μm, clavate. Cheilo 20–35 × 5–12 × 4–8 μm, lageniform or ventricose, tapering to a blunt top. CC a cutis, with hyphae 3–10 μm diameter.

Season: July–September.
Habitat: On peat, in the mountains.
Distribution: Alps and Northern Europe.

---

p.157 *Gymnopilus spectabilis* (Weinm.:Fr.) Sm.
Crepidotaceae, Cortinariales

Description: *Cap fibrillose, rufous orange. Gills rust. Stipe with a ring.*
Cap 3–20 cm (1.2–8 in), hemispherical then sometimes umbonate, fibrillose or with small fibrillose scales, dry, rufous orange to bright orange-tawny, slightly darker or dull when old. *Gills* broadly adnate, rather close, yellow-buff then rust to bright rust. *Flesh* yellowish to rather tawny, bitter. *Stipe* 4–20 × 0.5–5 cm (1.6–8 × 0.2–2 in), cylindric or fusoid, often clavate near the host, fibrillose, concolorous or paler under the gills and almost whitish at the very base. Ring ascending, membranous, pale then rust, rather fragile when old.
Microscopic features Sp 8–11 × 4.5–6.5 μm, elliptic to ovoid or slightly almond-shaped, warted, dextrinoid. Bas 20–35 × 6–9 μm, clavate. Cheilo 25–45 × 6–10 × 3–5 μm, lageniform with a cylindric neck, more or less capitate. Pleuro rare, 20–35 × 6–7 μm, lageniform to utriform. CC a cutis, the scales trichodermial, with hyphae 5–10 μm diameter.

Season: August–December.
Habitat: In clusters or gregarious, on broad-leaved trees, sometimes also on conifers.
Distribution: Widespread but frequency uneven.
Not edible. The only *Gymnopilus* in Europe with a ring; it resembles *Phaeolepiota aurea* (*see* p.679).

p.152 *Gymnopilus penetrans* (Fr.) Murrill
Crepidotaceae, Cortinariales

Description: *Cap rather tawny. Gills with rusty spots. Cortina fleeting.*
Cap 2–6 cm (0.8–2.4 in), soon shallowly convex to almost flat, rather rufous yellow then brighter or speckled, with rufous orange to brownish rufous zones in the centre. *Gills* adnate, rather narrow, pale yellow then with bright rust speckles when mature. *Flesh* bitter, white to pale yellowish. *Stipe* 2–6 × 0.3–0.7 cm (0.8–2.4 × 0.12–0.28 in), hollow, cylindric or curved, fibrillose, pale yellowish, spotted brownish to dingy rust from the base when old. Veil a cortina, fleeting, with a few whitish fibrils near the stipe base.
Microscopic features Sp 6.5–9.5 × 4–5.5 µm, elliptic to ovoid or slightly almond-shaped, rather coarsely warted, dextrinoid. Bas 25–30 × 6–7 µm, clavate. Cheilo 25–50 × 5–10 × 3–5 µm, fusoid-lageniform, the neck long cylindric, wavy, obtuse or more or less capitate. Pleuro quite rare, 20–35 × 4–8 µm, lageniform or fusiform ventricose. CC a cutis, with trichodermial tufts, the hyphae 4–8 µm diameter.
Season: Practically all year round.
Habitat: Trunks, fallen branches and twigs (conifers and also broad-leaved trees).
Distribution: Widespread. Common or quite common.

G. hybridus (Sow.:Fr.) Maire, a very similar species, has gills which do not spot rusty and remain pale yellow or rather rufous, while the white or silvery cortina is more persistent, forming a pale ring. In *G. sapineus* (Fr.:Fr.) Maire, the flesh is yellower in the middle.

p.126 *Phaeocollybia arduennensis* M. Bon
Crepidotaceae, Cortinariales

Description: *Cap dark brown, pointed. Stipe rooting, concolorous.*
Cap 2–6 cm (0.8–2.4 in), conical, with a sharp umbo, hygrophanous, fibrillose to

slightly silky, with fine radial wrinkles when old, slimy, dark chestnut to reddish brown or crimson-brown, paler towards the margin when dry. *Gills* narrowly adnate, rusty buff, rather dark when mature. *Flesh* thin. Taste slightly or slowly bitter. *Stipe* 5–10 × 0.2–0.4 cm (2–4 × 0.08–0.16 in), rather fragile, rooting, concolorous with the cap but paler below. *Microscopic features* Sp 5–6.5 × 3–4 µm, elliptic or almond-shaped, the apex slightly acute, with rather fine warts. Cheilo 20–40 × 3–5 × 5–8 µm, lageniform, the neck cylindric, more or less irregular or subcapitate. CC a cutis, with hyphae 2–5 µm diameter. Pigment mixed.

Season: September–November.
Habitat: Damp conifer woods, sometimes under mixed broad-leaved trees.
Distribution: Mainly northern.

The genus is well characterized by its pointed cap and strongly rooting stipe.

---

p.157 *Phaeocollybia lugubris* (Fr.:Fr.) Heim
Crepidotaceae, Cortinariales

Description: *Cap rather rufous tawny, umbonate. Stipe rooting, vinaceous brown below.*
Cap 3–7 cm (1.2–2.8 in), conico-convex then more expanded, with a rather blunt umbo, lubricious or shining, silky fibrillose when old, brownish tawny to rather rufous, often flecked orange-brown or reddish. Margin sometimes lobed. *Gills* narrowly adnate, rather bright rusty buff then more rust or dark. *Flesh* average, white. Smell of radish or honey. Taste slightly or slowly bitter, unpleasant. *Stipe* 6–12 × 0.5–1.5 cm (2.4–4.7 × 0.2–0.6 in), hollow, deeply rooting, smooth, often paler under the gills and reddish brown towards the base. *Microscopic features* Sp 7–9 × 4.5–5.5 µm, almond-shaped, the apex slightly elongated, rather coarsely warted, the SP smooth. Cheilo 25–50 × 5–10 × 2–5 µm, lageniform, the neck

cylindric, sometimes filiform or
subcapitate to capitulate, even forked.
CC a cutis, with hyphae 2–4 µm
diameter. Pigment mixed.
Clamps absent.

Season: August–November.
Habitat: Conifers, especially spruce, on damp, even peaty ground.
Distribution: Uneven. Rather rare to rare.

P. christinae (Fr.) Heim., a smaller species, is often brightly coloured. It occurs in the same habitats and smells more distinctly of honey.

---

p.147 *Phaeomarasmius erinaceus* (Fr.:Fr.) Kühn.
Crepidotaceae, Cortinariales

Description: *Cap with small bristling fibrillose scales, brown. Stipe spiky.*
Cap 0.5–1.5 cm (0.2–0.6 in), globose then hemispherical, rarely expanded, finely fibrillosely scaly and even shaggy, chestnut to rusty brown. *Gills* broadly adnate, rather distant, rather tawny-buff then more rust-coloured. *Flesh* yellow-buff to red-brown at the base.
Stipe 5–15 × 1–3 mm (0.2–0.6 × 0.04–0.12 in), rather fragile, sometimes with a small bulb, often curved, spiky like the cap beneath a rather high, clearly defined, concolorous zone, paler at the apex. *Microscopic features* Sp 9–14 × 6–10 µm, ovoid to broadly elliptic or slightly lozenge-shaped, smooth. Bas 2-spored (rarely 4-spored), 25–35 × 5–7.5 µm. Cheilo 25–50 × 1–6 µm, lageniform, the neck narrow subcapitate. CC a trichoderm, with septate hyphae, 400–700 × 6–8 µm, very thick-walled, brown.

Season: May–November.
Habitat: Branches of broad-leaved trees (willow), in scrub and coppice on damp or marshy ground.
Distribution: Uneven. Rather rare to rare.

Very characteristic. The genus is close to *Flammulaster*.

p.148 *Ramicola centunculus* (Fr.:Fr.) Watling
(= *Simocybe c.*)
Crepidotaceae, Cortinariales

Description: *Cap convex obtuse, olivaceous brownish. Gills concolorous.*
Cap 1–3 cm (0.4–1.2 in), hemispherical, then regularly convex or flat, hygrophanous, striate, glabrous, often finely wrinkled or sometimes cracked when old, olivaceous brownish, quite dark when fresh then paler when dry. *Gills* adnate, not very close, olivaceous cream then concolorous or more brownish.
Flesh concolorous. *Stipe* 0.5–2 × 0.1–0.4 cm (0.2–0.8 × 0.04–0.16 in), cylindric, slightly pruinose, concolorous, central or slightly excentric, often short and curved. *Microscopic features* Sp 6–9 × 3.5–5 µm, cylindro-elliptic to almost bean-shaped, smooth. Cheilo 30–70 × 4–12 µm, lageniform, the neck more or less wide and obtuse or cylindric with a clavate or subcapitate apex. CC a trichoderm, with septate interwoven hyphae 5–15 µm diameter; pileocystidia similar to the cheilocystidia.

Season: May–December.
Habitat: Rotting fallen branches of broad-leaved trees, in damp undergrowth.
Distribution: Widespread. Quite common.

F. *filopes* (Romagn) Bon, has a taller, more slender stipe, an often small cap and smaller spores.

---

p.104 *Tubaria autochtona* (Berk. & Br.) Saccardo
Crepidotaceae, Cortinariales

Description: *Cap pale yellowish buff. Gills pale yellow.*
Cap 0.5–2 cm (0.2–0.8 in), convex then expanded to faintly umbonate, not striate, fibrillose to slightly felted, yellowish buff or yellowish beige. *Gills* slightly decurrent, arching, not very close, straw-yellow. *Stipe* 2–5 × 0.1–0.3 cm (0.8–2 × 0.04–0.12 in), sometimes with a small bulb, faintly fibrillose, pale yellowish buff,

sometimes almost whitish. *Microscopic features* Sp 5–7.5 × 2.5–4 μm, ovoid or almond-shaped, the apex somewhat rounded and acute, smooth. Cheilo 20–30 × 9–13 μm, clavate to lageniform, the neck wavy. CC a subtrichodermial cutis, with hyphae 10–15 μm diameter.

Season: April–December.
Habitat: Under hawthorn.
Distribution: Uneven. Rather rare to rare.

Can be confused with some species of *Flammulaster*; however, *Tubaria* can be recognized in the field by the decurrent or subdecurrent gills.

---

p.104 *Tubaria conspersa* (Pers.:Fr.) Fayod
Crepidotaceae, Cortinariales

Description: *Cap rather rufous buff. Margin with white wisps of veil. Stipe silky fibrillose.*
Cap 0.5–3 cm (0.2–1.2 in), convex then fairly quickly expanded to almost flat, not striate, rather rufous buff, slightly paler when dry, felted to subsquamulose, the veil whitish and developing into characteristic wisps on the margin. *Gills* subdecurrent, arched, rather close, almost concolorous, sometimes with pinkish tones. *Stipe* 2–4 × 0.1–0.4 cm (0.8–1.6 × 0.04–0.16 in), sometimes pinched or compressed, with a veil or cortina of fleecy whitish fibrils then glabrous, but remaining slightly fibrillose below. *Microscopic features* Sp 6–10 × 3.5–5 μm, ovoid or almond-shaped, the apex obtuse or faintly rounded and acute, smooth. Cheilo 25–70 × 5–13 μm, often clavate but also lageniform, the neck wavy. CC subtrichodermial, with hyphae 6–15 μm diameter. Pigment almost smooth.

Season: May–December.
Habitat: On the ground; on a range of habitats.
Distribution: Widespread. Common or quite common.

T. furfuracea (Pers.:Fr.) Gillet has transient velar remains over the entire cap, and distinctly darker, rather tawny to warm-brown gills. *T. hiemalis* Rom. ex M. Bon is often larger, the cap soon glabrous, rusty brown, fading and the gills rather dark rust. It usually occurs at the end of the season or in winter.

---

p.104 *Tubaria romagnesiana* Arnolds
(= *T. pellucida* p.p.)
Crepidotaceae, Cortinariales

Description: *Margin with long striations. Cap rufous orange. Cap* 0.5–2 cm (0.2–0.8 in), convex then expanded, or almost flat, hygrophanous, with rather long striations, felted or with flecks or small scales in the centre, rufous orange to rather bright tawny, fading to pale ochraceous beige. Margin with fleecy fibrils, becoming glabrous.
Gills subdecurrent, arched, not very close, tawny to rather dark tawny brown.
Stipe 2–4 × 0.1–0.3 cm (0.8–1.6 × 0.04–0.12 in), cylindric, with a faint cortina then fibrillose or glancing, almost concolorous. *Microscopic features* Sp 6–8.5 × 3.5–5 µm, elliptic, smooth, slightly thick-walled. Cheilo 30–70 × 5–15 × 3–8 µm, lageniform, the neck wavy, sometimes subcapitate. CC a cutis, with septate hyphae 5–10 µm diameter.
Season: May–December.
Habitat: Grass in parks, wood edges, sometimes in woods among plant debris.
Distribution: Widespread but frequency uneven.

ENTOLOMATACEAE AND ALLIES
Species in the *Entolomataceae* have the following characters: spores pink (field character); spores distinctive, polygonal, ribbed or faceted (microscopic character); gills not absolutely free (field character); gill trama not distinctive (microscopic character).

Although this family is rather easy to recognize, it is much more difficult to attribute a collection to any one of the three genera, which are separated basically by spore shape (microscopic character): *Entoloma* have polygonal spores, *Clitopilus* have ribbed spores and *Rhodocybe* have faceted spores.

The great majority of members of this family are *Entoloma* species. Classification is complex and rather difficult, based in part on the silhouette of the fruit body, but particularly on microscopic details which can be difficult to see. In this book, species are arranged according to their silhouette: tricholomatoid, collybioid, mycenoid, clitocyboid and omphalinoid. Some taxa being intermediate between these silhouettes, the reader should be cautious in how they interpret the photographs and descriptions, especially since this is a huge genus and many rather common species are not illustrated.

'Allies' indicates that other closely related families (*Macrocystidiaceae* and *Rhodotaceae*) are included here.

---

p.163 *Entoloma lividum* Quélet (poisonous)
Entolomataceae, Entolomatales

Description: *Cap greyish cream. Gills notched, pale yellow then salmon. Smell mealy.*
Cap 5–20 cm (2–8 in), conico-convex then with a broad umbo, slightly viscid then dry and rather shining, smooth or with fine radial wrinkles, not hygrophanous, pale, ochraceous cream with greyish or yellowish beige tones. *Gills* adnate-notched, horizontal or slightly sinuate, butter yellow then salmon. *Stipe* 4–18 × 1–4 cm (1.6–7 × 0.4–1.6 in), solid, fibrillose then becoming glabrous, white to greyish cream. *Flesh* thick, white. Smell mealy then slightly sharp to nauseous. *Microscopic features* Sp 8–11 × 7–9.5 μm, with five or six angles. CC an ixocutis or a cutis, with hyphae 2–5 μm diameter. Pigment intracellular, brownish yellow.
Season: August–October.
Habitat: Broad-leaved trees, on calcareous clays, usually damp soils.
Distribution: Rather widespread but frequency uneven.

Causes severe gastro-enteritis, often requiring hospital treatment. Some people confuse this species with *Clitocybe nebularis* (edible, *see* p.821), which has more decurrent or arching gills, which are white and remain so, a greyer cap with slight radial streaking, and no smell of meal.

---

p.173 *Entoloma inopiliforme* M. Bon
Entolomataceae, Entolomatales

Description: *Cap umbonate, brownish, fibrillose. Stipe yellowish below. Smell mealy.*
Cap 4–7 cm (1.6–2.8 in), conical then convex or almost flat but retaining a distinct umbo, slightly viscid or dry, rather shining, with innate radial streaks to fissured like *Inocybe*, brownish buff, rather pale towards the margin. *Gills* notched, rather deep, not very close, slightly thick, white then a pure pink, edge rather thick, eroded.
Stipe 4–8 × 0.5–1.5 cm (1.6–3.2 × 0.2–0.6 in), rather short, tapering, concolorous with the cap or paler, generally yellowish at the base. *Flesh* rather thick, white or sepia-cream, yellowish at the base. Smell and taste mealy. *Microscopic features* Sp 6.5–8.5 × 6.5–8 μm, subglobose, with five to seven angles. CC an ixocutis, with hyphae 2–7 μm diameter. Pigment intracellular, brownish.
Season: September–November.
Habitat: Unimproved grassland.
Distribution: Rare to very rare, recorded in France.

Some authorities combine this species with *E. prunuloides* (Fr.:Fr.) Quélet, which occurs in the same habitat and has a smoother and paler cap. *E. bloxamii* (Berk. & Br.) Saccardo, with a very similar silhouette and habitat, differs in its blue or grey-blue colours, at least when young.

---

p.141 *Entoloma clypeatum* (L.) Kummer
Entolomataceae, Entolomatales

Description: *Cap brown, hygrophanous, umbonate. Smell mealy.*
Cap 4–10 cm (1.6–4 in), conical then convex with an obtuse umbo, sometimes in

a depression, the margin undulating, finely striate or splitting; slightly viscid then dry, rather shiny, smooth to innately streaky or radially fissured, sometimes even slightly scaly when old or in dry weather, hygrophanous, quite dark brown, slightly grey or olivaceous when dry. *Gills* notched, whitish or greyish then pink. *Stipe* 4–15 × 2–6 cm (1.6–6 × 0.8–2.4 in), fibrous, fibrillose or even grooved, white then brownish grey. *Flesh* white or greyish, brownish near the surface. Smell and taste mealy. *Microscopic features* Sp 9–11.5 × 7.5–10 µm, subglobose, with five to seven angles, distinctly angular. CC an ixocutis, with hyphae 2–7 µm diameter. Pigment intracellular, brown.

Season: April–June.
Habitat: Hedges, wood edges, sometimes coppiced areas, associated with shrubs in Rosaceae. Sometimes in clusters.
Distribution: Widespread but frequency uneven.

Quite a good edible species. The early fruiting season reduces the number of species with which it might be confused (*Inocybe patouillardii*, a very poisonous species, is one danger to avoid). The spring-fruiting *Entoloma* are difficult to identify and some have been linked to digestive disorders.

---

p.141 *Entoloma porphyrophaeum* (Fr.) P. Karsten
Entolomataceae, Entolomatales

Description: *Cap fibrillose, crimson-brown or the colour of porphyry. Stipe concolorous.*
Cap 4–15 cm (1.6–6 in), conical then spreading, with a large obtuse umbo, fibrillose to felted, sometimes with slight innate streaks or small scales when old, crimson-brown or brownish grey with a hint of porphyry (violaceous tones).
Gills notched to almost free, rather distant, white or pale cream then pinkish brown, edge thin, eroded to toothed. *Stipe* 5–16 × 0.5–2 cm (2–6.3 × 0.2–0.8 in), brittle, fusoid, fibrillose to subsquamulose, concolorous or with crimson tones on a pale

buff ground. *Flesh* rather thick at the umbo, white to concolorous. Smell insignificant. *Microscopic features* Sp 8–12 × 6–8 μm, with many angles, elliptic to irregular. Cheilo numerous, bottle-shaped subcapitate or capitate, 20–60 × 7–20 × 2.5–7 × 3–15 μm. CC a cutis, with hyphae 10–25 μm diameter, sometimes erect and forming a trichoderm, the terminal elements long fusiform. Pigment intracellular, brownish.

Season: August–November.
Habitat: Unimproved grassland and meadows, on poor soils.
Distribution: Rather widespread. Rare.

Other taxa with a fibrillose or almost felted cap are smaller and lack the crimson or lilaceous tones. The best known is *E. jubatum* (Fr.:Fr.) P. Karsten, which has a dark grey-brown cap, greyish brown gills and grey stipe.

---

p.136 *Entoloma sericeum* (Bull.:Fr.) Quélet
Entolomataceae, Entolomatales

Description: *Cap hygrophanous, brown, shining or silky. Smell mealy.*
Cap 2–6 cm (0.8–2.4 in), shallowly convex, remaining umbonate and sometimes depressed, hygrophanous, with rather long striations, lubricious, shining then slightly silky when dry, smooth, dark chocolate brown, fading markedly, greyish cream to pale ochraceous beige.
Gills ventricose, not very close, cream then quite dark brownish pink. *Stipe* 2–8 × 0.3–1 cm (0.8–3.2 × 0.12–0.4 in), fragile, almost concolorous or silvery beige, very fibrillose to striate and slightly silvery.
Flesh fragile, concolorous or paler when dry. Smell and taste mealy. *Microscopic features* Sp 7–10.5 × 6.5–9.5 μm, almost globose, generally with 5 angles. CC a cutis, with hyphae 2–12 μm diameter. Pigment epimembranal, encrusting, especially near the surface. Clamps present especially in the hymenium.

Season: August–December.
Habitat: Short turf, meadows, sometimes in towns.
Distribution: Very widespread. Common.

Very variable. F. *nolaniformis* (Kühner) Noordeloos is more slender.

---

p.136 *Entoloma alpicola* (Favre) Noordeloos
Entolomataceae, Entolomatales

Description: *Cap black-brown, hygrophanous. Smell mealy.*
Cap 2–6 cm (0.8–2.4 in), broadly convex, umbonate, irregular, hygrophanous, not very striate, slightly viscid, shining in dry weather, smooth but radially fissured or slightly scurfy, blackish brown when fresh, sometimes with reddish or greyish tones, scarcely fading when dry. *Gills* notched, rather close, white then pink to pinkish brown. *Stipe* 2–7 × 0.5–1.5 cm (0.8–2.8 × 0.2–0.6 in), hollow in the centre, tapering slightly, very pale, with silvery fibrils. *Flesh* firm then fragile, whitish or pale to concolorous. Smell and taste faintly mealy. *Microscopic features* Sp 8–10 × 7–9 µm, very shortly elliptic, with five to seven angles. CC a cutis, an ixocutis in some places, with hyphae 2–10 µm diameter. Pigment intracellular, brown, abundant near the surface.
Season: June–September.
Habitat: Areas of snow melt and adjacent turf, often with dwarf willow (especially *Salix herbacea*).
Distribution: Alpine or Arctic.

Probably one of the commonest Entolomas in this habitat.

---

p.137 *Entoloma sordidulum* (K. & R.) Orton
Entolomataceae, Entolomatales

Description: *Cap brownish grey, hygrophanous. Smell of water melon.*
Cap 1–5 cm (0.4–2 in), conico-convex umbonate then almost flat or irregular, hygrophanous, finely striate only at the

extreme edge, smooth or finely scurfy in the centre in dry weather, more or less dark brownish grey, fading to dingy whitish grey. *Gills* notched, often very narrow at the cap margin, almost distant, white then pale greyish pink. *Stipe* 2–6 × 0.3–0.5 cm (0.8–2.4 × 0.12–0.2 in), hollow, slightly clavate, pale or even whitish, slightly glancing but not fibrillose. *Flesh* firm then brittle, pale greyish beige. Smells and tastes strongly of water melon. *Microscopic features* Sp 8–11.5 × 7–8.5 µm, very shortly elliptic in profile, with 5–7 angles. CC a cutis, with hyphae 2–7 µm diameter. Pigment epimembranal, encrusting, brownish.

Season: June–December.
Habitat: Broad-leaved trees, coppice, sometimes urban parks.
Distribution: Widespread. Quite common.

---

p.115 *Entoloma rhodopolium* f. *nidorosum* (Fr.) Noordeloos
Entolomataceae, Entolomatales

Description: *Cap pale, hygrophanous. Stipe very pale. Smell nitrous or sharp.*
Cap 3–8 cm (1.2–3.2 in), convex then almost flat, with or without an umbo, hygrophanous, finely striate or striate, slightly silky, smooth, greyish brown to yellowish ochraceous brown, drying out in paler to whitish radial streaks.
Gills notched, rather narrow, not very close, white then pale pink or pale greyish pink. *Stipe* 4–10 × 0.5–1 cm (1.6–4 × 0.2–0.4 in), sometimes twisted, white to pale beige-grey, with silky fibrils. *Flesh* brittle, slightly watery in the stipe, white or very pale. Smell nitrous (bleach), sometimes merely sharp, unpleasant when rubbed. *Microscopic features* Sp 7–10 × 6–8 µm, shortly elliptic, with five to seven angles. CC an ixocutis, with hyphae 2–7 µm diameter. Pigment intracellular, very pale.

Season: August–November.
Habitat: Broad-leaved trees, cool or damp places.
Distribution: Widespread. Common.

Not very distinct from *E. rhodopolium* (Fr.:Fr.) Kummer, which is slightly more robust, darker and completely odourless. *E. subradiatum* (Kühn. & Romagn.) Moser has a similar shape but the cap is strongly striate. There are records of gastro-intestinal poisoning associated with these species.

p.173 *Entoloma lividoalbum* (K. & R.) Kubicka
Entolomataceae, Entolomatales

Description: *Cap hygrophanous, yellowish brown. Smell of cucumber.*
Cap 3–12 cm (1.2–4.7 in), conico-campanulate then expanded, broadly convex umbonate, not or not very striate, often wavy, smooth and slightly shining, sometimes very finely wrinkled or silky when dry, quite dark yellowish brown, pale yellowish grey when dry, dull and sometimes almost whitish or greyish. *Gills* notched, rather deep, not very close, white then pink. *Stipe* 4–10 × 0.5–2 cm (1.6–4 × 0.2–0.8 in), firm, very pale, sometimes slightly dingy buff below. *Flesh* white or almost concolorous. Smells and tastes strongly of cucumber. *Microscopic features* Sp 8–10.5 × 8–9 µm, very shortly elliptic, with five to seven angles. CC a cutis and tending towards an ixocutis, with hyphae 2–10 µm diameter. Pigment intracellular, brown.
Season: August–November.
Habitat: Broad-leaved trees, especially on calcareous soils.
Distribution: Uneven. Rather rare to rare.

E. griseoluridum (Kühner) Moser is more highly coloured, quite dark reddish brown initially and with greyish gills.

p.135 *Entoloma excentricum* Bresadola
Entolomataceae, Entolomatales

Description: *Cap very pale. Stipe can be slightly eccentric.*
Cap 2–6 cm (0.8–2.4 in), convex then almost flat or domed, sometimes asymmetrical, not hygrophanous, smooth

or finely velvety in places, whitish to pale tan. *Gills* notched, rather narrow, quite close, white then pinkish buff, edge eroded, sometimes brownish. *Stipe* 3–7 × 0.4–0.7 cm (1.2–2.8 × 0.16–0.28 in), fibrillose or striate, white or very pale, with yellowish or pale buff tones, sometimes slightly eccentric. *Flesh* relatively firm, very pale. Smell slightly mealy, sharp. *Microscopic features* Sp 10.5–13 × 7–9 µm, rather longer than usual, with five to eight angles. Cheilo numerous, 40–110 × 10–25 µm, the base inflated, the neck narrow, more or less long, sometimes slightly warted, occasionally with brownish contents. CC a cutis, with hyphae 5–10 µm diameter, or subtrichodermial, with erect wider or clavate hyphae. Pigment finely encrusting. Clamps present especially in the hymenium.

Season: June–November.
Habitat: Dry grassland, coppice woodland on calcareous soils.
Distribution: Uneven. Rare.

---

p.136 *Entoloma ameides* (Berk. & Br.) Saccardo
Entolomataceae, Entolomatales

Description: *Cap shining, brownish grey. Strong smell of pear drops.*
Cap 2–5 cm (0.8–2 in), conical then umbonate or even depressed to slightly domed, hygrophanous, striate, silky, greyish with brownish or buff tones, paler and shining when dry. *Gills* notched or almost free, ventricose, not very close, greyish then dark pink, with greyish or brownish tones. *Stipe* 3–8 × 0.3–0.8 cm (1.2–3.2 × 0.12–0.32 in), fibrillose, concolorous or paler to buff. *Flesh* brittle, pale greyish buff, almost concolorous near the surface. Smell very strong of amyl acetate (pear drops). *Microscopic features* Sp 9.5–10.5 × 7–8 µm, shortly elliptic, with five or six angles. Bas sometimes 2-spored. CC a cutis, with hyphae 1.5–6 µm diameter. Pigment epimembranal, finely encrusting, rather pale

and hard to see. Clamps present, especially in the hymenium.
Season: August–October.
Habitat: Meadows and unimproved grassland but also under broad-leaved trees.
Distribution: Widespread. Rather rare.

The smell is characteristic.

---

p.146 *Entoloma euchroum* (Pers.:Fr.) Donk
Entolomataceae, Entolomatales

Description: *Cap, gills and stipe violaceous blue.*
Cap 1–4 cm (0.4–1.6 in), convex, with a small umbo then almost flat or irregular, not hygrophanous, not striate, velvety or tomentose then more distinctly squamulose in the centre, violaceous blue, fading or becoming a dull purplish brown then dingy brown. *Gills* sinuate or ventricose, rather distant, concolorous then with pinkish tones. *Stipe* 2–6 × 0.2–0.5 cm (0.8–2.4 × 0.08–0.2 in), often curved or even eccentric, fibrillose to striate, slightly pruinose under the gills, concolorous then dull. *Flesh* rather thin, concolorous or paler. Smell aromatic, like violets.
Microscopic features Sp 9–11.5 × 6–8 µm, rather long, with five to seven angles. Cheilo 20–60 × 5–15 µm, cylindric or slightly clavate, sometimes tapering, thick- or thin-walled, with violaceous intracellular pigment. CC a trichoderm with erect hyphae 20–90 × 5–30 µm. Pigment mostly intracellular, violaceous brown.
Season: August–November.
Habitat: Stumps and rotten wood of broad-leaved trees, rarer on conifers. Lignicolous.
Distribution: Widespread. Quite common.

---

p.146 *Entoloma corvinum* (Kühner) Noordeloos
Entolomataceae, Entolomatales

Description: *Cap blue-black, slightly squamulose. Gills white. Stipe paler.*
Cap 1–4 cm (0.4–1.6 in), conical or truncate then very slightly convex or

slightly depressed, not hygrophanous, not striate, velvety or radially fibrillose, finely squamulose in the centre, blue-black or very dark. *Gills* almost free, not very close, white then pinkish. *Stipe* 2–6 × 0.2–0.5 cm (0.8–2.4 × 0.08–0.2 in), sometimes compressed or grooved, with darker fibrils on a dark blue to grey-blue or violaceous ground. *Flesh* fragile, bluish grey, darker and bluer near the surface. *Microscopic features* Sp 8–11.5 × 6–8 μm, distinctly long, with five to seven angles. Bas without clamps. Cheilo 25–65 × 5–15 μm, cylindro-clavate. CC a cutis, more hymenidermial near the centre, the terminal cells club-shaped 15–32 μm diameter. Pigment intracellular. Clamps absent.

Season: June–November.

Habitat: Meadows and lawns, more rarely broad-leaved or coniferous woodland.

Distribution: Widespread. Rather rare.

---

p.146 *Entoloma lazulinum* (Fr.) Noordeloos
Entolomataceae, Entolomatales

Description: *Cap blue-black, striate. Gills blue-grey. Stipe shining.*
Cap 1–4 cm (0.4–1.6 in), flat bell-shaped, then depressed, not or not very hygrophanous, with long striations and finely fibrillose-tomentose, weakly squamulose in the centre when old, dark blue, paler at the margin and fading when old. *Gills* almost free, rather distant, dark blue to grey-blue then with pinkish tones, edge uneven, often brown. *Stipe* 2–7 × 0.2–0.5 cm (0.8–2.8 × 0.08–0.2 in), smooth and shining, almost concolorous but soon fading, finally bluish or glaucous. *Flesh* greyish or dingy cream to almost concolorous. *Microscopic features* Sp 8.5–12 × 6.5–8.5 μm, rather long, with five to nine angles. Bas sometimes 2-spored, without clamps. Cheilo 30–75 × 5–20 μm, cylindro-clavate, with intracellular, brown pigment. CC almost trichodermial in the centre, with clavate cells 10–30 μm diameter. Pigment intracellular, dark blue. Clamps absent.

Season: June–November.
Habitat: Lawns and meadows.
Distribution: Very widespread. Quite common.

E. chalybaeum (Pers.:Fr.) Noordeloos, is a very similar species, characterized especially by its much less striate cap and less scaly centre.

---

p.147 *Entoloma mougeotii* (Fr.) Hesler
Entolomataceae, Entolomatales

Description: *Cap greyish blue or violaceous, subsquamulose. Gills pale.*
Cap 1–5 cm (0.4–2 in), convex then depressed, not hygrophanous, not striate, fibrillose-tomentose then slightly squamulose in the centre, sometimes cracking, violaceous blue then violaceous grey, sometimes darker in the centre. *Gills* adnate, rather close, pale cream then pinkish, edge eroded. *Stipe* 3–7 × 0.2–0.5 cm (1.2–2.8 × 0.08–0.2 in), fibrillose, greyish violet, with a metallic sheen, paler than the cap. *Flesh* cream or pale greyish, almost concolorous near the surface. Smell weakly aromatic. *Microscopic features* Sp 9–12 × 6–8.5 µm, rather long, with five to nine angles. Bas without clamps. Cheilo 25–70 × 10–20 µm, cylindric or clavate. CC a trichoderm, the terminal cells club-shaped 10–25 µm diameter. Pigment intracellular, violaceous. Clamps absent.
Season: August–November.
Habitat: Meadows, rarer under broad-leaved trees. Usually on calcareous soils.
Distribution: Widespread.

---

p.146 *Entoloma tjallingiorum* Noordeloos
Entolomataceae, Entolomatales

Description: *Cap brownish grey. Gills pale. Stipe blue with small dark scales.*
Cap 2–8 cm (0.8–3.2 in), conico-convex then with a broad umbo or depression around the centre, not hygrophanous, not striate, fibrillose or with small scales, especially in the centre, rather dark

brownish grey but violaceous at the margin. *Gills* notched, white or bluish then pinkish, edge eroded. *Stipe* 3–10 × 0.4–1 cm (1.3–4 × 0.16–0.4 in), dark violaceous blue to bluish grey, fibrillose and with small dark violaceous brown scales, especially towards the apex. *Flesh* fragile, whitish to concolorous. *Microscopic features* Sp 9–11 × 6–7 µm, rather long, with six to nine angles. Cheilo 20–60 × 6–15 µm, cylindric or tapering, thin- or slightly thick-walled and brownish. CC rather trichodermial towards the centre, with inflated cells 10–30 µm diameter. Pigment intracellular, violaceous brown, sometimes also epimembranal and encrusting in places.

Season: August–November.
Habitat: Broad-leaved trees, on oak or birch stumps or near rotten wood.
Distribution: Not well known, but widespread.

---

p.147 *Entoloma querquedula* (Rom.) Noord.
Entolomataceae, Entolomatales

Description: *Cap olivaceous blue. Edge blue-black. Stipe pale olivaceous blue or glaucous.*
Cap 1–5 cm (0.4–2 in), conical then flat to depressed, sometimes with a small umbo, not hygrophanous, not or not very striate, fibrillose then slightly squamulose in the centre, rather dark brown with olivaceous blue tones, especially towards the margin. *Gills* adnate, not very close, whitish to pale greyish blue then pinkish grey, edge eroded and edged with blue-black. *Stipe* 2–6 × 0.2–0.5 cm (0.8–2.4 × 0.08–0.2 in), pale olivaceous blue-grey, fibrillose or with blue dots at the apex, smooth and almost polished elsewhere. *Flesh* whitish or greyish, almost concolorous near the surface. *Microscopic features* Sp 8.5–10.5 × 7–8 µm, rather long, with five to seven angles. Bas without clamps. Cheilo 40–110 × 5–20 µm, cylindric or clavate, with abundant intracellular blue pigment. CC trichodermial near the centre, with cells 10–25 µm diameter. Pigment intracellular, blue. Clamps absent.

Season: September–October.
Habitat: Grass or moss, in damp, mainly broad-leaved woodland.
Distribution: Uneven. Rare.

E. serrulatum (Pers.:Fr.) Hesler, with a blue-black to dark blue-grey cap, pale bluish gills and an almost concolorous stipe, has an aromatic smell.

---

p.146 *Entoloma lampropus* (Fr.:Fr.) Hesler
Entolomataceae, Entolomatales

Description: *Cap dark brown. Gills white. Stipe steely blue with dark blue fibrils.*
Cap 1–4 cm (0.4–1.6 in), obtusely conical, rarely depressed, not hygrophanous, not striate, finely fibrillose or with small scales in the centre, rather dark brown, at least when young. *Gills* adnate, rather distant, white or greyish, very pale then pinkish. *Stipe* 4–7 × 0.2–0.5 cm (1.6–2.8 × 0.08–0.2 in), with dark blue fibrils on a steely blue-grey ground. *Flesh* greyish white to almost concolorous. *Microscopic features* Sp 8.5–11.5 × 6–8.5 µm, rather long, with six to nine angles. CC trichodermial near the centre, with broadly clavate terminal cells 10–30 µm diameter. Pigment intracellular, brown. Clamps present, especially in the hymenium.
Season: August–November.
Habitat: Broad-leaved woodland (mainly clearings or wood edges), unimproved grassland. Usually on calcareous soils.
Distribution: Appears to be widespread. Rare.

---

p.147 *Entoloma poliopus* (Rom.) Noordeloos
Entolomataceae, Entolomatales

Description: *Cap brownish grey. Gills white with a brownish edge. Stipe blue, shining.*
Cap 2–5 cm (0.8–2 in), conical then with a depression or a small umbo, not very hygrophanous, not or not very striate, fibrillose to squamulose in the centre, rather dull brownish grey, almost black in

the centre when young. *Gills* adnate, not very close, white or pale grey then dingy pink, edge brown to brownish.
Stipe 2–7 × 0.2–0.5 cm (0.8–2.8 × 0.08–0.2 in), with a polished appearance, a beautiful metallic blue then fading to bluish grey. *Flesh* whitish to greyish but almost concolorous near the surface.
Microscopic features Sp 10–13.5 × 6–9 µm, rather long, with six to eight angles. Bas without clamps. Cheilo 30–100 × 5–20 µm, cylindric or clavate, with brownish intracellular pigment. CC trichodermial to hymenidermal in the centre, the terminal cells club-shaped 10–25 µm diameter. Pigment intracellular, brown. Clamps absent.

Season: July–November.
Habitat: Various. Usually on damp acid soils.
Distribution: Widespread. Rare.

---

p.130 *Entoloma roseum* (Longyear) Hesler
Entolomataceae, Entolomatales

Description: *Cap reddish pink. Gills white or pinkish. Stipe pinkish, shining.*
Cap 1–4 cm (0.4–1.6 in), hemispherical then with a depression, not hygrophanous, not or not very striate, fibrillose, sometimes slightly scaly in the centre, reddish pink to old rose. *Gills* adnate to subdecurrent, not very close, white or pinkish then pink.
Stipe 2–6 × 0.2–0.5 cm (0.8–2.4 × 0.08–0.2 in), glabrous and shining, concolorous or paler. *Flesh* pinkish or pink near the surface. *Microscopic features* Sp 8.5–10.5 × 7–8 µm, rather long, with 6–9 angles. Bas without clamps. Cheilo sometimes rare, 20–60 × 5–20 µm, cylindric or clavate, sometimes tapering. CC mainly a trichoderm but a hymeniderm in the centre, with inflated terminal cells 20–40 µm diameter. Pigment intracellular, pink. Clamps absent.

Season: August–November.
Habitat: Unimproved meadows or grassy places. Usually on calcareous soils.
Distribution: Rare to very rare.

p.129 *Entoloma incanum* (Fr.:Fr.) Hesler
Entolomataceae, Entolomatales

Description: *Cap yellowish olivaceous brown. Stipe greenish yellow with a dark green base.*
Cap 1–4 cm (0.4–1.6 in), with a flattened top then depressed and sometimes with a slight umbo, slightly hygrophanous, with rather long striations, smooth then fibrillose, disrupted or slightly scaly in the centre, yellowish green or olive green, sometimes more brownish, reddish or yellowish. *Gills* subdecurrent, rather distant, pale greenish then pink to brownish pink. *Stipe* 2–7 × 0.2–0.4 cm (0.8–2.8 × 0.08–0.16 in), sometimes pruinose at the apex, yellowish green to ochraceous yellow but dark blue-green at the base. *Flesh* pale or concolorous with the outside. Smell of mouse urine, or some kinds of biscuit. *Microscopic features* Sp 10–13.5 × 7.5–10 µm, rather long, with six to nine angles. Bas without clamps. CC a trichoderm near the cap centre, with club-shaped terminal cells 10–30 µm diameter. Pigment intracellular, yellowish brown. Clamps absent.
Season: June–November.
Habitat: Unimproved grassland, grassy rides, sandy to stony places. Usually on calcareous soils.
Distribution: Widespread. Rather rare or very rare.

This species has striking fluorescent colours and is easy to identify.

p.128 *Entoloma icterinum* (Fr.:Fr.) Moser
Entolomataceae, Entolomatales

Description: *Cap yellow. Gills subdecurrent. Smell of pear drops.*
Cap 1–3 cm (0.4–1.2 in), bell-shaped then plano-convex, sometimes with a small umbo, hygrophanous, striate, smooth then slightly felted in the centre, lemon yellow to buff, paler and yellowish when dry.
Gills subdecurrent, rather close, yellowish then a pretty salmon pink. *Stipe* 2–6 ×

0.2–0.5 cm (0.8–2.4 × 0.08–0.2 in), smooth and glabrous, concolorous or paler. Base white. *Flesh* whitish or concolorous with the outer surfaces. Smell of amyl acetate (pear drops). *Microscopic features* Sp 8–11 × 6.5–8.5 µm, rather long, with five or six angles. CC a cutis with hyphae 4–7 µm diameter. Pigment intracellular, yellow. Clamps present especially in the hymenium.

Season: August–November.
Habitat: Damp broad-leaved woodland, sometimes coppice woodland, urban parks, disturbed sites.
Distribution: Quite widespread. Rather rare or very rare.

The colour and smell are very characteristic.

---

p.89 *Entoloma sericellum* (Fr.:Fr.) Kummer
Entolomataceae, Entolomatales

Description: *Cap white, silky, yellowing slightly. Gills decurrent.*
Cap 0.5–4 cm (0.2–1.6 in), hemispherical or convex, the centre flattened or depressed, not hygrophanous, not striate, finely tomentose in the centre, silky and slightly shining, white but sometimes slightly ochraceous yellow, yellowing slightly. *Gills* decurrent, rather close, white then pink. *Stipe* 2–6 × 0.2–0.4 cm (0.8–2.4 × 0.08–0.16 in), silky, white then yellowing slightly. *Flesh* white or yellowing slightly. *Microscopic features* Sp 8–11.5 × 6–9 µm, slightly elongated, with five to eight angles. Bas sometimes 2-spored. Cheilo in clusters, 30–85 × 8–15 µm, cylindric or fusoid, sometimes tapering. CC almost a trichoderm in the centre, with slightly clavate terminal hyphae 5–15 µm diameter, sometimes more.

Season: July–December.
Habitat: Meadows, broad-leaved trees (damp woods, etc.).
Distribution: Very widespread. Quite common.

E. cephalotrichum (Orton) Noordeloos, which is more frequent in very damp and rather disturbed sites, is a small version of *E. sericellum*, with long, distinctly capitate hairs on the cap and stipe. *E. olorinum* (Romagn. & Favre) Noordeloos, which yellows rather more strongly, differs (under the microscope) by the absence of cheilocystidia.

---

p.120 *Entoloma cetratum* (Fr.:Fr.) Moser
Entolomataceae, Entolomatales

Description: *Cap honey brown, hygrophanous, with a dark eye.*
Cap 1–4 cm (0.4–1.6 in), flat bell-shaped, remaining convex for a long time, hygrophanous, with long striations, slightly viscid then somewhat shining, smooth or very finely wrinkled, honey-brown, fading to ochraceous cream when dry; the centre is marked by a characteristic brown spot. *Gills* almost free, not very close, yellowish honey then rather dark salmon-pink. *Stipe* 3–8 × 0.2–0.4 cm (1.2–3.2 × 0.08–0.16 in), often twisted, ochraceous honey, silky with silvery fibrils. *Flesh* fragile, pale or almost concolorous. Smell slightly mealy. *Microscopic features* Sp 9.5–13.5 × 7–9.5 µm, rather long, with five to eight angles. Bas without clamps. CC a cutis with hyphae 4–8 µm diameter. Pigment epimembranal smooth, often faint. Clamps absent.
Season: August–December.
Habitat: Damp conifer woods, sometimes heathland or peaty acid places.
Distribution: Widespread. Rather common.

The dark spot in the cap centre is a constant character, and resembles an eye in the middle of a lighter striate band.

---

p.122 *Entoloma hebes* (Romagnesi) Trimbach
Entolomataceae, Entolomatales

Description: *Cap papillate, warm-brown, striate. Stipe rather stiff.*
Cap 0.5–3 cm (0.2–1.2 in), sharply conical, gradually flattening but retaining a small rather pointed umbo, slightly hygrophanous, slightly striate, smooth, sometimes silky or slightly micaceous,

especially in the centre, warm-brown to beige-grey. *Gills* almost free, rather close, pale greyish beige then pinkish or grey-pink. *Stipe* 3–10 × 0.1–0.4 cm (1.2–4 × 0.04–0.16 in), rather stiff, sometimes twisted, with silvery fibrils on brownish grey ground. Base white. *Flesh* pale or almost concolorous. Smell faint, of rancid meal. *Microscopic features* Sp 8.5–13 × 5.5–7.5 µm, rather long, with five to eight angles. Cheilo 25–60 × 5–12 µm, cylindric or more often tapering into an obtuse neck. CC a cutis, with hyphae 2–10 µm diameter. Pigment epimembranal, encrusting, in fine irregular bands, brown, also intracellular.

Season: July–December.
Habitat: In humus on wood edges, coppice woodland, damp broad-leaved woodland. Usually on calcareous soils.
Distribution: Widespread but frequency uneven.

E. hirtipes (Schum.:Fr.) Moser, which differs in its more robust build, smells strongly of the sea or putty (it has larger, wider spores: 10–15 × 8–9.5 µm).

---

p.132 *Entoloma araneosum* (Quélet) Moser
Entolomataceae, Entolomatales

Description: *Cap very fibrillose, silvery. Gills grey. Stipe silvery.*

Cap 1–4 cm (0.4–1.6 in), conical then convex, not hygrophanous, not striate, strongly fibrillose or almost shaggy squamulose in the centre, silvery grey, sometimes with a hint of brown.
Gills almost free, rather distant, grey then dark pinkish grey, edge finely floccose.
Stipe 3–6 × 0.2–0.5 cm (1.2–2.4 × 0.08–0.2 in), brittle, silky or fibrillose, almost woolly or strigose below, concolorous or paler.
Flesh greyish or darker near the surface. Smell slightly mealy. *Microscopic features* Sp 10–14 × 7–8.5 µm, rather long, with six to eight angles. Bas without clamps. Cheilo 50–100 × 10–30 µm, with an inflated base and long, narrow, quite well-defined neck. CC a trichoderm, with long cells 10–25 µm

diameter. Pigment intracellular brownish grey, also epimembranal, finely encrusting, in irregular bands in deep tissue. Clamps absent.
Season: August–November.
Habitat: Broad-leaved coppice woodland and damp woods, on rich or slightly disturbed ground.
Distribution: Uneven. Quite rare.

E. dysthales (Peck) Saccardo has a small blackish grey-brown to brownish grey cap, grey-brown gills (under broad-leaved trees on damp soils).

---

p.93 *Entoloma politum* (Pers.:Fr.) Donk
Entolomataceae, Entolomatales

Description: *Cap brown, striate, hygrophanous. Gills arching, subdecurrent.*
Cap 1–5 cm (0.4–2 in), convex, then depressed or umbilicate, hygrophanous, with rather long striations, slightly viscid, then shining when dry, smooth or very finely wrinkled, brown, with yellowish, greyish or reddish tones. *Gills* almost decurrent, not very close, white or pale cream then pink. *Stipe* 2–7 × 0.2–0.6 cm (0.8–2.8 × 0.08–0.24 in), often shining, concolorous with the cap or paler.
Flesh pale or concolorous. *Microscopic features* Sp 8–10 × 7–8 µm, rather shortly elliptic, with five or six angles. CC a cutis or ixocutis, with variable hyphae 4–10 µm diameter. Pigment intracellular, brown.
Season: August–November.
Habitat: Damp woodland, sometimes in mud.
Distribution: Widespread. Quite common.

E. pernitrosum (Orton) Trimbach, with a distinct nitrous smell, is very similar.

---

p.93 *Entoloma rhodocylix* (Lasch.:Fr.) Moser
Entolomataceae, Entolomatales

Description: *Cap buff, umbilicate, striate. Gills white. Stipe cream.*
Cap 0.4–1 cm (0.16–0.4 in), convex with a flat top then depressed, hygrophanous,

with long striations, slightly viscid then dry, smooth or slightly fibrillose, light brown to ochraceous cream.
Gills decurrent, distant, white then pink. *Stipe* 1–4 × 0.1–0.2 cm (0.4–1.6 × 0.04–0.08 in), shining, pale cream or yellowish. *Flesh* concolorous. *Microscopic features* Sp 8–10 × 7–9 µm, rather shortly elliptic. Cheilo in clusters, 20–70 × 2–8 µm, cylindric or tapering into an obtuse neck. CC a cutis, with hyphae 2–7 µm diameter. Pigment epimembranal, encrusting, brownish. Clamps present, especially in the hymenium.

Season: August–October.
Habitat: Moss, especially in woods.
Distribution: Not yet determined. Rare.

---

p.97 *Entoloma undatum* (Fr.→Gillet) Moser
Entolomataceae, Entolomatales

Description: *Cap greyish, silky, with concentric zones. Smell mealy.*
Cap 1–4 cm (0.4–1.6 in), soon flattened, umbilicate, not hygrophanous, not striate, silky, with radial fibrils that form concentric, shining, glancing, brown to greyish zones. Margin initially inrolled, wavy. *Gills* decurrent, rather close, brownish grey then pinkish brown. *Stipe* 1–3 × 0.2–0.5 cm (0.4–1.2 × 0.08–0.2 in), pruinose at the apex and rather glabrous elsewhere, almost concolorous with the cap. *Flesh* pale or concolorous. Smell mealy. *Microscopic features* Sp 7.5–10 × 6–7 µm, rather long, with six to eight angles. CC a cutis, with slender or more clavate hyphae. Pigment epimembranal, encrusting, yellowish brown.

Season: July–November.
Habitat: Grassy places, moss on sandy soils.
Distribution: Widespread. Quite common.

The glancing, shiny appearance of the cap and the concentric zonation are good diagnostic characters for this species. A few rather similar taxa occur.

p.94 *Entoloma rusticoides* (Gillet) Noordeloos
Entolomataceae, Entolomatales

Description: *Cap brown, striate. Gills distant, brown.*
Cap 0.5–2 cm (0.2–0.8 in), remaining convex for a long time, sometimes depressed, hygrophanous, with long striations, smooth or slightly roughened to scurfy in the centre, warm-brown, paler when dry. *Gills* decurrent, very distant, brown then pinkish brown, edge thick. *Stipe* 0.5–3 × 0.1–0.2 cm (0.2–1.3 × 0.04–0.08 in), sometimes pruinose under the gills, almost concolorous. *Flesh* pale or concolorous near the surface. *Microscopic features* Sp 8–10.5 × 7–9.5 µm, subglobose, with five or six angles. Bas 4- or 2-spored, without clamps. CC a trichoderm near the centre, with cylindric or clavate hyphae, the terminal cells sometimes broadly inflated, 20–40 µm diameter. Pigment epimembranal, encrusting, in irregular bands, dark brown, sometimes also intracellular in underlying layers. Clamps absent.

Season: September–December.
Habitat: Moss in short turf, often in dunes or on sandy substrates.
Distribution: Widespread. Rather rare to very rare.

E. phaeocyathus Noordeloos differs in the presence of abundant clavate or bladder-like cheilocystidia and a mealy smell.

p.91 **The Miller**
Clitopilus prunulus (Scop.:Fr.) Quélet
Entolomataceae, Entolomatales

Description: *Cap white or dirty cream, velvety, soon soft. Strong mealy smell.*
Cap 3–10 cm (1.2–4 in), soon depressed but sometimes umbonate, white, sometimes dingy cream to pinkish beige, matt, felted or velvety, soft. Margin smooth or occasionally grooved at the extreme edge. *Gills* decurrent, rather close, soft, white then pink.
Stipe 2–5 × 0.5–1 cm (0.8–2 ×

0.2–0.4 in), slightly fibrillose, concolorous. *Flesh* soon soft, white. Smells strongly mealy (cucumber). Taste mealy. *Microscopic features* Sp 9–13 × 4.5–7 μm, fusoid with a suprahilar plage, and six to eight longitudinal ribs. CC a cutis, with hyphae 2–7 μm diameter. Clamps absent.

Season: August–December.
Habitat: Broad-leaved and also coniferous trees.
Distribution: Widespread but frequency uneven.

A very good edible species. Avoid confusion with the highly poisonous white *Clitocybe* spp., some of which occur in woods (*C. phyllophila* p.823, for example) and are rather similar, apart from the spore colour (white) and the smell (not typically mealy).

---

p.85 *Clitopilus hobsonii* var. *daamsii* (Noordeloos) Courtecuisse
Entolomataceae, Entolomatales

Description: *Very small crepidotoid species. Cap white. Gills pale pinkish cream.*
Cap 2–8 mm (0.08–0.3 in), kidney-shaped, silky felted, slightly more strigose close to the host, white. Margin fibrillose or pubescent. *Gills* confluent, rather distant, white then pale pinkish cream. *Stipe* absent or rudimentary, reduced to a small eccentric stub. *Flesh* thin, white. Smell absent. *Microscopic features* Sp 8–12 × 5–7 μm, elliptic or slightly elongated, with six to ten longitudinal ribs. Bas sometimes 2-spored. CC subtrichodermial, with hyphae 2–5 μm diameter. Clamps absent.

Season: September–December.
Habitat: On broad-leaved trees.
Distribution: France and Holland. Rare.

A variety of *C. hobsonii* (Berk.) Orton, which is a much commoner species. The spore size, which is slightly smaller (7–9 × 5–6 μm) in the type, does not justify specific distinction.

p.100 *Rhodocybe gemina* (Fr.) Kuyper &
Noordeloos (= *R. truncata*)
Entolomataceae, Entolomatales

Description: *Cap pinkish brown, matt to pruinose. Gills pinkish. Smell rancid mealy.*
Cap 2–12 cm (0.8–4.7 in), soon convex or even depressed, often wavy, matt and pruinose, sometimes felted to subsquamulose, flesh beige, tan, pinkish brown, warm-brown, reddish brown, fading when old. *Gills* notched to subdecurrent, rather close, white or pale cream then pinkish cream to flesh-beige. *Stipe* 2–6 × 0.3–3 cm (0.8–2.4 × 0.12–1.2 in), pruinose to fibrillose streaky, white then pinkish buff to beige-tan, finally almost concolorous. *Flesh* slightly or slowly bitter, white or almost concolorous. Smell mealy or rancid.
Microscopic features Sp 5–7 × 4–5.5 µm, elliptic to subglobose, somewhat angular. Cheilo filiform, 20–60 × 2–6 µm. Pigment epimembranal, encrusting. Clamps absent.
Season: June–December.
Habitat: Broad-leaved and coniferous trees.
Distribution: Rather widespread but frequency uneven.

Varieties have been described which differ in stature, smell, taste, etc.

---

p.100 *Rhodocybe popinalis* (Fr.:Fr.) Singer
Entolomataceae, Entolomatales

Description: *Cap pruinose or cracked, dingy brownish grey. Gills dingy. Smell rancid. Taste bitter.*
Cap 2–8 cm (0.4–3.2 in), conico-convex then expanded, umbonate or wavy, pruinose frosted then smooth or more often rivulose to cracked, brownish grey to dingy crimson-brown, fading when old, often mottled. *Gills* decurrent, not very close and rather thick, greyish cream then pinkish beige-grey to dingy pinkish brown-grey, becoming dingy. *Stipe* 1.5–5 × 0.5–1.5 cm (0.6–2 × 0.2–0.6 in), pruinose to fibrillose,

concolorous, becoming dingy.
Flesh slightly or slowly bitter, greyish then brownish grey, blackening slightly or darkening when cut or bruised. Smell rancid mealy. *Microscopic features* Sp 4.5–7.5 × 4–6 μm, elliptic to subglobose, somewhat angular. Pigment epimembranal, encrusting. Clamps absent.

Season: June–December.
Habitat: Grassy places, coppices and undergrowth.
Distribution: Widespread. Rather rare to rare.

---

p.84 *Rhodocybe fallax* (Quélet) Singer
Entolomataceae, Entolomatales

Description: *Cap white, matt or cracked. Smell mealy. Taste bitter.*
Cap 1–5 cm (0.4–2 in), convex then clitocyboid or omphalinoid, pruinose, becoming glabrous, matt and soon cracked, rivulose or fissured, white or very pale. *Gills* decurrent, rather close, whitish then pinkish yellow.
Stipe 1–5 × 0.2–0.7 cm (0.4–2 × 0.08–0.28 in), pruinose then fibrillose, white or concolorous. *Flesh* bitter, white. Smell mealy to fruity, pleasant.
Microscopic features Sp 5–8.5 × 3.5–5.5 μm, elliptic, somewhat angular. Clamps absent.

Season: June–December.
Habitat: Lawns, coppices, woods.
Distribution: Uneven. Rare.

R. mundula (Lasch:Fr.) Sing. is often larger, blackens and usually occurs in woods.

---

p.132 *Rhodocybe caelata* (Fr.) Maire
Entolomataceae, Entolomatales

Description: *Cap dark brownish grey, cracked. Stipe silvery.*
Cap 0.5–3 cm (0.2–1.2 in), globose then convex, sometimes depressed, often lobed, pruinose or frosted, smooth then finely wrinkled or cracked, dark grey to brownish grey. *Gills* adnate, rather

distant, greyish cream then pinkish brownish grey, edge thick.
Stipe 1–4 × 0.2–0.5 cm (0.4–1.6 × 0.08–0.2 in), pruinose-fibrillose then slightly shiny, silvery grey on a brownish grey then brownish ground.
Flesh brownish to concolorous near the surface. Smell slightly mealy. *Microscopic features* Sp 7–9 × 4–5 µm, fusiform or elliptic, somewhat angular. Cheilo and Pleuro 35–90 × 4–10 µm, cylindro-fusoid but sometimes ventricose to clavate or lageniform, with refractive granular contents. CC a cutis. Pigment epimembranal, encrusting. Clamps absent.

Season: July–November.
Habitat: Heathland on sandy acid soils, sometimes coppice woodland or low alpine shrub zone; rarer in mature woodland.
Distribution: Uneven. Quite rare.

---

p.150 *Rhodotus palmatus* (Bull.:Fr.) Maire
Rhodotaceae, Entolomatales (?)

Description: *Cap with reticulate wrinkles, apricot pink, with an elastic gelatinous skin.*
Cap 2–10 cm (0.8–4 in), globose then convex, apricot pink, sometimes with rose red droplets when young, pruinose or glabrous, with reticulate wrinkles, and a separable elastic gelatinous skin.
Gills notched, rather distant, rather thick, concolorous or deeper pink.
Stipe 2–7 × 0.5–1.5 cm (0.8–2.8 × 0.2–0.6 in), rather leathery, often curved, pruinose and longitudinally striate, concolorous, in clusters or gregarious.
Flesh white to concolorous. Smell of apricot. *Microscopic features* Sp 5–7 µm diameter, subglobose, warted. Cheilo 35–60 × 5–8 µm, fusoid. CC an hymeniderm, the hyphae clavate or pyriform, with a filiform appendage. Clamps present.

Season: September–November.
Habitat: Stumps and low down on the trunks of broad-leaved trees (elm).
Distribution: Uneven. Rare to very rare.

p.127 *Macrocystidia cucumis* (Pers.:Fr.) Joss.
Macrocystidiaceae, Entolomatales (?)

Description: *Cap dark red-brown with an orange margin. Stipe almost black below.*

Cap 1–6 cm (0.4–2.4 in), conico-convex then expanded, pruinose, red-brown to date brown, edged with yellowish brown, orange or yellow. *Gills* almost free, cream then pinkish beige. *Stipe* 3–8 × 0.2–0.5 cm (1.2–3.2 × 0.08–0.2 in), rather stiff, pruinose to velvety, cream to pinkish beige at the apex, darker, red-brown to black-brown or almost black at the base. *Flesh* brownish to brown. Smell of putty, cod-liver oil or cucumber. *Microscopic features* Sp 7–9 × 3.5–4.5 μm, elliptic, smooth. Cheilo and Pleuro 50–100 × 10–25 μm, broadly fusoid and tapering to a point. CC a trichoderm, with filiform hyphae 2–5 μm diameter, and dermatocystidia similar to hymenial cystidia. Caulocystidia abundant. Clamps present.

Season: July–December.

Habitat: Broad-leaved and coniferous trees, sometimes in coppice woodland, often in slightly damp and/or disturbed sites.

Distribution: Uneven.

Several varieties have been described. They differ in gill shape (deeper in var. *latifolia* Lange), spore colour (white in var. *leucospora* Lange) or size (very small in f. *minor* Josserand).

GOMPHIDIACEAE

The *Gomphidiaceae* is a very small family of gilled Boletales. It is easily defined, with several distinct characters: mycorrhizal on conifers; gills decurrent, distant; spore print dark brown to black; spores fusiform with a suprahilar plage; large, more or less cylindric, encrusted cystidia; viscid veil, leaving the cap viscid, greasy or waxy.

Characters common to all species are: cap convex, smooth; margin inrolled then expanded, not striate; surface at least greasy to waxy; gills decurrent, thick, distant; stipe tapering; spores fusoid, with a pronounced suprahilar plage, very dark when mature; cap cuticle an ixocutis.

p.103 *Gomphidius roseus* (Nees.:Fr.) Gillet
Gomphidiaceae, Boletales

Description: *Cap rose red, viscid. Gills white then blackish grey.*
Cap 2–7 cm (0.8–2.8 in), soon depressed, wavy, viscid, shining, smooth, red then reddish pink to pink, the margin often paler, even whitish. *Gills* white then blackish grey. *Stipe* 2–6 × 0.5–1.5 cm (0.8–2.4 × 0.2–0.6 in), white to brownish fibrillose, often pink at the base (mycelium), viscid to lubricious. Ring or thick fibrillose belt rather high.
Flesh white or red near the surface.
Microscopic features Sp 15–25 × 5–6 μm, very narrow. Bas 45–55 × 8–12 μm, clavate, without clamps. Cheilo and Pleuro 75–130 × 10–15 μm, cylindric to cylindro-clavate, with a band of crystals mid-way or at the apex, the wall about 0.5 μm diameter. CC an ixocutis, with hyphae 4–8 μm diameter. Clamps often absent.

Season: July–November.
Habitat: Pine, especially in damp, somewhat open places, in grass, under bracken, etc. Appears always to be associated with *Suillus bovinus*.
Distribution: Uneven.

G. glutinosus (Sch.:Fr.) Fr., which has a glutinous, buff-brown cap with crimson or greyish tones and a stipe with a bright yellow base, occurs under spruce and pine.
G. maculatus (Scop.) Fr., which has an pinkish buff-brown cap with black-brown patches and a fibrillose stipe with crimson-brown blackening dots, grows under larch (rarer).

p.104 *Chroogomphus rutilus* (Sch.:Fr.) Miller
(= *Gomphidius viscidus*)
Gomphidiaceae, Boletales

Description: *Cap coppery brown to reddish. Gills olivaceous buff then blackish brown.*
Cap 2–12 cm (0.8–4.7 in), eventually depressed, wavy, sticky, smooth to more or

less innately streaky, buff brown to reddish coppery brown, sometimes several colours together. *Gills* olivaceous buff then olive-brown to blackish. *Stipe* 4–12 × 0.4–2 cm (1.6–4.7 × 0.16–0.8 in), fibrillose or mottled with olivaceous brown zones then concolorous, with reddish tones, sticky near the base. Ring zone rather high.
Flesh yellowish cream to reddish buff below, reddening slightly when cut.
Microscopic features Sp 16–22 × 6–8 μm. Bas 50–70 × 12–14 μm, clavate, without clamps. Trama pigmented. Cheilo and Pleuro 60–170 × 12–20 μm, cylindric or the base slightly inflated, with a band of crystals mid-way or at the apex, rather thick-walled 1–2.5 μm. CC an ixocutis, with hyphae 3–10 μm diameter. Pigment intracellular and epimembranal, often also extracellular. Clamps absent.

Season: July–January.
Habitat: Woods with two-needle pines.
Distribution: Widespread but frequency uneven.

C. fulmineus (Heim) Courtecuisse is very similar but the colours are much brighter and warmer, vermilion or very bright orange.

HYGROPHORACEAE: THE WAX CAP FAMILY

The *Hygrophoraceae* are a fairly homogenous group with the following characters: flesh often somewhat watery, generally terrestrial, saprotrophic or mycorrhizal; gills thick, distant and with a waxy consistency; spore print white; spores non-amyloid; brightly coloured, frequently with a viscid or greasy cuticle.

Familiarity with the general appearance of wax caps is soon acquired. It is, however, possible to confuse the dry-capped *Hygrophorus* species with *Tricholoma* or *Clitocybe* and the *Cuphophyllus* species with similarly shaped *Omphalina*. The distant gills and waxy feel should be sufficient to overcome these pitfalls.

Characters common to all species are: surface glabrous, smooth; gills thick and often interveined, distant; gill edge thick, concolorous; stipe cylindric or tapering below, glabrous; flesh watery to fibrous; smell and taste insignificant; spores hyaline, elliptic, smooth, non-amyloid; basidia 4-spored, narrowly cylindro-clavate, with clamps; cystidia absent; pigment intracellular; clamps abundant; terrestrial.

p.98 *Cuphophyllus grossulus* (Pers.) M. Bon
Hygrophoraceae, Tricholomatales

Description: *Cap yellowish beige. Gills pale yellow. Stipe curved.*
Cap 1–3 cm (0.4–1.2 in), convex then depressed, striate, sometimes toothed or fluted at the edge, often moist or slightly shiny, hygrophanous, yellowish brown with olivaceous tones, fading with age and at the margin to lemon yellow or even whitish. *Gills* decurrent, pale lemon yellow. *Stipe* 1–4 × 0.2–0.7 cm (0.4–1.6 × 0.08–0.28 in), curved or flexuous, almost concolorous to pale olivaceous yellow or whitish with age. *Flesh* thin, pale yellow. *Microscopic features* Sp 6–10 × 4–6 µm. Bas 35–50 × 4–8 µm. Clamps absent. CC subtrichodermial, with hyphae 5–15 µm diameter, the terminal cell cylindric to clavate.
Season: August–December.
Habitat: Very damp, rotten conifer wood.
Distribution: Usually continental to montane.

p.95 **Meadow Wax Cap**
Cuphophyllus pratensis (Pers.:Fr.) M. Bon
Hygrophoraceae, Tricholomatales

Description: *Cap more or less dull orange. Gills pale.*
Cap 2–7 cm (0.8–2.8 in), convex then cup-shaped, often retaining a large umbo, regular or lobed, not striate, orange, often dull or dingy, fading with age to pale orange-beige. *Gills* decurrent, pale orange-cream. *Stipe* 2–7 × 0.5–1.5 cm (0.8–2.8 × 0.2–0.6 in), smooth to fibrillose, concolorous or paler. *Flesh* firm, cream or whitish. *Microscopic features* Sp 5–7.5 ×

4–5.5 µm, elliptic to ovoid. Bas 40–65 × 5–7 µm. CC subtrichodermial, with hyphae 2–8 µm diameter, the terminal cell cylindric or slightly clavate.
Season: September–December.
Habitat: Unimproved grassland, lawns, etc.
Distribution: Widespread but frequency uneven.

Quite a good edible species (thick fleshed, which is a rather rare feature in the genus) but some cause indigestion.

---

p.90 *Cuphophyllus berkeleyi* (Ort. & Watl.) M. Bon
Hygrophoraceae, Tricholomatales

Description: *Cap ochraceous cream with a darker umbo, dry. Gills and stipe pale.*
Cap 2–5 cm (0.8–2 in), domed then expanded, remaining umbonate, with short striations or entire, slightly shiny, not greasy, cream near the edge, darker towards the brownish buff centre. *Gills* decurrent, cream to pale ochraceous beige.
Stipe 4–7 × 0.5–1.2 cm (1.6–2.8 × 0.2–0.47 in), almost concolorous or pale but, if so, with a brownish buff base.
Flesh whitish or dingy cream. *Microscopic features* Sp 5.5–8 × 3.5–5.5 µm, elliptic rather tear-shaped. Bas sometimes 2-spored (spores are then larger), 35–60 × 5–8 µm. CC a cutis, with hyphae 2–5 µm diameter.
Season: September–December.
Habitat: Lawns and unimproved grassland.
Distribution: Not fully known.

C. ochraceopallidus (Orton) M. Bon, a smaller, more striate and more hygrophanous species, has larger spores (up to 10 µm).

---

p.90 *Cuphophyllus borealis* (Peck) M. Bon
Hygrophoraceae, Tricholomatales

Description: *Cap domed, pure white, dry. Gills white. Stipe fairly robust.*
Cap 3–8 cm (1.2–3.2 in), domed then expanded umbonate, with or without faint striations, dry, smooth to slightly silky,

white. *Gills* decurrent, white. *Stipe* 4–8 × 0.3–1 cm (1.6–3.2 × 0.12–0.4 in), fleshy, white. *Flesh* quite thick, white. *Microscopic features* Sp 7.5–9.5 × 4.5–6.5 µm, elliptic. Bas 40–65 × 5–8 µm. CC a dry cutis, with hyphae 3–5 µm diameter, the terminal cell cylindric.

Season: September–December.
Habitat: Lawns and unimproved grassland, sometimes wood pasture or wood edges. Often on calcareous soils.
Distribution: Not fully known.

Confusion is possible with robust forms of *C. virgineus* (*see below*), which has a lubricious cuticle and is in a different group.

---

p.90 *Cuphophyllus virgineus* (Wulf.:Fr.) Kovalenko
Hygrophoraceae, Tricholomatales

Description: *Cap lubricious, white, more or less hygrophanous. Gills and stipe white.*
Cap 1–5 cm (0.4–2 in), hemispherical then almost flat or depressed, hygrophanous, more or less striate, lubricious, shiny, white but often tinged pale cream. *Gills* decurrent, white or cream. *Stipe* 2–7 × 0.2–1 cm (0.8–2.8 × 0.08–0.4 in), white or cream. *Flesh* white or very pale cream. *Microscopic features* Sp 7–12 × 4–7 µm, elliptic to cylindro-elliptic or tear-shaped. Bas 2- or 4-spored, 35–65 × 5–10 µm, with or without clamps (2-spored forms). CC an ixocutis, with hyphae 2–4 µm diameter.

Season: August–January.
Habitat: Lawns, unimproved grassland.
Distribution: Widespread. Still fairly common.

This species is declining. Small forms have long been called *C. niveus* (Scop.:Fr.) M. Bon but intermediates of all sizes can be found. Var. *fuscescens* (Bres.) Arnolds, is distinctly brownish to rather strong brownish grey in the centre (*see also C. cereopallidus*, p.566). Quite a good edible species, but avoid confusion with the highly poisonous species of white *Clitocybe* which sometimes share the same habitat.

p.90 *Cuphophyllus russocoriaceus* (Berk. & Miller) M. Bon
Hygrophoraceae, Tricholomatales

Description: *Cap yellowish beige. Smell of Russian leather.*
Cap 1–3 cm (0.4–1.2 in), hemispherical then almost flat, slightly hygrophanous, finely striate, lubricious, shiny, pale buff with yellowish tones, fading. *Gills* decurrent, arching, whitish to almost concolorous. *Stipe* 1–5 × 0.2–0.4 cm (0.4–2 × 0.08–0.16 in), concolorous or paler. *Flesh* pale, fragile. Smell very strong and penetrating, like Russian leather (or cedar wood). *Microscopic features* Sp 7–12 × 4–7 µm, elliptic to rather shortly ovoid. Bas 40–55 × 4.5–8 µm, with clamps. CC an ixocutis, with hyphae 1.5–4.5 µm diameter.
Season: September–December.
Habitat: Unimproved lawns, and sometimes wood pasture.
Distribution: Uneven.

Can easily be identified by the smell.

p.90 *Cuphophyllus cereopallidus* (Clç.) M. Bon
Hygrophoraceae, Tricholomatales

Description: *Cap ochraceous cream, hygrophanous, lubricious. Stipe pale.*
Cap 1.5–4 cm (0.6–1.6 in), hemispherical then convex, sometimes umbonate or almost flat, hygrophanous, very finely striate, slightly viscid or lubricious, shiny, ochraceous beige or pale ochraceous cream, fading when dry to a whitish zone. *Gills* decurrent, whitish to cream. *Stipe* 3–5 × 0.2–0.5 cm (1.2–2 × 0.08–0.2 in), pale to whitish except at the almost concolorous base. *Flesh* thin, whitish or very pale. *Microscopic features* Sp 6–8.5 × 4–5.5 µm, elliptic. Bas 25–40 × 5–8 µm. CC an ixocutis, with hyphae 1.5–4 µm diameter.
Season: September–December.
Habitat: Unimproved or semi-improved lawns.
Distribution: Not fully known. Quite rare.

See comments for *C. virgineus* (p.565).

p.91 *Cuphophyllus colemannianus* (Blox.) Bon
Hygrophoraceae, Tricholomatales

Description: *Cap reddish brown, not very striate. Gills almost concolorous. Stipe beige.*
Cap 3–8 cm (1.2–3.2 in), conico-convex then umbonate, hygrophanous, not or not very striate, lubricious, shiny, red-brown then fading to reddish brown or pinkish brown. *Gills* decurrent, interveined, pinkish beige. *Stipe* 2–7 × 0.4–1 cm (0.8–2.8 × 0.16–0.4 in), pinkish to pinkish beige, often pale. *Flesh* quite thick, pale pinkish beige. *Microscopic features* Sp 6.5–10 × 5–7 µm, elliptic. Bas 40–65 × 6–10 µm. CC an ixocutis, with hyphae 2–5 µm diameter.
Season: September–December.
Habitat: Short unimproved calcareous grassland.
Distribution: Uneven. Very rare.

C. radiatus Arnolds (= *C. subradiatus* (Schum.) M. Bon ss. auct.) is comparable but the cap is smaller and distinctly striate (very rare, in the same habitat).

p.106 *Hygrocybe punicea* (Fr.:Fr.) Kummer
Hygrophoraceae, Tricholomatales

Description: *Cap conico-convex, poppy red. Innermost flesh white.*
Cap 5–12 cm (2–4.7 in), fleshy, conico-convex then campanulate or expanded, slightly hygrophanous, smooth to very finely wrinkled, sometimes fading in zones, blood-red, with brownish or red-brown, sometimes scarlet tones, fading slightly to yellowish red-brown, the extreme edge orange-yellow from the start.
Gills notched, yellow to orange-yellow. *Stipe* 6–12 × 1–3 cm (2.4–4.7 × 0.4–1.2 in), soon hollow, tapering, coarsely fibrillose, dry, orange-red, more yellow or even whitish near the base. *Flesh* yellow but innermost flesh in the stipe whitish. *Microscopic features* Sp 8–11 × 4.5–6 µm, elliptic to cylindro-elliptic, sometimes slightly curved. Bas 45–65 × 6–10 µm.

Cheilo occasional, slightly clavate or wavy, 20–30 × 3–6 µm. CC a cutis or weakly gelatinized, sometimes subtrichodermial, with hyphae 2–5 µm diameter.

Season: October–December.
Habitat: Short unimproved calcareous grassland.
Distribution: Formerly widespread.

A good edible species but it should be treated with the utmost respect on account of its increasing rarity.

---

p.106 *Hygrocybe splendidissima* (Orton) Moser
Hygrophoraceae, Tricholomatales

Descriptions: *Cap conico-convex, dark orange-red. Innermost flesh bright yellow.*
Cap 3–8 cm (1.2–3.2 in), conico-convex then often umbonate, very slightly viscid or dry, with faint fibrillose wrinkles, very bright orange-red, discolouring slightly, extreme edge orange-yellow. *Gills* notched, an average distance apart, orange or orange-red. *Stipe* 4–8 × 0.5–2 cm (1.6–3.2 × 0.2–0.8 in), soon stuffed, less red, often orange-yellow below. *Flesh* chrome yellow, including the innermost flesh in the stipe. *Microscopic features* Sp 7.5–9.5 × 4.5.5–5 µm, elliptic or sometimes slightly almond-shaped. Bas 40–65 × 6–10 µm. CC subtrichodermial sometimes almost gelatinized, with hyphae 3–5 µm diameter.
Season: September–December.
Habitat: Lawns and unimproved grassland. Tends to be on acid soils.
Distribution: Uneven. Very rare.

---

p.107 *Hygrocybe aurantiosplendens* f. *luteosplendens* M. Bon
Hygrophoraceae, Tricholomatales

Description: *Cap orange-yellow, zoned. Gills pale yellow. Stipe fusiform.*
Cap 3–8 cm (1.2–3.2 in), conico-convex, then broadly umbonate, not or not very striate, slightly viscid, smooth or very faintly wrinkled, zoned, egg-yellow or

orange-yellow and paler yellow in the centre. *Gills* almost free, rather pale yellow, edge sometimes paler. *Stipe* 5–10 × 0.5–1.5 cm (2–4 × 0.2–0.6 in), soon hollow, fusiform, dry, orange-yellow, fading. *Flesh* fragile, yellow, innermost flesh whitish. *Microscopic features* Sp 7–9 × 3.5–5 µm, cylindro-elliptic but often constricted and wider near the apiculus. Bas 30–55 × 7–10 µm. CC an ixocutis, with hyphae 1.5–5 µm diameter, the terminal cells slender and more or less erect.

Season: September–December.
Habitat: Lawns, grassy clearings or wood pasture. Often on calcareous soils.
Distribution: Rare to very rare, usually continental to montane.

F. *luteosplendens* occurs mainly in lowland areas, and is replaced in the mountains by the type *H. aurantiosplendens* Haller which is more red.

---

p.106 **Scarlet Hood**
Hygrocybe coccinea (Fr.:Fr.) Kummer
Hygrophoraceae, Tricholomatales

Description: *Cap obtuse, scarlet, smooth. Stipe concolorous. Gills orange-yellow.*
Cap 2–6 cm (0.8–2.4 in), hemispherical then convex, rather regular, sometimes umbonate, slightly viscid or greasy, smooth to finely grained, hygrophanous, scarlet, fading from the centre in orange to yellowish patches or zones. *Gills* broadly adnate, orange-yellow with salmon tones. *Stipe* 3–8 × 0.4–1 cm (1.2–3.2 × 0.16–0.4 in), concolorous or more orange, pale below. *Flesh* orange-red in the cap, innermost flesh in the stipe quite bright yellow. *Microscopic features* Sp 8–11 × 4.5–6 µm, elliptic or slightly almond-shaped. Bas 40–65 × 5–10 µm. CC a cutis or ixocutis, with hyphae 2–8 µm diameter, some clavate and erect.

Season: August–January.
Habitat: Unimproved grassland, grassy places.
Distribution: Widespread. Fairly common.

H. phaeococcinea (Arnolds) M. Bon, which has a dark red to brownish cap, salmon gills and red flesh, is a rare species of short, usually acid, grassland.

---

p.108 *Hygrocybe quieta* (Kühner) Singer
Hygrophoraceae, Tricholomatales

Description: *Cap orange-yellow, dry, sometimes slightly disrupted. Smell of soap.*
Cap 3–7 cm (1.2–2.8 in), hemispherical then convex to lobed, dry, smooth or scurfy to disrupted, slightly hygrophanous, orange-yellow sometimes reddish, fading from the centre to a paler yellow.
Gills deeply notched, broad. *Stipe* 2–7 × 0.5–1 cm (0.8–2.8 × 0.2–0.4 in), cylindric or fusiform, often compressed or irregular, concolorous or paler. *Flesh* fragile, pale yellow to whitish. Smell soapy or of wet laundry. *Microscopic features* Sp 7–10.5 × 4.5–6.5 μm, cylindro-elliptic, constricted in the middle. Bas sometimes 2-spored, 35–60 × 5–10 μm. CC a cutis (sometimes somewhat gelatinized) sometimes subtrichodermial, with hyphae 2–7 μm, the terminal cells tapering or constricted.
Season: August–December.
Habitat: Lawns, meadows, grassy sites, etc.
Distribution: Widespread but frequency uneven.

---

p.105 *Hygrocybe ceracea* var. *vitellinoides* (M. Bon) M. Bon
Hygrophoraceae, Tricholomatales

Description: *Cap viscid, striate, orange-yellow. Gills decurrent. Stipe dry.*
Cap 1–3 cm (0.4–1.2 in), subglobose then almost flat, with rather long striations, viscid, egg-yellow to orange-yellow, fading slightly.
Gills decurrent, edge concolorous. *Stipe* 3–6 × 0.3–0.5 cm (1.2–2.4 × 0.12–0.2 in), often indented, dry, orange-yellow, pale below. *Flesh* fragile, almost concolorous.
Microscopic features Sp 6–8.5 × 3.5–4.5 μm, cylindro-elliptic or bean-shaped to constricted in the middle. Bas 30–50 ×

5–8 μm. CC an ixocutis with trichodermial areas, with hyphae 2–3 μm diameter.
Season: September–December.
Habitat: Short unimproved mossy grass.
Distribution: Not fully known. Rather common.

F. rubella (M. Bon) M. Bon is a variant with more red in its colouring.

---

p.107 *Hygrocybe insipida* (Lange) Moser
Hygrophoraceae, Tricholomatales

Description: *Cap orange-red with a yellow margin, slightly viscid. Stipe dry. Taste mild.*
Cap 1–3 cm (0.4–1.2 in), hemispherical then convex, finally irregular, striate, lubricious, smooth or finely veined, bright red to orange-red, often more yellow at the margin. *Gills* broadly adnate, pale then more or less orange to reddish, edge paler. *Stipe* 2–4 × 0.2–0.4 cm (0.8 1.6 × 0.08–0.16 in), slightly viscid then dry, twisted or indented, concolorous or more yellow, especially below. *Flesh* concolorous or paler. *Microscopic features* Sp 6–9.5 × 3–5 μm, cylindro-elliptic, sometimes constricted. Bas sometimes 2-spored, 25–40 × 5–10 μm. CC an ixocutis with trichodermial patches, the hyphae 2–5 μm diameter, sometimes branched.
Season: September–December.
Habitat: Lawns, grassy places.
Distribution: Widespread. Fairly common.

H. reai (Maire) J.E. Lange, a very similar species, differs mainly in its slightly smaller size, more viscid stipe and especially in its very bitter flesh.

---

p.105 *Hygrocybe subminutula* (Murrill) Pegler
Hygrophoraceae, Tricholomatales

Description: *Cap red, discolouring, viscid. Gills white. Stipe bright red above, viscid.*
Cap 0.5–3 cm (0.2–1.2 in), globose then almost flat, finely striate, viscid and shiny, bright red, discolouring to orange-red,

orange-yellow or even dull yellow.
Gills decurrent, white or very pale.
Stipe 2–6 × 0.1–0.4 cm (0.8–2.4 × 0.04–0.16 in), viscid, indented, almost concolorous with the cap, fading but always bright red under the pale contrasting gills.
Flesh almost concolorous or pale, bright red at the stipe apex. *Microscopic features* Sp 6–8 × 3–4 µm, cylindro-elliptic, curved or constricted. Bas 25–40 × 5–8 µm. CC an ixocutis, with hyphae 2–4 µm diameter, sometimes branched.

Season: October–December.
Habitat: Grassy places. Often on acid soils.
Distribution: Uneven. Quite rare.

---

p.106 *Hygrocybe reidii* Kühner
Hygrophoraceae, Tricholomatales

Description: *Cap orange, not very striate, dry. Stipe almost concolorous. Smell of honey.*
Cap 2–5 cm (0.8–2 in), hemispherical then expanded to wavy, hygrophanous, very faintly striate, dry, smooth to subsquamulose or irregularly disrupting with age, orange then fading to dingy orange or orange-yellow.
Gills subdecurrent, almost concolorous or sometimes whitish. *Stipe* 2–6 × 0.2–0.5 cm (0.8–2.4 × 0.08–0.2 in), dry, sometimes indented, almost concolorous, soon fading to quite a bright yellow from the base.
Flesh almost concolorous but innermost layers yellowish. Smells strongly of honey.
Microscopic features Sp 6.5–9 × 4.5–6 µm, elliptic to shortly ovoid-elliptic, sometimes slightly constricted. Bas 40–60 × 6–8 µm. CC a trichoderm, with elements up to 80 × 7 µm, tapering.

Season: September–December.
Habitat: Lawns, grassy places, etc.
Distribution: Uneven, from sea level to the alpine zone.

---

p.105 *Hygrocybe cantharellus* (Schw.:Fr.) Murr.
(= *H. lepida*)
Hygrophoraceae, Tricholomatales

Description: *Cap orange-red. Gills decurrent.*

Cap 1.5–4 cm (0.6–1.6 in), convex then becoming depressed, slightly hygrophanous, very faintly striate, often crenate, dry, smooth to subsquamulose, orange-red, fading to reddish orange-yellow. *Gills* decurrent, pale yellowish with orange to reddish tones near the base. *Stipe* 2–6 × 0.2–0.4 cm (0.8–2.4 × 0.08–0.16 in), dry, almost concolorous or yellowish at the base. *Flesh* orange-yellow to almost concolorous, the innermost layers whitish. *Microscopic features* Sp 8–11.5 × 5–7.5 µm, elliptic or bean-shaped, sometimes constricted.

Season: August–December.
Habitat: Grassy or damp mossy places.
Distribution: Widespread. Quite rare.

---

p.107 Hygrocybe coccineocrenata (Orton) Moser
Hygrophoraceae, Tricholomatales

Description: *Cap finely crenate, orange-red with small brown scales.*
Cap 1–3 cm (0.4–1.2 in), globose then hemispherical to almost depressed, not or very shortly striate, scurfy then squamulose, orange-red (margin remaining red for longer), with brownish then brown or even blackish scales. Margin finely crenate or scalloped. *Gills* subdecurrent, yellowish white to yellow or more or less orange (and slightly interveined). *Stipe* 3–7 × 0.1–0.4 cm (1.2–2.8 × 0.04–0.16 in), dry, orange-red, paler below. *Flesh* concolorous or innermost flesh yellowish orange. *Microscopic features* Sp 9–13 × 6–8 µm, elliptic to somewhat tear-shaped or slightly constricted. Bas 35–65 × 8–12 µm. Trama with hyphae, 10–35 µm diameter. CC a trichoderm, with elements 10–30 µm diameter. Pigment intracellular, brownish.

Season: July–November.
Habitat: Sphagnum in peat bogs, marshy sites.
Distribution: Quite widespread but frequency uneven.

In similar habitats, *H. turunda* (Fr.:Fr.) P. Karsten is a more dingy orange with brownish scales and smooth margin.

H. miniata (Fr.:Fr.) Kummer, a bright red species with small concolorous scales and adnate or notched, concolorous gills, occurs on lawns and heaths, usually on acid soils. Its counterpart on calcareous soils is *H. calciphila*, which may be more of an orange-red.

---

p.105 **Hygrocybe laeta** (Pers.:Fr.) Kummer
Hygrophoraceae, Tricholomatales

Description: *Wholly viscid. Cap salmon pink or orange. Gills decurrent.*
Cap 1–4 cm (0.4–1.6 in), hemispherical then sometimes depressed, striate, very viscid, smooth and shiny, salmon pink or with orange tones, sometimes with mixed colours. *Gills* decurrent, almost concolorous or flushed pink to pinkish. *Stipe* 3–6 × 0.2–0.4 cm (1.2–2.4 × 0.08–0.16 in), slightly indented, very viscid, concolorous or paler. *Flesh* almost concolorous or paler. *Microscopic features* Sp 6–9 × 4–5.5 μm, elliptic or cylindro-elliptic. Bas 30–50 × 5–8 μm, with a medallion clamp at the base. Gill edge with closely packed, slender, gelatinized and sometimes slightly branched hyphae, 25–80 × 1–3 μm. CC an ixotrichoderm with contorted hyphae 1–3 μm diameter. Clamps quite rare.
Season: September–January
Habitat: Lawns and grassy places.
Distribution: Widespread but frequency uneven.

H. luteolaeta Arnolds is similar but distinctly yellow and *H. perplexa* (A.H. Smith & Hesler) Arnolds (= *H. sciophana* p.p.) is more dull or dingy reddish like an onion skin.

---

p.108 **Parrot Wax Cap**
Hygrocybe psittacina (Sch.:Fr.) Kummer
Hygrophoraceae, Tricholomatales

Description: *Very viscid. Cap, gills and stipe dark green, discolouring.*

Cap 1–5 cm (0.4–2 in), conico-convex then expanded umbonate, striate, hygrophanous, very viscid and shining in dry weather, dark green but soon discolouring, orange, reddish, yellow, pinkish, sometimes bluish.
Gills notched, concolorous with the cap near the base but discolouring to orange or yellowish at the thick edge. *Stipe* 2–8 × 0.2–0.7 cm (0.8–3.2 × 0.08–0.28 in), often grooved, viscid, concolorous and discolouring but remaining dark green at the apex. *Flesh* brittle, concolorous or paler. *Microscopic features* Sp 7–9 × 4–6 µm, elliptic to cylindro-elliptic. Bas 30–60 × 6–10 µm, with a wide medallion clamp. CC an ixocutis, with hyphae 1.3–5 µm, sometimes branched. Clamps quite rare.

Season: August–December.
Habitat: Rather ubiquitous.
Distribution: Uneven.

The Parrot Wax Cap is easily recognized by its beautiful and vivid colours.

---

p.109 *Hygrocybe unguinosa* (Fr.:Fr.) P. Karsten
Hygrophoraceae, Tricholomatales

Description: *Very viscid. Cap and stipe beige-grey. Gills pale.*
Cap 1.5–6 cm (0.6–2.4 in), conico-campanulate then convex, striate, very viscid, shiny in dry weather, grey to greyish beige, sometimes chamois.
Gills notched, white or pale beige-grey. *Stipe* 3–6 × 0.3–0.5 cm (1.2–2.4 × 0.12–0.2 in), long clavate, often indented, very viscid, almost concolorous or paler. *Flesh* quite thin, fragile, brittle, pale. *Microscopic features* Sp 6–9 × 4.5–6 µm, elliptic. Bas 40–60 × 5–8 µm, with a broad medallion clamp. CC an ixocutis, with hyphae 1.5–4 µm diameter, slightly branched. Clamps quite rare.

Season: September–December.
Habitat: Lawns, meadows, grassy heaths and grassy places. Usually on acid soils.
Distribution: Uneven. Quite rare to rare.

p.108 *Hygrocybe fornicata* var. *streptopus* (Fr.) Arn.
Hygrophoraceae, Tricholomatales

Description: *Cap brownish beige. Gills whitish. Stipe yellowing below.*
Cap 2–7 cm (0.8–2.8 in), conico-convex or ovoid then umbonate or rounded-pointed to pointed, not striate, splitting, slightly lubricious or slightly viscid, soon dry, fibrillose to innately striate, brownish beige to beige-grey, fading from the margin.
Gills narrowly adnate, rather close, whitish to greyish or tinged very pale beige.
Stipe 5–10 × 0.4–1.2 cm (2–4 × 0.16–0.47 in), fibrillose, concolorous or paler and tending to yellow from the base.
Flesh rather fragile, white or very pale. Smell none or slightly earthy. *Microscopic features* Sp 6.5–9 × 4–6.5 µm, elliptic to somewhat tear-shaped. Bas 35–45 × 5–9 µm. CC a radial, weakly gelatinized or dry cutis, with hyphae 2–5 µm.

Season: September–December

Habitat: Unimproved lawns, sometimes coppice woodland, set-aside land, etc. Usually on calcareous soils.

Distribution: Uneven. Rare or very rare.

H. fornicata (Fr.) Singer has an almost whitish cap and a stipe that browns slightly on handling. Var. *clivalis* (Fr.) M. Bon, has a pink flush at the base and in the gills.

---

p.107 *Hygrocybe chlorophana* var. *aurantiaca* Bon
Hygrophoraceae, Tricsholomatales

Description: *Wholly viscid. Cap yellow to orange. Stipe concolorous.*
Cap 2–6 cm (0.8–2.4 in), ovoid to convex, not umbonate, not or not very striate, very viscid, shiny, orange, the centre sometimes almost reddish (the type is yellow, sometimes egg-yellow in the centre). *Gills* deeply notched, orange-yellow with a paler, yellow edge. *Stipe* 4–7 × 0.4–0.8 cm (1.6–2.8 × 0.16–0.32 in), irregular, very viscid, concolorous or slightly paler. *Flesh* yellowish to whitish. *Microscopic features* Sp 6.5–9.5 ×

4–6 μm, elliptic. Bas 30–45 × 6–10 μm. Trama parallel. CC an ixotrichoderm, with hyphae 2–4 μm diameter, sometimes diverticulate or branched.

Season: August–December.
Habitat: Unimproved lawns, coppice, wood pasture.
Distribution: Very widespread. Quite rare.

---

p.109 *Hygrocybe conica* (Scop.:Fr.) Kummer
Hygrophoraceae, Tricholomatales

Description: *Cap conical, bright red. Stipe red and yellow. Blackening.*
Cap 1–4 cm (0.4–1.6 in), pointed like an elf's cap then conical or umbonate, not or not very striate, slightly viscid, bright red, fading to orange-red then darkened by the overall blackening of the fruit body. *Gills* notched, whitish to orange-yellow. *Stipe* 2–6 × 0.2–0.5 cm (0.8–2.4 × 0.08–0.2 in), tall, smooth to fibrillose, concolorous but soon fading, sometimes bright yellow below. Blackening strongly from the base. *Flesh* concolorous or more yellow, blackening. *Microscopic features* Sp 9–15 × 5–9 μm, cylindro-elliptic or cylindric, sometimes slightly constricted. Bas 2- or 4-spored, 30–55 × 6–15 μm, with or without clamps. Gill edge with protruding, cylindric, tapering hyphae, up to 100 × 5–15 μm, often with intracellular blackish necropigment. CC a cutis, with fusiform hyphae, and some hyphae 2–3 μm diameter. Clamps present or absent.

Season: June–January.
Habitat: Rather varied.
Distribution: Widespread.

H. conica is the best known of a group of taxa or entities with more or less orange-red caps, all inedible. Three other species in this complex are described below.

---

p.110 *Hygrocybe tristis* (Pers.) Moeller
Hygrophoraceae, Tricholomatales

Description: *Rather tall. Cap dull reddish orange. Gills yellowish. Blackening.*

Cap 1.5–4 cm (0.6–1.6 in), dull or dingy orange-red, slightly viscid, sometimes slightly radially fibrillose. *Gills* yellow or yellowish then greying. *Stipe* 4–10 × 0.3–0.5 cm (1.6–4 × 0.12–0.2 in), concolorous, greying then blackening from the base. *Flesh* yellowish, blackening. *Microscopic features* as for *H. conica* but basidia more frequently 4-spored.

Season: July–December.
Habitat: Lawns, meadows, coppice, wood edges.
Distribution: Widespread.

---

p.109 *Hygrocybe cinereifolia* Courtecuisse & Priou
Hygrophoraceae, Tricholomatales

Description: *Tall. Cap orange-red. Gills grey. Stipe bright yellow above. Blackening.*
Cap 1.5–4 cm (0.6–1.6 in), slightly viscid, sometimes slightly fibrillose, somewhat dingy or dirty orange-red. *Gills* more or less dark ash-grey and remaining so to the end. *Stipe* 4–10 × 0.3–0.5 cm (1.6–4 × 0.12–0.2 in), very bright yellow in the upper part and long remaining so, a more dull orange-yellow below, blackening strongly. *Flesh* pale, blackening. *Microscopic features* as for *H. conica* but basidia 4-spored and cheilocystidia better differentiated, clavate, sometimes septate, 25–45 × 5–10 μm.

Season: September–December.
Habitat: Grassy places, lawns, coppice.
Distribution: Quite widespread. Not fully known.

---

p.109 *Hygrocybe conicoides* (Ort.) Ort. & Watl.
Hygrophoraceae, Tricholomatales

Description: *Cap bright red. Gills salmon.*
Cap 2–5 cm (0.8–2 in), bright red, scarcely blackening, often dry, slightly fibrillose. *Gills* a beautiful salmon colour, sometimes bright red, especially well away from the gill edge. *Stipe* 4–10 × 0.3–0.6 cm (1.6–4 × 0.12–0.24 in), concolorous or paler, scarcely blackening. *Flesh* concolorous, scarcely blackening. *Microscopic features* as for

H. conica, but spores more often constricted and basidia more often 4-spored.

- Season: October–December.
- Habitat: Coastal dunes, short grassland on sandy soils, grassy coppice.
- Distribution: Rather widespread; all around the coast of Europe.

---

p.109 *Hygrocybe olivaceonigra* (Orton) Moser
Hygrophoraceae, Tricholomatales

Description: *Cap shallowly blunt, pale olivaceous grey. Gills pale greenish.*
Cap 1.5–5 cm (0.6–2 in), shallowly conico-convex then almost flat, more or less fibrillose, rather dry, greenish grey to olivaceous, rather pale then becoming grey and finally entirely black. *Gills* pale greenish, becoming grey. *Stipe* 5–10 × 0.3–0.6 cm (2–4 × 0.12–0.24 in), deeply buried in the sand, whitish to pale greenish or olivaceous grey then greying strongly. *Flesh* watery and fragile, greenish, greying. *Microscopic features* as for *H. conica* but spores rather small for the group and basidia usually 4-spored.

- Season: October–December.
- Habitat: Coastal dunes.
- Distribution: Rare to very rare.

---

p.110 *Hygrocybe spadicea* (Scop.) P. Karsten
Hygrophoraceae, Tricholomatales

Description: *Cap brown. Gills yellow. Stipe pale yellow.*
Cap 3–7 cm (1.2–2.8 in), conico-convex then broadly umbonate, not striate, slightly viscid or soon dry, sometimes innately striate at the margin, brown, chestnut to milky coffee-coloured. *Gills* free or almost free, quite bright yellow. *Stipe* 4–7 × 0.5–1.2 cm (1.6–2.8 × 0.2–0.47 in), solid, fibrillose or striate, quite bright yellow. *Flesh* slightly watery in the stipe, yellowish white or lemon yellow. *Microscopic features* Sp 9–12 × 5–8 µm, elliptic or slightly oval. Bas sometimes 2-spored, 35–55 × 8–12 µm, with or without clamps.

CC a cutis, with hyphae 2–5 μm diameter, above a layer with broader fusiform hyphae.
Season: July–December.
Habitat: Lawns, meadows, open coppice.
Distribution: Submontane or continental.

Easy to recognize from its unusual colours.

---

p.108 *Hygrocybe calyptriformis* (Berk.) Fayod
Hygrophoraceae, Tricholomatales

Description: *Cap shaped like an elf cap, old rose. Gills, stipe and flesh pinkish.*
Cap 2–5 cm (0.8–2 in), elf cap-shaped (up to 8 cm (3.2 in) tall in the young adult), expanding but always conical, not striate but splitting deeply, with silky radial fibrils, old rose, fading slightly. *Gills* not quite free, whitish to pinkish. *Stipe* 5–15 × 0.5–1.5 cm (2–6 × 0.2–0.6 in), long, fusoid, white or pinkish, dry. *Flesh* very fragile, whitish to pinkish. *Microscopic features* Sp 6.5–9 × 5–7 μm, shortly elliptic to subglobose. Bas 35–50 × 5–8 μm. Gill edge with protruding hyphae up to 150 × 20 μm, tapering fusiform. CC a cutis, with hyphae 2–10 μm diameter, tapering fusiform. Clamps rare.
Season: September–December.
Habitat: Short unimproved grassland on calcareous soils.
Distribution: Uneven. Rare.

Remarkable for its unique colour; a wonderful species that should be fully protected.

---

p.108 *Hygrocybe konradii* Haller
Hygrophoraceae, Tricholomatales

Description: *Cap bluntly conical, yellow to orange-yellow, lubricious. Stipe yellow.*
Cap 2–6 cm (0.8–2.4 in), conico-convex then expanded, remaining umbonate or obtuse, not striate, lubricious then with slight radial fibrils, yellow to orange-yellow, sometimes flushed radially with reddish or

red streaks. *Gills* notched, yellow to yellowish with a paler edge. *Stipe* 5–8 × 0.5–1 cm (2–3.2 × 0.2–0.4 in), sometimes compressed, fibrillose, dry. *Flesh* yellowish. *Microscopic features* Sp 8–13 × 6–9 µm, broadly elliptic. Bas 30–50 × 8–15 µm, sometimes subcapitate. Trama parallel. CC subtrichodermial, with slender or tapering fusiform hyphae 2–10 µm diameter.

Season: September–December.
Habitat: Lawns and meadows, open grassy scrub. Usually on calcareous soils.
Distribution: Widespread. Quite rare.

H. aurantiolutescens Orton (coastal turf) has a paler, more dingy centre to the cap. *H. intermedia* (Pass.) Fayod (usually continental, grassy places and lawns) is fairly robust and has a dry, fibrillose or somewhat innately striate, radially splitting cap, flushed red on an orange or orange-yellow ground.

---

p.91 *Hygrophorus penarius* Fr.
Hygrophoraceae, Tricholomatales

Description: *Fleshy. Cap white or very pale, dry. Gills decurrent. Stipe thick.*
Cap 5–18 cm (2–7 in), domed and finally slightly convex with a broad umbo, very slightly lubricious then almost felted, sometimes slightly squamulose in the centre, white to pale cream, with a wavy, sometimes fluted margin. *Gills* decurrent, white or cream. *Stipe* 3–10 × 1–3 cm (1.2–4 × 0.4–1.2 in), fusoid ventricose, pruinose or fleecy at the apex. *Flesh* thick, quite firm, white. Smell faint, more or less oily.
Chemical test yellow on the stipe with KOH.
Microscopic features Sp 6–8 × 4–6 µm, elliptic to slightly ovoid. Bas 40–55 × 6–9 µm. Trama rather bilateral. CC a subtrichodermial cutis, the hyphae 3–5 µm diameter, with a slightly inflated apex.

Season: August–November.
Habitat: Woods on calcareous soils, especially under beech. Usually in warm sites.
Distribution: Widespread. Rather rare.
A good edible species.

Next to be described is the large mycorrhizal genus *Hygrophorus*.

---

p.91 ***Hygrophorus fagi*** Becker & M. Bon
Hygrophoraceae, Tricholomatales

Description: *Cap convex subumbonate, cream. Stipe with small scales below.*
Cap 4–10 cm (1.6–4 in), subglobose then broadly umbonate, very slightly lubricious then slightly felted in dry weather, whitish to pinkish cream. *Gills* subdecurrent, whitish to pinkish cream. *Stipe* 5–12 × 0.5–1.5 cm (2–4.7 × 0.2–0.6 in), with a pointed base, almost scaly at the very bottom, slightly pruinose fleecy at the apex, white or very pale. *Flesh* white or pale cream. *Microscopic features* Sp 7–10 × 4–5 µm, elliptic or slightly bean-shaped. Bas 45–60 × 5–8 µm. CC subtrichodermial, sometimes slightly gelatinized, the hyphae 4–8 µm diameter, more or less thick at the apex.
Season: August–November.
Habitat: Woods and coppice on calcareous soils, especially under beech.
Distribution: Germany, France and former Czechoslovakia.

H. piceae Kühner is smaller and rather slender with pinkish or rufous tones, especially in the gills, and is found under conifers. Under broad-leaved trees on acid soils, *H. penarius* var. *barbatulus* (Becker) M. Bon, is a medium-sized species, with a felted to fibrillose or bearded margin when young.

---

p.101 ***Hygrophorus arbustivus*** (Fr.) Fr.
Hygrophoraceae, Tricholomatales

Description: *Cap orange-tawny, fibrillose to innately streaky. Stipe white, dry.*
Cap 3–6 cm (1.2–2.4 in), convex then with a broad, sometimes faint umbo, quite soon dry, fibrillose or with innate radial streaks, orange-tawny or more dull, tawny beige, especially in the centre. *Gills* subdecurrent, white. *Stipe* 3–8 × 0.3–0.8 cm (1.2–3.2 ×

0.12–0.32 in), cylindric, fleecy under the gills, white. *Flesh* quite soft with age, white. *Microscopic features* Sp 7–9 × 4.5–5.5 μm, long elliptic. Bas 40–55 × 5–8 μm. CC an ixotrichoderm, with hyphae 2–6 μm diameter. Pigment mostly intracellular.

Season: September–December.
Habitat: Woods on calcareous soils (beech and also oak).
Distribution: Widespread. Quite rare.

---

p.100 *Hygrophorus leucophaeus* (Scop.) Fr.
Hygrophoraceae, Tricholomatales

Description: *Cap rufous brown with a pale cream margin. Stipe pruinose above.*
Cap 2–5 cm (0.8–2 in), campanulate convex then umbonate or slightly domed, viscid to slightly viscid in dry weather, with faint radial fibrils, zoned, cream to whitish with a rufous to rufous orange centre. *Gills* decurrent, whitish or slightly pinkish. *Stipe* 4–7 × 0.4–0.8 cm (1.6–2.8 × 0.16–0.32 in), lubricious only at the base, slightly fibrillose elsewhere and pruinose under the gills, whitish.
Flesh not very thick, white. *Microscopic features* Sp 7–9 × 5–6.5 μm, elliptic. Bas 30–45 × 6–8 μm. CC an ixocutis or trichodermial, the hyphae 2–4 μm diameter, the apex regular or clavate.

Season: September–November.
Habitat: Broad-leaved woodland on calcareous soils.
Distribution: Uneven.

H. leucophaeo-ilicis M. Bon & Chevassut, a slightly larger, fleshy and less pale species of warm sites, occurs under oaks in the south of Europe. Other taxa are very similar to *H. leucophaeus* but have a viscid stipe (*see H. lindtneri*, p.587).

---

p.102 *Hygrophorus russula* (J.C. Sch.:Fr.) Quélet
Hygrophoraceae, Tricholomatales

Description: *Cap soon becoming spotted vinaceous crimson-brown.*

Cap 5–15 cm (2–6 in), hemispherical or domed, then flat, sometimes umbonate, slightly viscid but quite soon dry, fibrillose to slightly disrupted, almost white or pinkish but covered with vinaceous pink then crimson-brown spots, finally wholly this colour. Margin sometimes almost grooved or undulating. *Gills* adnate, pinkish white then with vinaceous pink to crimson-red, finally red-brown spots or dots. *Stipe* 4–10 × 1–3 cm (1.6–4 × 0.4–1.2 in), stocky, slightly dotted at the apex, concolorous with the cap. *Flesh* thick, slightly or slowly bitter, pale but flushed or becoming reddish pink to vinaceous. Smell slightly herbaceous. *Microscopic features* Sp 6–8.5 × 4–5.5 μm, elliptic. Bas 40–55 × 5–8 μm. CC a somewhat gelatinized trichoderm, with hyphae 1–4 μm diameter. Pigment intracellular on superficial hyphae, parietal and encrusting on underlying layers.

Season: September–November.
Habitat: Especially under oak on calcareous soils. Usually in warm sites.
Distribution: Widespread; especially away from cold regions.

Best known of a group of more or less reddening species.

---

p.102 *Hygrophorus camarophyllus* (Alb. & Schw.:Fr.) Dumée *et al.*
Hygrophoraceae, Tricholomatales

Description: *Cap dark brown. Gills pale.*
Cap 5–13 cm (2–5.1 in), hemispherical then almost flat and slightly depressed, slightly viscid then dry, nearly smooth or radially fibrillose, dark brown, fading near the margin with age, the extreme edge sometimes almost white. *Gills* subdecurrent, soft, whitish to pale yellowish grey. *Stipe* 5–10 × 1–1.5 cm (2–4 × 0.4–0.6 in), pruinose above, pale, almost white at the base and apex but sometimes almost concolorous in the middle. *Flesh* white or very pale. Smell faint or like

honey. Taste nutty. *Microscopic features* Sp 7–10 × 4–6 μm, elliptic. Bas 40–55 × 5–8 μm. CC a trichodermial ixocutis, with hyphae 3–7 μm diameter. Pigment intracellular, brown.

Season: August–November.
Habitat: Damp conifer woods.
Distribution: Montane areas. Quite rare.

H. atramentosus (Alb. & Schw.) Haas & Haller ex M. Bon, a rather similar species but usually with a dark bluish grey to blackish cap, occurs in mixed woodland on calcareous soils.

---

p.172 *Hygrophorus marzuolus* (Fr.:Fr.) Bresadola
Hygrophoraceae, Tricholomatales

Description: *Cap whitish to blackish grey. Gills white.*
Cap 3–12 cm (1.2–4.7 in), hemispherical then convex, slightly viscid to fibrillose or felted, from white to grey-black according to age and light levels (the darker forms are in the light). *Gills* subdecurrent, white to greyish. *Stipe* 2–8 × 1–3.5 cm (0.8–3.2 × 0.4–1.4 in), fleecy above, fibrillose elsewhere, white or greyish. *Flesh* white or greyish near the surface. *Microscopic features* Sp 5–7.5 × 4–6 μm, elliptic. Bas 50–75 × 6–8 μm. CC an ixocutis, with hyphae 2–5 μm diameter. Pigment intracellular, brown.

Season: February–April.
Habitat: Broad-leaved and coniferous trees, on calcareous soils.
Distribution: Montane zones. Rare everywhere.

A fairly good edible species. *H. calophyllus* P. Karsten, which occurs in the autumn, has a brownish grey cap but bright salmon pink gills.

---

p.92 *Hygrophorus agathosmus* (Fr.) Fr.
Hygrophoraceae, Tricholomatales

Description: *Cap grey. Stipe apex granular. Smell of bitter almonds.*

Cap 3–8 cm (1.2–3.2 in), hemispherical then somewhat umbonate, slightly viscid or sticky, finally dry or even scurfy to subsquamulose in the centre, more or less dark grey, sometimes with pinkish or lilaceous tones. *Gills* subdecurrent, white. *Stipe* 5–8 × 1–1.5 cm (2–3.2 × 0.4–0.6 in), smooth to slightly fibrillose below but distinctly pruinose, fleecy under the gills. *Flesh* white. Smells strongly of bitter almonds. *Microscopic features* Sp 7.5–10 × 4.5–6 μm, elliptic to cylindro-elliptic. Bas 40–60 × 6–9 μm. CC an ixotrichoderm with hyphae 2–3 μm diameter. Pigment intracellular, pale.

Season: August–December.
Habitat: Conifer woods. Usually on calcareous soils.
Distribution: Widespread. More frequent in continental areas.

---

p.91 *Hygrophorus eburneus* (Bull.:Fr.) Fr.
Hygrophoraceae, Tricholomatales

Description: *Very viscid. Cap white. Stipe tapering, fleecy under the gills.*
Cap 2–8 cm (0.8–3.2 in), hemispherical then sometimes recurved, very viscid, shiny, pure white or very pale cream.
Gills decurrent, white or cream.
Stipe 5–12 × 0.3–0.7 cm (2–4.7 × 0.12–0.28 in), tapering, viscid, fleecy at the apex. *Flesh* soft, white. Smell pleasant, rather faint. *Chemical tests* orange-brown at the base of the stipe, pale yellowish buff on the cap with KOH. *Microscopic features* Sp 7–9 × 4.5–6 μm, elliptic to cylindro-elliptic. Bas 30–50 × 7–9 μm. CC an ixotrichoderm, the hyphae slender 2–5 μm diameter, with a long clavate apex.

Season: August–December.
Habitat: Broad-leaved trees, especially beech.
Distribution: Widespread but not fully understood.

H. cossus (Sow.:Fr.) Fr. is white or very pale and has a clinging smell of Jerusalem artichokes or cooking shellfish.
H. discoxanthus (Fr.) Rea (= *H. chrysaspis*)

has a distinct tendency to redden with age and has an orange-yellow to bright rufous reaction to KOH over the entire fruit body; it smells often of mandarin oranges but varies in intensity.

---

p.100 *Hygrophorus lindtneri* Moser
Hygrophoraceae, Tricholomatales

Description: *Very viscid. Cap orange-tawny in the centre. Gills pale orange-cream.*
Cap 2–5 cm (0.8–2 in), conico-convex then domed to umbonate, very viscid, shiny, zoned, rufous orange or bright rufous brown in the centre and almost whitish or cream towards the margin. *Gills* subdecurrent, white or pale orange-cream. *Stipe* 5–10 × 0.3–1.5 cm (2–4 × 0.12–0.6 in), tapering, viscid, very pale, sometimes flushed rufous orange, pruinose at the apex.
Flesh white to almost concolorous just below the surface. Smell faintly aromatic.
Microscopic features Sp 6.5–10 × 4–6 µm, elliptic. Bas 30–50 × 6–8 µm. CC an ixotrichoderm, with hyphae 2–5 µm diameter, slightly branched. Pigment faint, intracellular.
Season: September–December.
Habitat: Coppice woodland, often under hazel.
Distribution: Not fully known. Quite rare.

Differs mainly from *H. leucophaeus* (see p.583) by its viscid stipe and slightly more robust stature. Var. *carpini* (Gröger) M. Bon, is a somewhat more slender, less zoned species with less definite colours that occurs especially under hornbeam. *H. discoideus* (Pers.:Fr.) Fr. has a similarly bicoloured or zoned cap but has a distinct, thick, viscid ring zone, above which the stipe is dry.

---

p.101 *Hygrophorus pudorinus* (Fr.:Fr.) Fr.
Hygrophoraceae, Tricholomatales

Description: *Cap orange-buff, paler towards the margin. Stipe apex fleecy.*

Cap 5–13 cm (2–5.1 in), subglobose then convex or slightly umbonate, viscid, becoming somewhat radially fibrillose or speckled in the centre, rather pale orange-buff, paler towards the margin which is whitish to pinkish and sometimes fluted or slightly crenate. *Gills* subdecurrent, white or pale ochraceous cream.
Stipe 6–10 × 2–6 cm (2.4–4 × 0.8–2.4 in), tapering cylindric or fusiform, viscid or sticky, fibrillose and browning below, fleecy at the apex, white to ochraceous cream. *Flesh* white to pinkish. Smell slightly aromatic. Taste unpleasant.
Microscopic features Sp 7–11 × 4.5–6.5 µm, elliptic. Bas 45–65 × 6–10 µm. CC an ixotrichoderm, the hyphae 2–6 µm diameter, with a more or less clavate or irregular apex. Pigment intracellular.

Season: August–November.
Habitat: Conifers, often fir. Usually on calcareous soils.
Distribution: Continental and submontane.

The more robust equivalent, on acid soils, of the *H. lindtneri* (see p.587) group which occurs with deciduous trees.

---

p.102 *Hygrophorus olivaceoalbus* (Fr.:Fr.) Fr.
Hygrophoraceae, Tricholomatales

Description: *Cap viscid, olive brown. Stipe with snake-like bands of olive brown.*
Cap 2–6 cm (0.8–2.4 in), conico-convex then sometimes depressed, the umbo persistent, viscid then sticky, slightly radially fibrillose, brownish sepia to olivaceous brown, sometimes almost black at the umbo. *Gills* subdecurrent, soft, white. *Stipe* 8–12 × 0.5–1 cm (3.2–4.7 × 0.2–0.4 in), tall, with a thick high viscid ring zone, with distinct irregular olivaceous brown banding beneath. *Flesh* white to yellowish at the base. *Microscopic features* Sp 10–15 × 6.5–8.5 µm, rather long elliptic. Bas 60–80 × 8–15 µm. CC a trichodermial ixocutis, the hyphae 3–6 µm diameter,

with an obtuse cylindric apex. Pigment mostly intracellular.
Season: August–November.
Habitat: Forests of spruce or other conifers. Often on damp acid soils.
Distribution: Especially submontane or montane.

Var. *gracilis* R. Maire, which can be misidentified as it grows under broad-leaved trees, is much smaller, often not umbonate and lacks olivaceous tones. *H. fuscoalbus* (Lasch:Fr.) Fr. may also cause confusion: it has a greyish sepia to blackish cap and occurs in grassy places under a light canopy in mixed coniferous and broad-leaved woodland.

p.101 *Hygrophorus persoonii* Arnolds
Hygrophoraceae, Tricholomatales

Description: *Cap dull rufous brown, viscid. Stipe mottled with rufous brown.*
Cap 3–6 cm (1.2–2.4 in), hemispherical then flat, sometimes with a rather broad umbo, viscid then shiny, dirty rufous brown, darker in the centre (occasionally with faint olivaceous tones).
Gills adnate or subdecurrent, white.
Stipe 5–10 × 0.5–1.5 cm (2–4 × 0.2–0.6 in), tapering, with a viscid ring zone; irregular brownish to pale rufous brown bands beneath. *Flesh* quite thick, firm, white. *Chemical test* Dark green on the cap with ammonia. *Microscopic features* Sp 8–12 × 5–7 µm, elliptic. Bas 50–70 × 8–12 µm. CC an ixotrichoderm, the hyphae 3–5 µm diameter with an obtuse, clavate to branched apex. Pigment intracellular, extracellular patches dark blue-green in ammonia.

Season: August–December.
Habitat: Broad-leaved trees or coppice on rather warm calcareous sites.
Distribution: Widespread. Quite rare to very rare.

The ammonia reaction and habitat are good diagnostic characters since the other species in the group usually grow under conifers.

p.101 *Hygrophorus latitabundus* Britz.
(= *H. limacinus*)
Hygrophoraceae, Tricholomatales

Description: *Cap bronze-sepia, viscid. Stipe fusiform, mottled.*
Cap 5–15 cm (2–6 in), conico-convex then with a rather low umbo, viscid or very slimy, brownish sepia, lighter at the margin, sometimes with faint olivaceous tones. *Gills* adnate, soft, white. *Stipe* 5–15 × 1–4 cm (2–6 × 0.4–1.6 in), viscid, fusiform but tapering below, white and mottled pale sepia to brownish under a rather indistinct ring zone. *Flesh* thick, firm, white. *Chemical test* More or less orange on the cap (sometimes faintly) with ammonia. *Microscopic features* Sp 8–11.5 × 5–7.5 µm, elliptic. Bas 45–70 × 7–12 µm. CC an ixotrichoderm, the hyphae 2–6 µm diameter, with a more or less cylindric apex. Pigment intracellular.
Season: August–December.
Habitat: Pine woods, both standing and cleared, on calcareous soils; slight preference for warmer sites.
Distribution: Widespread but frequency uneven.

The largest *Hygrophorus* in this group. A good edible species.

---

p.92 *Hygrophorus chrysodon* (Batsch:Fr.) Fr.
Hygrophoraceae, Tricholomatales

Description: *Cap white, with bright yellow fleece on the margin.*
Cap 2–6 cm (0.8–2.4 in), hemispherical umbonate then convex or flat, viscid but quite soon dry and almost felted, white, the margin with golden yellow fleece sometimes reaching half-way to the centre. *Gills* decurrent, white or tinged pale lemon-yellow, edge sometimes more yellow. *Stipe* 3–8 × 0.5–1.2 cm (1.2–3.2 × 0.2–0.47 in), whitish, fibrillose under a very high ring zone, picked out by fine golden-yellow fleece. *Flesh* white to pale lemon-yellow. *Microscopic features* Sp 7–10.5 × 4–5.5 µm, elliptic to narrowly

fusiform-elliptic. Bas 35–50 × 5–8 μm. CC a trichodermial ixocutis, with hyphae 3–6 μm diameter. Pigment in bright yellow extracellular masses.

Season: August–November.
Habitat: Broad-leaved woodland. Usually in warm sites, on calcareous soils.
Distribution: Widespread. Quite rare.

---

p.101 *Hygrophorus hypothejus* (Fr.:Fr.) Fr.
Hygrophoraceae, Tricholomatales

Description: *Cap viscid, olive brown with a rather orange margin. Gills yellow.*
Cap 2–5 cm (0.8–2 in), hemispherical then convex or depressed, viscid, dark brown with olive tones in the centre; margin soon orange-yellow or rather orange.
Gills decurrent, yellowish to orange-cream.
Stipe 4–8 × 0.5–1.2 cm (1.6–3.2 × 0.2–0.47 in), hollow, viscid under a thick ring zone; slightly mottled below, whitish to yellowish beige, yellowing from the base. *Flesh* whitish but yellow or orange just below the surface. *Microscopic features* Sp 7–9.5 × 4–6 μm, elliptic. Bas 40–60 × 6–8 μm. CC an ixotrichoderm, with regular or branched hyphae 2–5 μm diameter. Pigment intracellular, extracellular encrustations more or less refractive.

Season: November–February.
Habitat: Grassy areas in standing and cleared conifer woods.
Distribution: Widespread but frequency uneven.

Other yellow or orange species occur in the mountains, usually under larch.

LEPIOTACEAE

Species in *Lepiotaceae* have the following characters: saprotrophic; gills free; spore print white, rarely red or green; veil forming a ring or fleecy, fibrillose, sometimes sheathing; spores usually dextrinoid; cap often disrupted into small scales.

The reaction of the endosporium to cresyl blue is diagnostic in separating the two subfamilies: *Lepiotoideae* and *Leucocoprinoideae* (no reaction and metachromatic reaction, respectively).

The genus *Lepiota* in the strict sense is divided into several groups, according to spore shape and, in some cases, to the reaction of microscopic characters to various chemical reagents. The three main groups of *Lepiota* differ in the following way: *Ovisporae* have ovoid spores; *Fusisporae* have fusoid spores; and *Stenosporae* have spurred spores.

Species in this family have the following characters in common: margin usually appendiculate or overhanging, not striate; gills free, ventricose, rather close, white or very pale, edge thin, entire, concolorous; flesh white to almost concolorous just below the surface; smell and taste insignificant; spores slightly thick-walled; basidia 4-spored, clavate; pleurocystidia absent; clamps present.

---

p.197 *Echinoderma echinaceum* (Lange) M. Bon
Lepiotaceae, Agaricales

Description: *Cap dark brown, with erect fragile scales. Sheath granular, concolorous.*
Cap 2–5 cm (0.8–2 in), convex to subumbonate, with erect, pyramidal scales, more or less concentric towards the margin, fleeting, dark brown on a brownish, slightly fibrillose or finely grooved ground. *Stipe* 2.5–6 × 0.3–0.7 cm (1–2.4 × 0.12–0.28 in), concolorous, with a sheath of erect fleecy scales, under a thick ring zone, soon splitting apart or shattering. *Flesh* rather thin. Smell sharp. *Microscopic features* Sp 4–6 × 2.5–3 µm, cylindro-elliptic. Cheilo very rare. CC in septate chains, with spherical or elliptic cells 20–50 × 15–30 µm, the terminal cell regular or bottle-shaped. Pigment mostly parietal, smooth.
Season: September–December.
Habitat: Mixed broad-leaved woods, on slightly disturbed or damp, enriched sites.
Distribution: Widespread but frequency uneven.

Echinoderma species are characterized by the erect, more or less fleeting scales (very difficult in the field).

---

p.197 *Echinoderma perplexum* (Knudsen) M. Bon
Lepiotaceae, Agaricales

Description: *Cap with small fibrillose, pointed scales, more or less fleeting, brown.*
Cap 5–10 cm (2–4 in), umbonate or flat, with fine erect scales, which eventually spread out and more or less disappear; brown on a paler ground. *Gills* rather narrow, edge frosted. *Stipe* 3–6 × 0.3–1 cm (1.2–2.4 × 0.12–0.4 in), beige to brownish, fibrillose under a rather wide ring, edged with scales like those on the cap, fragile and quite soon breaking up. *Flesh* rather thin. Smell sharp. *Microscopic features* Sp 5–7 × 2.5–3.5 μm, fusoid elliptic. Cheilo 20–50 × 5–15 μm, cylindro-clavate to fusoid ventricose. CC with chains of cylindric to subglobose cells 25–80 × 15–40 μm. Pigment parietal, smooth.
Season: August–December.
Habitat: Disturbed or humus-rich undergrowth. Usually on calcareous soils.
Distribution: Widespread but frequency uneven.

E. asperum (Pers.:Fr.) M. Bon, with extremely close forking gills is slightly more robust and more common.

---

p.194 *Cystolepiota adulterina* (Moeller) M. Bon
Lepiotaceae, Agaricales

Description: *Cap micaceous, cream to beige. Stipe pruinose, becoming somewhat rufous.*
Cap 1–4 cm (0.4–1.6 in), hemispherical then umbonate, pruinose, with fleecy scales that are sometimes quite dense towards the centre, fleeting; whitish to pale ochraceous beige. *Gills* greying a little, edge slightly eroded. *Stipe* 2–6 × 0.2–0.6 cm (0.8–2.4 × 0.08–0.24 in), whitish or slightly buff to dingy vinaceous below, pruinose to fleecy under a poorly defined ring zone.

Microscopic features Sp 4.5–7.5 × 2.5–3.5 μm, elliptic, cylindric or slightly curved. Cheilo 15–35 × 6–15 μm, lageniform, the neck sometimes with several slight constrictions. CC with sphaerocytes 25–50 μm diameter.

Season: August–December.
Habitat: Damp disturbed thickets and coppice woodland.
Distribution: Uneven. Quite rare to rare.

Cystolepiota, which is similar to the preceding genus, comprises species with an epithelial or subcellular cuticle.

---

p.195 *Cystolepiota hetieri* (Boudier) Singer
Lepiotaceae, Agaricales

Description: *Cap pruinose to micaceous, cream, reddening. Stipe concolorous.*
Cap 1–3 cm (0.4–1.2 in), hemispherical then slightly umbonate, pruinose, white to pinkish cream, reddening. *Gills* cut off sharply, not very close, reddening, edge eroded, with rust spots. *Stipe* 2.5–4.5 × 0.2–0.4 cm (1–1.8 × 0.08–0.16 in), pruinose or fleecy under a fragile mealy ring zone, concolorous and reddening.
Flesh reddening. Smell slightly sharp.
Microscopic features Sp 4–5.5 × 2.3 μm, trapezoidal from a slightly truncate base. Pleuro and Cheilo 20–45 × 6–12 μm, lageniform, the neck with many constrictions, sometimes misshapen or diverticulate. CC with sometimes elliptic or pedunculate sphaerocytes 20–60 μm diameter.

Season: August–December.
Habitat: Damp coppice, wood edges; sometimes in disturbed places.
Distribution: Uneven. Quite rare to rare.

---

p.194 *Cystolepiota seminuda* (Lasch) M. Bon
Lepiotaceae, Agaricales

Description: *Cap white, with a dentate margin. Stipe pinkish then vinaceous at the base.*

Cap 0.5–1 cm (0.2–0.4 in), conico-convex then more or less umbonate, with fleeting, overhanging triangular veil remains, micaceous fleecy then mealy, white or very pale cream to yellowish. *Stipe* 2–5 × 0.1–0.2 cm (0.8–2 × 0.04–0.08 in), filiform, finely mealy, flushing slightly pink or becoming vinaceous from the base. *Microscopic features* Sp 3.5–5 × 2–2.5 µm, elliptic or ovoid. CC an epithelium.

Season: June–December.
Habitat: Damp coppice, undergrowth; sometimes in disturbed places.
Distribution: Widespread. Common or fairly common.

C. sistrata (Fr.:Fr.) M. Bon, which is very similar but larger, has more broadly elliptic spores.

---

p.201 *Cystolepiota bucknallii* (Berk. & Br.) Singer & Clemençon
Lepiotaceae, Agaricales

Description: *Cap granular, violaceous lilac. Gills lemon yellow. Strong sulphurous smell.*
Cap 1–5 cm (0.4–2 in), hemispherical then expanded or dented, fleecy or mealy then pruinose, violaceous lilac to pale lilac, sometimes yellowish when bruised then fading to greyish yellow or pale reddish grey. Margin with whitish veil remains, mealy or becoming glabrous. *Gills* a pretty lemon or sulphur yellow, fading. *Stipe* 3–6 × 0.2–0.5 cm (1.2–2.4 × 0.08–0.2 in) yellowish, but fleecy under a fragile mealy ring zone, concolorous or more violaceous below. Smells strongly of coal gas or sulphurous. *Microscopic features* Sp 7.5–10.5 × 2.5–4 µm, subfusoid, slightly trapezoidal from a truncated base. CC with sometimes irregular sphaerocytes 15–25 µm diameter. Pigment slightly encrusting, dotted.

Season: August–November.
Habitat: Cool or damp disturbed coppice.
Distribution: Uneven.

p.194 *Lepiota ignivolvata* Josserand
Lepiotaceae, Agaricales

Description: *Centre warm-brown. Ring oblique, with a brownish edge. Base with orange colours.*
Cap 4–12 cm (1.6–4.7 in), conico-convex then umbonate, smooth in the centre (quite well-defined central area), disrupting outwards into small fine scales or fibrillose tufts, reddish brown to rufous buff on a cream to whitish ground. *Gills* white to pale greyish cream, edge slightly eroded. *Stipe* 6–12 × 0.5–1.2 cm (2.4–4.7 × 0.2–0.47 in), whitish with a pinkish or brownish veil. Ring rather low, the edge pinkish or brownish, whitish fibrillose above and with concolorous bands below. Base tinged rather rufous orange with age. *Microscopic features* Sp 9–12.5 × 5–7 µm, fusiform, quite broad. Cheilo 20–35 × 8–10 µm, clavate to obtusely tapering. CC a trichoderm, with hyphae 120–400 × 10–12 µm, the underlying hymeniform layer not very developed. Pigment parietal, smooth.

Season: September–November.
Habitat: Coniferous or broad-leaved trees; sometimes in disturbed places.
Distribution: Uneven. Quite rare to rare.

In *Lepiota* in the strict sense, classification depends very much on spore shape (i.e. microscope required).

---

p.193 *Lepiota laevigata* (Lange) Lange
Lepiotaceae, Agaricales

Description: *Cap with a low umbo, almost smooth, pale tawny buff. Stipe fleecy.*
Cap 2–5 cm (0.8–2 in), conico-convex then umbonate to almost flat, nearly smooth or slightly disrupted around the umbo, rather pale tawny-buff, especially at the margin. *Stipe* 2–5 × 0.2–0.5 cm (0.8–2 × 0.08–0.2 in), pale to almost concolorous or slightly more rufous below. Ring floccose, fragile and not very pronounced, the sheath with faint fleecy bands. *Microscopic features* Sp 13–18 × 5–7 µm, fusiform.

Cheilo faint, 15–25 × 5–12 μm, clavate to obtusely tapering. CC a trichoderm, with variable hyphae 100–250 × 5–10 μm. Pigment mostly intracellular.

Season: September–November.
Habitat: Turf or thickets on sandy, usually calcareous soils.
Distribution: Usually coastal or Atlantic. Quite rare.

An example of a very homogenous group of *Lepiota* spp. with fusiform spores, only slightly disrupted cap cuticle and more or less sun-loving. *L. sublaevigata* M. Bon & Boiffard has a less well-defined ring.

---

p.192 *Lepiota ochraceodisca* M. Bon
Lepiotaceae, Agaricales

Description: *Cap umbonate, with a brownish centre and white margin, smooth.*
Cap 3–6 cm (1.2–2.4 in), campanulate then umbonate or almost flat, smooth or with faint fibrillose scales at the margin, almost white at the edge but the centre with a brownish rufous 'eye'. *Stipe* 3–6 × 0.4–1 cm (1.2–2.4 × 0.16–0.4 in), white or buff below. Sheath floccose or woolly, up to a rather short-lived ring zone. Smell faint or sharp. *Microscopic features* Sp 11–16 × 5–7.5 μm, fusiform or fusiform-elliptic. Cheilo 25–35 × 8–16 μm, clavate or slightly obtusely tapering. CC a trichoderm, with hyphae 100–250 × 8–12 μm, the hymeniform layer beneath incomplete. Pigment faint.

Season: September–January.
Habitat: Short coastal grassland or thickets.
Distribution: Uneven. Rare or not fully known.

This is a variant of *L. alba* (Bres.) Saccardo, a white or very pale taxon.

---

p.196 *Lepiota clypeolaria* (Bull.:Fr.) Kummer
Lepiotaceae, Agaricales

Description: *Cap with a buff-brown central disk. Stipe woolly felted.*

Cap 2–7 cm (0.8–2.8 in), conico-campanulate then slightly umbonate, smooth in the centre (well-defined disk), finely fibrillosely scaly elsewhere, orange-brown to buff-brown on a white or cream ground. White veil remains on the margin forming triangular teeth. *Gills* close, edge finely frosted. *Stipe* 5–12 × 0.3–1 cm (2–4.7 × 0.12–0.4 in), white or pale cream, fusiform, woolly fibrillose under a floccose, diffuse and fragile ring zone. *Flesh* white or almost concolorous just below the surface. *Microscopic features* Sp 12–16 × 5–6 µm, fusiform. Cheilo quite hard to find, 20–40 × 5–15 µm, clavate or cylindro-clavate. CC a trichoderm in the centre, with hyphae 200–400 × 10–15 µm, and an hymeniform layer beneath with hyphae 25–30 × 5–10 µm. Pigment mixed or parietal at the base of the hyphae.

Season: August–December.
Habitat: Broad-leaved trees.
Distribution: Widespread. Fairly common.

L. ochraceosulfurescens (Locquin) M. Bon, differs in the less well-defined disk, less separable stipe and especially in the strong yellowing that occurs when handled.

---

p.196 *Lepiota ventriosospora* var. *fulva* M. Bon
Lepiotaceae, Agaricales

Description: *Cap with small scales and an orange-brown centre. Stipe with yellow fleece.*
Cap 3–6 cm (1.2–2.4 in), conico-campanulate then convex umbonate, almost smooth in the orange-brown centre, with small concentric scales or fibrillosely scaly elsewhere, orange-buff to tawny on a somewhat rufous to cream or yellow ground. Margin with yellow veil remains, forming triangular teeth. *Gills* yellowish to whitish. *Stipe* 4–12 × 0.4–1.2 cm (1.6–4.7 × 0.16–0.47 in), with somewhat rufous buff fleecy fibrils under a fragile ring zone. *Flesh* yellowish or buff. Smell slightly aromatic. *Microscopic features* Sp 17–25 × 4–6 µm, long fusiform.

Cheilo 15–30 × 5–15 µm, clavate or sublageniform and tapering. CC a trichoderm in the centre, with cylindric or undulating hyphae 150–400 × 5–15 µm, lacking a well-defined hymeniform layer beneath. Pigment mixed or mostly parietal towards the base of the hyphae.

Season: August–December.
Habitat: Broad-leaved trees on warm sites.
Distribution: Probably widespread; usually southern.

L. ventriosospora Reid (under conifers) is more brightly coloured and has a bright yellow veil with more orange or yellow at the margin.

---

p.198 *Lepiota cristata* (Bolt.:Fr.) Kummer
Lepiotaceae, Agaricales

Description: *Cap with small brown scales on a pale ground. Stipe vinaceous below. Smell strong.*
Cap 1–5 cm (0.4–2 in), campanulate then umbonate, more or less well defined and smooth in the centre and with short-lived concentric brown, crimson brown to light buff scales on a whitish to cream or pinkish ground. *Gills* quite close, edge finely eroded. *Stipe* 2–6 × 0.2–0.7 cm (0.8–2.4 × 0.08–0.28 in), pale towards the apex, dingy brown or vinaceous brown at the base. Ring ascending, quite wide but fragile and short-lived. Smell strong, rubbery. *Microscopic features* Sp 7–8.5 × 3–4 µm, triangular on account of the dorsal spur, faintly dextrinoid. Cheilo 15–25 × 8–14 µm, clavate. CC an hymeniderm, with clavate hyphae 30–50 × 10–25 µm. Pigment parietal, smooth or slightly encrusting in underlying layers.

Season: June–January.
Habitat: Various, usually disturbed places.
Distribution: Widespread. Common.

L. saponella Priou & Bodin, a very similar species occurring on the Atlantic coast of France, has dingy buff gills, a less regularly disrupted cuticle, more distinctly triangular spores and a soapy smell.

p.198 *Lepiota audreae* (Reid) M. Bon
Lepiotaceae, Agaricales

Description: *Cap with a rufous central disk and small scales. Stipe base with concolorous bands.*
Cap 2–7 cm (0.8–2.8 in), hemispherical then subumbonate, dark chestnut or rufous on a yellowish cream ground, the central area more or less well-defined, and with irregular or more or less fleeting small scales, especially at the margin which then appears radially grooved on account of those scales that remain. *Stipe* 3–6 × 0.5–1.5 cm (1.2–2.4 × 0.2–0.6 in), with floccose fibrils or mottled scaly below. Sheath low, with bands of scales concolorous with the cap, under a short-lived fibrillose ring zone. Smell aromatic or with a hint of geranium. *Microscopic features* Sp 7–9.5 × 3–3.5 µm, triangular or fusoid with a slightly truncated base, the dorsal spur not very developed, dextrinoid. Cheilo 20–30 × 8–12 µm, clavate to sphaeropedunculate. CC a trichoderm, with long fusoid or cylindric hyphae 200–350 × 10–15 µm, and with a few shorter hyphae. Pigment mostly parietal, smooth.
Season: September–November.
Habitat: Broad-leaved coppice; especially in warm sites.
Distribution: Uneven; continental

Looks like an *Echinoderma* because of the more or less erect and irregularly disintegrating scales on the cap.

---

p.200 *Lepiota boudieri* Bresadola
Lepiotaceae, Agaricales

Description: *Cap orange with bronze tones. Gills pinkish. Stipe with irregular bands below.*
Cap 1.5–4 cm (0.6–1.6 in), conico-convex then subumbonate, fibrillose to fibrillosely scaly in the centre when young, orange-brown then orange-buff, with the centre more olivaceous bronze. *Gills* pale cream or pinkish, edge finely eroded. *Stipe* 2.5–6 × 0.2–0.5 cm (1–2.4 × 0.08–0.2 in), yellowish

cream at the apex, with irregular orange-brown bands under a poorly defined, rather low ring zone. *Flesh* yellowish cream to pinkish when old. Smell rather aromatic. *Microscopic features* Sp 7.5–9.5 × 3–4 µm, triangular on account of the dorsal spur. Cheilo 20–35 × 7–12 µm, clavate or ventricose fusoid. CC a septate trichoderm, with cylindric cells, clamps present. Pigment mostly intracellular.

Season: September–November.

Habitat: Broad-leaved trees on somewhat disturbed sites; usually on calcareous soils and in warm places.

Distribution: Not fully assessed. Quite rare to very rare.

L. fulvella Rea, which differs in lacking the olive or pinkish tones, is more orange-tawny, and has ornament higher on the stipe. The pigment is intracellular but the elements in the cap cuticle lack clamps.

---

p.200 *Lepiota acerina* Peck
Lepiotaceae, Agaricales

Description: *Cap tawny chestnut, felted to finely fibrillosely scaly. Stipe with irregular zones.*
Cap 1.5–4 cm (0.6–1.6 in), conico-campanulate then subumbonate, felted or with small fibrillose scales, quite dark in the centre, brownish chestnut to somewhat tawny, paler towards the margin.
Stipe 2.5–7 × 0.2–0.6 cm (1–2.8 × 0.08–0.24 in), pale, with a few irregular, concolorous zones below, and a rather high, short-lived ring zone. *Microscopic features* Sp 8–10 × 3–4.5 µm, triangular on account of the dorsal spur and almost constricted. Cheilo 20–45 × 5–10 µm, clavate or fusiform. CC a septate trichoderm, with irregular, rather fusiform cells, clamps present. Pigment intracellular but also parietal, smooth towards the base.

Season: September–November.

Habitat: Broad-leaved trees on disturbed and sometimes slightly damp ground. Also with conifers.

Distribution: Not fully known. Quite rare.

p.199 *Lepiota ignipes* Locquin ex M. Bon
Lepiotaceae, Agaricales

Description: *Cap brick red, finely squamulose. Stipe base with a few scales.*
Cap 1–4 cm (0.4–1.6 in), conico-convex then subumbonate or almost flat, nearly smooth in the centre to finely scaly, chestnut then orange-brown to orange–brick-red, paler at the margin. *Gills* more or less reddening. *Stipe* 2–5 × 0.1–0.4 cm (0.8–2 × 0.04–0.16 in), fibrillose, the extreme base with bands of scales, concolorous with the cap or brighter, and a very short-lived ring zone. *Flesh* thin, white to somewhat tawny. Smell aromatic or like pencils. *Microscopic features* Sp 10–13.5 × 4–6 µm, triangular and slightly constricted. Cheilo 25–40 × 8–15 µm, clavate or cylindro-fusoid. CC a trichoderm, with cylindro-fusoid hyphae 50–250 × 10–20 µm. Pigment mostly parietal, smooth.
Season: September–December.
Habitat: Coppice or undergrowth. Usually in disturbed, warm places.
Distribution: Not fully known, usually southern.

L. castanea Quélet is very similar but has a slightly browner cap on a rufous chestnut ground and hyphae with clamps in the cap cuticle.

---

p.196 *Lepiota pseudofelina* Lange ex Lange
Lepiotaceae, Agaricales

Description: *Cap blackish grey, with small scales. Stipe with irregular concolorous banding.*
Cap 1–2 cm (0.4–0.8 in), conico-convex then subumbonate, floccose or with erect fibrils in the centre, sometimes more scaly, grey or black in the centre, on a greyish to whitish or sepia ground. *Gills* slightly greyish. *Stipe* 2–5 × 0.1–0.3 cm (0.8–2 × 0.04–0.12 in), with broken zigzag concolorous bands under a rather transient ring zone. Smell fruity or

pleasantly sharp. *Microscopic features* Sp 6.5–10 × 3–4 µm, triangular on account of the dorsal spur. Cheilo 15–30 × 5–10 µm, clavate fusiform or slightly lageniform. CC a trichoderm, with cylindro-fusoid hyphae 200–300 × 10–15 µm, lacking a well-defined lower layer. Pigment mostly parietal, smooth.

Season: September–December.
Habitat: Broad-leaved trees on slightly disturbed sites.
Distribution: Uneven. Rare to very rare.

Similar to *L. felina*, in the *Ovisporae* group of *Lepiota* (see p.607).

---

p.199 *Lepiota grangei* (Eyre) Lange
Lepiotaceae, Agaricales

Description: *Cap with small dark green scales, becoming bright orange at the margin. Stipe orange when mature.*
Cap 1–3 cm (0.4–1.2 in), conico-convex then subumbonate, felted or disrupted into rather poorly defined, dark greenish blue, greenish brown or coppery scales on a somewhat rufous cream ground, becoming rust or bright orange from the margin. *Gills* whitish then slightly rufous, edge finely eroded, concolorous then edged or dotted with rufous orange. *Stipe* 3–6 × 0.2–0.5 cm (1.2–2.4 × 0.08–0.2 in), with incomplete, broken, concolorous bands, under a rather well-defined but fleeting ring zone, flushing rufous orange from the base.

Flesh reddening. Smell sharp. *Microscopic features* Sp 10–14 × 3–5 µm, triangular or curved on account of the dorsal spur. Cheilo not very abundant. CC a trichoderm, with cylindro-fusoid or undulating hyphae, more or less septate towards the bottom. Pigment mostly intracellular, blue-green.

Season: September–December.
Habitat: Broad-leaved trees on slightly disturbed sites.
Distribution: Widespread but frequency uneven.

p.198 *Lepiota josserandii* M. Bon & Boiffard
(deadly poisonous)
Lepiotaceae, Agaricales

Description: *Cap with small pinkish buff scales. Ring appressed. Smell fruity.*
Cap 2–5 cm (0.8–2 in), globose then almost flat, with more or less concentric, small fibrillose scales, sometimes with a central disk, pinkish buff to flesh beige on a whitish to pinkish cream ground. *Gills* white or slightly pinkish. *Stipe* 3–6 × 0.3–0.8 cm (1.2–2.4 × 0.12–0.32 in), with a floccose or irregularly banded sheath, concolorous on a whitish ground, under slightly appressed or woolly, rather fragile ring. *Flesh* white or pinkish to almost concolorous. Smells strongly of mandarin oranges. *Microscopic features* Sp 5.5–7.5 × 3–4.5 µm, elliptic to cylindro-elliptic. Cheilo not obvious, 15–30 × 5–12 µm. CC a trichoderm with cylindric, obtusely tapering hyphae 8–12 µm diameter, not or not very septate at the base, lacking an hymeniform lower layer. Pigment mostly parietal, smooth.
Season: August–November.
Habitat: Coppice, parks, wood edges, gardens. In somewhat disturbed and often warm sites.
Distribution: Uneven. Quite rare.

L. helveola Bresadola is more robust, with a well-defined, starry, slightly darker vinaceous brown disk. It smells less strongly and typically grows in warm southern sites. Both species are deadly poisonous: they contain the same toxins as *Amanita phalloides*. Every year people die after collecting them, mistaking them for the Fairy Ring Champignon or other edible species.

p.201 *Lepiota brunneoincarnata* Chodat & Martin
(deadly poisonous)
Lepiotaceae, Agaricales

Description: *Cap scaly, vinaceous blackish brown in the centre on a pinkish brown ground.*
Cap 2–6 cm (0.8–2.4 in), hemispherical then almost flat, not or not very umbonate,

with irregular and rather coarse scales, sometimes with a smooth central area, vinaceous brown to blackish in the centre, paler towards the edge, on a pinkish brown ground, almost whitish towards the margin. *Stipe* 2–5 × 0.4–1.2 cm (0.8–2 × 0.16–0.47 in), stocky, white at the apex and with concolorous scales below. Ring narrow and not very thick, rather fragile, with an edging of vinaceous brown, the sheath below with bands or spirals of small brownish grey or concolorous scales. Smell like mandarin oranges. *Microscopic features* Sp 8–10.5 × 4–6 µm, elliptic. Cheilo 20–30 × 8–12 µm, clavate or fusiform-clavate. CC a trichoderm, with cylindro-fusoid hyphae 150–350 × 10–15 µm, the lower hymeniform layer loose, with variable hyphae 20–50 × 5–15 µm. Pigment mixed.

Season: August–December.

Habitat: Lawns or various kinds of coppice; on fairly warm sites.

Distribution: Widespread. Quite rare.

Deadly poisonous like the preceding and following species.

---

p.200 *Lepiota brunneolilacea* M. Bon & Boiffard (deadly poisonous)
Lepiotaceae, Agaricales

Description: *Cap lilaceous brown on a pinkish russet ground. Ring oblique, greenish sepia.*
Cap 3–7 cm (1.2–2.8 in), hemispherical then more or less umbonate, with felted, lilaceous pink then darker fibrillose scales, lilaceous brown in the centre, on a pinkish russet ground, paler towards the margin. *Gills* white or whitish with faint pinkish or pale lilac tones, sometimes with rusty spots. *Stipe* 5–9 × 0.5–1.2 cm (2–3.5 × 0.2–0.47 in), often covered in sand, fibrillose to satiny. Ring slightly woolly, edged with greenish sepia. *Flesh* white then slightly lilac or vinaceous in the stipe base.
Microscopic features Sp 8.5–12 × 5–6.5 µm, elliptic or slightly almond-shaped. Cheilo 20–35 × 5–10 µm, cylindro-clavate or

slightly tapering. CC a trichoderm, with hyphae 100–220 × 10–18 µm, the lower hymeniform layer rather dense. Pigment mostly parietal, smooth.

Season: October–January.
Habitat: Dunes.
Distribution: Mediterranean and Atlantic coasts. Rare.

---

p.199 *Lepiota pseudohelveola* var. *sabulosa* M. Bon (deadly poisonous)
Lepiotaceae, Agaricales

Description: *Cap finely scaly, brownish. Ring with a brownish edge.*
Cap 2–4 cm (0.8–1.6 in), conico-convex then subumbonate, finely scaly, the central area poorly defined, brown with pinkish or crimson tones, on a brownish beige to pinkish cream ground. *Stipe* 2–4 × 0.3–0.5 cm (0.8–1.6 × 0.12–0.2 in), with small scales in a slightly zigzag pattern towards the base, concolorous with the cap. Ring relatively low, membranous, with a brownish edge on the underside; another silky fibrillose whitish zone nearer the gills. Smell slightly sharp. *Microscopic features* Sp 6.5–9.5 × 4–5 µm, elliptic or ovoid. Cheilo 30–50 × 5–12 µm, fusi-clavate or with a long neck. CC a trichoderm, with hyphae 150–250 × 10–18 µm, the lower hymeniform layer more or less complete. Pigment mostly parietal.

Season: September–January.
Habitat: Fixed dunes; rarer inland.
Distribution: Not fully assessed.

L. pseudohelveola Kühner ex Hora often occurs under broad-leaved trees on disturbed ground, and is less coastal in its distribution. The cap has small, rather dense, dark, more pinkish brown scales on a rather pale ground.

---

p.199 *Lepiota rhodorrhiza* (Rom. & Locq.) ex Orton
Lepiotaceae, Agaricales

Description: *Cap with fine pinkish brown speckles. Stipe with more or less red rhizoids.*

Cap 0.5–2.5 cm (0.2–1 in), hemispherical then subumbonate with small, more or less erect to erect scales in the centre, pinkish brown to reddish brown or mahogany, becoming dull or fading with age. *Stipe* 2–6 × 0.1–0.3 cm (0.8–2.4 × 0.04–0.12 in), with fine white to bright red rhizoids at the base, and more or less conspicuous, irregular bands below, thinning out to a poorly defined zone. *Flesh* sometimes red at the stipe base. Smell like *L. cristata* (p.599). *Microscopic features* Sp 5–7 × 3–4 µm, ovoid. Cheilo 20–35 × 5–10 µm, ventricose fusoid to sublageniform. CC a trichoderm, with straight hyphae 150–350 × 12–18 µm, the lower layer hymeniform. Pigment mostly parietal, smooth.

Season: August–December.
Habitat: Under broad-leaved trees on slightly disturbed ground; much rarer under conifers.
Distribution: Rather uneven.

---

p.196 *Lepiota felina* (Pers.) P. Karsten
Lepiotaceae, Agaricales

Description: *Cap with small blackish scales on a pale ground. Stipe with irregular banding below.*
Cap 1–4 cm (0.4–1.6 in), hemispherical then subumbonate, the central disk sometimes smooth, scaly elsewhere, blackish to dark sepia brown on a cream to whitish ground. *Stipe* 2.5–6 × 0.2–0.5 cm (1–2.4 × 0.08–0.2 in), very pale but with small concolorous scales in irregular bands at the base. Ring relatively high, submembranous and black-edged. Smell of geranium. *Microscopic features* Sp 6–8 × 3.5–4 µm, elliptic to cylindro-elliptic. Cheilo 20–35 × 5–10 µm, clavate to fusiform-clavate. CC a trichoderm, with rather short fusoid hyphae 200–350 × 8–15 µm, the lower hymeniform layer with hyphae 30–40 × 10–20 µm. Pigment parietal, smooth to encrusting in the lower layers.

Season: August–December.

Habitat: Conifers or dense broad-leaved coppice, usually on calcareous soils.
Distribution: Uneven.

---

p.201 *Lepiota fuscovinacea* Moeller & Lange
Lepiotaceae, Agaricales

Description: *Cap woolly or with small felted lilaceous brown scales. Stipe woolly.*
Cap 2–5 cm (0.8–2 in), hemispherical then subumbonate, felted or with small appressed scales forming a slight star in the centre and more or less disrupted around it; lilaceous brown to vinaceous, on a lilaceous pink to pinkish cream ground. *Gills* white to very pale lilaceous cream. *Stipe* 3–6 × 0.2–0.6 cm (1.2–2.4 × 0.08–0.24 in), yellowish cream at the apex. Sheath concolorous or more pinkish, felted woolly, breaking up into irregular bands under a thick shaggy ring zone. *Flesh* white or pinkish to lilac. Smell slightly sharp. *Microscopic features* Sp 4–6 × 2–3 µm, cylindro-elliptic to cylindric. Cheilo 15–35 × 5–15 µm, clavate or quite broadly inflated. CC a trichoderm, with cylindric septate hyphae, the elements 50–100 × 7–15 µm. Pigment parietal or mixed.
Clamps absent.
Season: September–December.
Habitat: Coppice, more or less disturbed wood edges.
Distribution: Uneven. Rare.

This taxon is very easily recognized by its distinctive colour.

---

p.197 *Lepiota ochraceofulva* P.D. Orton
Lepiotaceae, Agaricales

Description: *Cap with a reddish brown disk and fine rust scales. Yellowing.*
Cap 3–6 cm (1.2–2.4 in), hemispherical then subumbonate, the central disk well defined, almost smooth, with small concentric red-brown to rust brown,

sometimes more rufous scales, contrasting with the ochraceous cream ground, which is whitish at the margin. Yellowing moderately, especially towards the margin where there are fleeting white velar remains. *Gills* cream to pale buff. *Stipe* 4–8 × 0.5–1 cm (1.6–3.2 × 0.2–0.4 in), sometimes fusoid, fibrillose to fleecy. Sheath with small concolorous scales towards the base and ochraceous yellow or yellow, woolly or fleecy fibrils under a short-lived fragile ring. Yellowing moderately or quite strongly below. *Flesh* ochraceous cream, yellowing slightly. Smell rather strongly aromatic, of fruit or pastries. *Microscopic features* Sp 6–8 × 3.5–4.5 µm, elliptic. Cheilo 20–30 × 5–8 µm, clavate or like basidioles. CC a rather loose hymeniderm, with uneven, cylindric or clavate hyphae, 25–55 × 10–25 µm. Pigment mostly parietal.

Season: September–November.
Habitat: Coppice, disturbed wood edges.
Distribution: Uneven. Rare to very rare.

Easily recognized by the fine, rather regular scales on the cap, the generalized, moderate yellowing and the aromatic smell.

---

p.194 *Sericeomyces sericatellus* (Mal. & Bert.) Bon
(= *Leucoagaricus s.*)
Lepiotaceae, Agaricales

Description: *A slender, white species. Cap silky. Ring fleeting.*
Cap 1.5–3 cm (0.6–1.2 in), narrowly ovoid then rounded and acute to umbonate, with silky fibrils, the fibrils rather more disrupted with age, white, shiny and sometimes cream to pinkish buff in the centre. Margin with fleeting fleecy velar remains, later splitting to finely striate. *Stipe* 4–8 × 0.2–0.5 cm (1.6–3.2 × 0.08–0.2 in), very fragile, white or slightly coloured and sometimes slightly fleecy at the base. Ring thin, slightly membranous but soon breaking up. *Microscopic features* Sp 9–15 × 4–6 µm, fusiform or

almond-shaped, with a conical or umbonate apex and a metachromatic endosporium. Bas mostly 2-spored. Cheilo 25–50 × 8–15 µm, lageniform, the neck sometimes wavy. CC a cutis, with hyphae 5–12 µm. Clamps absent.

Season: September–December.
Habitat: Broad-leaved coppice woodland. Warm sites.
Distribution: Usually southern.

Sericeomyces are characterized by the glabrous to silky cuticle (cutis not very distinctive).

---

p.191 *Leucoagaricus macrorrhizus* (Locquin) ex Horak
Lepiotaceae, Agaricales

Description: *Cap with small fibrillose scales, greyish beige. Stipe rooting. Ring rather fragile.*
Cap 3–10 cm (1.2–4 in), hemispherical then broadly convex, with fibrillose or shaggy to felted erect scales in the centre, sometimes simply fibrillose, beige-grey, brownish, greyish brown on a whitish cream ground. *Gills* sometimes tinged grey or beige. *Stipe* 5–10 × 0.5–1.5 cm (2–4 × 0.2–0.6 in), fusiform or swollen ventricose at the base but rooting, white or pale beige grey to brownish below, smooth or slightly fibrillose. Ring narrow and rather fleeting. *Flesh* whitish or brownish beige. *Microscopic features* Sp 6.5–10 × 4.5–6 µm, elliptic or slightly depressed on the ventral face, with a metachromatic endosporium. Cheilo 25–40 × 5–15 µm, clavate or slightly tapering into an obtuse neck. CC a septate trichoderm, with cylindric elements 20–50 × 5–10 µm. Pigment mixed, mostly intracellular towards the apex of the hyphae. Clamps absent.
Season: August–December.
Habitat: Lawns, parks, coppice. Usually in nitrogen-rich sites.
Distribution: Widespread, at least in warmer regions.

Abundant where it occurs in some years.

p.191 *Leucoagaricus pinguipes* (Pearson) M. Bon
Lepiotaceae, Agaricales

Description: *Cap fibrillose, pinkish brown to beige. Stipe slightly lubricious.*
Cap 2–7 cm (0.8–2.8 in), globose then convex, subumbonate, nearly smooth, slightly punctate in the centre, pinkish brown on an ochraceous beige to greyish ground. *Gills* not very close, whitish. *Stipe* 4–8 × 0.5–1 cm (1.6–3.2 × 0.2–0.4 in), white or pale beige-grey, lubricious and slightly sticky when fresh. Ring narrow, rather fleeting. *Microscopic features* Sp 6–8.5 × 3.5–5 µm, ovoid or almond-shaped with a metachromatic endosporium. Cheilo, 20–25 × 10–18 µm, clavate or utriform. CC a septate (ixo)trichoderm with cylindric elements 25–50 × 5–12 µm. Pigment mixed, mainly parietal. Clamps absent.
Season: October–December.
Habitat: Short grass on dunes and coppice near the coast.
Distribution: France, Holland, Italy, UK.

Very similar to the preceding species from which it differs mainly in the less shaggy cap, less rooting and slightly lubricious stipe.

---

p.199 *Leucoagaricus gauguei* M. Bon & Boiffard
Lepiotaceae, Agaricales

Description: *Cap with a central disk and concentric, pinkish brown scales. Stipe fusoid.*
Cap 1.5–5 cm (0.6–2 in), conico-campanulate then bell-shaped, often umbonate, with a more or less well-defined central disk, surrounded by small, somewhat short-lived scales, on a fibrillose or fleecy ground; pinkish brown to ochraceous cream, the ground colour pale. Margin striate, sometimes even splitting. *Stipe* 6–12 × 0.2–0.8 cm (2.4–4.7 × 0.08–0.32 in), cylindric to fusoid, white or pinkish cream at the base. Ring very fragile, white. *Flesh* thin, brittle, white. *Microscopic features* Sp 6.5–9 × 3–5 µm,

almond-shaped, sometimes umbonate, with a faintly metachromatic endosporium. Cheilo 30–60 × 5–15 µm, fusiform-lageniform to clavate or ventricose with a short neck. CC a trichoderm lying more or less horizontally, the terminal cell cylindric-fusoid, 50–150 × 8–15 µm, and the lower layer with the hyphae short or in septate chains. Pigment mostly intracellular. Clamps absent.

Season: September–January.
Habitat: Damp disturbed undergrowth.
Distribution: Rather warm coastal or Atlantic areas.

---

p.192 *Leucoagaricus littoralis* (Menier) M. Bon & Boiffard
Lepiotaceae, Agaricales

Description: *Cap pinkish buff, nearly smooth. Stipe covered in sand, bulbous.*
Cap 4–10 cm (1.6–4 in), globose then convex, nearly smooth or felted to finely subsquamulose towards the margin, pinkish, sometimes more beige or buff. *Gills* close. *Stipe* 5–8 × 2–6 cm (2–3.2 × 0.8–2.4 in), bulbous, finely fleecy below. Ring hanging, narrow and rather fragile. *Microscopic features* Sp 7.5–10 × 4.5–5.5 µm, almond-shaped or ovoid or, alternatively, umbonate, with a metachromatic endosporium. Cheilo 30–45 × 8–15 µm, clavate to utriform, often encrusted with small refractive crystals. CC a trichoderm lying more or less horizontally, with cylindric or undulating hyphae 50–150 × 5–12 µm. Pigment mixed, pale. Clamps absent.

Season: October–January.
Habitat: Dunes on the Atlantic and Mediterranean coasts.
Distribution: France and Italy.

L. wychanskyi (Pilát) M. Bon & Boiffard is a species of more continental, warm sites; it has a white universal veil (rather faint velar remains on the cap centre). *L. sublittoralis* (Kühner ex Hora) Bon is rather similar but a taller, continental species of enriched ground.

p.201 *Leucoagaricus purpureorimosus* M. Bon & Boiffard
Lepiotaceae, Agaricales

Description: *Cap with a star-shaped vinaceous to vinaceous pink central disk. Ring fragile.*
Cap 4–10 cm (1.6–4 in), conico-convex then subumbonate, smooth to slightly viscid in the centre, the arms of the star-like disk sometimes extending to the margin, vinaceous pink or reddish crimson, on a white ground. *Gills* distant from the stipe. *Stipe* 5–10 × 0.5–1.2 cm (2–4 × 0.2–0.47 in), fusoid, white or slightly pinkish to flesh-beige below. Ring membranous, rather narrow and fragile. Smell slightly sharp.
Microscopic features Sp 7.5–9 × 4–5.5 µm, almond-shaped, the apex regular or almost papillate, with a metachromatic endosporium. Cheilo 20–35 × 7–12 µm, clavate or fusoid with a wide obtuse neck, the top slightly dusted with fine crystals. CC rather like an ixocutis in the centre, with slender hyphae 3–5 µm diameter. Pigment mainly parietal. Clamps absent.

Season: September–December.
Habitat: Lawns, coppice or open scrub.
Distribution: Atlantic and Mediterranean coasts. Rare.

Paler or less pinkish variants may cause difficulties in identification.

---

p.200 *Leucoagaricus croceovelutinus* (M. Bon & Boiffard) M. Bon & Boiffard
Lepiotaceae, Agaricales

Description: *Cap fibrillose-squamulose, vinaceous, yellowing, reddening then blackening.*
Cap 2–5 cm (0.8–2 in), hemispherical then almost flat, nearly smooth to fibrillose then with more or less finely grooved, radial scales, distant towards the margin, red-brown then soon yellowing (bright orange-yellow), reddening and finally browning to blackening. *Gills* reddening, especially from the edge. *Stipe* 5–12 × 0.3–1 cm (2–4.7 × 0.12–0.4 in), with irregular fibrillose white zones, staining

saffron red, red then vinaceous brown or black from the base. Ring hanging, membranous but soon breaking up, sometimes with a brown edge when young. *Flesh* strongly reddening. *Microscopic features* Sp 7–9 × 3.5–4.5 µm, almond-shaped, rounded and acute or umbonate at the apex, with a metachromatic endosporium. Cheilo 25–50 × 10–15 µm, lageniform, the base ventricose and the neck wavy or with many constrictions, red-brown when old. CC a trichoderm lying more or less horizontally, with variable, cylindro-fusoid hyphae 50–300 × 10–25 µm. Pigment mostly intracellular. Clamps absent.

Season: September–December.
Habitat: Coppice, open scrub, thickets on dunes; warm and slightly nitrogen-rich sites.
Distribution: Uneven. Rare to very rare.

---

p.193 *Leucoagaricus badhamii* (Berk. & Br.) Singer
Lepiotaceae, Agaricales

Description: *Cap finely speckled, pale. Reddening very rapidly then blackening.*
Cap 3–10 cm (1.2–4 in), hemispherical then subumbonate, velvety or finely disrupted subsquamulose, cream to pale brownish on a white or pale cream ground. When touched, turns instantly orange-red or bright red-brown then blackens.
Gills reddening. *Stipe* 4–15 × 0.5–1.5 cm (1.6–6 × 0.2–0.6 in), sometimes slightly swollen to nearly bulbous, velvety fibrillose below, white or cream to beige, reddening then browning to blackening like the rest of the fruit body. Ring ascending, membranous but fragile. *Flesh* fragile, reddening vividly when cut then becoming brown to black.
Chemical test Deep green with ammonia.
Microscopic features Sp 6.5–9 × 4.5–6 µm, ovoid or almond-shaped, the apex slightly conical or elongated, with a metachromatic endosporium. Cheilo 30–70 × 10–20 µm, lageniform, the neck wavy or with many constrictions, and

brown intracellular ageing pigment. CC a trichoderm, with variable cylindro-fusoid hyphae 100–300 × 8–15 μm. Pigment parietal, smooth in underlying layers and intracellular, blackish brown with age. Clamps absent.

Season: September–December.
Habitat: Undergrowth, wood edges, coppice; especially on nitrogen-rich and fairly warm sites.
Distribution: Uneven. Rare to very rare.

---

p.192 *Leucoagaricus leucothites* (Vittadini) Wasser (= *L. naucinus*)
Lepiotaceae, Agaricales

Description: *Cap white. Gills pinkish. Ring thin.*
Cap 3–10 cm (1.2–4 in), hemispherical then shallowly convex, smooth, slightly felted or velvety, white to pale beige.
Gills flushing pink with age.
Stipe 3–7 × 0.5–1.5 cm (1.2–2.8 × 0.2–0.6 in), white or concolorous, more or less clavate. Ring membranous, narrow and quite soon appressed, concolorous or slightly cream. *Flesh* quite thick, white, unchanging. *Microscopic features*
Sp 8–10 × 5–6 μm, elliptic, the apex not elongated, with a metachromatic endosporium and a distinct GP with a metachromatic plug. Cheilo 40–60 × 12–25 μm, clavate to fusiform-lageniform. CC a trichoderm, with cylindric or fusoid hyphae 50–120 × 10–15 μm, sometimes with a few shorter hyphae at the base. Clamps absent.
Season: August–January.
Habitat: Unimproved lawns and meadows.
Distribution: Very widespread. Fairly common.

L. carneifolius (Gillet) Wasser, has pink gills and a greyish or violaceous grey cap.
L. subcretaceus M. Bon, a large robust species whose cap is scaly when mature, browns slightly. *L. holosericeus* (Gillet) Moser is white, yellowing when handled, as does *L. cinerascens* (Q.) M. Bon & Boiffard, which has a grey to quite dark sepia cap.

616 *Lepiotaceae*

p.195 *Leucocoprinus brebissonii* (Godey) Locquin
Lepiotaceae, Agaricales

Description: *Cap bell-shaped, with a blackish grey central disk. Ring fragile.*
Cap 1–5 cm (0.4–2 in), cylindro-conical then umbonate, with a smooth to scurfy disk, surrounded by small speckled to fleecy scales, short-lived at the edge, blackish to greyish on a white ground. Margin distinctly striate. *Gills* rather narrow, not very close. *Stipe* 3–8 × 0.2–0.5 cm (1.2–3.2 × 0.08–0.2 in), slightly scurfy below, white. Ring membranous, fragile, white or greyish, soon appressed and breaking up. *Flesh* very fine, white.
Microscopic features Sp 8–13 × 5–7 µm, ovoid or almond-shaped, sometimes lemon-shaped, with a metachromatic endosporium and slightly protruding GP with a faint callus. Bas sphaeropedunculate. Cheilo 20–50 × 10–15 µm, fusiform to lageniform, with a wide obtuse neck. Pseudoparaphyses globose. CC with occasional clavate hyphae 20–60 × 5–10 µm, over chains of sphaerocytes 10–30 µm diameter. Pigment mostly intracellular. Clamps absent.
Season: July–December.
Habitat: Undergrowth, coppice, wood edges. Often on nitrogen-rich and fairly warm sites.
Distribution: Widespread but frequency uneven.

Leucocoprinus (a primarily tropical genus, with very few European species) embraces those Lepiotas with fragile flesh, a striate margin and subglobose pseudoparaphyses inserted among the basidia, as in *Coprinus*. They are more frequently found in greenhouses or in a few types of warm habitat.

---

p.190 *Macrolepiota fuliginosa* (Barla) M. Bon
Lepiotaceae, Agaricales

Description: *Cap umbonate with fleeting scales. Stipe with small fleeting scales.*
Cap 8–30 cm (3.2–11.8 in), conico-convex then umbonate, with a nearly smooth centre,

the surrounding area disrupting into fine, short-lived scales; often leaving a star with more or less upturned arms on a fibrillose to fleecy ground; dull brown to sooty brown, the ground colour beige to cream, even whitish at the margin. *Gills* sometimes slightly pinkish. *Stipe* 10–30 × 1.5–5 cm (4–11.8 × 0.6–2 in), quite firm then fragile, with a bulbous base, felted then with irregular zones (looks like scribbling), that disappear when handled; concolorous with the cap or paler. Ring complex and thick, moveable, brown or concolorous beneath. *Flesh* cream to beige, flushing pink then slightly vinaceous near the edge. *Microscopic features* Sp 13–16 × 8–11 µm, elliptic, with a metachromatic endosporium, and a distinct GP, edged with a refractive callus; thick-walled. Cheilo 15–40 × 6–18 µm, fusoid or cylindro-lageniform. CC a trichoderm, with hyphae 100–200 × 8–12 µm. Pigment parietal, slightly encrusting at the base of the hyphae. Clamps only present at the base of young basidia.

Season: August–November.
Habitat: Broad-leaved trees, rarer under conifers.
Distribution: Widespread but frequency uneven.

M. procera (Scop.:Fr.) Singer (**Parasol Mushroom**) has much more pronounced and well defined zigzag banding on the stipe and practically unchanging flesh. It may be even larger. All *Macrolepiota* spp. with a mottled stipe are edible.

---

p.190 *Macrolepiota rickenii* (Vel.) Bellù & Lanzoni
(= *M. gracilenta*)
Lepiotaceae, Agaricales

Description: *Cap umbonate, squamulose, brownish. Stipe faintly zoned. Ring double.*
Cap 5–10 cm (2–4 in), conico-convex then with a conspicuous umbo, the central disk more or less well-defined, smooth to finely squamulose, surrounded by small irregular scales, sometimes radiating towards the margin and more or less fleeting; scales brown or brownish beige, paler at the

margin, on a cream or whitish ground. *Gills* pale cream, edge concolorous or with a pale brownish edge. *Stipe* 10–25 × 0.8–3 cm (4–10 × 0.32–1.2 in), very tall and slender, the base more or less bulbous, finely zoned or fleecy then becoming glabrous, concolorous with the cap or paler. Ring complex, rather thick, moveable at maturity, beige or whitish. *Flesh* cream to pale brownish. *Microscopic features* Sp 10–16 × 7–10 μm, elliptic, with a metachromatic endosporium and a conspicuous GP, closed by a refractive callus; thick-walled. Cheilo 25–40 × 10–20 μm, septate, the terminal cell clavate or ventricose utriform. CC a trichoderm, with hyphae 100–200 × 8–10 μm, septate at the base. Pigment mixed. Clamps only at the base of young basidia.

Season: August–December.
Habitat: Broad-leaved trees, grassy places.
Distribution: Widespread. Quite rare.

This remarkably elegant species is similar to *M. mastoidea* (Fr.:Fr.) Singer, which is more stocky and also has a distinctly umbonate cap. Both species are edible.

---

p.190 **Shaggy Parasol**
Macrolepiota rhacodes (Vittadini) Singer
Lepiotaceae, Agaricales

Description: *Cap with upturned scales. Stipe not mottled. Flesh reddening.*
Cap 5–15 cm (2–6 in), globose then hemispherical or domed, sometimes broadly umbonate, the central disk often fairly distinct but small, large, coarsely fibrillose, quite long-lasting, more or less upturned scales elsewhere, dark brown to brownish beige on a whitish, cream or greyish beige ground. *Gills* close, thick, cream to pale greyish beige. *Stipe* 6–12 × 1–3 cm (2.4–4.7 × 0.4–1.2 in), firm, with a round bulbous base, smooth, whitish to greyish beige or brownish from the base, bruising red then orange to vinaceous. Ring thick and complex, moveable at maturity.

Flesh fibrous, whitish or cream, reddish orange then vinaceous when cut, finally browning. *Microscopic features* Sp 10–14 × 6–9 μm, elliptic, thick-walled, with a metachromatic endosporium and truncate GP 1.5–3 μm diameter, lacking a conspicuous callus. Cheilo 30–50 × 12–20 μm, clavate. CC subhymeniform with hyphae 30–50 × 5–15 μm. Pigment in underlying layers mostly parietal. Clamps only at the base of young basidia.

Season: August–December.
Habitat: Conifers, especially spruce; rarer under broad-leaved trees or in coppice woodland.
Distribution: Widespread but frequency uneven.

A good edible species but beware of var. *bohemica* (Wich.) Bellù & Lanz. (= var. *hortensis*), which is more robust and darker, with a short stipe, very marginate bulb (up to 4 cm (1.6 in) across at the bulb) and has less reddening flesh. This taxon, of more disturbed sites with dappled shade, appears to be more or less indigestible, especially when gathered in very disturbed sites.

---

p.195 *Melanophyllum haematospermum* (Bull.:Fr.) Kreisel (= *M. echinatum*)
Lepiotaceae, Agaricales

Description: *Cap pruinose, brownish grey. Gills bright red then red-brown.*
Cap 1–4 cm (0.4–1.6 in), hemispherical then subumbonate, granular or pruinose fleecy, dull dark brown or yellowish brown, brownish grey then dingy beige, fading. Margin often with broad velar remains when young. *Gills* quite bright red then rather crimson-brown, sometimes dark when old. Spore print olivaceous green, drying red-brown. *Stipe* 3–6 × 0.2–0.5 cm (1.2–2.4 × 0.08–0.2 in), pruinose, concolorous or paler. PV forming a very fleeting floccose zone. *Flesh* pale then brownish, reddening when cut. Smell rather faint, sharp. *Microscopic features* Sp 4.5–7 × 2.5–3.5 μm, elliptic, very finely punctate. Cheilo quite rare, 15–30 × 5–8 μm, clavate

or fusoid. CC with sphaerocytes 25–50 μm diameter. Pigment mixed, intracellular and parietal, roughened.

Season: August–December.
Habitat: Thickets, coppice, wood edges in damp disturbed places.
Distribution: Widespread but frequency uneven.

---

p.195 *Melanophyllum eyrei* (Massee) Singer
Lepiotaceae, Agaricales

Description: *Cap pruinose, whitish or buff. Gills bluish green. Cap* 0.5–3 cm (0.2–1.2 in), globose then convex, pruinose or finely granular, becoming glabrous, white then slightly buff or somewhat pale tawny. Margin with fleeting veil remains. *Gills* white then greenish to bluish green, more grey when old. Spore print greenish then greenish brown. *Stipe* 2–4 × 0.1–0.4 cm (0.8–1.6 × 0.04–0.16 in), pruinose, concolorous, pale tawny buff below. PV visible as a vague edge to the pruina near the stipe apex. *Microscopic features* Sp 3–5 × 2.5–3 μm, elliptic, faintly roughened with dots. CC with sphaerocytes 15–35 μm diameter. Pigment very faint.
Season: September–December.
Habitat: Damp disturbed places, broad-leaved coppice or wood edges, etc.
Distribution: Uneven. Very rare.

The green gills are unique among gilled Agarics in the temperate zone.

MARASMIACEAE AND ALLIES

This group, which comprises the *Marasmiaceae* and *Dermolomataceae*, has the following characters: species are fleshy to very slender, the texture fibrous or sometimes somewhat cartilaginous in the stipe; gills decurrent to notched, never free; spores white, rarely slightly coloured; silhouette very seldom tricholomatoid or clitocyboid, usually omphalinoid, mycenoid, collybioid, marasmioid or sometimes pleurotoid; cap cuticle ordinary (cutis) or differentiated (trichoderm, hymeniderm or even epithelium). The distinction between *Marasmiaceae* and *Dermolomataceae* is a rather fine one. The latter family usually has a more differentiated cap cuticle; it is transitional to the Agaricales.

The fairly numerous genera exhibit great variation. Consequently, there are not many constant characters: gill edge thin, entire, concolorous; stipe cylindric, hollow at maturity; flesh pale to nearly concolorous near the surface; taste and smell insignificant; spores elliptic, hyaline, smooth, not amyloid or dextrinoid, thin-walled, lacking a germ pore; basidia 4-spored, clavate, with clamps; pleurocystidia absent; clamps present.

---

p.114 *Collybia cookei* (Bres.) J.D. Arnold
Marasmiaceae, Tricholomatales

Description: *Cap white. Gills close. Stipe on an orange-yellow sclerotium.*
Cap 2–10 mm (0.08–0.4 in), very soon flat or slightly depressed, smooth to very finely wrinkled, white to cream in the centre. *Gills* broadly adnate, close, white. *Stipe* 1–5 × 0.03–0.15 cm (0.4–2 × 0.012–0.06 in), filiform, whites, fibrillose, arising from a round, yellow, orange or yellow-buff, sometimes brownish or warm brown sclerotium, up to 5 mm (0.2 in) long. *Flesh* very thin, slightly membraneous. *Microscopic features* Sp 4–6 × 2.5–3.5 µm, ovoid-elliptic. Bas 15–20 × 3–5 µm. Cheilo scattered, 20–30 × 2–4 µm cylindric or wavy, obtuse or tapering. CC an (ixo)cutis with hyphae 3–5 µm diameter.
Season: August–December.
Habitat: On plant remains or old rotting toadstools.
Distribution: Uneven. Generally quite rare.

C. cirrhata (Pers.) Kummer, which is quite common and often occurs in large troops, has no sclerotium. *C. tuberosa* (Bull.:Fr.) Kummer is closer because it has a black sclerotium (up to 1 cm (0.4 in)).

p.114 *Collybia racemosa* (Pers.) Quélet
Marasmiaceae, Tricholomatales

Description: *Cap grey. Gills greyish. Stipe with short lateral branches.*
Cap 2–10 mm (0.08–0.4 in), conico-convex then slightly umbonate, smooth to slightly wrinkled, grey, more or less dark or with brownish tones. Margin slightly denticulate or scalloped, sometimes striate or grooved. *Gills* notched, quite close, grey or greyish beige. *Stipe* 1–7 × 0.03–0.15 cm (0.4–2.8 × 0.012–0.06 in), like a pale club, filiform, with many, quite short branches, often arising from a black, globular sclerotium. *Microscopic features* Sp 4–5.5 × 2.5–3 µm. Bas 15–20 × 3–4 µm. Cheilo absent. CC a cutis, with hyphae 1–5 µm diameter. Pigment parietal, encrusting.

Season: August–November.
Habitat: Like *C. cookei* but more often in moss.
Distribution: Not fully assessed. Rare.

---

p.137 **Butter Cap**
Collybia butyracea (Bull.:Fr.) Kummer
Marasmiaceae, Tricholomatales

Description: *Cap umbonate, greasy in the centre. Stipe hollow.*
Cap 3–8 cm (1.2–3.2 in), conico-convex then with a blunt umbo, hygrophanous, smooth, lubricious, especially on the umbo which is greasy and slippery to the touch, dark red-brown to brownish buff, fading when dry; light forms may fade to whitish with a coloured centre.
Gills notched, quite close, white.
Stipe 4–8 × 0.5–1.5 cm (1.6–3.2 × 0.2–0.6 in), compressible, soon smooth and shiny, nearly concolorous or paler.
Microscopic features Sp 5.5–7.5 × 3–4 µm, elliptic to somewhat tear-shaped, slightly dextrinoid. Bas 25–35 × 5–8 µm. Cheilo scattered, 20–35 × 5–12 µm, cylindric to fusoid or irregularly lageniform. CC an ixocutis, with hyphae 2–5 µm diameter. Pigment parietal, encrusting in a zigzag pattern.

Season: September–November.
Habitat: Broad-leaved and coniferous woodland.
Distribution: Widespread. Common.

---

p.137 *Collybia distorta* (Fr.) Quélet
Marasmiaceae, Tricholomatales

Description: *Cap umbonate, red-brown to tan. Gills close. Stipe paler, cartilaginous.*
Cap 4–10 cm (1.6–4 in), umbonate like the preceding species but not very hygrophanous and not lubricious, dark red-brown to warm buff-brown when dry.
Gills notched, fairly narrow, close, white or slightly speckled reddish or brownish.
Stipe 4–10 × 0.4–1 cm (1.6–4 × 0.16–0.4 in), cartilaginous, fibrillose-pruinose, nearly concolorous or paler, quite often twisted. *Chemical test* $FeSO_4$ pinkish.
Microscopic features Sp 3.5–4.5 × 3–4.5 µm, ovoid to subglobose, slightly dextrinoid. Bas 20–30 × 5–7 µm. Cheilo scattered, 30–45 × 3–5 µm, cylindric to flexuous or sometimes forked. CC a cutis, with hyphae 2–5 µm diameter. Pigment parietal, smooth to encrusted with zigzag banding. Caulocystidia similar to the cheilocystidia.
Season: August–November.
Habitat: Conifers; rarer under broad-leaved trees.
Distribution: Continental or submontane.

---

p.148 *Collybia fusipes* (Bull.:Fr.) Quélet
Marasmiaceae, Tricholomatales

Description: *Cap irregular, reddish brown. Gills distant. Stipe fusiform.*
Cap 2–12 cm (0.8–4.7 in), globose then often irregular, slightly hygrophanous, quite dark red-brown to buff or pinkish cream when dry, developing dark brown spots. *Gills* strongly notched, irregular, distant, cream then spotted brown.
Stipe 5–15 × 0.5–3 cm (2–6 × 0.2–1.2 in), quite leathery, elastic, fusiform and rooting, deeply grooved, whitish or ochraceous cream, later spotted and flushed dark red-brown. In clusters. *Flesh* firm

and slightly elastic. *Microscopic features*
Sp 4–6.5 × 3–4.5 μm. Bas 20–25 × 4–5 μm, cylindro-clavate. Cheilo scattered, 30–35 × 3–5 μm, cylindric to fusoid. CC a cutis, with more or less branching hyphae 3–12 μm diameter. Pigment parietal, smooth.

Season: August–December.
Habitat: Stumps, base of standing broad-leaved trees.
Distribution: Widespread but frequency uneven.

People may be tempted to eat this species but attacks of gastro-enteritis have been reported. However, the fungi were probably too old and contaminated by various bacteria. Indeed, because the flesh is quite leathery, the fungus does last quite a long time before rotting.

---

p.162 **Foxy Spot**
Collybia maculata (Alb. & Schw.:Fr.) Quélet
Marasmiaceae, Tricholomatales

Description: *Cap white then developing small reddish spots. Gills very close.*
Cap 4–12 cm (1.6–4.7 in), hemispherical then convex, smooth to silky, white, soon developing small, confluent, reddish or rust then red-brown spots, sometimes entirely dirty reddish brown.
Gills notched, fairly narrow, very close, soon with rust or reddish brown spots.
Stipe 7–15 × 0.5–2 cm (2.8–6 × 0.2–0.8 in), quite firm, concolorous, reddening later. *Flesh* spotted rust when old. *Microscopic features* Sp 5–7 × 4.5–5.5 μm, ovoid subglobose, slightly dextrinoid. Bas 25–35 × 5–8 μm. Cheilo scattered, 20–50 × 3–8 μm, flexuous cylindric or lobed to diverticulate. CC an (ixo)cutis, subtrichodermial in places, the hyphae 5–15 μm diameter, with a more or less erect and branching apex. Pigment parietal smooth.

Season: August–November.
Habitat: Coniferous but sometimes also broad-leaved woodland.
Distribution: Widespread but frequency uneven.

Var. *scorzonera* (Fr.) Gillet, has a less spotted cap and, in particular, more or less yellow gills.

---

p.165 *Tricholompsis platyphylla* (Pers.:Fr.) Singer (= *Collybia p.* = *Megacollybia p.*)
Dermolomataceae (?), Tricholomatales

Description: *Cap grey-brown. Gills distant, very wide. Stipe with rhizoids.*
Cap 3–15 cm (1.2–6 in), convex then faintly depressed to very slightly umbonate, with fine innate radial streaks, more or less brown, often with rather pronounced greyish tones, the centre often darker. Margin shortly overhanging and up-turned, finely striate, splitting.
Gills strongly notched, very wide and soft, distant, white. *Stipe* 5–15 × 0.5–2.5 cm (2–6 × 0.2–1 in), hollow at maturity, white to greyish cream. Rhizoids well developed, forming white branching cords attached to the litter. *Flesh* very thin in the cap.
Microscopic features Sp 6–10 × 5–8 µm, elliptic to subglobose. Bas 40–55 × 8–12 µm. Cheilo abundant, 30–70 × 10–20 µm, clavate to inflated elliptic. CC an (ixo)cutis with some trichodermial tendencies, the hyphae 4–10 µm diameter with a more or less clavate apex. Pigment intracellular, brown.
Season: June–December.
Habitat: Broad-leaved woods; rarer under conifers.
Distribution: Very widespread. Quite common.

---

p.136 *Collybia dryophila* (Bull.:Fr.) Kummer
Marasmiaceae, Tricholomatales

Description: *Cap orange-buff or yellowish. Gills white. Stipe hollow.*
Cap 1–6 cm (0.4–2.4 in), convex then flat-convex, hygrophanous, smooth, more or less bright orange-buff to pale yellowish cream. *Gills* notched, close, white. *Stipe* 3–8 × 0.2–0.5 cm (1.2–3.2 × 0.08–0.2 in), compressible, cylindric, smooth and shiny, concolorous with the

cap or slightly paler. *Microscopic features* Sp 4.5–6.5 × 2–3 µm, elliptic or somewhat tear-shaped. Bas 15–25 × 5–8 µm. Cheilo 18–40 × 3–10 µm, cylindric to clavate, often wavy, sometimes branching or diverticulate. CC with interwoven branching hyphae 2–10 µm diameter. Pigment parietal, encrusting.

Season: April–December.
Habitat: Broad-leaved and coniferous woods; rather varied.
Distribution: Very widespread. Common.

C. ocior (Pers.) Vilgalys & Miller (= *C. succinea*) is very similar (cap tan to reddish brown and spores wider). *C. luteifolia* Gill. (= *C. dryophila* var. *funicularis*) occurs mainly in the spring (cap darker but gills distinctly yellow). *C. kuehneriana* Singer is also quite frequent (cap pale or medium coloured but stipe bright red-brown).

---

p.114 *Collybia confluens* (Pers.:Fr.) Kummer
Marasmiaceae, Tricholomatales

Description: *Tufted. Cap beige-grey. Gills very close. Stipe pruinose.*
Cap 1–4 cm (0.4–1.6 in), hemispherical then flat-convex, matt, grey to greyish beige sometimes more buff or brownish, almost white when old and dry.
Gills notched, narrow, extremely close, pale greyish beige. *Stipe* 6–12 × 0.2–0.5 cm (2.4–4.7 × 0.08–0.2 in), cylindric, with a greyish white pruina, the ground concolorous or crimson-brown below. *Microscopic features* Sp 5.5–10 × 2.5–3.5 µm, elliptic to somewhat tear-shaped. Bas 25–35 × 3–8 µm. Cheilo 25–60 × 3–8 µm, knobbly diverticulate, sometimes lobed or constricted. CC a subtrichodermial cutis, the hyphae 2–7 µm diameter, sometimes with a clavate apex. Pigment parietal, encrusting.

Season: June–December.
Habitat: Broad-leaved trees, in tufts.
Distribution: Widespread but frequency uneven.

Very easy to recognize, except when it grows singly. The pruinose stipe and very close gills are always the best diagnostic characters.

p.117 Wood Woolly-foot
Collybia peronata (Bolt.:Fr.) Kummer
(= *Marasmius urens*)
Marasmiaceae, Tricholomatales

Description: *Gills very distant. Stipe base with long, often yellow or yellowish hairs.*
Cap 1–8 cm (0.4–3.2 in), hemispherical then sometimes depressed or umbonate, slightly hygrophanous, often finely wrinkled to wrinkled radially, yellowish ochraceous beige but sometimes dark brown to bright golden yellow.
Gills strongly notched, distant, dingy ochraceous beige to bright or pale yellow.
Stipe 3–10 × 0.3–0.5 cm (1.2–4 × 0.12–0.2 in), compressible, pruinose, the base bright golden yellow (rarely) to ochraceous beige but with long, erect, more yellow to yellowish buff hairs.
Flesh smelling of vinegar when squashed. Taste very acrid. *Microscopic features*
Sp 7–9 × 3.5–4.5 µm, elliptic to somewhat tear-shaped. Bas 30–45 × 3–8 µm, cylindro-clavate. Cheilo numerous, 40–100 × 5–10 µm, cylindro-clavate to wavy, the apex obtuse, tapering or sometimes constricted mucronate. CC a cutis or subtrichodermial, with hyphae 2–5 µm diameter. Pigment parietal, encrusting.
Season: June–December.
Habitat: Broad-leaved and coniferous woodland.
Distribution: Widespread. Common or quite common.

The only really constant character is the smell of vinegar when squashed.

p.116 *Marasmiellus ramealis* (Bull.:Fr.) Singer
Marasmiaceae, Tricholomatales

Description: *Trooping on wood. Cap cream, Stipe granular and red-brown below.*

Cap 0.3–2 cm (0.12–0.8 in) (generally about 0.5–1 cm (0.2–0.4 in)), hemispherical then flat, often pruinose, finely wrinkled to wrinkled radially, sometimes shrivelled in the centre, pale ochraceous beige, the centre rufous buff to mid-brown or reddish brown and the margin often finely striate to grooved, paler, sometimes white.
Gills adnate, fairly narrow, white then pinkish cream to pale beige. *Stipe* 4–20 × 0.3–1 mm (0.16–0.8 × 0.012–0.04 in), not compressible, tapering below, floccose or granular towards the base, whitish to ochraceous beige, darker, brown to dark red-brown, sometimes black below. *Microscopic features* Sp 7.5–11.5 × 2.5–5 μm, cylindro-elliptic to subfusiform, sometimes slightly curved. Bas 12–30 × 3–7 μm, cylindro-clavate. Cheilo 20–40 × 7–20 μm, clavate to subglobose, with finger-like and diverticulate, sometimes branching outgrowths, 2–6 × 1–4 μm. CC a subtrichodermial cutis, with thick branching hyphae 6–15 μm diameter, bearing a few loose brush-like diverticulate outgrowths, intermixed with more slender hyphae. Pigment parietal, encrusting.

Season: June–December.
Habitat: On wood, usually on twigs from broad-leaved trees, frequently on dead bramble, rarely on conifers.
Distribution: Common everywhere.

M. vaillantii (Fr.:Fr.) Singer is similar but the cap is often pale, the gills less close and the stipe more red-brown and less granular-pruinose.

---

p.112 *Marasmiellus candidus* (Bolt.) Singer
Marasmiaceae, Tricholomatales

Description: *On wood. Cap irregular, white. Gills poorly formed.*
Cap 0.5–2 cm (0.2–0.8 in), hemispherical then umbonate, sometimes flattened, finely wrinkled to wrinkled radially or fluted, white or very pale. Margin thin, more or less striate, even pectinate. *Gills* adnate to

decurrent, often poorly formed, distant and irregular, interveined, white. *Stipe* 4–15 × 0.3–1 mm (0.16–0.6 × 0.012–0.04 in), pruinose, white, becoming grey to grey-brown from the base. *Microscopic features* Sp 11–16.5 × 3.5–6.5 µm, elliptic to fusiform. Bas 20–40 × 4–10 µm, clavate to fusoid. Cheilo 40–90 × 3–12 µm, lageniform, the base slightly ventricose and the neck straight or wavy. CC subtrichodermial, with hyphae 4–12 µm diameter, coralloid and interwoven, with some erect pileocystidia similar to the cheilocystidia. Pigment parietal, encrusting, or mixed, very pale.

Season: June–November.
Habitat: In troops on twigs from broad-leaved trees.
Distribution: Widespread. Rather rare to rare.

p.95 *Marasmiellus omphaliformis* (Kühner) Noordeloos
Marasmiaceae, Tricholomatales

Description: *On wood. Cap tawny brown. Gills decurrent. Stipe base dark brown.*
Cap 0.8–1.5(2) cm (0.32–0.6(0.8) in), subglobose then sometimes depressed or umbilicate, radially wrinkled, finely pubescent, ochraceous beige to more or less dark brown from the centre, paler when dry. Margin striate to fluted and often lobed scalloped. *Gills* decurrent, curved, distant, more or less interveined below, whitish then cream. *Stipe* 8–15 × 0.5–1 mm (0.32–0.6 × 0.02–0.04 in), quite leathery or flexible, often thinner below, finely pubescent, white to yellowish cream and red-brown to brownish grey from the base. *Microscopic features* Sp 7–10 × 3.5–4.5 µm, elliptic to somewhat tear-shaped. Bas 20–35 × 4–7 µm, cylindro-clavate. Cheilo rare or absent, sometimes hair-like and tapering. CC a cutis, the hyphae 5–15 µm diameter, with numerous, more or less erect pileocystidia, 10–90 × 5–12 µm, with a ventricose or cylindric base and an apex tapering to a narrow flagellum. Pigment parietal, encrusting.

Season:	September–December.
Habitat:	Twigs and small branches, especially oak and sometimes cypress.
Distribution:	Very rare, described from Morocco, recorded in France and Portugal; locally abundant.

M. trabutii (R. Maire) Singer has a similar shape but the cap is white and the stipe almost black.

---

p.84 *Campanella caesia* Singer
Marasmiaceae, Tricholomatales

Description: *Pleurotoid, gelatinized. Greenish blue. Gills very distant, interveined.*
Cap 3–10 mm (1.2–4 in), attached by its top, globose then more or less reniform convex, slightly hygrophanous, smooth or finely wrinkled, sometimes more or less fluted, rather greenish blue, sometimes quite dark, fading as it ages and when dry, greyish to whitish. *Gills* convergent, often poorly formed, forked and interveined, very distant, whitish to pale greenish or bluish. Edge often frosted. *Microscopic features* Sp 7.5–10 × 5.5–6.5 µm, elliptic or almond-shaped. Cheilo 30–60 × 4–9 µm, cylindro-lageniform, capitate, the base irregular or with a few small outgrowths. Trama gelatinized. CC a subtrichodermial ixocutis, the hyphae with brush-like outgrowths; numerous pileocystidia similar to the cheilocystidia. Pigment parietal, encrusting in a zigzag manner.

Season: October–December.
Habitat: Grass stalks; especially in coastal dunes.
Distribution: Insufficient information. Rare or very rare.

The genus *Campanella*, which has a reduced fruit body, is placed for anatomical and structural reasons in the tribe *Collybieae*.

---

p.84 *Campanella inquilina* Romagnesi
Marasmiaceae, Tricholomatales

Description: *Like* C. caesia *(see above), but the gills more numerous and lacking blue or green colours.*

Cap 3–10 mm (0.12–0.4 in), with the same characters as *C. caesia*, but sometimes irregular, with a lobed or scalloped margin, greyish beige to grey.
Gills like *C. caesia* but often more close, like folds, forked, sometimes forming an irregular honeycomb, reticulately veined at the base, pale grey to whitish. Gill edge distinctly frosted or fringed.
Microscopic features Sp 7.5–9.5 × 4.5–5.5 μm, elliptic or almond-shaped, rhomboidal. Cheilo 35–60 × 3.5–10 μm, cylindro-lageniform, the neck obtuse or subcapitate, the base ventricose, often with diverticulate outgrowths. CC an (ixo)cutis, sometimes subtrichodermial, the hyphae 1–5 μm diameter, often diverticulate or branching into a rather dense brush, with numerous pileocystidia similar to the cheilocystidia. Pigment parietal, encrusting.

Season: October–December.
Habitat: Dead grass stalks; especially on coastal dunes.
Distribution: Very little information. Rare to very rare.

May look like some *Arrhenia*, but these never have a gelatinous texture.

---

p.111 *Crinipellis stipitarius* (Fr.:Fr.) Pat.
Marasmiaceae, Tricholomatales

Description: *Cap pale with concentric bands of appressed rufous brown fibrils. Stipe fibrillose.*
Cap 0.5–1(1.5) cm (0.2–0.4(0.6) in), hemispherical then flat to convex, ochraceous beige to whitish, with stiff appressed hairs, arranged in concentric dark brown to rufous brown zones, covering the pale ground. Often with greenish tones near the umbo. Margin often undulating or scalloped to grooved.
Gills notched, not very close, pale.
Stipe 1.5–5 × 0.05–0.15 cm (0.6–2 × 0.02–0.06 in), rather tough, covered, like the cap, in appressed hairs which are more dense towards the darker base.
Flesh rather tough, thin. *Microscopic*

features Sp 6–10 × 4.5–6 µm, elliptic. Bas 35–45 × 3–9 µm, clavate. Cheilo inconspicuous, 25–50 × 5–10 µm, cylindro-fusiform or wavy to deformed. CC with stiff hairs 100–250 × 5–10 µm, the apex obtusely tapering, thick-walled, brown, turning greenish yellow in KOH.

Season: July–December.
Habitat: Dead grass stalks.
Distribution: Widespread but frequency uneven.

---

p.111 *Setulipes androsaceus* (L.:Fr.) Antonin
(= *Marasmius a.*)
Marasmiaceae, Tricholomatales

Description: *Cap pinkish beige or violaceous, striate. Gills brown. Stipe black. Smell none.*
Cap 0.2–1.5 cm (0.08–0.6 in), hemispherical and soon flat, slightly hygrophanous, finely radially wrinkled, pinkish brown to rufous brown, sometimes with violaceous tones, fading to flesh beige-brown when dry, the centre often more russet. Margin sometimes fluted. *Gills* notched, not very close to distant, pale then flesh brown.
Stipe 2–6 × 0.03–0.1 cm (0.8–2.4 × 0.012–0.04 in), stiff and cartilaginous, smooth and shiny, black-brown, twisting as it dries. Smell none. Mycelium forming very fine rhizomorphs which run through the substrate. *Microscopic features* Sp 6.5–9.5 × 3–4.5 µm, oblong, elliptic to somewhat tear-shaped. Bas 20–40 × 6–10 µm. Basidioles fusiform. Cheilo 20–55 × 5–20 µm, cylindro-clavate to bilobed, with diverticulate outgrowths (2–7 µm diameter). CC an irregular trichoderm, with irregular or branching to coralloid hyphae 3–25 µm diameter, bearing rather short loose brush-like outgrowths. Pigment parietal, encrusting, making a zigzag pattern. Dextrinoid hyphae present, especially in the stipe.

Season: May–December.
Habitat: Conifer needles; rare under broad-leaved trees or in heathland.
Distribution: Very widespread but frequency uneven.

S. quercophilus (Pouzar) Antonin (= *S. splachnoides*), which has a very pale cap with a brown 'eye' in the centre, occurs on leaves of oak and other broad-leaved trees.
Micromphale perforans (Hoffm.:Fr.) S.F. Gray, which is very similar, smells of rotten cabbage, and often has a wrinkled or fluted cap, a slightly gelatinized texture and non-dextrinoid hyphae.

p.111 *Marasmius epiphylloides* (Rea) Sacc. & Trott. (= *M. hederae*)
Marasmiaceae, Tricholomatales

Description: *Cap white. Gills very distant, like folds. Stipe brownish below, pruinose.*
Cap 1–6 mm (0.04–0.24 in), hemispherical then convex, sometimes flat in the centre, undulating to lobed, indented or wrinkled radially, white to very pale yellowish cream. *Gills* adnate to decurrent, poorly formed and rather like folds, not always reaching the margin, very distant, sometimes forked, white. *Stipe* 4–12 × 0.1–0.3 mm (0.16–0.47 × 0.04–0.012 in), tough, filiform, pruinose below, white to brownish or reddish brown from the base.
Microscopic features Sp 9–17 × 2–4 µm cylindro-elliptic. Bas 25–40 × 6–12 µm. Basidioles cylindro-clavate to cylindro-fusoid. Cheilo and Pleuro 40–70 × 5–10 µm, fusoid-lageniform, with an obtuse or subcapitate apex. Trama not dextrinoid. CC an hymeniderm, the hyphae clavate 15–40 × 8–25 µm, with diverticulate outgrowths, 2–8 µm long. Pileocystidia numerous, similar to the cheilocystidia, sometimes thick-walled.
Season: September–March.
Habitat: On fallen ivy leaves.
Distribution: Widespread. Quite rare.

M. epiphyllus (Pers.:Fr.) Fr. is rather similar but often less small, with a wider range of habitat (various leaves and woody twigs). It has more or less dextrinoid hyphae.

p.110 *Marasmius rotula* (Scop.:Fr.) Fr.
Marasmiaceae, Tricholomatales

Description: *Cap like a parachute, white or cream. Gills distant, attached to a collar. Stipe black.*
Cap 0.5–1.5(2) cm (0.2–0.6(0.8) in), globose then like a parachute, sometimes umbilicate, wrinkled pectinate, whitish to pale cream, sometimes slightly more coloured in the centre. *Gills* attached to a collar, quite wide, rather distant, white. *Stipe* 2–6 × 0.02–0.06 cm (0.8–2.4 × 0.008–0.02 in), cartilaginous, filiform, smooth and shiny, yellowish cream just under the gills, red-brown to black below. *Microscopic features* Sp 8–11 × 3.5–5 µm, elliptic to somewhat tear-shaped. Bas 25–35 × 4–8 µm. Basidioles cylindro-clavate to fusoid-lageniform. Cheilo 12–30 × 5–18 µm, broadly clavate or pyriform, with short, fairly regular diverticulate outgrowths. Trama dextrinoid. CC an hymeniderm, with clavate pyriform to utriform hyphae 15–40 × 10–20 µm, diverticulate like cheilocystidia.

Season: June–November.
Habitat: Woody debris; in troops.
Distribution: Widespread but frequency uneven.

M. limosus Boudier & Quélet (cap up to 2 mm (0.08 in)) occurs on herbaceous marsh plants. *M. bulliardii* Quélet (cap about 2–6 mm (0.08–0.24 in)) often has a dark central spot and occurs mainly on leaves and needles. The following species also belongs in this group with collars.

---

p.110 *Marasmius curreyi* Berk. & Br.
Marasmiaceae, Tricholomatales

Description: *Cap like a parachute, coral then paler. Gills attached to a collar, cream.*
Cap 0.4–1 cm (0.16–0.4 in), like *M. rotula* (see p.634) but red-brown or coral, fading and becoming dull when old. *Gills* attached to a narrow collar, quite wide, horizontal or slightly ventricose,

distant, white or pale cream.
Stipe 1.5–3 × 0.02–0.05 cm
(0.6–1.2 × 0.008–0.02 in), same
characters as for *M. rotula*. *Microscopic features* Sp 8–11.5 × 4–6 μm, cylindro-elliptic or more or less almond-shaped. Bas 25–40 × 7–9 μm. Basidioles cylindro-clavate to fusoid. Cheilo 10–20 × 6–12 μm, clavate, with fairly regular diverticulate outgrowths. Trama dextrinoid. CC an hymeniderm, with clavate to cylindro-clavate, diverticulate hyphae 10–25 × 8–15 μm, the wall yellowish brown under the microscope.

Season: June–October.
Habitat: Growing on dead or rotting grass stalks and other herbaceous plants.
Distribution: Uneven. Quite common to rare.

---

p.137 *Marasmius alliaceus* (Jacq.:Fr.) Fr.
Marasmiaceae, Tricholomatales

Description: *Cap ochraceous beige, matt. Stipe brown to black, very pruinose. Strong smell of garlic.*
Cap 1–5 cm (0.4–2 in), conico-convex then with a low umbo, hygrophanous, pruinose, smooth to wrinkled, grey then grey-brown to more or less warm or dull brown, fading when dry to greyish beige, the centre remaining more coloured. *Gills* narrowly adnate, not very close, white or greyish cream, sometimes spotted brownish or reddish. *Stipe* 5–15 (20) × 0.1–0.3 cm (2–6(8) × 0.04–0.12 in), rather stiff, very pruinose, whitish then yellowish cream and red-brown to almost black from the base. Smell of garlic, very strong and persistent. *Microscopic features* Sp 8–13 × 5.5–9 μm, elliptic or almond-shaped ventricose, sometimes somewhat lozenge-shaped. Bas 30–55 × 5–10 μm, clavate to subcapitate. Basidioles clavate to fusiform. Cheilo 30–70 × 6–18 μm, fusoid-lageniform. Trama not dextrinoid. CC an hymeniderm, the hyphae 15–40 × 8–20 μm, pyriform, cylindro-clavate to utriform, not diverticulate. Caulocystidia cylindro-lageniform.

Season: June–November.
Habitat: Almost exclusively a beech species, on woody debris.
Distribution: Widespread but frequency uneven.

M. scorodonius (Fr.:Fr.) Fr., another garlic-smelling species, occurs mainly under conifers (cap pinkish brown, stipe reddish brown to almost black below, but shiny).

p.126 *Marasmius undatus* (Berk.) Fr.
Marasmiaceae, Tricholomatales

Description: *Cap grey to brown, pruinose. Stipe with a dense pale grey pruina, crimson below.*
Cap 0.5–2.5 cm (0.2–1 in), globose then shallowly convex, very pruinose, becoming glabrous, sometimes finely wrinkled in the centre, whitish or greyish then more brownish in places, finally with yellow, olivaceous to warm brown tones. *Gills* broadly adnate, rather distant, white then brownish cream in places. Gill edge sometimes with small clear droplets. *Stipe* 4–10 × 0.1–0.3 cm (1.6–4 × 0.04–0.12 in), tough, very pruinose or velvety, the apex white or cream, often with small clear droplets when young, soon tinged brownish to red-brown or crimson-brown below. *Flesh* thin, sometimes with a mealy smell or a suggestion of garlic when bruised. *Microscopic features* Sp 8.5–13 × 4.5–6.5 µm, broadly fusoid or lemon-shaped. Bas 30–45 × 5–10 µm. Basidioles cylindro-clavate to slightly fusoid. Cheilo and Pleuro 25–70 × 5–13 µm, fusoid-lageniform subcapitate. Trama not dextrinoid. CC an hymeniderm, the hyphae 5–40 × 5–15 µm, pyriform to broadly fusiform, smooth. Pileocystidia abundant, 30–100 × 5–12 µm, lageniform to cylindric, the neck straight, wavy or subcapitate.
Season: June–December.

Habitat: Where ferns, especially bracken, occur, in woods, heaths, acid dunes, etc.
Distribution: Uneven. Quite rare to very rare.

---

p.110 *Marasmius anomalus* Lasch
Marasmiaceae, Tricholomatales

Description: *Cap shallowly convex. Gills distant. Stipe red-brown below.*
Cap 1–4 cm (0.4–1.6 in), conico-convex then sometimes almost flat but with a faint umbo, scalloped, smooth to faintly grained, tawny buff, sometimes more brown in the centre, fading from the margin.
Gills nearly free, not attached to a collar, ventricose, distant, white or pale.
Stipe 4–10 × 0.1– 0.5 cm (1.4–4 × 0.04–0.2 in), cartilaginous, shiny, yellowish cream at the apex, more brown to dark red-brown towards the base. *Microscopic features* Sp 7–10 × 3.5– 5.5 µm, elliptic to tear-shaped. Bas 25–40 × 4– 8.5 µm. Basidioles cylindro-clavate. Cheilo 9–30 × 5–10 µm, cylindro-clavate, with quite short, wavy or straight, finger-like processes (2–10 × 0.5–1.5 µm), with thick yellowish brown walls. Pleuro 35–60 × 6– 15 µm, fusoid-clavate or ventricose fusoid, thin-walled. Trama dextrinoid. CC an hymeniderm, with hyphae like the cheilocystidia.
Season: June–December.
Habitat: Mossy or grassy turf.
Distribution: Uneven. Common or quite rare.

Quite easy to recognize but variable under the microscope. Several taxa have been described but it is best to unite them under a single epithet.

---

p.116 *Marasmius cohaerens* (Pers.:Fr.) Cke. & Q.
Marasmiaceae, Tricholomatales

Description: *Cap brownish buff. Stipe cartilaginous, dark red-brown and shiny.*
Cap 1–3 cm (0.4–1.2 in), convex then expanded, often smooth, yellowish brown to

buff brown. *Gills* notched, not very close, white or yellowish cream. *Stipe* 2–8 × 0.1–0.3 cm (0.8–3.2 × 0.04–0.12 in), cartilaginous and stiff, shiny, yellowish cream at the apex, red-brown to black-brown from the base. *Microscopic features* Sp 8–11 × 3.5–5 µm, elliptic to somewhat tear-shaped. Bas 25–35 × 4–8 µm, cylindro-clavate. Basidioles cylindro-clavate to cylindro-fusiform. Cheilo 10–20 × 3–9 µm, clavate or trapezoidal, with stiff, erect finger-like processes (1–10 µm), and thick yellowish brown walls. Pleuro resembling hymenial setae, arising from deep origins, 30–100 × 4–12 µm, fusiform, with brown walls up to 3 µm diameter. Trama and cystidia dextrinoid. CC an hymeniderm, with hyphae 12–25 × 4–10 µm, like the cheilocystidia or stiffer and more strongly coloured.

Season: June–November.
Habitat: Broad-leaved trees (dead leaves or humus).
Distribution: Uneven but widespread. Quite rare.

M. torquescens Quélet has a similar appearance (especially shape and colour) but the stipe is matt, velvety or pruinose.

---

p.114 *Marasmius wynneae* Berk. & Br.
(= *M. globularis*)
Marasmiaceae, Tricholomatales

Description: *Cap pinkish beige to lilaceous grey. Stipe blackish violaceous from the base.*
Cap 1–4 cm (0.4–1.6 in), globose then convex, hygrophanous, finely striate, smooth to finely wrinkled, often dull, with greyish or brownish beige tones often with a hint of lilac or violet, pale to whitish when dry. *Gills* notched, not very close, lilaceous grey or greyish cream. *Stipe* 3–8 × 0.02–0.04 cm (1.2–3.2 × 0.008–0.016 in), not snapping when twisted, shiny at the apex, pruinose or velvety below, whitish to ochraceous cream at the very apex, reddish brown to violaceous black from the base. *Microscopic features* Sp 6–8.5 × 3–4.5 µm. Bas 25–40 × 4–8 µm. Basidioles clavate to cylindro-fusiform. Cheilo scattered, 15–55 × 5–15 µm, clavate

or fusiform, sometimes very irregular, thin- or thick-walled. Trama dextrinoid. CC an hymeniderm, the hyphae 15–30 × 5–20 μm, clavate to pyriform. Caulo 20–35 × 5–15 μm, wavy cylindric, in clusters.

Season: June–December.
Habitat: Broad-leaved trees, especially beech; rare under conifers.
Distribution: Widespread but frequency uneven.

p.116 **Fairy Ring Champignon**
Marasmius oreades (Bolt.:Fr.) Fr.
Marasmiaceae, Tricholomatales

Description: *Cap umbonate, light buff. Stipe not snapping when twisted.*
Cap 1–6(8) cm (0.4–2.4(3.2) in), conico-convex then umbonate, hygrophanous, smooth to finely wrinkled, light buff, sometimes more brownish on the umbo or paler at the margin, often undulating. *Gills* notched, not very close, pale ochraceous cream. *Stipe* 4–8 × 0.03–0.07 cm (1.4– 3.2 × 0.012–0.028 in), leathery and not snapping when twisted (it can be twisted several times), ochraceous cream, slightly browner from the base. *Microscopic features* Sp 7.5–10.5 × 4–6 μm, almond-shaped or almost lemon-shaped. Bas 30–40 × 6–10 μm, cylindro-clavate. Basidioles cylindro-fusiform. Trama dextrinoid. CC an hymeniderm, the hyphae 10–40 × 5–15 μm, cylindro-clavate. Caulo 30–70 × 2–10 μm, cylindro-clavate.
Season: Almost all the year.
Habitat: Grassy places, often in fairy rings.
Distribution: Widespread. Common.

Edible (remove leathery stipe). Often grows in towns or beside roads (so remember advice concerning the possible pollution of fungi). Avoid confusion with the very poisonous white *Clitocybe* spp. (gills more or less decurrent, close, colours very pale), which occur on the same kinds of grassland and sometimes in rings (care is required as the rings may interlock).

p.48 *Calyptella capula* (Holmskj.) Quélet
Marasmiaceae, Tricholomatales

Description: *Stipe short and cap cup-shaped, white.*
Frb cyphelloid, i.e. reduced. Fertile part cup-shaped, campanulate or asymmetrically hooded, on a small lateral stipe. White or pale cream, sometimes greying near the host when old. Outer surface shortly pubescent near the stipe. Inner surface smooth or rather finely wrinkled. *Microscopic features* Sp 6–9 × 4–6 µm, elliptic somewhat tear-shaped. Bas 20–25 × 4–7 µm. CC a cutis, the hyphae 2–5 µm diameter, sometimes erect, with diverticulate lateral outgrowths.
Season: June–November.
Habitat: Litter, in damp or marshy places.
Distribution: Widespread. Quite common.

C. campanula (Nees) Cooke is very similar but sulphur yellow.

---

p.119 *Mycenella bryophila* (Voglino) Singer
Marasmiaceae, Tricholomatales

Description: *Cap beige grey to brownish. Stipe pruinose, rooting.*
Cap 1–4 cm (0.4–1.6 in), conico-convex then umbonate or pointed, smooth then finely wrinkled, slightly pruinose, beige grey to grey-brown or sometimes warmer brown. Margin with long striations. *Gills* notched, close, white to greyish or pale beige-grey. Gill faces pruinose under the hand lens. *Stipe* 3–8 × 0.1–0.3 cm (1.2–3.2 × 0.04–0.12 in), rather fragile, rooting, velvety then pruinose, pale at the apex and brownish grey below *Microscopic features* Sp 6–9.5 × 5–8 µm, elliptic to subglobose, with a very conspicuous apiculus and large hemispherical warts. Bas 2-spored, 25–35 × 7–12 µm. Cyst 35–80 × 8–15 × 3–6 µm, lageniform, with an obtuse apex. CC an (ixo)cutis, with interwoven hyphae 1–3.5 µm diameter, and lageniform pileocystidia 25–45 × 3–6 µm. Pigment parietal. Clamps absent.
Season: September–November.
Habitat: Moss or bare ground, in open undergrowth.

Distribution: Uneven. Quite rare to very rare.

The remarkable cystidia and the spores which often have large warts and a very conspicuous apiculus are characteristic for the genus.

---

p.115 *Mycena galericulata* (Scop.:Fr.) S.F. Gray
Marasmiaceae, Tricholomatales

Description: *Variable. Gills interveined at the base, becoming pink at maturity.*
Cap 1–6 cm (0.4–2.4 in), bell-shaped or conical then convex, often retaining an umbo, slightly viscid then dry, often slightly shiny, smooth then wrinkled, whitish to dark greyish brown. *Gills* adnate, quite distant when old, white then pink to bright pink when old (spore print white). Gill faces smooth then interveined at the base. *Stipe* 4–10 × 0.2–0.5 cm (1.6–4 × 0.04–0.2 in), fairly tough to cartilaginous, often with a groove, shiny but frequently hairy at the base, concolorous with the cap or paler. Taste mealy if chewed. *Microscopic features* Sp 10–12.5 × 7–8.5 µm, elliptic to tear-shaped, amyloid. Bas sometimes 2-spored, 30–50 × 7–10 µm. Cheilo 20–50 × 5–15 µm, clavate, covered with slender, more or less branching outgrowths. Trama vinaceous in iodine. CC a cutis, the hyphae 1.5–2.5 µm diameter, covered with filiform but densely branching and interwoven processes. With or without clamps.
Season: Practically all year round.
Habitat: Especially on stumps and rotten wood.
Distribution: Widespread. Common.

Very variable.

---

p.115 *Mycena polygramma* (Bull.:Fr.) S.F. Gray
Marasmiaceae, Tricholomatales

Description: *Stipe silvery, with parallel longitudinal grooves.*
Cap 1–5 cm (0.4–2 in), bell-shaped then expanded, striate or sometimes denticulate at the edge, smooth to radially wrinkled, often grey but sometimes with a hint of

brown, yellowish, blackish in the centre, etc. *Gills* notched, quite close, white to dingy cream, sometimes pinkish or spotted pinkish brown. *Stipe* 5–20 × 0.2–0.3 cm (2–8 × 0.04– 0.12 in), quite brittle, distinctly grooved from top to bottom, silvery or grey, often hairy below.
Flesh relatively elastic, pale. *Microscopic features* Sp 8.5–10.5 × 6.5–7 µm, elliptic to tear-shaped, amyloid. Bas 25– 35 × 8–10 µm. Cheilo long, fusoid or ventricose, the neck more or less differentiated, the apex sometimes with scattered, filiform or very irregular outgrowths. Trama vinaceous brown in iodine. CC a cutis, the hyphae 1.5–3.5 µm diameter, with occasional clusters of branching or entangled outgrowths.

Season: Practically all year round.
Habitat: Broad-leaved woods, often in contact with or near stumps or roots.
Distribution: Widespread. Common.

---

p.115 *Mycena inclinata* (Fr.) Quélet
Marasmiaceae, Tricholomatales

Description: *Cap buff. Stipe orange at the base. Smell of candles.*
Cap 1–4 cm (0.4–1.6 in), conical or convex umbonate, slightly viscid in damp weather, smooth to finely wrinkled, greyish cream, then darkening to ochraceous grey-brown, darker in the centre, sometimes with yellowish or reddish tones. Margin with short striations, sometimes denticulate.
Gills adnate, quite close, white, sometimes dingy or spotted vinaceous pink. Faces smooth or finely wrinkled to interveined at the base. *Stipe* 4–12 × 0.2–0.5 cm (1.6–4.7 × 0.08–0.2 in), slightly elastic then brittle, soon shiny, pale but rapidly becoming beige to rather bright orange from the base. Usually tufted. Smells strongly of candle wax. *Microscopic features* Sp 7–9 × 5–6.5 µm, tear-shaped, amyloid. Bas 25–35 × 6–8 µm. Cheilo 20–35 × 6–13 µm, clavate to cylindric, with a few filiform, slightly branched outgrowths. Trama vinaceous

brown in iodine. CC a cutis, with hyphae 1.5–3.5 μm diameter, the largest with scattered, rather short, unbranched processes.

Season: August–December.
Habitat: Stumps of broad-leaved trees.
Distribution: Widespread.

---

p.115 *Mycena maculata* P. Karsten
Marasmiaceae, Tricholomatales

Description: *Stipe fragile. Fruit body with red-brown spots when old.*
Cap 1–4 cm (0.4–1.6 in), campanulate then almost flat, slightly hygrophanous, often quite pale or dull, the centre often darker. Red-brown spots develop when old. Margin striate, sometimes slightly grooved. *Gills* decurrent with a tooth, not very close, pale greyish to pinkish then spotted like the cap. *Stipe* 4–10 × 0.2–0.3 cm (1.6–4 × 0.08–0.12 in), slightly cartilaginous but brittle, shiny, sometimes hairy below, beige grey to brownish, spotting like the rest of the fruit body. *Microscopic features* Sp 8–10 × 5–6.5 μm, tear-shaped, amyloid. Bas 30–40 × 6–9 μm. Cheilo in clusters, 20–40 × 5–12 μm, clavate or irregular, with simple or diverticulate outgrowths. Trama vinaceous brown in iodine. CC a cutis, the hyphae 1.5–4.5 μm diameter, the widest with often short, slender outgrowths.

Season: August–November.
Habitat: Rotten wood and litter in broad-leaved woodland.
Distribution: Widespread. Quite common.

M. zephyrus (Fr.:Fr.) Kummer, is a rather similar species occurring mainly under conifers but the stipe and flesh are more fragile (cystidia lacking outgrowths).

---

p.134 *Mycena pura* (Pers.:Fr.) Kummer (poisonous)
Marasmiaceae, Tricholomatales

Description: *Cap fragile. Gills pale. Bluish, pink, violaceous. Smell of radish.*

Cap 1–4 cm (0.4–1.6 in), hemispherical then shallowly convex, striate, the edge hygrophanous, bluish, pink, violaceous, lilaceous, fading. *Gills* adnate, not very close, white or very pale, sometimes more or less tinged with the cap colour.
Stipe 3–8 × 0.2–0.6 cm (1.2–3.2 × 0.08–0.24 in), fragile, concolorous with the cap or paler. Smells strongly of radish.
Microscopic features Sp 7.5–8.5 × 3.5–5 μm, cylindric to tear-shaped, amyloid. Bas 25–35 × 5–8 μm. Pleuro and Cheilo 35–80 × 8–16 μm, fusiform or ventricose with a wide obtuse neck, usually smooth. Trama vinaceous brown in iodine. CC a cutis, with hyphae 1.5–6.5 μm diameter, smooth.

Season: June–December.
Habitat: Broad-leaved trees and coniferous woods.
Distribution: Widespread. Common.

Poisonous. Avoid confusion with *Laccaria amethystina* (Huds.) Cooke (smell none, gills thick, distant, concolorous with the cap, much more intensely violet, edible). *M. pura* is variable. There are white, yellow, bicoloured or tricoloured forms (or varieties).

---

p.133 *Mycena rosea* (Bull.) Gramberg (poisonous)
Marasmiaceae, Tricholomatales

Description: *Cap a beautiful pink, umbonate. Stipe rather robust. Smell of radish.*
Cap 3–7 cm (1.2–2.8 in), bell-shaped, then convex, often wrinkled around the large persistent umbo, hygrophanous, typically old rose, sometimes paler or dull.
Gills notched, quite distant, pinkish.
Stipe 5–10 × 0.5–1 cm (2–4 × 0.2–0.4 in), quite robust but fragile, hollow, nearly free, concolorous with the cap or paler.
Flesh rather fragile, pale pink. Smell strong, of radish. *Microscopic features* Sp 6.5–9 × 4.5–5.5 μm, tear-shaped, amyloid. Bas 25–32 × 6–9 μm. Pleuro and Cheilo 40–90 × 10–35 μm, fusoid, cylindric, or broadly inflated or clavate, smooth. Trama vinaceous brown in iodine. CC a cutis, the

hyphae rather slender 1.5–5.5 μm diameter, smooth.
Season: September–December.
Habitat: Mainly in broad-leaved woodland.
Distribution: Widespread. Quite common.

M. rosea is in the same group as *M. pura*. It is more poisonous.

---

p.134 *Mycena pelianthina* (Fr.:Fr.) Quélet
Marasmiaceae, Tricholomatales

Description: *Cap dull lilaceous grey. Gills dark, with a crimson-black edge.*
Cap 2–5 cm (0.8–2 in), hemispherical, then almost flat to faintly umbonate, often darker at the edge, smooth to finely wrinkled, hygrophanous, violaceous brown to pale lilac-grey when dry. *Gills* notched, quite wide, rather distant, dull but relatively dark violaceous, edge frosted and with a crimson-black line. *Stipe* 3–8 × 0.4–1 cm (1.2–3.2 × 0.16–0.4 in), hollow, smooth, lilaceous grey like the gills. *Flesh* slightly cartilaginous then fragile. Smells rather strongly of radish. *Microscopic features* Sp 6–7.5 × 3–4 μm, tear-shaped, amyloid. Bas 20–25 × 4–6 μm. Pleuro numerous and Cheilo in clusters, 40–70 × 8–14 μm, fusiform or tapering at the apex, often (especially the cheilocystidia) filled with violaceous brown pigment. Trama vinaceous in iodine. CC a cutis, with hyphae 1.5–4.5 μm diameter, smooth.
Season: August–December.
Habitat: Especially under beech.
Distribution: Widespread. Quite common.

---

p.133 *Mycena seynii* Quélet
Marasmiaceae, Tricholomatales

Description: *Cap pinkish brown, striate. Gill edge crimson-brown.*
Cap 1–4 cm (0.4–1.6 in), conical, smooth to finely wrinkled, chestnut to crimson-brown, fading from the margin to pinkish brown or pinkish beige, sometimes with

vinaceous tones. Margin striate.
Gills adnate, not very close, pale pinkish grey but edged with crimson-brown.
Stipe 4–6 × 0.2–0.5 cm (1.6–2.4 × 0.04–0.2 in), hollow, often curved, finally slightly shiny, concolorous or slightly paler.
Flesh thin, fragile, concolorous. *Microscopic features* Sp 10–14.5 × 5.5–7.5 µm, tear-shaped or long elliptic, amyloid. Bas 30–40 × 9–12 µm. Cheilo 25–45 × 7–15 µm, club-shaped, sometimes fusiform or with a few digitate and often stalked outgrowths, the pigment intracellular, reddish brown. Trama vinaceous in iodine. CC a sometimes slightly gelatinized cutis, the hyphae 1.5–4 µm diameter, smooth or with scattered processes. Pigment intracellular, red-brown.

Season: August–January.
Habitat: Fallen pine cones.
Distribution: Especially in coastal regions.

M. rubromarginata (Fr.:Fr.) Kummer is similar (cap greyish beige or with pinkish tones, gill edge reddish brown) and found on rotting wood (especially coniferous).

---

p.116 *Mycena capillaripes* Peck
Marasmiaceae, Tricholomatales

Description: *Cap reddish brown. Gill edge reddish. Smell alkaline.*
Cap 0.5–2 cm (0.2–0.8 in), conical, campanulate, hygrophanous, slightly wrinkled, dark brown to reddish brown when fresh, fading markedly when dry to pale greyish beige or even whitish. Margin striate. *Gills* nearly free, not very close, whitish but with a pinkish brown or reddish edge. *Stipe* 3–6 × 0.1–0.2 cm (1.2–2.4 × 0.04–0.08 in), fragile, with long pale hairs below, nearly free, concolorous. *Flesh* pale or concolorous, fragile. Smell strongly alkaline. *Microscopic features* Sp 8–11.5 × 4.5–6 µm, cylindric to long tear-shaped, amyloid. Bas 25–30 × 6–9 µm. Cheilo 30–75 × 5–15 µm, fusiform or cylindric, with more

or less red intracellular pigment. Trama red-brown in iodine. CC a cutis, the hyphae 2–8 μm diameter, with relatively short and rather dense or sometimes long diverticulate processes, the latter also branching.

Season: September–December.
Habitat: Conifer needles; in troops.
Distribution: Widespread. Relatively common.

M. roseofusca (Kühner) M. Bon, an often small species with a pinkish brown cap and a stipe that is less hairy at the base, also has a red-brown edge to the gills and a slight nitrous smell but occurs in mossy grass.

---

p.130 *Mycena rosella* (Fr.:Fr.) Kummer
Marasmiaceae, Tricholomatales

Description: *Cap pink to brownish salmon. Gills pink with a darker edge.*
Cap 0.5–2 cm (0.2–0.8 in), bell-shaped, sometimes umbonate, hygrophanous, smooth or finely wrinkled, bright pink to a more dull salmon-pink, sometimes brownish in the centre, fading when dry. Margin striate. *Gills* adnate, not very close, pale pink but with a reddish or pinkish red edge. *Stipe* 2–5 × 0.05–0.2 cm (0.8–2 × 0.02–0.08 in), hollow, fibrillose towards the base, reddish pink, fading rapidly to yellowish pink. *Flesh* very thin, concolorous. *Microscopic features* Sp 8–9.5 × 4.5–5 μm, tear-shaped, amyloid. Bas 25–35 × 6–9 μm. Cheilo 30–80 × 5–20 μm, club-shaped with more or less well-developed outgrowths. Pleuro 30–60 × 10–15 μm, scattered, more fusiform or bottle-shaped, with a smooth apex, (cystidia with red intracellular pigment). Trama pale vinaceous brown in iodine. CC a cutis, the hyphae 2.5–10 μm diameter, with occasional or more dense, long or short outgrowths. Pigment intracellular.

Season: August–December.
Habitat: Pine and spruce needles.
Distribution: Quite common.

p.121 *Mycena pterigena* (Fr.:Fr.) Kummer
Marasmiaceae, Tricholomatales

Description: *Minute. Cap reddish pink. Gill edge bright coral pink.*
Cap 1–5 mm (0.04–0.2 in), tiny, bell-shaped, hygrophanous, reddish pink to pinkish, sometimes more orange at the edge, which does not fade. Margin straight to upturned, striate and sometimes grooved. *Gills* adnate, distant, pale pinkish but the edge outlined or dotted with bright coral pink. *Stipe* 1–4 × 0.02–0.03 cm (0.4–1.6 × 0.008–0.012 in), extremely slender, fragile, brownish but fading to pinkish, with orange to pale pink fibrils at the base. *Microscopic features* Sp 9–12.5 × 4.5–5.5 µm, tear-shaped, amyloid. Bas 20–25 × 8–10 µm. Cheilo 20–40 × 7–20 µm, club-shaped or fusiform, sometimes very irregular, with very slender, more or less long processes of varying density. Trama vinaceous brown in iodine. CC a cutis, the hyphae 5–15 µm diameter, covered with slender, roughened outgrowths.

Season: August–December.
Habitat: Dead fern fronds.
Distribution: Widespread. Usually very rare.

---

p.131 *Mycena renatii* Quélet
Marasmiaceae, Tricholomatales

Description: *Cap lilaceous brown or pinkish. Stipe yellow. Smell slightly nitrous.*
Cap 1–2 cm (0.4–0.8 in), campanulate then faintly umbonate, smooth or faintly wrinkled, lilaceous brown with pinkish or yellowish tones, the centre darker, browning when old. Margin striate. *Gills* adnate, quite close, pale cream or with a hint of pink, edge thin and often reddish or red-brown, at least near the cap margin or when old. *Stipe* 2–6 × 0.1–0.3 cm (0.8–2.4 × 0.04–0.12 in), hollow, often curved, a beautiful bright yellow, but gradually vinaceous brown from the base.
Flesh concolorous or brownish. Smell often

slightly nitrous. *Microscopic features* Sp 11.5–13 × 6.5–7 μm, elliptic, amyloid. Bas 25–30 × 6–10 μm (rarely 2-spored and then lacking clamps). Cheilo 20–60 × 2–15 μm, clavate or bottle-shaped, the neck sometimes branching or diverticulate, with brown intracellular pigment. Pleuro rare, similar. Trama vinaceous brown in iodine. CC a cutis, the hyphae 2–8 μm diameter, smooth or with quite rare or occasional, swollen diverticulate or very irregular, thick outgrowths.

Season: August–December.
Habitat: Rotten wood of broad-leaved trees, in tufts.
Distribution: Rare, usually continental.

---

p.127 *Mycena chlorantha* (Fr.:Fr.) Kummer
Marasmiaceae, Tricholomatales

Description: *Cap brownish olive-yellow. Gill edge more strongly coloured.*
Cap 0.5–2 cm (0.2–0.8 in), bell-shaped then shallowly convex, pruinose, smooth or finely wrinkled, olive-brown to olivaceous yellow, fading on drying. Margin finely striate. *Gills* narrowly adnate, quite close, very pale greenish yellow grey, edge greenish lemon. *Stipe* 2–5 × 0.1–0.2 cm (0.8–2 × 0.04–0.08 in), often curved, olivaceous brown-grey, often pale but darker below. *Flesh* pale greenish, with a strong smell of iodoform on drying. *Microscopic features* Sp 9.5–11.5 × 5.5–7 μm, cylindro-elliptic, amyloid. Bas 25–40 × 8–12 μm. Cheilo 20–50 × 8–20 μm, clavate, with short slender outgrowths. Pleuro sporadic. Trama vinaceous in iodine. CC a cutis, the hyphae 2–4.5 μm diameter, covered with slender, sometimes branching and clustered outgrowths.

Season: September–January.
Habitat: Dead grass and plant remains.
Distribution: Widespread. Rare, mainly coastal.

M. olivaceomarginata (Massee) Massee (cap olivaceous brown, margin striate, gills pale with an olivaceous brown edge, smell faint), like *M. thymicola* Vel. (darker, smell alkaline, nitrous), occurs in mossy grass.

---

p.118 *Mycena arcangeliana* Bresadola
Marasmiaceae, Tricholomatales

Description: *Cap olivaceous brownish with a yellow margin. Smell of iodine when dry.*
Cap 0.5–2.5 cm (0.2–2 in), conico-convex then sometimes umbonate, smooth, olivaceous brownish, often yellow to yellowish at the margin, then greyish or whitish. *Gills* nearly free, quite close, pale. *Stipe* 3–10 × 0.1–0.3 cm (1.2–4 × 0.04–0.12 in), rather fragile, pruinose at the apex, violaceous grey to olivaceous brownish, fading above. *Flesh* thin. Smell faint, then smelling strongly of iodine when dry. *Microscopic features* Sp 8–11 × 4.5–5.5 µm, somewhat tear-shaped to broadly elliptic, amyloid. Bas 25–35 × 6–10 µm. Cheilo 20–65 × 10–25 µm, clavate, with slender, very short, brush-like processes. Pleuro quite numerous, identical. Trama vinaceous in iodine. CC a cutis, the hyphae 2–5 µm diameter, with slender outgrowths.
Season: August–November.
Habitat: Litter under broad-leaved trees.
Distribution: Widespread but frequency uneven.

Similar in overall appearance to *M. amicta* (*see* p.656), which usually occurs on conifer needles.

---

p.129 *Mycena aurantiomarginata* (Fr.:Fr.) Quélet
(= *M. elegans*)
Marasmiaceae, Tricholomatales

Description: *Cap brown with a pale or orange margin. Gill edge bright orange.*
Cap 0.5–2 cm (0.2–0.8 in), campanulate, often umbonate, smooth or with fine radial wrinkles, date brown in the centre, the edge lighter, sometimes bright orange. Margin striate. *Gills* narrowly adnate, not very close, yellowish orange to pale beige.

Gill edge with a fiery, then orange line. *Stipe* 3–7 × 0.1–0.2 cm (1.2–2.8 × 0.04–0.08 in), fragile, hairy below, yellowish brown to orange-yellow. *Microscopic features* Sp 8–10.5 × 4.5–5.5 µm, elliptic to cylindro-elliptic, amyloid. Bas 22–32 × 5–7 µm. Cheilo 20–40 × 5–15 µm, cylindro-clavate to broadly clavate, the apex with very short outgrowths and the pigment intracellular, orange. Pleuro less abundant, identical. Trama vinaceous brown in iodine. CC a somewhat gelatinized cutis, the hyphae 2–3.5 µm diameter, with slender, distant or densely clustered outgrowths.

Season: September–December.
Habitat: Usually on conifer needles.
Distribution: Widespread but frequency uneven.

---

p.119 *Mycena galopus* (Pers.:Fr.) Kummer
Marasmiaceae, Tricholomatales

Description: *Cap brownish with a darker centre. Milk white. Cap* 0.5–2 cm (0.2–0.8 in), campanulate then smooth to finely wrinkled, typically rather warm brown, darker in the centre. Margin with quite long striations. *Gills* narrowly adnate, fairly close, pale. *Stipe* 4–8 × 0.1–0.3 cm (1.6–3.2 × 0.04–0.12 in), fragile, exuding a white milk when broken, greyish beige, often paler than the cap, sometimes hairy at the base. *Microscopic features* Sp 10–14.5 × 5–6.5 µm, elliptic tear-shaped, amyloid. Bas 25–35 × 6–10 µm. Cheilo abundant, cylindric or fusiform to lageniform, the apex sometimes papillate or with thick outgrowths. Pleuro cylindric, protruding. Trama vinaceous brown in iodine. CC a cutis, the hyphae slender 1.5–3.5 µm diameter, with simple or branching, more or less dense outgrowths.

Season: June–January.
Habitat: In woods, on the ground or on litter.
Distribution: Widespread. Common.

Various colour forms have been recorded.

Mycena haematopus (Pers.:Fr.) Kummer
Marasmiaceae, Tricholomatales

p.131

Description: *Cap reddish brown. Milk vinaceous red-brown.*
Cap 1–3 cm (0.4–1.2 in), conical then convex, often umbonate, pruinose in dry weather, reddish brown to pinkish beige. Margin dentate appendiculate then entire, striate. *Gills* adnate, quite close, whitish to pinkish beige. *Stipe* 3–7 × 0.2–0.3 cm (1.2–2.8 × 0.08–0.12 in), rather fragile, exuding a dark red-brown to dark vinaceous latex, pinkish brown to dark crimson from the base where there may be some erect reddish mycelium. *Flesh* quite thin, concolorous, watery. *Microscopic features* Sp 8.5–9.5 × 5.5–6 µm, elliptic to somewhat tear-shaped, amyloid. Bas 30–40 × 7–11 µm. Cheilo 35–70 × 4–15 µm, often fusiform or lageniform, sometimes almost pointed. Pleuro similar. Trama vinaceous in iodine. CC a cutis, the hyphae 1.5–4.5 µm diameter, with a few rather thick, diverticulate outgrowths, themselves sometimes branching. Caulo in clusters, very diverticulate to very irregular.

Season: July–December.
Habitat: On wood, mainly of broad-leaved trees.
Distribution: Widespread but frequency uneven.

Can be confused with the following species but the habitat (*M. haematopus* occurs on wood) and gill edge (concolorous with the faces) are different. *M. haematopus* is also usually much larger than *M. sanguinolenta*.

Mycena sanguinolenta (Alb. & Schw.:Fr.) Kummer
Marasmiaceae, Tricholomatales

p.131

Description: *Cap reddish crimson. Gill edge vinaceous brown. Latex vinaceous crimson.*
Cap 0.5–1(1.5) cm (0.2–0.4(0.6) in), conical then convex, dark red-brown to reddish crimson, the centre long remaining darker. Margin striate. *Gills* narrowly adnate, rather close, whitish to pale pinkish

red, edge with a fine vinaceous brown line. *Stipe* 3–8 × 0.015–0.04 cm (1.2–3.2 × 0.006–0.016 in), fragile, nearly free, concolorous or darker below. *Flesh* pale brownish, exuding a vinaceous crimson latex when broken (especially in the stipe). *Microscopic features* Sp 8–10.5 × 5.5–6 μm, tear-shaped, amyloid. Bas 25–35 × 8–10 μm. Cheilo with red-brown contents, fusiform or the upper part more or less abruptly contracted into a sharp point. Pleuro similar but often quite rare. Trama vinaceous in iodine. CC a cutis, the hyphae 1.5–4.5 μm diameter, with dense, complex branching outgrowths. Caulo in clusters, similar to the cheilocystidia or more readily diverticulate to branching. Clamps numerous.

Season: July–December.
Habitat: Plant debris, mainly on the ground.
Distribution: Widespread. Common.

See *M. haematopus*.

---

p.130 *Mycena crocata* (Schrad.:Fr.) Kummer
Marasmiaceae, Tricholomatales

Description: *Cap brownish. Latex bright orange when broken.*
Cap 0.8–2.5 cm (0.32–1 in), conico-campanulate then expanded, greyish beige to orange-brown, paler at the margin, staining orange in patches when old. Margin striate. *Gills* narrowly adnate, rather close, white then spotted orange. *Stipe* 5–10 × 0.1–0.3 cm (2–4 × 0.04–0.12 in), fragile, with tufts of whitish to orange mycelium below, concolorous or orange-beige to somewhat russet, brighter or dark below. *Flesh* fragile, exuding an orange-yellow to bright orange-red latex when broken (especially in the stipe). *Microscopic features* Sp 8.5–10.5 × 5.5–6.5 μm, cylindric to tear-shaped, amyloid. Bas 30–40 × 8–10 μm. Cheilo 15–35 × 6–12 μm, clavate, covered in the upper half with slender outgrowths (up to 6 × 2 μm). Pleuro not very numerous,

simply clavate and often lacking outgrowths. Trama faintly vinaceous in iodine. CC a cutis, the hyphae 1.5–3.5 μm diameter, with small slightly branching outgrowths. Hyphae of the stipe cuticle with sparse, small outgrowths.

Season: August–November.
Habitat: Usually under beech.
Distribution: Widespread. Quite rare, usually in warm places.

---

p.119 *Mycena epipterygia* (Scop.:Fr.) S.F. Gray
Marasmiaceae, Tricholomatales

Description: *Cap viscid, brownish buff. Stipe yellow, viscid.*
Cap 0.5–2 cm (0.2–0.8 in), conico-campanulate, covered with a separable gelatinous skin, pale yellowish grey to yellow-buff or rather light brownish. Margin striate. *Gills* adnate, rather close, white then yellowish to pinkish, edge separable as a gelatinous thread.
Stipe 4–8 × 0.1–0.2 cm (1.6–3.2 × 0.04–0.08 in), fairly tough then more fragile, viscid, bright yellow, then yellowish white, sometimes with reddish zones at the base when old.
Flesh concolorous or paler. *Microscopic features* Sp 8–10.5 × 4.5–6 μm, slightly tear-shaped, amyloid. Bas 25–35 × 5–8 μm. Cheilo 25–50 × 5–10 μm, cylindro-clavate, with slender sometimes branching outgrowths. Trama vinaceous in iodine. CC a cutis, the hyphae 2–3 μm diameter, finely diverticulate.

Season: July–December.
Habitat: Wood edges and woods. Often on damp acid soils.
Distribution: Widespread. Quite common.

There are three 2-spored varieties: var. *epipterygioides* (Pearson) Kühner (cap and stipe greenish yellow sometimes reddish below; clamps present; in moss under conifers), var. *badiceps* M. Lange (cap blackish brown, smell rancid mealy; clamps absent; in moss) and var. *brunneola* Favre ex Maas G. (cap yellowish brown, smell none;

clamps present; under conifers in subalpine areas). All the other varieties are 4-spored and have clamps: var. *viscosa* (R. Maire) Ricken (cap rather pale, smell rancid, staining red-brown in places when old; on rotten stumps and conifer needles), var. *atroviscosa* Malençon ex Maas G. (cap dark brown, flesh reddening and stipe lemon yellow; on wood of Mediterranean conifers), var. *fuscopurpurea* (Arnolds) Maas G. (cap dark violaceous red-brown, stipe red-brown, gills reddish grey, smell rancid; grassland on sandy acid soils) and var. *pelliculosa* (Fr.) Maas G. (cap blackish olive-brown, stipe pale grey, gills greyish, smells slightly of iodine; mossy grass).

---

p.118 *Mycena erubescens* v Höhnel
Marasmiaceae, Tricholomatales

Description: *Cap reddish brown, whitish at the extreme edge. Taste bitter.*
Cap 0.5–1.2 cm (0.2–0.47 in), obtusely conical then bell-shaped, lubricious in damp weather, dark brown to reddish beige, the margin paler and the extreme edge often whitish. Margin striate.
Gills narrowly adnate, not very close, whitish to pale beige-grey. *Stipe* 2–4 × 0.05–0.15 cm (0.8–1.6 × 0.02–0.06 in), fragile, often curved, pruinose, often paler than the cap but with distinct reddish brown tones. *Flesh* brownish, fragile, exuding a colourless watery latex (especially in the stipe). Taste very definitely bitter.
Microscopic features Sp 9–11.5 × 6.5–7.5 µm, elliptic to somewhat tear-shaped, rather faintly amyloid. Bas 2-spored (sometimes 4-spored and then with clamps), 25–35 × 6–10 µm. Cheilo 15–65 × 5–20 µm, pointed fusiform or very irregular and partly covered with more or less branching filiform outgrowths. Pleuro rather long fusiform. Trama vinaceous in iodine. CC a cutis, the hyphae 1.5–3.5 µm, with filiform outgrowths. Clamps absent in the 2-spored forms.

Season: September–November.

Habitat: Mossy base of standing broad-leaved trees.
Distribution: Widespread. Quite rare.

---

p.118 *Mycena clavicularis* (Fr.:Fr.) Gillet
Marasmiaceae, Tricholomatales

Description: *Cap grey-brown, slightly viscid. Gill edge not gelatinized. Stipe glutinous.*
Cap 0.5–2 cm (0.2–0.8 in), hemispherical then bell-shaped, smooth, slightly viscid when fresh, dry and shiny in dry weather, dark brown to grey-brown, the centre and marginal striations remaining darker. *Gills* adnate, not very close, white. Gill edge not gelatinized. *Stipe* 3–7 × 0.1–0.2 cm (1.2–2.8 × 0.04–0.08 in), fairly tough, glabrous and smooth, glutinous and slippery, shiny, nearly free, concolorous or often paler, beige-grey, with a fibrillose base. *Flesh* thin, brownish beige. *Microscopic features* Sp 7.5–10 × 4.5–5.5 µm, cylindric to tear-shaped, amyloid. Bas 20–30 × 5–8 µm. Cheilo 15–40 × 8–15 µm, rather broadly clavate, the apex with often short, slender diverticulate outgrowths. Pleuro similar, often not very numerous. Trama vinaceous in iodine. CC an ixocutis, the hyphae 2–5 µm diameter, with more or less long, sometimes interwoven, filiform outgrowths.
Season: August–December.
Habitat: Conifer, especially pine, needles (in groups, sometimes in large troops).
Distribution: Widespread. Quite common.

M. vulgaris (Pers.:Fr.) Kummer, which is difficult to collect in damp weather because the stipe can be slippery, has a gelatinized gill edge separable as a gelatinous thread (cheilocystidia embedded in a thick slime, cylindric, the apex branching or forked, with long filiform branching outgrowths).

---

p.119 *Mycena amicta* (Fr.:Fr.) Quélet
Marasmiaceae, Tricholomatales

Description: *Cap pale, often bluish at the margin. Stipe pruinose, blue-green below.*

Cap 0.5–2 cm (0.2–0.8 in), conico-campanulate then expanded, with a separable gelatinized skin (in damp weather), pruinose, brownish grey to chamois beige, sometimes slightly olivaceous in the centre, the margin paler, often bluish or greenish, but sometimes also more yellowish.
Gills narrowly adnate, rather close, pale greyish, edge finely frosted, often paler.
Stipe 4–8 × 0.1–0.2 cm (1.6–3.2 × 0.04–0.08 in), somewhat rooting, pruinose, grey-brown to whitish, glaucous or blue green, sometimes distinctly blue below. *Microscopic features* Sp 8–10.5 × 4.5–5.5 μm, tear-shaped, amyloid. Bas 30–40 × 5–8 μm. Cheilo cylindro-clavate or fusoid-clavate, smooth. Trama vinaceous or dark crimson in iodine. CC a cutis, the hyphae 2–5 μm diameter, simple or very slightly diverticulate. Stipe covered with long cylindric caulocystidia tapering at the apex, up to 150 μm long.

Season: August–November.
Habitat: Leaf and needle litter, in woods.
Distribution: Widespread but frequency uneven.

This species may be confused with *M. arcangeliana* (see p.650), which tends to have rather olivaceous yellow tones on the margin and stipe base.

p.118 *Mycena filopes* (Bull.:Fr.) Kummer
Marasmiaceae, Tricholomatales

Description: *Cap brown, lighter at the margin. Stipe filiform. Smell of iodine when dry.*
Cap 0.5–2 cm (0.2–0.8 in), conico-campanulate then convex, fibrillose to innately streaky, dark brown in the centre, fading towards the margin to pale grey-brown or even whitish. Margin striate. *Gills* narrowly adnate, close, white or cream. *Stipe* 5–15 × 0.05–0.2 cm (2–6 × 0.02–0.08 in), grey-brown, quite pale but darker and often fibrillose towards the base. Smell of

iodine developing on drying. *Microscopic features* Sp 8.5–12 × 5–6.5 μm, tear-shaped, amyloid. Bas 2-spored (sometimes 4-spored: spores then small), 20–30 × 5–12 μm. Cheilo 15–30 × 8–20 μm, subglobose to clavate, the apex covered with rather distant, usually short, slender outgrowths. Trama vinaceous in iodine. CC a cutis, the hyphae 1.5–3 μm diameter, with slender, sometimes branching, often rather dense processes.

Season: August–December.
Habitat: Various kinds of litter, sometimes bark.
Distribution: Widespread. Quite common.

M. metata (Fr.:Fr.) Kummer (= *M. phyllogena* (Pers.) Singer), which has a brownish beige cap and pale margin, often with a pinkish flush, has whitish then distinctly pinkish gills when mature.

---

p.118 *Mycena leptocephala* (Pers.:Fr.) Gillet
Marasmiaceae, Tricholomatales

Description: *Cap brownish grey. Smell strongly alkaline.*
Cap 0.5–2 cm (0.2–0.8 in), conical to almost flat or umbonate, grey-brown, often darker in the centre and in the marginal striations. *Gills* adnate, not very close, whitish then greyish at maturity. *Stipe* 3–6 × 0.1–0.2 cm (1.2–2.4 × 0.04–0.08 in), fragile, nearly free, concolorous, grey-brown to brownish below where it is often fibrillose.
Flesh brittle, nearly concolorous or paler. Smell alkaline or nitrous (very like crushed poppy petals). *Microscopic features*

Sp 8.5–11 × 4.5–6.5 μm, tear-shaped, amyloid. Bas 25–35 × 7–10 μm, cylindro-clavate. Cheilo 25–80 × 5–20 μm, cylindro-lageniform or fusiform, the base thick and the neck obtusely tapering, sometimes bifid. Pleuro rare, identical. Trama vinaceous in iodine. CC a cutis, the hyphae 2–5 μm diameter, with filiform or finger-like outgrowths. Caulo narrow at the base but the apex broadly clavate or

irregular, sometimes diverticulate to misshapen.

Season: August–December.
Habitat: In large troops, especially on the ground and under conifers.
Distribution: Widespread. Common.

M. stipata Maas G & Schwöbel (= *M. alcalina* pp), often slightly darker, is a caespitose lignicolous species on conifer wood (cystidia smaller, often more fusoid-lageniform and stipe lacking clavate hyphae). *M. silvae-nigrae* Maas G & Schwöbel, which has a blackish brown then brown to grey-brown or dull pinkish beige cap, a dark metallic blue stipe when young, soon whitish or paler below, occurs on rotten conifer stumps or in tall mosses (basidia 2-spored and cheilocystidia diverticulate).

---

p.119 *Mycena vitilis* (Fr.) Quélet
Marasmiaceae, Tricholomatales

Description: *Cap dull, often finely wrinkled. Gills white. Stipe shiny, smooth.*
Cap 0.5–2 cm (0.2–0.8 in), conico-campanulate then almost flat, more or less grooved when old, greyish beige to quite dark brownish, often paler at the margin. *Gills* narrowly adnate, quite close, quite often pure white.
Stipe 4–10 × 0.05–0.2 cm (1.6–4 × 0.02–0.08 in), fairly stiff and resistant to pulling (finally breaking with a little snap), often shiny, paler than the cap, pale beige-grey to whitish or greyish.
Flesh firm then fragile, pale. *Microscopic features* Sp 9.5–11.5 × 5.5–7 µm, elliptic tear-shaped, amyloid. Bas 25–35 × 7–10 µm, lacking clamps. Cheilo 20–45 × 2–15 µm, clavate with a few finger-like or thick, more or less misshapen outgrowths. Trama vinaceous brown in iodine. CC a cutis, with hyphae 2–8 µm diameter, more or less thick and covered with slender, simple or complexly branched processes, slightly gelatinized when old. Hyphae on the

stipe surface with a fusoid, simple apex or with finger-like or varied outgrowths. Clamps absent.

Season: June–December.
Habitat: Debris from broad-leaved trees, especially oak twigs.
Distribution: Widespread. Common or quite rare in the absence of oaks.

---

p.132 *Mycena aetites* (Fr.) Quélet
Marasmiaceae, Tricholomatales

Description: *Cap dark, blackish then beige-grey. Gills dark grey.*
Cap 0.8–3 cm (0.32–1.2 in), hemispherical then expanded, often umbonate, blackish then fading to grey-brown or pale greyish beige, the grey tones often dominant. Margin striate. *Gills* narrowly adnate, not very close, a characteristic dark grey. *Stipe* 2–8 × 0.1–0.3 cm (0.8–3.2 × 0.04–0.12 in), fragile, grey and concolorous with the gills above, brownish towards the base.
Flesh brownish to grey-brown. Smells slightly of radish. *Microscopic features* Sp 8.6–10.5 × 5.5–7 µm, elliptic to tear-shaped, amyloid. Bas 25–35 × 5–10 µm. Cheilo cylindro-clavate often with a long neck or a few finger-like, rarely complex outgrowths. Trama vinaceous brown in iodine. CC a cutis, the hyphae 1.5–5 µm diameter, with slender finger-like, sometimes bushy processes.

Season: September–December.
Habitat: Lawns and grassy places.
Distribution: Widespread. Quite rare to rare.

The grey gills are a striking feature of this species, which is easy to identify in the field.

---

p.117 *Mycena flavoalba* (Fr.) Quélet
Marasmiaceae, Tricholomatales

Description: *Cap pale yellow, the margin whitish. Gills and stipe whitish.*

Cap 0.5–1.5 cm (0.2–0.6 in), conical then campanulate or almost flat, smooth, slightly viscid in damp weather, fairly pale yellow, especially at the edge which may become whitish. Margin striate.
Gills narrowly adnate, not very close, whitish. *Stipe* 1–4 × 0.05–0.1 cm (0.4–1.6 × 0.02–0.04 in), fragile, tapering, white or very pale yellow. *Flesh* white and slightly watery. *Microscopic features* Sp 6–8 × 3–4.5 μm, tear-shaped, not amyloid. Bas 20–25 × 5–7 μm, cylindro-clavate. Cheilo 35–60 × 3–13 μm, fusiform or sometimes clavate. Pleuro quite numerous, similar. Trama not reacting in iodine. CC a cutis, the hyphae 2–5 μm diameter, slightly gelatinized, with digitate or wavy, sometimes branched to coralloid outgrowths. Caulo rather long fusiform and almost pointed.

Season: July–December.
Habitat: Open places, sometimes also in woods.
Distribution: Widespread. Fairly uncommon.

---

p.131 *Mycena acicula* (Sch.) Kummer
Marasmiaceae, Tricholomatales

Description: *Cap orange. Stipe yellow, very slender. Gills whitish.*
Cap 2–8 mm (0.08–0.32 in) (sometimes tiny), hemispherical then bell-shaped, finely pruinose, then smooth, bright orange, fading to yellow or even whitish at the margin. *Gills* narrowly adnate, not very close, white or pale yellowish. *Stipe* 1–5 × 0.02–0.1 cm (0.4–2 × 0.008–0.04 in), fragile, very slender, orange-yellow to yellowish, fading. *Microscopic features* Sp 9–10.5 × 3–4 μm, fusoid to tear-shaped, not amyloid. Bas 15–25 × 5–8 μm. Cheilo 15–30 × 3–8 μm, fusiform or clavate to sublageniform. Pleuro rare, similar. Trama not reacting in iodine. CC a cutis, the hyphae 2–3.5 μm diameter, with short outgrowths, over a layer of broadly inflated cells. Hyphae of the stipe surface erect, with broadly cylindric, more or less

Season: September–December.
Habitat: Litter, especially under broad-leaved trees.
Distribution: Widespread. Quite common.

M. oregonensis A.H. Smith, a similar American species also occurring in Europe, has an orange-red or bright orange cap (the colour appears to remain unchanged to the end while it fades quite quickly in *M. acicula*). The gills may have a yellow or pale orange edge. The spores are more stocky (slightly shorter and slightly wider). The cheilocystidia have a yellow pigment and the stipe has true caulocystidia, which are clavate to fusiform or sublageniform.

---

p.131 *Mycena adonis* (Bull.:Fr.) S.F. Gray
Marasmiaceae, Tricholomatales

Description: *Cap pointed, coral red then salmon. Stipe white.*
Cap 0.5–2 cm (0.2–0.8 in), like an elfin cap then expanding, striate, slightly lubricious, coral red, fading to orange-red or salmon. *Gills* narrowly adnate, quite distant, white or pale pinkish. *Stipe* 1–3 × 0.1–0.2 cm (0.4–1.2 × 0.04–0.08 in), fragile, finely pruinose, pinkish then pure white or sometimes slightly yellowish. *Flesh* slightly watery, white or pale.
Microscopic features Sp 8–11.5 × 4.5–6.5 µm, elliptic somewhat tear-shaped, not amyloid. Bas 2-spored 20–35 × 5–7 µm, cylindro-clavate, without clamps. Cheilo 35–80 × 3–14 µm, fusoid-lageniform. Pleuro fairly abundant, similar. Trama not reacting in iodine. CC a cutis, the hyphae 2–5 µm diameter, with occasional, slender or finger-like branching outgrowths. Stipe cuticle with fusiform or more or less clavate to capitate caulocystidia. Clamps absent.

Season: August–November.
Habitat: Moss and plant remains in peat bogs or damp, slightly acid places.
Distribution: Widespread, but very rare in some places.

---

p.130 *Mycena meliigena* (Berk. & Cke.) Saccardo
(= *M. corticola*)
Marasmiaceae, Tricholomatales

Description: *Cap red-brown, striate. Stipe curved.*
Cap 4–8 mm (0.16–0.32 in), globose then bell-shaped, striate, pruinose, reddish brown, sometimes pale or also more violaceous. *Gills* subdecurrent, distant, pinkish beige. *Stipe* 5–15 × 0.5–1 mm (0.2–0.6 × 0.02–0.04 in), thicker at the base, curved, concolorous or darker then fading, often with pale hyphae below. *Flesh* thin, concolorous or paler. *Microscopic features* Sp 9–12 µm diameter, spherical or ellipsoid somewhat tear-shaped, amyloid. Bas 2- or 4-spored, 25–35 × 10–15 µm, clavate, without clamps (2-spored) or with clamps (4-spored). Cheilo 20–40 × 5–15 µm, cylindro-clavate to clavate, with filiform, more or less wavy outgrowths. Trama vinaceous in iodine. CC a cutis, the hyphae 3–10 µm diameter, covered with variable, short or branching outgrowths. Stipe cuticle with long terminal hyphae (20–80 µm), the apex clavate or inflated, the finger-like processes not very dense, more or less filiform.
Season: September–December.
Habitat: Mossy bark on living broad-leaved trees.
Distribution: Widespread. Quite rare.

p.111 *Mycena stylobates* (Pers.:Fr.) Kummer
Marasmiaceae, Tricholomatales

Description: *Cap buff to whitish. Small disc at the stipe base. Cap* 0.5–1 cm (0.2–0.4 in), hemispherical then expanded, finely striate or slightly grooved, with a separable gelatinous layer, whitish to buff, often quite pale. *Gills* nearly free, quite close, whitish. *Stipe* 1–4 × 0.05–0.1 cm (0.4–1.6 × 0.02–0.04 in), slightly pubescent below, whitish to nearly free, concolorous, with a flat basal disc (up to 2 mm (0.08 in)), ciliate at the edge. *Microscopic features* Sp 7–11 × 3.5–5.5 µm, cylindric to tear-shaped, amyloid. Bas 20–25 × 5–9 µm. Cheilo 20–60 × 5–12 µm, clavate but deformed at the apex by large elliptic or contorted outgrowths. Trama vinaceous in iodine. CC an ixocutis, the hyphae 2–5 µm diameter, branched and variously ornamented with complex slender or filiform outgrowths. Stipe cuticle with clusters of lageniform or variable, sometimes large caulocystidia.

Season: July–December.
Habitat: Dead leaves, conifer needles and various plant remains.
Distribution: Widespread. Quite common.

M. bulbosa (Cejp) Kühner, which occurs among tall grasses in damp places, has a greyish to beige cap, whitish stipe, and a distinct hemispherical basal bulb. *Resinomycena saccharifera* (Berk. & Br.) Redhead which is rather similar to *M. bulbosa* and occurs in the same habitats has a less pronounced basal disc and very distant gills.

p.111 *Mycena tenerrima* (Berk.) Quélet
Marasmiaceae, Tricholomatales

Description: *Cap pruinose, white. Stipe pruinose. Base bulbous.*
Cap 1–5 mm (0.04–0.2 in), globose then expanded, striate, pruinose then becoming glabrous, white or dull and slightly dingy, greyish beige, especially in the centre.
Gills nearly free when old, distant, white.
Stipe 5–20 × 0.2–0.5 mm (0.2–0.8 × 0.008–0.02 in), fragile, swelling into a disc at the base, pruinose, almost translucent to whitish or pale greyish.
Microscopic features Sp 8–10 × 5–6 µm, elliptic to tear-shaped, amyloid. Bas 2-spored, 15–20 × 7–10 µm. Cheilo 15–35 × 5–10 µm, clavate to lageniform, the rather narrow neck sometimes with short slender brush-like outgrowths. Trama vinaceous in iodine. CC a cutis, the hyphae 2–15 µm diameter, more or less clavate at the apex, with slender, more or less long, often rather dense outgrowths. Caulo lageni-fusiform, smooth.
Season: All year round.
Habitat: Plant debris.
Distribution: Widespread. Quite rare.

M. alphitophora (Berk.) Saccardo lacks the basal disc and is much more pruinose under the hand lens (under the microscope, cap and stipe cuticles have dense brush-like processes; basidia 4-spored).

p.112 *Mycena corynephora* Maas Geesteranus
Marasmiaceae, Tricholomatales

Description: *Small white species, very strongly mealy. Margin grooved.*
Cap 2–5 mm (0.08–0.2 in), globose then bell-shaped, mealy to coarsely powdered, white. Margin opaque, often grooved.
Gills broadly adnate, distant, white, edge slightly fleecy. *Stipe* 5–10 × 0.2–0.5 mm (0.2–0.4 × 0.008–0.02 in), often curved, very pruinose-mealy, white. *Microscopic features* Sp 7–9 × 6.5–8 µm, subglobose,

amyloid. Bas 20–30 × 8–10 μm. Cheilo 15–30 × 5–15 μm, clavate or sphaeropedunculate, the apex covered with very short, slender, warty, brush-like outgrowths. Trama vinaceous brown in iodine. CC subtrichodermial, the hyphae 2–10 μm diameter, the terminal elements erect and inflated like a balloon or sphaeropedunculate, 15–30 μm diameter, covered with short, slender processes. Stipe surface with comparable hyphae.

Season: July–November.
Habitat: Bark, usually moss-covered, of broad-leaved trees.
Distribution: Very rare, recently described, discovered in Italy and recorded in France, Germany and Switzerland.

---

p.112 *Hemimycena tortuosa* (Orton) Redhead
Marasmiaceae, Tricholomatales

Description: *Small white species, the pruinose cuticle retaining pinkish droplets.*
Cap 0.1–1 cm (0.04–0.4 in), hemispherical then expanded, slightly kidney-shaped in eccentrically stiped forms, pruinose, holding pinkish droplets in damp sites, white. *Gills* adnate, rather close, white, edge finely frosted.
Stipe 1–40 × 0.2–1 mm (0.04–1.6 × 0.008–0.04 in), sometimes eccentric or almost lateral, pruinose and bearing pinkish droplets, white. *Microscopic features* Sp 8.5–11 × 2.5–3.5 μm, fusiform, not amyloid. Bas 10–20 × 5–8 μm, narrowly clavate. Cheilo fusiform with a ventricose base, the apex pointed, sometimes with many constrictions. Trama not reacting in iodine. CC a trichoderm, the hyphae twisting like a corkscrew, 1.5–4.5 μm diameter, with a clavate or globose apex, 3–10 μm diameter. Underlying hyphae with finger-like and twisted outgrowths, which are sometimes present at the base of the cuticular hyphae.

Season: All year round.
Habitat: On damp wood of broad-leaved trees.
Distribution: Fairly widespread. Quite common.

H. cephalotricha (Joss.) Singer only differs in its boat-shaped spores (slightly lozenge-shaped) and merely capitate hyphae (not cork-screwing). Apparently more rare.

---

p.112 *Hemimycena lactea* (Pers.:Fr.) Singer
Marasmiaceae, Tricholomatales

Description: *Cap campanulate, pure white. Gills close, narrow.*
Cap 0.4–1.5 cm (0.16–0.6 in), conico-campanulate then expanded, very finely pruinose then smooth, sometimes faintly wrinkled, white. *Gills* narrowly adnate, often narrow, quite close, white.
Stipe 1–4 × 0.05–0.15 cm (0.4–1.6 × 0.02–0.06 in), fragile, pruinose, white.
Microscopic features Sp 7.5–10 × 2.5–3.5 µm, fusoid to cylindric, not amyloid. Bas 2- or 4-spored, 20–30 × 4–6 µm. Cheilo 20–32 × 5–6 µm, lageniform, the neck quite wide, slightly contorted and clavate at the apex (2.5–3.5 µm diameter). Trama not reacting in iodine. CC a trichoderm, the hyphae similar to the cheilocystidia or more slender, erect, 20–30 × 2–5 µm.
Season: August–December.
Habitat: Conifer needles.
Distribution: Widespread. Common.

The fruit body size of this species is very variable. *H. cucullata* (Pers.) Singer, with a sometimes larger cap and stipe, is its look-alike under broad-leaved trees. It is also common.

---

p.128 *Mycena speirea* (Fr.:Fr.) Gillet
Marasmiaceae, Tricholomatales

Description: *Cap umbonate, brownish. Gills decurrent.*
Cap 0.4–1 cm (0.16–0.4 in), conico-campanulate then often umbonate, smooth or finely wrinkled towards the centre, grey-brown to brownish beige, sometimes yellowish beige, paler at the striate margin. *Gills* decurrent, white to pale beige.

Stipe 1–5 × 0.05–0.1 cm (0.4–2 × 0.02–0.04 in), sometimes curved, finely pruinose, whitish to yellowish cream below. *Microscopic features* Sp 8.5–10.5 × 4–5.5 μm, elliptic somewhat tear-shaped, not amyloid. Bas 2-spored, 20–30 × 4–7 μm, without clamps. Cheilo 15–45 × 3–8 μm, cylindric to wavy sublageniform, sometimes coarsely diverticulate. Trama not reacting in iodine. CC a cutis, the hyphae 2–5 μm diameter, occasionally diverticulate or with more or less branching, finger-like outgrowths. Caulo cylindric or tapering.

Season: All year round.
Habitat: Woody debris under broad-leaved trees.
Distribution: Widespread. Quite common.

The photograph illustrates young fruit bodies of var. *camptophylla* (Berk.) Courtec., which is characterized by an initially bright yellow stipe.

---

p.94 *Mycena belliae* (Johnston) Orton
Marasmiaceae, Tricholomatales

Description: *Gills very decurrent. Cap pinkish brown, slightly viscid. On Phragmites.*
Cap 0.5–2 cm (0.2–0.8 in), hemispherical then bell-shaped, umbilicate when old, slightly viscid, smooth and shiny, pinkish brown, slightly paler towards the striate, often scalloped or lobed margin.
Gills deeply decurrent, very distant, pinkish white. *Stipe* 2–6 × 0.1–0.3 cm (0.8–2.4 × 0.04–0.12 in), powdery to pruinose, the apex whitish or very pale and slightly translucent, the base more pinkish or nearly concolorous, pinkish brown.
Flesh thin, pale brownish or watery.
Microscopic features Sp 12–15 × 5.5–6.5 μm, cylindric or narrowly tear-shaped, amyloid. Bas 30–45 × 7–10 μm. Cheilo 30–55 × 3–12 μm, clavate but with finger-like or branching outgrowths. Trama vinaceous in iodine. CC an ixocutis, the hyphae 1–5 μm diameter, smooth or with occasional slender processes.

Season: September–November.
Habitat: Marshy places; on *Phragmites* stems.
Distribution: Very rare.

The shape may suggest looking for a match among the *Omphalina* spp. but there are no species in that genus which occur in this particular habitat.

---

p.89 *Mycena rorida* (Fr.:Fr.) Quélet
Marasmiaceae, Tricholomatales

Description: *Cap brownish beige. Gills decurrent. Stipe with a thick coating of slime.*
Cap 0.5–1 cm (0.2–0.4 in), globose then convex, sometimes slightly depressed, often finely wrinkled in the centre, pale beige-grey to brownish, darker in the centre. Margin sometimes scalloped, striate.
Gills broadly adnate to decurrent, not very close, whitish. *Stipe* 1–4 × 0.05–0.1 cm (0.4–1.6 × 0.02–0.04 in), sheathed in a very thick coating of transparent slime, shiny in dry weather. *Flesh* thin, nearly free, concolorous or pale. *Microscopic features* Sp 10–14 × 4–6 µm, elongated cylindro-elliptic, amyloid. Bas 2-spored, 20–35 × 5–8 µm. Cheilo 20–40 × 5–10 µm, fusoid-lageniform, the neck quite wide, sometimes clavate, often exuding slightly refractive drops. Trama vinaceous in iodine. CC hymeniform, the hyphae clavate to broadly capitate, 10–25 µm diameter. Pigment intracellular, brown.
Season: August–December.
Habitat: Twigs (often on bramble).
Distribution: Widespread. Quite rare.

---

p.88 *Delicatula integrella* (Pers.:Fr.) Fayod
Marasmiaceae, Tricholomatales

Description: *Cap irregular, white. Gills fold-like.*
Cap 0.5–1 cm (0.2–0.4 in), hemispherical then convex or irregular, almost translucent, often lobed, slightly grooved or fluted, white. *Gills* subdecurrent, poorly formed, forked or anastomosing,

not reaching the margin, distant, white, edge rather thick. *Stipe* 0.5–2 × 0.05–0.1 cm (0.2–0.8 × 0.02–0.04 in), tapering, white. *Microscopic features* Sp 6.5–9 × 4–5.5 µm, lemon-shaped, ventricose, papillate, amyloid. Bas 25–50 × 5–9 µm. Trama not reacting in iodine. CC a cutis, the hyphae 2–5 µm diameter, smooth. Stipe base with a few thick-walled fusoid-clavate cells.

Season: August–December.
Habitat: Bare ground in ruts in woodland.
Distribution: Widespread but frequency uneven.

---

p.93 *Hemimycena ochrogaleata* (Favre) Moser
Marasmiaceae, Tricholomatales

Description: *Cap light buff. Gills distant, decurrent. Stipe pale.*
Cap 2–12 mm (0.08–0.47 in), conico-campanulate then convex, sometimes depressed, with short striations, smooth, ochraceous cream or pale yellowish beige. *Gills* decurrent, quite distant, white to very pale yellowish. *Stipe* 1–5 × 0.1–0.2 cm (0.4–2 × 0.04–0.08 in), the apex pruinose, white or faintly yellowish buff, especially below, almost translucent. *Microscopic features* Sp 7.5–10 × 4.5–6 µm, lemon-shaped or rhomboidal, often papillate, not amyloid. Bas 20–30 × 5–7 µm, cylindro-clavate. Cheilo 35–65 × 3–6 µm lageniform or with a very long, contorted filiform neck. Pleuro rarer, identical. Trama not reacting in iodine. CC a trichoderm, the hyphae similar to the cheilocystidia or more slender, erect, 20–35 × 2–5 µm.

Season: July–September.
Habitat: Clumps of spiny thistle (*Cirsium spinosissimum*) in upper mountain zones.
Distribution: The Alps and mountains in Scandinavia.

H. candida (Bres.) Singer is its look-alike in lowland areas, where it occurs on comfrey; however, it is whiter, often pure white. Similar species occur on lawns: *H. mairei* (Gilbert) Singer, is more stocky and possibly slightly larger, white to greyish cream and has no smell; *H. delectabilis* (Pk.) Singer, is often more fragile and smells strongly nitrous.

---

p.132 *Hydropus scabripes* (Murrill) Singer
Marasmiaceae, Tricholomatales

Description: *Cap brown, pruinose. Gills greyish white. Stipe pruinose.*
Cap 1.5–3 cm (0.6–1.2 in), conico-campanulate then convex, often umbonate, pruinose, finely wrinkled to rugose in the centre, dark brown to dark brownish grey, sometimes blackish in the centre. Margin inrolled then expanded, striate.
Gills nearly free, often interveined at the base, distant, white to greyish or greyish beige. *Stipe* 3–8 × 0.1–0.3 cm (1.2–3.2 × 0.04–0.12 in), cylindric or with a groove, strongly pruinose, or even granular above, beige-grey to tawny brown under the pruina, sometimes longitudinally striate.
Microscopic features Sp 8–12 × 4.5–6 µm, elliptic, amyloid. Bas (sometimes 2-spored) 35–45 × 7–10 µm, cylindro-clavate. Cyst numerous (facial and marginal), 60–120 × 10–20 µm, lageniform or fusoid-lageniform, the neck not always well differentiated. Trama not reacting in iodine. CC a cutis, the hyphae 2–5 µm diameter, overlying more inflated hyphae. Caulo abundant, variable. Pigment intracellular.
Season: September–December.
Habitat: Broad-leaved trees, sometimes mixed with conifers, grassy places, etc.
Distribution: Uneven. Rare.

---

p.98 *Hydropus marginellus* (Pers.:Fr.) Singer
Marasmiaceae, Tricholomatales

Description: *Cap striate, ochraceous beige to yellow-brown. Gill edge often darker.*
Cap 0.8–1.5 cm (0.32–0.6 in), convex then umbonate or depressed, finely pruinose, smooth, quite dark brownish

grey then greyish beige or yellowish buff. Margin with long striations.
Gills subdecurrent, straight or more often arching, quite close, white to whitish, edge strongly pruinose or with a brownish to brownish grey line. *Stipe* 1–3 × 0.1–0.2 cm (0.4–1.2 × 0.04–0.08 in), fragile, finely pruinose and matt, nearly free, concolorous or ochraceous grey then pale beige. *Flesh* quite distinctly watery. *Microscopic features* Sp 6–7.5 × 3–4.5 μm, elliptic, amyloid. Bas 20–30 × 5.5–8 μm, cylindro-clavate. Cheilo numerous 40–110 × 7–15 μm, lageniform or fusoid-lageniform, the neck sometimes subcapitate. Trama not reacting in iodine. CC subhymenidermal, the hyphae cylindro-clavate or fusoid-utriform, 15–50 × 10–20 μm. Caulo abundant, variable. Pigment intracellular.

Season: August–September.
Habitat: Rotten stumps or debris of firs.
Distribution: Mainly continental or in submontane areas. Rare to very rare.

---

p.137 **Rooting Shank**
Oudemansiella radicata (Relh.:Fr.) Singer
(= *Collybia r.*)
Dermolomataceae, Tricholomatales

Description: *Cap viscid, umbonate, wrinkled. Gills distant. Stipe rooting.*
Cap 1–15(20) cm (0.4–6(8) in), convex then almost flat to depressed, often umbonate, viscid then shiny, wrinkled and sometimes crumpled around the umbo, light ochraceous tan but sometimes brown or whitish beige, often rather uniformly coloured.
Gills adnate, wide and soft, rather distant, white, edge sometimes edged brownish or brown (var. *marginata*). *Stipe* 4–30 × 0.3–1.5 cm (1.6–11.8 × 0.12–0.6 in), rather tough, compressible, thickened at ground level, deeply rooting below, glabrous, often grooved, buff. *Microscopic features* Sp 12–16 × 9–11 μm, elliptic, slightly thick-walled. Bas 60–70 × 12–16 μm, cylindro-clavate. Cyst (facial and marginal)

50–130 × 15–35 µm, club-shaped, lageniform to ventricose fusiform. CC an ixohymeniderm, with hyphae 25–60 × 10–25 µm. Pigment intracellular.

Season: July–December.
Habitat: Broad-leaved trees and sometimes conifers, often arising from woody debris.
Distribution: Widespread. Common.

Size ranges widely. *O. pudens* (Pers.) Pegler & Young (= *O. longipes*), with a brownish grey to dark brown cap and beige-grey to ochraceous grey stipe is wholly covered with stiff erect hairs.

---

p.112 **Poached Egg Fungus, Porcelain Fungus**
Oudemansiella mucida (Schrad.:Fr.) v Höhn
(= *Mucidula m.*)
Dermolomataceae, Tricholomatales

Description: *Cap convex, viscid, white. Ring brownish grey. Cap* 1.5–10 cm (0.6–4 in), hemispherical then striate, translucent, Extremely slippery to viscid, shiny when dry, smooth or finely wrinkled, often white, sometimes dingy cream or ochraceous beige from the centre. *Gills* notched, wide and soft, distant, white, edge sometimes brownish or brown. *Stipe* 2–8 × 0.3–1 cm (0.8–3.2 × 0.12–0.4 in), compressible, often curved and often swollen at the base, white or brownish grey at the base. Ring ascending, rather high, membranous and relatively wide then breaking up, white but grey to blackish beneath. *Flesh* very thin, the cap translucent. *Microscopic features* Sp 13–19 µm diameter, more or less spherical, with a conspicuous apiculus, thick-walled. Bas 40–80 × 12–20 µm, cylindric. Cyst (facial and marginal) 60–140 × 15–40 µm, bottle-shaped stocky and ventricose, sometimes clavate. CC an ixohymeniderm, the hyphae 15–40 × 3–8 µm, clavate or very irregular and diverticulate.
Season: August–November.
Habitat: In clusters or groups on dead beech wood.
Distribution: Widespread wherever beech occurs.

p.151 *Flammulina velutipes* (Curt.:Fr.) Karsten
(= *Collybia v.*)
Dermolomataceae, Tricholomatales

Description: *Cap lubricious, orange-brown. Stipe velvety, black-brown.*
Cap 2–7 cm (0.8–2.8 in), convex to almost flat, viscid or lubricious, smooth to finely wrinkled, orange-brown, the margin orange-yellow or yellow and the centre sometimes darker. *Gills* notched, quite close, yellowish or cream. *Stipe* 1–8 × 0.2–1 cm (0.4–3.2 × 0.08–0.4 in), quite tough, often curved, velvety, pale yellow, becoming brown to black-brown from the base. *Flesh* rather tough in the stipe. *Microscopic features* Sp 6.5–9 × 3–5 µm, subcylindric, sometimes slightly tear-shaped. Bas 25–30 × 5–6 µm, cylindro-clavate. Cyst (facial quite rare, marginal more abundant) 40–60 × 10–15 µm, fusiform to fusoid-lageniform. CC an ixotrichoderm, the hyphae slender, interwoven 1–3 µm diameter, the numerous pileocystidia long conico-fusoid, pedicellate, 60–150 × 5–15 µm, thick-walled. Pigment mainly parietal. SC with cells similar to the pileocystidia but more variable or longer.

Season: October–February.
Habitat: Tufted or gregarious, on dead wood and stumps of broad-leaved trees.
Distribution: Widespread but frequency uneven.

Strong tendency to appear in the winter (often a sign of the end of the season).

---

p.140 *Dermoloma atrocinereum* (Pers.) Orton
Dermolomataceae, Tricholomatales

Description: *Cap grey-brown, the umbo very dark. Gills greyish. Smell mealy.*
Cap 2–5 cm (0.8–2 in), conico-convex then almost flat, often with an umbo, matt and pruinose, smooth to finely wrinkled or crumpled in the centre, sometimes cracked, blackish grey-brown, fading from the margin, grey-brown to beige-grey, the

umbo remaining very dark. Margin more or less finely striate, splitting. *Gills* notched, not very close to distant, white or greyish. *Stipe* 2–5 × 0.3–0.7 cm (0.8–2 × 0.12–0.28 in), fragile and brittle, tapering, fibrillose to pruinose, white or pale greyish. *Flesh* brittle. Smell strongly mealy.
Microscopic features Sp 6–8 × 3.5–5 µm, broadly elliptic, not amyloid. Bas 20–30 × 5–8 µm. CC a regular hymeniderm. Pigment parietal, smooth to encrusting at the base of the superficial cells.

Season: September–November.
Habitat: Dry calcareous grassland, sometimes in somewhat disturbed calcareous coppice woods.
Distribution: Uneven but quite widespread. Rare.

D. cuneifolium (Fr.:Fr.) Singer ex M. Bon is probably the most common species in the genus (cap convex, grey brown; gills very deep, triangular and distant; frequent on lawns).

---

p.94 *Camarophyllopsis foetens* (Phill.) Arnolds
Dermolomataceae, Tricholomatales

Description: *Cap brownish beige. Gills distant and thick. Stipe granular.*
Cap 0.5–2(3) cm (0.2–0.8(1.2) in), hemispherical then convex, matt, dry, coarsely wrinkled or cracked when old, grey-brown or milk-coffee to ochraceous beige. *Gills* decurrent, thick and distant, concolorous or sometimes dark grey-brown. *Stipe* 1.5–4 × 0.2–0.3 cm (0.6–1.6 × 0.08–0.12 in), brittle, fibrillose to granular but quite soon becoming glabrous, nearly free, concolorous or with warmer colours, pinkish beige to tawny buff, sometimes whitish below and at the apex.
Flesh brittle. Smell of moth balls.
Microscopic features Sp 4.5–6 × 4–5 µm, broadly elliptic to subglobose, not amyloid. Bas 35–40 × 5–7 µm, cylindro-clavate. CC an hymeniderm. Pigment parietal, encrusting at the base of the superficial cells. Clamps absent.

Season: August–October.
Habitat: Under broad-leaved trees, coppice woodland, especially on calcareous soils.
Distribution: Uneven. Rare.

C. atropuncta (Pers.:Fr.) Arnolds, a very similar species but with almost no smell, has large black-brown granules near the middle and upper half of the stipe.

---

p.143 *Cystoderma amianthinum* (Scop.) Fayod
Dermolomataceae, Tricholomatales

Description: *Cap granular, rufous orange. Sheath fleecy above.*
Cap 1–4 cm (0.4–1.6 in), hemispherical then shallowly convex, wrinkled, granular scurfy, rufous buff to rufous orange or orange-yellow. Margin appendiculate with white or pale, sometimes abundant velar remains. *Gills* nearly free, quite close, whitish to pale yellowish buff. *Stipe* 2–8 × 0.2–0.7 cm (0.8–3.2 × 0.08–0.28 in), hollow, whitish at the apex and concolorous, granular to fleecy under the ring. *Flesh* whitish to pale buff. Smell strong and unpleasant, like mouldy earth. *Microscopic features* Sp 4.5–6 × 2.5–4 µm, narrowly elliptic to cylindric, amyloid. CC subcellular, the hyphae septate, with elliptic or globose elements, 15–50 × 7–20 µm. Pigment parietal, smooth.
Season: June–December.
Habitat: Broad-leaved trees and conifers, mossy lawns.
Distribution: Widespread but frequency uneven.

C. amianthinum has a number of varieties and quite a few satellite species.

---

p.117 *Cystoderma carcharias* (Pers.) Fayod
Dermolomataceae, Tricholomatales

Description: *Cap granular scurfy, whitish to pinkish. Sheath concolorous.*
Cap 3–6 cm (1.2–2.4 in), hemispherical then broadly umbonate or wrinkled,

granular scurfy, whitish to pinkish, sometimes pinkish buff over the umbo. Margin appendiculate with white velar remains. *Gills* narrowly adnate, close, white, whitish to pinkish or yellowish. *Stipe* 3–7 × 0.3–0.8 cm (1.2–2.8 × 0.12–0.32 in), hollow, slightly clavate, white at the apex. Sheath well developed, concolorous, ornamented like the cap, with a fragile, rather flaring ring. *Flesh* white. Smell unpleasant, earthy. *Microscopic features* Sp 4–5.5 × 3–4 µm, shortly elliptic, amyloid. CC subcellular or an epithelium, the elements very short, 25–35 × 15–20 µm, the terminal cells more or less separable. Pigment mixed or parietal, smooth, intracellular in the terminal cells.

Season: July–November.
Habitat: Conifers.
Distribution: Mainly continental but sometimes near the sea. Quite rare.

In the *Cystoderma* group with a wide submembranous ring are: *C. fallax* Smith & Singer (rusty brown, strongly granular, with small spores, 4–5 × 3–3.5 µm), *C. tuomikoskii* Harmaja (fairly robust, tawny brown, with large spores, 5–7 × 3.5–4.5 µm) and *C. intermedium* Harmaja (more slender, buff, with medium-sized spores, 5–6 × 2.5–3.5 µm). The spores of all these species are amyloid.

p.144 *Cystoderma terreyi* (Berk. & Br.) Harmaja
Dermolomataceae, Tricholomatales

Description: *Cap granular, red-brown or bright rust. Sheath fleecy.*
Cap 3–8 cm (1.2–3.2 in), hemispherical, sometimes umbonate, slightly wrinkled, with fine, sometimes scaly, rather fleeting granules, cinnabar red, red-brown or bright rust, then fading slightly. *Gills* nearly free, quite close, pale cream, sometimes with a pinkish tone. *Stipe* 3–6 × 0.5–1 cm (1.2–2.4 × 0.2–0.4 in), slightly stuffed, clavate, the apex pale buff. Sheath fleecy,

nearly concolorous or fading, ending in a rather fleeting ring zone, the fleece pale. *Flesh* pinkish ochraceous cream. Smell rather faint. *Microscopic features* Sp 3–4 × 2.5–3 μm, elliptic, not amyloid. Cyst (facial and marginal), 30–45 × 5–12 μm, ventricose, tapering to a rather slender neck, the apex with barbed crystals. CC subcellular, the hyphae septate, with elliptic to spherical elements, 20–50 × 15–30 μm, the most superficial sometimes separating. Pigment mixed, mostly parietal, forming a zigzag encrustation.

Season: July–November.
Habitat: Conifers but also broad-leaved trees.
Distribution: Mainly northern or continental. Quite rare to very rare.

---

p.191 *Chamaemyces fracidus* (Fr.) Donk
Dermolomataceae, Tricholomatales

Description: *Cap ochraceous beige, finely wrinkled. Sheath with amber droplets.*
Cap 2–7 cm (0.8–2.8 in), globose then almost flat, smooth or finely wrinkled in the centre, slightly viscid, dry later, ochraceous cream to rufous beige, the centre sometimes darker. Margin with amber droplets, forming small brownish spots when dry. *Gills* nearly free, whitish to yellowish cream, spotting rust, edge thin, weeping small amber droplets which dry as brown spots. *Stipe* 2–7 × 0.3–0.8 cm (0.8–2.8 × 0.12–0.32 in), the apex pruinose or finely fleecy. Sheath granular or nearly smooth, ending in a fleeting, narrow ring, with amber droplets when young. *Flesh* whitish to pale cream. Smell faint, sharp. *Microscopic features* Sp 4–5 × 2–3.5 μm, ovoid. Cyst (facial and marginal) clavate, cylindro-clavate or lageniform-utriform, 40–70 × 12–18 μm. CC an hymeniderm, the hyphae broadly clavate or pyriform, 30–60 × 10–25 μm, with a few pileocystidia up to 70 × 30 μm. Pigment intracellular.

Season: August–November.
Habitat: Wood edges, open woodland; mainly on

	warm calcareous soils.
Distribution:	Widespread but uneven.

p.165 *Phaeolepiota aurea* (Matt.) R. Maire
Dermolomataceae, Tricholomatales

Description:	*Cap granular, golden brown, wrinkled. Sheath granular and wrinkled with a ring.*
	Cap 5–20 cm (2–8 in), globose then convex, granular or even pruinose, often radially wrinkled, golden brown to tawny orange, sometimes more dull when old. Margin appendiculate. *Gills* nearly free, quite close, cream then rather rufous to golden yellow. Spore print rufous brown. *Stipe* 6–15 × 2–3 cm (2.4–6 × 0.8–1.2 in), quite firm, cylindric-clavate, with a membranous, concolorous sheath similar to the cap, wrinkled or longitudinally grooved, ending in a rather persistent membranous ring. *Flesh* thick, cream to ochraceous yellow. *Microscopic features* Sp 10–13 × 4.5–5.5 µm, fusiform or long almond-shaped, with a slight suprahilar depression, brownish, finely warted at maturity. Bas 25–40 × 6–10 µm. CC an epithelium, the hyphae septate with rounded or elliptic, separating cells, 15–60 × 15–30 µm, brown and sometimes warty. Pigment parietal.
Season:	September–October.
Habitat:	Grass, disturbed undergrowth, town parks, etc.
Distribution:	Uneven. Occurs in sites disturbed by Man but quite rare.

PAXILLACEAE AND RELATED FAMILIES

This group comprises the gilled Boletales, with the exception of *Gomphidiaceae* (*see* p.560). It is composed of three families, *Paxillaceae*, *Hygrophoropsidaceae* and *Omphalotaceae*. The species are characterized by: a fibrous texture; decurrent gills, more or less separable from the flesh; and a clitocyboid, rarely pleurotoid silhouette. These gilled fungi are placed in the Boletales on the basis of their microscopic characters, anatomy or chemistry.

The three families are separated by the following characters:
- *Hygrophoropsidaceae*: spore print white; gills forked; terrestrial.
- *Omphalotaceae*: spore print white; gills forked; lignicolous; luminescent; very poisonous.
- *Paxillaceae*: spore print brown; gills very separable; terrestrial; sometimes very poisonous.

Characters common to this group are: cap depressed at maturity; margin inrolled then expanded; gills decurrent, close; taste and smell insignificant; spores smooth, not amyloid; basidia 4-spored; pleurocystidia absent; clamps present.

p.105 **False Chanterelle**
Hygrophoropsis aurantiaca (Wulf.:Fr.) Maire
Hygrophoropsidaceae, Boletales

Description: *Cap orange. Gills bright orange, decurrent, forked.*
Cap 2–8 cm (0.8–3.2 in), soon depressed, finely velvety or glabrous, orange, sometimes with more yellow or very pale tones, sometimes brownish to dingy buff. Margin wavy to lobed when old.
Gills decurrent, narrow, close, forked, often slightly crimped along the stipe, bright orange but sometimes yellowish cream. Spore print white. *Stipe* 1–5 × 0.3–1 cm (0.4–2 × 0.12–0.4 in), flexible, glabrous, concolorous or brighter, quite often blackening from the base.
Flesh rather thin, concolorous or paler.
Microscopic features Sp 5.5–8 × 3–5 μm, elliptic. Bas 25–40 × 5–8 μm. CC a trichoderm, with hyphae 4–15 μm diameter. Pigment intracellular.
Season: July–December.
Habitat: Conifers, broad-leaved trees. Fairly ubiquitous.
Distribution: Widespread. Common to very common.

Edible but mediocre, often confused with the chanterelle, from which it differs in its soft flesh, thin, forked gills and often brighter colours. *H. fuscosquamulosa* Orton has a velvety cap or fine fibrillose dark brown scales on a dull orange ground. The gills, however, are bright orange.

---

p.103 *Omphalotus olearius* (De Cand.:Fr.) Fayod
Omphalotaceae, Boletales

Description: *Clustered, lignicolous, bright orange. Gills forked. Stipe eccentric.*
Cap 5–15 cm (2–6 in), soon funnel-shaped, fibrillose to innately streaky or radially rimose, bright orange to orange-red, sometimes darkened or more dull from brown fibrils near the centre. Margin wavy to lobed and splitting. *Gills* very decurrent, narrow, close and forked, bright orange to orange-yellow. Spore print white. *Stipe* 5–18 × 0.5–2 cm (2–7 × 0.2–0.8 in), central (then curved) to eccentric, tapering, fibrillose to striate, concolorous or dark red-brown from the base. *Flesh* saffron or marbled orange. *Microscopic features* Sp 5–7 × 4.5–6 µm, subglobose. Bas 35–50 × 7–10 µm. Cheilo almost absent or sterile cells, 20–45 × 5–8 µm. CC a radial cutis, with hyphae 1.5–7.5 µm. Pigment epimembranal.
Season: September–December.
Habitat: Stumps and roots of broad-leaved trees (especially oak); warm places.
Distribution: Southern, extending to southern England, northern France, etc.

A highly poisonous luminescent species, often confused with the chanterelle.

---

p.105 **Brown Roll-rim** (poisonous)
Paxillus involutus (Batsch:Fr.) Fr.
Paxillaceae, Boletales

Description: *Margin inrolled, grooved. Gills separable, bruising brown.*
Cap 4–12(20) cm (1.6–4.7(8) in), soon funnel-shaped, sometimes umbonate in the bottom of the funnel, viscid in the centre, shiny when dry, felted, tomentose or glabrous at the margin, brown to brownish, sometimes olivaceous, dirty putty. Margin

inrolled, ribbed or grooved, gradually expanding. *Gills* decurrent, soft, quite narrow, close, yellowish buff, darkening with age and bruising brown, easily separable from the flesh (push up along the stipe with a finger nail). *Stipe* 3–8 × 0.5–1.5 cm (1.2–3.2 × 0.2–0.6 in), somewhat brittle, yellowish to buff, becoming brown or red-brown from the base. *Flesh* quite thick, firm then soft, ochraceous yellow, red-brown below, becoming slightly reddish when cut. *Microscopic features* Sp 7–10 × 5–7 μm, ovoid-elliptic. Bas 45–60 × 7–11 μm. Cyst. 30–70 × 8–12 μm, fusoid to sublageniform. CC an ixocutis, sometimes subtrichodermial, with hyphae 5–15 μm diameter. Pigment mixed.

Season: June–December.
Habitat: Various, especially birch, sometimes conifers.
Distribution: Widespread. Very common.

Can be deadly (*see* p.20). It is abundant on urban grassland and therefore potentially very dangerous. *P. filamentosus* (Scop.) Fr., a smaller species under alder, has a fibrillose to finely scaly cap.

p.88 *Paxillus atrotomentosus* (Batsch:Fr.) Fr.
Paxillaceae, Boletales

Description: *Cap olivaceous brown. Stipe eccentric, black and velvety.*
Cap 5–20 cm (2–8 in), convex then depressed, velvety to felted, sometimes cracking, brown to brownish, with buff, yellowish or rufous tones. *Gills* decurrent, arched, often crinkled along the stipe, pale orange-yellow, darkening with age, separable. *Stipe* 3–8 × 2–5 cm (1.2–3.2 × 0.8–2 in), laterally eccentric, robust, felty, velvety or hispid from the black-brown hairs. *Flesh* slightly or slowly bitter, yellowish cream or buff, becoming somewhat reddish or violaceous when cut.

Microscopic features Sp 4.5–7 × 3–5 μm, ovoid-elliptic, pale. Bas 35–45 × 5–8 μm.

CC subtrichodermial, with hyphae 5–10 μm diameter. Pigment mixed.
Season: August–November.
Habitat: Conifer stumps or roots.
Distribution: Uneven. Quite rare to rare.

Easy to identify on account of the black tomentum on the stipe.

---

p.85 *Paxillus panuoides* (Fr.:Fr.) Fr.
Paxillaceae, Boletales

Description: *Pleurotoid. Cap olivaceous buff. Gills orange.*
Cap 1–10 cm (0.4–4 in), hooded then spathulate or reniform, convex or depressed, often wavy, velvety to felted, white to ochraceous beige or yellowish buff. *Gills* converging towards the point of attachment, narrow, crinkled where inserted, yellow or orange-yellow, more cinnamon when old, separable. *Stipe* rudimentary, whitish or sometimes violaceous to lilac where inserted (var. *ionipus* Q). *Flesh* thin, soft, pale. *Microscopic features* Sp 4.5–6 × 3–4 μm, ovoid-elliptic, pale. Bas 25–40 × 5–7 μm. CC subtrichodermial, with hyphae 2–10 μm diameter. Pigment mixed.
Season: August–December.
Habitat: Conifer stumps.
Distribution: Uneven. Quite rare to rare.

Could be confused with *Phyllotopsis nidulans* (Pers.:Fr.) Singer, which has orange gills, but the cap is more strigose, nearly concolorous or orange pink. It gives off a fetid smell of garlic.

PLEUROTACEAE

The *Pleurotaceae* includes those species which usually have an eccentric or even lateral or absent stipe. Species still have the following traditional characters: fibrous but sometimes leathery at maturity; gills decurrent or confluent towards the point of attachment of the fruit body; spore print white or pinkish; often lignicolous.

Although the illustrated section includes a certain number of pleurotoid species, (morphological type C1 in the visual key), there are not many *Pleurotaceae*, in the strict sense. Constant characters are: cap domed then hollowed or depressed in the centre; surface smooth; margin inrolled then thin; gills decurrent, sinuate, not very close, white, edge thin, concolorous; stipe eccentric; flesh leathery in the stipe or near the point of attachment when mature, white to nearly concolorous near the surface; taste and smell insignificant; spores hyaline, smooth, not amyloid; basidia 4-spored, rather long cylindro-clavate, with clamps; pleurocystidia absent; trama interwoven; clamps present; on wood.

---

p.87 *Lentinellus cochleatus* (Pers.:Fr.) P. Karsten
Lentinellaceae, Hericiales

Description: *Cap split down one side, buff. Gill edge strongly toothed.*
Cap 2–10 cm (0.8–4 in), split and spiralling asymmetrically, ochraceous yellow to brownish, sometimes with pinkish tones. *Gills* extending well down the stipe as a thread, cream to buff, edge finely toothed. *Stipe* 1–7 × 0.3–1 cm (0.4–2.8 × 0.12–0.4 in), leathery, blackish brown at the base, often slightly strigose. *Flesh* rather elastic to leathery. Smell very strong of anise (there is a form with no smell). *Microscopic features* Sp 4–5.5 × 3.5–4.5 μm, globose to subspherical, finely warty spiny, amyloid. Bas 20–25 × 4– 6.5 μm. CC a cutis, with hyphae 2–6 μm diameter.
Season: July–December.
Habitat: In clusters, especially on broad-leaved trees.
Distribution: Widespread but frequency uneven.

---

p.87 *Lentinellus micheneri* (Berk. & Curt.) Pegler
(= *L. omphalodes*)
Lentinellaceae, Hericiales

Description: *Cap pinkish buff. Gill edge toothed. Taste acrid.*
Cap 1–3 cm (0.4–1.2 in), hygrophanous,

sometimes lobed, pinkish buff to pinkish brown, sometimes quite dark, fading when dry. *Gills* decurrent, quite wide, white then greying or greyish beige, edge coarsely toothed. *Stipe* 0.5–1.5 × 0.1–0.5 cm (0.2–0.6 × 0.04–0.2 in), short, central to lateral, leathery, concolorous with the gills and darker at the base. *Flesh* quite thin, very acrid. *Microscopic features* Sp 4.5–6 × 3–4.5 µm, rather broadly elliptic, finely warty punctate, amyloid. Bas 15–25 × 5–7 µm. Cyst 20–40 × 3–7 µm, fusiform, the apex sometimes tapering, papillate. CC a cutis, with hyphae 2–5 µm diameter.

Season: August–November.
Habitat: On broad-leaved trees, rarer on conifers.
Distribution: Widespread but frequency uneven.

Some, often smaller, specimens have a mild taste and may be called *L. bisus* (Quélet) Kühner & Maire.

---

p.88 *Lentinus tigrinus* (Bull.:Fr.) Fr.
Pleurotaceae, Tricholomatales

Description: *Cap pale, streaked with small grey scales. Gills close, pale yellow.*
Cap 1–10 cm (0.4–4 in), soon funnel-shaped, mottled or streaked with small appressed, grey to black, more or less dense scales, on a whitish to yellowish beige ground. *Gills* quite narrow, close, butter yellow. Edge finely toothed. *Stipe* 1–5 × 0.2–0.8 cm (0.4–2 × 0.08–0.32 in), central or slightly eccentric, finely scaly under a ring zone close under the gills. *Flesh* quite thin. Sometimes clustered. *Microscopic features* Sp 6–9 × 2.5–3.5 µm, cylindro-elliptic. Bas 25–30 × 4–8 µm. Cheilo 20–30 × 3–6 µm, cylindric. CC subtrichodermial in the scales, with hyphae 2–7 µm diameter. Pigment epimembranal.

Season: Practically all year round.
Habitat: Stumps and trunks of broad-leaved trees, often on willow, in damp places.
Distribution: Widespread. Very common to common.

L. tigrinus is a good edible if collected young. Later it becomes too leathery.

---

p.87 *Panellus stipticus* (Bull.:Fr.) P. Karsten
Pleurotaceae, Tricholomatales

Description: *Cap pale buff-brown. Stipe lateral, distinct.*
Cap 1–4 cm (0.4–1.6 in), kidney-shaped to spathulate, faintly zoned, finely fibrillose to felted, ochraceous beige to brownish buff. *Gills* quite narrow and close. Edge entire. *Stipe* reduced and lateral, 0.3–0.5(2) × 0.2–1 cm (0.12–0.2(0.8) × 0.08–0.4 in), often wider than long, concolorous, finely pubescent to scurfy, delimited by a smooth curve where it meets the gills. *Flesh* concolorous, very styptic when chewed (attacks the throat). *Microscopic features* Sp 3–5.5 × 1.5–3 µm, cylindro-elliptic to faintly allantoid. Bas 15–25 × 2.5–5 µm. Cheilo 20–80 × 5–10 µm, cylindric or clavate, with resinous crystals on the apex. Pleuro very rare. CC a trichoderm, with hyphae 2–6 µm diameter.
Season: Practically all year round.
Habitat: Stumps and large branches of broad-leaved trees.
Distribution: Widespread but frequency uneven.

P. mitis (Pers.:Fr.) Kühner is white to very pale cream when old, fibrillose to silky, with a mild taste and very narrow, cylindric to allantoid spores (3.5–6.5 × 1–1.3 µm). It is quite rare or very rare, generally continental or submontane.

---

p.87 *Panellus serotinus* (Pers.:Fr.) Kühner
Pleurotaceae, Tricholomatales

Description: *Cap dark green then yellow to reddish brown. Cuticle gelatinous.*
Cap 2–10 cm (0.8–4 in), domed, sometimes minutely scaly or felted near the substrate, covered with an elastic,

gelatinous layer, dark olive-green, discolouring reddish brown or ochraceous yellow. *Gills* quite close, pale yellow to ochraceous cream, edge entire. *Stipe* robust but short and lateral, 1–4 cm (0.4–1.6 in), a rounded cushion, orange or yellow-brown, finely scaly or punctate. *Flesh* quite thick. *Microscopic features* Sp 4–6 × 1–1.5 µm, cylindric to allantoid. Cheilo 35–70 × 7–12 µm, abundant, cylindric fusoid or clavate. Pleuro rarer. CC an ixocutis, with hyphae 2–8 µm diameter.

Season: October–April.
Habitat: Stumps and large branches of broad-leaved trees.
Distribution: Widespread but frequency uneven.

---

p.87 **Oyster Fungus**
Pleurotus ostreatus (Jacq.:Fr.) Kummer
Pleurotaceae, Tricholomatales

Description: *Cap grey or beige to brownish. Gills white. Stipe very short.*
Cap 3–20 cm (1.2–8 in), circular then asymmetrical or spathulate, grey, beige, brownish to brown, sometimes with bluish or violaceous tones. *Gills* decurrent, white. Edge entire. *Stipe* very short, often eccentric 1–3 × 0.5–1 cm (0.4–1.2 × 0.2–0.4 in), tomentose to strigose below, white. *Flesh* quite thick. *Microscopic features* Sp 7–11 × 2.5–4 µm, cylindro-elliptic to allantoid. Bas 25–35 × 5–7 µm. CC a cutis, with hyphae 2–5 µm diameter.

Season: September–December.
Habitat: Stumps and trunks or fallen branches of broad-leaved trees; rarer on conifers.
Distribution: Widespread but frequency uneven.

Edible if collected fairly young (its relationship to the polypores is made evident by the development of skeletal hyphae toughening the stipe and flesh of old specimens). It is also grown commercially and is widely available in the shops.

p.88 *Pleurotus cornucopiae* (Pers.) Rolland
Pleurotaceae, Tricholomatales

Description: *Cap beige to pale buff. Gills decurrent as a thread right to the base.*
Cap 2–12 cm (0.8–4.7 in), circular then asymmetrical or spathulate, beige to brownish buff, often pale. *Gills* very decurrent with a long thread right to the base of the stipe, white, edge entire. *Stipe* fairly well-developed, usually eccentric, 1–6 × 0.5–2 cm (0.4–2.4 × 0.2–0.8 in), white. *Flesh* quite thin. *Microscopic features* Sp 9–12 × 4–5 µm, cylindro-elliptic. CC a cutis, with hyphae 3–7 µm diameter.

Season: August–November.

Habitat: Stumps, trunks or fallen or dead branches of various broad-leaved trees.

Distribution: Widespread but frequency uneven.

This is the only species in the genus with gills which reach so far down the stipe.

---

p.88 *Pleurotus eryngii* (De Cand.:Fr.) Quélet
Pleurotaceae, Tricholomatales

Description: *Cap innately streaky or finely scaly, brown to dark rufous.*
Cap 5–10 cm (2–4 in), circular then asymmetrical or spathulate, innately streaky to radially fibrillose to wrinkled, finally minutely scaly or scaly, variable, white, grey, brownish to brown, sometimes very dark or with violaceous tones. *Gills* decurrent, white, edge entire. *Stipe* more or less eccentric, 1–5 × 2–6 cm (0.4–2 × 0.8–2.4 in), pale. *Flesh* quite thick. *Microscopic features* Sp 10–13 × 4.5–5.5 µm, cylindro-elliptic. Bas 35–50 × 6–10 µm. CC a cutis, with hyphae 4–7 µm diameter. Pigment epimembranal, in zigzag encrustations.

Season: August–December.

Habitat: Grassland, on roots of *Eryngium*.

Distribution: Widespread. Rather rare.

P. nebrodensis (Inz.) Q. (also restricted to umbellifer roots) is paler and occurs in the mountains.

PLUTACEAE

The *Pluteaceae* is one of the easiest families to recognize in the field because it has two major diagnostic characters: free gills and pink spores. Furthermore, the gill trama is convergent (microscopic character).

In Europe, this family contains two very distinct genera: *Volvariella*, which has a universal veil remaining as a membranous volva and *Pluteus*, which has no veil and a virtually naked stipe (sometimes fibrillose or pruinose).

There are many constant characters in this very homogeneous family: gills free, ventricose, quite close, initially white then quite bright pink to ochraceous pink, edge thin, often slightly frosted, concolorous; stipe cylindric; flesh white or nearly concolorous; taste and smell insignificant; spores not absolutely hyaline under the microscope, somewhat thick-walled, smooth, not amyloid; basidia 4-spored, distinctly clavate and pinched in under the sterigmata; trama convergent; pigment intracellular, varying in proportion to the intensity of the cap colour; clamps absent.

---

p.181 *Pluteus cervinus* (Sch.:Fr.) Kummer
(= *P. atricapillus*)
Pluteaceae, Pluteales

Description: *Cap brown. Stipe with brown fibrils.*
Cap 2–15 cm (0.8–6 in), globose then convex, quite often umbonate, glabrous or slightly fibrillose to minutely scaly in the centre, sometimes wrinkled or crumpled over the umbo, with fine innate radial streaks, pale beige to dark brown, sometimes even almost black in the centre. *Gills* distant from the stipe, close. *Stipe* 3–15 × 0.3–3 cm (1.3–6 × 0.12–1.2 in), sometimes nearly bulbous, slightly pruinose to floccose under the gills, white or very pale but with brown fibrils or at least browning when old. Smell of radish. *Microscopic features* Sp 6–9 × 4–6.5 µm, shortly ovoid to elliptic ovoid.

Pleuro abundant, projecting markedly, 40–100 × 10–25 µm, ventricose fusoid, the neck topped with one to five curving hooks, thick-walled (up to 5 µm). Cheilo 20–70 × 10–25 µm clavate to sphaeropedunculate. CC subtrichodermial in the centre, with hyphae 5–25 µm diameter.

Season: April–December.
Habitat: Stumps and dead wood, sawdust, woody debris.
Distribution: Widespread. Very common.

P. tricuspidatus Velenovsky (no smell, on stumps and rotten wood of conifers), has a blackish brown edge to the gills and a more scaly centre to the cap. *P. salicinus* (Pers.:Fr.) Kummer, is another distinctive species with fibrils or fine scales in the centre of the cap and pronounced greenish or quite dark green tones, at least in the centre (clamps very frequent).

---

p.180 *Pluteus petasatus* (Fr.) Gillet
Pluteaceae, Pluteales

Description: *Cap white with a pale buff fibrillose to minutely scaly centre. Stipe white.*
Cap 4–15 cm (1.6–6 in), hemispherical then convex, sometimes umbonate, smooth, silky to fibrillose but sometimes minutely scaly or breaking into scales in the centre, white, sometimes with beige or buff tones in the centre later. *Gills* distant from the stipe, close. *Stipe* 5–15 × 0.5–2 cm (2–6 × 0.2–0.8 in), sometimes clavate, smooth to fibrillose, white then with buff or brownish or even blackish brown fibrils at the base. Often clustered or in groups. *Flesh* thick. Smell of radish, or earthy to fruity.
Microscopic features Sp 5–9 × 3.5–6 µm, long ovoid or cylindro-elliptic. Pleuro numerous, 45–80 × 10–20 µm, fusoid-lageniform, the apex crowned with one to five curving hooks, thick-walled (up to 2.5 µm). Cheilo 30–70 × 10–30 µm, clavate to cylindro-clavate. CC subtrichodermial in the centre, with hyphae 5–20 µm diameter.

Season: June–November.
Habitat: Stumps, debris and sawdust of broad-leaved trees.
Distribution: Quite widespread and requiring confirmation.

P. pellitus (Pers.:Fr.) Kummer, which is also pale, is often less robust and has no smell. The cap is fibrillose in the centre, white or pale dingy greyish buff. It occurs on stumps of broad-leaved trees (no clamps).

---

p.182 *Pluteus leoninus* (Sch.:Fr.) Kummer
(= *P. fayodii*)
Pluteaceae, Pluteales

Description: *Cap golden yellow to ochraceous yellow, fibrillose to fibrillosely scaly.*
Cap 1.5–8 cm (0.6–3.2 in), conico-convex then faintly umbonate, hygrophanous, finely fibrillose or with small fibrillose scales in the centre, bright tawny-yellow, sometimes more golden or more buff, fading to pale yellowish buff. Margin striate. *Gills* yellowish then salmon. *Stipe* 4–10 × 0.2–1 cm (1.6–4 × 0.08–0.4 in), fibrillose or with long striations, white then yellowish to yellowish buff from the base. *Flesh* white or yellowish to buff. *Microscopic features* Sp 6–8.5 × 5–6.5 µm, ovoid elliptic to subglobose. Pleuro few, 40–100 × 10–35 µm, fusoid utriform, with an obtuse, sometimes clavate apex. Cheilo numerous, 25–80 × 10–25 µm, fusiform clavate to fusiform, tapering to a slender obtuse apex. CC a trichoderm, with hyphae 10–30 µm diameter.
Season: June–November.
Habitat: Wood, debris or sawdust of broad-leaved trees.
Distribution: Uneven. Quite rare.

P. luteomarginatus Rolland, a more robust and darker species, with a yellow edge to the gills, occurs mainly on conifers in the mountains.

p.181 *Pluteus umbrosus* (Pers.:Fr.) Kummer
Pluteaceae, Pluteales

Description: *Cap velvety, reticulately wrinkled, dark brown. Edge with dark dots.*
Cap 2–10 cm (0.8–4 in), hemispherical then almost flat to umbonate, with an erect velvety tomentum in the centre, fibrillose elsewhere, reticulately wrinkled from the umbo or centre, dark date-brown on a lighter, buff-brown, yellowish brown ground, sometimes light beige at the margin. *Gills* close. Edge lined with dark brown floccose dots. *Stipe* 4–12 × 0.3–1 cm (1.6–4.7 × 0.12–0.4 in), thicker below, often curved, floccose or punctate, the ornament forming more or less vertical lines, pale above and progressively concolorous from the base. *Flesh* thick, yellowish grey. *Microscopic features* Sp 5–8 × 4–6.5 µm, ovoid elliptic. Pleuro few, 50–85 × 10–35 µm, broadly fusiform to utriform, sometimes with a rather narrow apex, obtuse, sometimes pigmented. Cheilo abundant, 25–80 × 10–30 µm, clavate, cylindro-clavate or fusiform, the apex obtuse, pigment abundant, intracellular, brown. CC a trichoderm, with hyphae 5–30 µm diameter.

Season: June–November.
Habitat: On stumps, rotten wood or debris of broad-leaved trees, especially beech and ash.
Distribution: Uneven. Quite rare to very rare.

A beautiful species, easy to identify.

---

p.182 *Pluteus seticeps* (Atk.) Singer
(= *P. minutissimus*)
Pluteaceae, Pluteales

Description: *Cap velvety, warm date-brown. Stipe finely granular below.*
Cap 0.5–2 cm (0.2–0.8 in), hemispherical then convex or with a low umbo, markedly velvety, matt and glancing, faintly cracking when old, especially near the margin, red-brown to date-brown then rosewood to warm buff-brown, the underlying flesh very pale. Margin quite thick, finely striate.

Gills often deep. *Stipe* 0.5–3 × 0.1–0.2 cm (0.2–1.2 × 0.04–0.08 in), glassy, faintly fibrillose but often scurfy towards the base or minutely scaly at the extreme base, white to greyish. *Flesh* thin, fragile, glassy, greyish. *Microscopic features* Sp 5–8 × 5–7 µm, subglobose. Cyst numerous, 25–70 × 10–25 µm clavate, fusiform clavate to broadly utriform. CC with two kinds of hyphae, the longest (trichodermial) pyriform, cylindric fusiform or lageniform, 70–180 × 10–30 µm, the shortest (hymenidermal) clavate to sphaeropedunculate, 25–80 × 12–50 µm.

Season: June–October.

Habitat: On soil or near rotten plant debris, in damp places.

Distribution: Requires further study. Quite rare.

Strikingly velvety on the cap owing to the nature of the cuticle (mixed). *P. podospileus* Saccardo & Cuboni, has the same characteristic but is usually lignicolous, slightly larger and, in particular, has distinct, dark chestnut-brown dots on the lower half of the stipe.

p.182 *Pluteus aurantiorugosus* (Trog.) Saccardo
Pluteaceae, Pluteales

Description: *Cap bright orange-red. Gills white or yellowish then pink.*
Cap 1–5 cm (0.4–2 in), conico-convex, with a low umbo, smooth to finely wrinkled in the centre, bright orange-red but orange or orange-yellow when old. *Gills* sometimes yellow at the base. *Stipe* 2–8 × 0.2–0.6 cm (0.8–3.2 × 0.08–0.24 in), slightly inflated, often curved, finely fibrillosely striate, whitish to rather bright yellow below. *Flesh* thin, white to yellow. *Microscopic features* Sp 5–7.5 × 4–5.5 µm, ovoid to subglobose. Pleuro scattered, 40–80 × 15–30 µm, broadly fusiform-clavate or utriform with a wide obtuse apex. Cheilo abundant 25–70 × 10–35 µm, sphaeropedunculate, clavate, utriform, broadly lageniform, sometimes with a

mucronate apex. CC an hymeniderm, with hyphae 20–65 × 15–45 μm. Pigment abundant, intracellular, orange.

Season: September–November.
Habitat: Stumps and rotten wood of broad-leaved trees.
Distribution: Uneven. Rare to very rare.

The first in the *Pluteus* group that have a cellular cuticle, where it is unique on account of its extraordinary colour.

---

p.181 *Pluteus phlebophorus* (Ditm.:Fr.) Kummer
Pluteaceae, Pluteales

Description: *Cap mid-brown or dark brown, with distinct fine wrinkles in the centre.*
Cap 1–4 cm (0.4–1.6 in), hemispherical to convex, wrinkled to somewhat reticulate or crimped in the centre, mid- or rather dark and warm brown, with no trace of cinnamon or yellow. Margin sometimes finely striate. *Stipe* 2–8 × 0.2–0.5 cm (0.8–3.2 × 0.04–0.2 in), glassy, fibrillose, white or pale ochraceous cream from the base. *Flesh* fragile, greyish white. Smell sharp. *Microscopic features* Sp 5.5–8.5 × 5–7 μm, subglobose to shortly elliptic or ovoid. Pleuro quite abundant, 50–120 × 15–30 μm, obtusely fusiform to lageniform-utriform, with a wide obtuse apex 5–11 μm. Cheilo numerous 25–80 × 10–25 μm, clavate to utriform or sometimes lageniform. CC an hymeniderm, with hyphae 25–70 × 15–45 μm.

Season: August–December.
Habitat: On the soil or on woody debris of broad-leaved trees.
Distribution: Widespread but not fully recorded. Quite rare.

---

p.181 *Pluteus romellii* (Britz.) Saccardo
(= *P. lutescens*)
Pluteaceae, Pluteales

Description: *Cap date-brown. Stipe bright yellow below.*
Cap 1–5 cm (0.4–2 in), hemispherical then

slightly umbonate, smooth or finely wrinkled in the centre, rather dark date-brown then cinnamon brown with underlying yellow. Margin finely striate. *Gills* yellowish then salmon. *Stipe* 2–5 × 0.2–0.6 cm (0.8–2 × 0.04–0.24 in), brittle, fibrillose, pale yellow to bright yellow below like the flesh. *Microscopic features* Sp 6–7.5 × 5.5–7 µm, subglobose. Pleuro rather rare, 40–90 × 20–40 µm, clavate to broadly utriform, with a broad obtuse neck. Cheilo 20–80 × 10–45 µm, sphaeropedunculate, fusiform-clavate or shortly utriform. CC an hymeniderm, with hyphae 30–60 × 20–35 µm.

Season: May–December.
Habitat: Wood or woody debris of broad-leaved trees.
Distribution: Uneven.

P. chrysophaeus (Sch.:Fr.) Quélet is intermediate, resembling *P. phlebophorus* (see p.694) in its slightly more finely wrinkled, yellow-brown cap but the outer surface of the stipe is only faintly yellowish.

p.181 *Pluteus thomsonii* (Berk. & Br.) Dennis
Pluteaceae, Pluteales

Description: *Cap wrinkled or veined, brown. Margin finely striate. Stipe grey.*
Cap 1–3 cm (0.4–1.2 in), campanulate to slightly umbonate, strongly reticulately wrinkled, sometimes to the edge, date-brown to light brown, more greyish on the veins. Margin finely striate. *Gills* greyish then dull pinkish. *Stipe* 2–6 × 0.1–0.4 cm (0.8–2.4 × 0.04–0.16 in), rather fragile, pruinose above and fibrillose elsewhere, greyish to dark grey from the base. *Flesh* thin, grey to dark grey below. *Microscopic features* Sp 6–9 × 5–7 µm, broadly ovoid to subglobose. Pleuro absent or very rare. Cheilo 30–65 × 8–20 µm, with an awl-shaped appendage, up to 50 × 1–3 µm. CC an hymeniderm, with hyphae 25–100 × 10–40 µm, clavate or variable, sometimes with a filiform appendage.

Season: May–December.
Habitat: Wood of broad-leaved trees
Distribution: Uneven.

The f. *evenosus* (Kühn.) Wuilbaut, with an almost smooth cap, is difficult to recognize in the field and could be confused with *P. griseopus* Orton, another small hymeniform *Pluteus* which has a date-brown or grey-brown cap and grey flesh in the stipe, sometimes bluish at the apex.

---

p.182 *Volvariella bombycina* (Sch.:Fr.) Singer
Pluteaceae, Pluteales

Description: *Cap white, silky to fibrillose floccose. Volva large.*
Cap 5–20 cm (2–8 in), globose then convex, silky to fibrillose fleecy, white or cream to pale buff, sometimes also with a hint of pink or, more rarely, yellow. Margin finely striate. *Gills* dark salmon at maturity. *Stipe* 6–20 × 0.5–2 cm (2.4–8 × 0.2–0.8 in), often clavate, pruinose but fibrillose or floccose near the base, white or concolorous. Volva large, loosely sheathing (up to 10 cm (4 in) at the top), whitish but soon beige to brownish. *Microscopic features* Sp 7–10 × 4.5–6.5 µm, elliptic to ovoid. Cheilo 30–140 × 10–25 (45) µm, clavate to fusiform-lageniform. Pleuro often narrower. CC a trichoderm, with hyphae 5–35 µm diameter, very long (up to 1500 µm long).
Season: April–November.
Habitat: Stumps or standing trunks of broad-leaved trees.
Distribution: Uneven. Rare to very rare.

Var. *flaviceps* (Murrill) R. Schaffer, is rather bright yellow.

---

p.183 *Volvariella hypopithys* (Fr.) Moser
Pluteaceae, Pluteales

Description: *Cap white or pale cream, fibrillose-floccose. Stipe pubescent.*

Cap 1–5 cm (0.4–2 in), globose then expanded, silky to fibrillose-fleecy, white or pale ochraceous cream. Margin fibrillose, overhanging, finely striate. *Stipe* 3–7 × 0.2–0.8 cm (1.2–2.8 × 0.04–0.32 in), pruinose or fibrillose-silky, white. Volva quite large, two- to four-lobed, white or pale ochraceous beige, silky or felted on the outer surface. *Microscopic features* Sp 5.5–8 × 3.5–5 µm, elliptic to long ovoid. Cheilo numerous, 30–120 × 10–25 µm, clavate to fusiform lageniform. Pleuro similar. CC a trichoderm, with hyphae up to 450 µm × 5–30 µm, scarcely tapering at the apex.

Season: July–November.

Habitat: Humus in woods, sometimes in grassy places.

Distribution: Uneven. Quite rare to very rare.

V. pusilla (Pers.:Fr.) Singer (= *V. parvula*), is smaller, with a finely pruinose to glabrous stipe, usually occurring on lawns.

---

p.182 *Volvariella surrecta* (Knapp) Singer
(= *V. loveiana*)
Pluteaceae, Pluteales

Description: *Cap white or cream, fibrillose to silky. On old* Clitocybe nebularis.
Cap 3–8 cm (1.2–3.2 in), convex, slightly viscid then silky to fibrillose-scaly, white or greyish cream to pale dingy beige. Margin fibrillose, slightly overhanging.
Stipe 3–9 × 0.4–1 cm (1.2–3.5 × 0.16–0.4 in), often curved, pruinose at the apex, white or concolorous with the cap. Volva large, simple or lobed, whitish to cream or dingy beige, pruinose on the outer surface. *Microscopic features* Sp 5–7.5 × 3–4 µm, elliptic to cylindro-elliptic. Cheilo numerous, 35–90 × 7–30 µm, fusiform-lageniform. Pleuro similar. CC subtrichodermial, sometimes slightly gelatinized, with hyphae 5–10 µm diameter, rather short and scarcely tapering at the apex.

Season: September–November.

Habitat: On old *Clitocybe nebularis* (see p.821).
Distribution: Uneven. Quite rare to very rare.

---

p.183 ***Volvariella gloiocephala*** (De Cand.:Fr.) Boekhout & Enderle (= *V. speciosa*)
Pluteaceae, Pluteales

Description: *Cap viscid or lubricious, cream to brown. Volva fragile.*
Cap 5–15 cm (2–6 in), conico-convex then broadly umbonate, smooth, slightly viscid to lubricious, often fibrillose to innately streaky, white to dark brown, sometimes with grey, greenish or violaceous tones.
Gills pinkish to brownish pink.
Stipe 8–20 × 0.7–2 cm (3.2–8 × 0.28–0.8 in), often thickened towards the base, white. Volva large, fragile with a somewhat creamy texture, loosely sheathing, white or greyish beige.
Microscopic features Sp 10–18 × 6–10 µm, long elliptic. Cheilo numerous, 40–110 × 20–45 µm, often fusiform-lageniform, tapering at the apex to a narrow neck. Pleuro with a shorter or no neck. CC an ixocutis, with hyphae 2–7 µm diameter.
Season: July–December.
Habitat: Enriched places. Sometimes on wood edges.
Distribution: Widespread but frequency uneven.

Can be eaten but care should be taken because the sites where this fungus grows are likely to contain various pollutants, which may explain why it has occasionally been blamed for making people ill.

RUSSULACEAE

The *Russulaceae* is a remarkably homogenous family with two dominant features: flesh composed mainly of sphaerocysts (rounded or more or less spherical cells), which give it a brittle granular texture; spores light coloured, their ornament amyloid (blue-black in Melzer's iodine solution).

Lactarius and *Russula* are very easy to recognize at generic level (in the field, the stipe snaps cleanly like a piece of chalk, leaving no threads or fibres). *Lactarius* species usually have an elegant outline, decurrent gills and an often depressed cap. *Russula* species are often more squat, with adnate gills and a less depressed cap. The presence or absence of milk must be checked by breaking the edge of the cap and watching for any possible flow of latex (a character confined to *Lactarius*). Once the genus has been determined, the problem of identifying the very numerous and sometimes very similar species remains.

The classification of these genera is complex and based mainly on microscopic characters.

Russula and *Lactarius* species are mycorrhizal with a very wide range of trees, and are sometimes narrowly host specific. These fungi do not therefore usually occur in grassland. They have the following constant characters: cap convex then expanded or depressed (especially *Lactarius*); margin obtuse then fully expanded; gills adnate to subdecurrent; stipe cylindric, brittle; spores elliptic to subglobose, with amyloid ornamentation, the suprahilar plage more or less amyloid, the apiculus prominent, often slightly ventral; basidia 4-spored; lactifers frequent; clamps absent. In addition, *Lactarius* has the following constant characters: cap depressed at maturity; gills adnate to decurrent, sloping or arched; stipe often compressible, the flesh pitted in the centre; spores with an indistinct suprahilar plage (*see* below), poorly defined or only slightly amyloid.

The term suprahilar plage (SP) describes a small area of the spore above the apiculus that differs from the rest of the spore wall (smoother, more uniformly amyloid, etc). The abbreviation SA is used for sulphoaldehyde reagents (sulphovanillin, sulphobenzaldehyde, etc).

Spore ornament is much less varied in *Lactarius* than in *Russula*. Therefore, drawings of microscopic features are provided only for the first few species of *Lactarius*. Subsequently, illustrations are provided only when there is a departure from the norm.

p.202 *Russula nigricans* (Bull.) Fr.
Russulaceae, Russulales

Description: *Large brown to black species, with very distant gills.*
Cap 5–20 cm (2–8 in), convex then depressed, rather dry, smooth or finely wrinkled, sometimes cracked, whitish, soon with brownish to dark brown patches then entirely black. *Gills* adnate, horizontal, very thick and distant, brittle, white then greyish cream, edge thick, reddening and blackening. Spore print white. *Stipe* 3–8 × 1–4 cm (1.3–3.2 × 0.4–1.6 in), short and thick, soon hollow, pruinose, sometimes cracked, white then reddish to brownish grey and black. *Flesh* hard, white but reddening then quickly blackening. Taste mild, insipid. *Chemical test* $FeSO_4$ greenish on the reddening flesh. *Microscopic features* Sp 6.5–8.5 × 5.5–7 µm, with a fine partial reticulum, SP smooth. Bas 45–60 × 6.5–8 µm. Cyst cylindric, 50–100 × 5–7 µm, faintly greyish at the apex with SA. CC hyphae wavy, 2–6 µm diameter, the pigment intracellular, brown, pileocystidia absent.
Season: June–November.
Habitat: Woods (broad-leaved and coniferous).
Distribution: Widespread. Very common.

This large species with distant gills is easy to identify. Other taxa in this group have thinner, close gills.

---

p.203 *Russula acrifolia* Romagnesi
Russulaceae, Russulales

Description: *Cap depressed, reddish brown. Flesh reddening and blackening. Gills acrid.*
Cap 5–15 cm (2–6 in), soon depressed, the depression acutely angled at the base, smooth or slightly cracked, lubricious to slightly viscid, brownish to dull reddish brown. *Gills* adnate to decurrent, close, smaller intermediate gills present, white then ochraceous cream. Gill edge browning. Spore print white.

Stipe 3–7 × 1–3 cm (1.2–2.8 × 0.4–1.2 in), soon stuffed and finely pruinose or cracked, whitish then dingy brownish. *Flesh* white or whitish, reddening then quickly or simultaneously dirty brown to blackish. Taste acrid, especially in the gills (faintly in the cap). *Chemical test* FeSO$_4$ greenish on the reddening flesh. *Microscopic features* Sp 7–10 × 6–7.5 μm, with slender warts linked by an often incomplete reticulum. SP smooth. Bas 45–55 × 7–11 μm. Cyst cylindric to narrowly fusoid, 55–120 × 5–8 μm, rostrate, the contents blackening with SA. CC hyphae slender, wavy 3–5 μm diameter, with a few narrow pileocystidia 5–7 μm diameter (contents refractive).

Season: July–October.
Habitat: Woods; often on neutral soils.
Distribution: Widespread. Common.

R. densifolia (Secrétan) ex Gillet, which has a less depressed, soon dry cap, and a mild or sharp taste, tends to occur on acid soils under deciduous trees.

p.202 *Russula delica* Fr.
Russulaceae, Russulales

Description: *Cap pale. Smell of fruit then fish.*
Cap 7–15 cm (2.8–6 in), with a deep obtuse depression, dry, finely wrinkled or wrinkled, white then dingy reddish buff. *Gills* decurrent, arched, at least as wide as the cap flesh is thick, not very close, whitish to pale cream, edge becoming brown or rufous. Spore print white. *Stipe* 2–6 × 1.5–4 cm (0.8–2.4 × 0.6–1.6 in), white, rarely with a greenish tinge under the gills. *Flesh* white, flushing slightly rufous. Smell fruity then fishy. Taste mild or slightly acrid in the gills. *Chemical test* FeSO$_4$ slowly pale orange pink. *Microscopic features* Sp 9–12 × 7–8.5 μm, with prominent warts, partially linked by zebroid ridges. SP slightly amyloid. Bas 45–70 × 10–15 μm. Cyst

fusoid, 70–150 × 7–12 µm, rostrate, blackening with SA. CC hyphae interwoven, 3–5 µm diameter; pileocystidia elongate 5–8 µm diameter, with contents blackening with SA.

Season: May–October.
Habitat: Open woods or wood pasture, calcareous, often warm sites.
Distribution: Widespread; rarer in the north.

---

p.202 *Russula chloroides* (Krombholz) Bresadola
Russulaceae, Russulales

Description: *Cap pale. Greenish ring frequent under the gills. Cap* 5–13 cm (2–5.1 in), convex then with a very acute, narrow depression, slightly felty, white then yellowish to somewhat rufous. *Gills* decurrent, arched (narrow by comparison with the thickness of the flesh in the cap), close, white with a bluish green tinge, edge browning slightly. Spore print white. *Stipe* 3–8 × 1–3 cm (1.2–3.2 × 0.4–1.2 in), quite tall, white, green or greenish ring often present at the apex. *Flesh* white, yellowing slightly to browning. Smell fruity, strong. Taste mild, but acrid in the gills. *Chemical test* $FeSO_4$ dingy reddish. *Microscopic features* Sp 8–11 × 6.5–9 µm, with conical spiny warts, joined by connectives forming an incomplete reticulum. SP amyloid. Bas 50–70 × 10–13 µm. Cyst fusoid, 50–150 × 6–12 µm, mucronate or rostrate, dark grey with SA. CC hyphae cylindric, 3–6 µm diameter; the typical pileocystidia absent.

Season: August–November.
Habitat: Broad-leaved trees (especially beech). Tends to be on neutral to acid soils.
Distribution: Widespread

The green ring under the gills may be missing in some collections. It is then easy to confuse this species with *R. delica* (see p.701), which may also have a greenish or glaucous tint at the stipe apex. Identification (microscopic characters apart), should be based on habitat and gill depth in relation to the flesh of the cap.

p.203 *Russula farinipes* Romell
Russulaceae, Russulales

Description: *Cap somewhat rufous yellow, margin sulcate, gills decurrent.*
Cap 3–8 cm (1.2–3.2 in), convex or slightly depressed, dry, sometimes almost minutely scaly, yellow-buff then somewhat russet brown. Margin finely sulcate tuberculate, tough, elastic.
Gills decurrent, arched, quite distant, white then pale yellowish, edge sometimes with clear droplets. Spore print white.
Stipe 3–7 × 1–2 cm (1.2–2.8 × 0.4–0.8 in), tapering, cavities soon developing, mealy at the apex, white then slightly spotted towards the base. *Flesh* somewhat tough or elastic, whitish, yellowing slightly. Smell fruity. Taste acrid. *Microscopic features* Sp 6–8 × 5–6.5 µm, with isolated conical warts. SP nearly smooth. Bas 40–55 × 8–10 µm. Cyst numerous, 50–120 × 6–10 µm, rostrate, black with SA. CC hyphae septate, wavy 4–8 µm diameter and pileocystidia cylindro-clavate, 5–10 µm diameter, the apex often with several constrictions, greying with SA.
Season: July–October.
Habitat: Broad-leaved trees, on damp clays. Often on calcareous soils.
Distribution: Widespread. Relatively rare.

R. pectinata Fr. (very rare) has a rather bright yellow-buff, often slightly viscid cap, with a more sulcate tuberculate margin, is sometimes reddish at the stipe base and smells slightly of burnt horn. In particular, it has a cream spore print.

p.204 *Russula foetens* Pers.:Fr.
Russulaceae, Russulales

Description: *Large species with a convex, somewhat rufous tawny cap. Smell strong, of burnt horn.*
Cap 12–20 cm (4.7–8 in), globose then convex depressed, glutinous then dry and shiny, somewhat tawny buff to bright rufous brown, sometimes marbled. Margin very

sulcate tuberculate. *Gills* adnate, not very close, interveined at the base, whitish then cream, finally rust spotted, edge sometimes with somewhat rufous dots. Spore print cream. *Stipe* 6–13 × 1.5–5 cm (2.4–5.1 × 0.6–2 in), fusoid, cavities soon developing, white then with a somewhat rufous flush or spotted below. *Flesh* rather fragile, soon rufous brown in the hollows in the stipe and where damaged. Smell strong and unpleasant of burnt feathers. Taste very acrid. *Chemical test* KOH reddish brown in the flesh of the stipe. *Microscopic features* Sp 8–12 × 6.5–8.5 µm, subglobose, with conspicuous, isolated, conical warts, the SP faint, not or faintly amyloid. Bas 50–65 × 10–14 µm. Cyst fusoid, rostrate, 60–120 × 8–10 µm, blackening with SA. CC terminal hyphae 1.5–4 µm diameter, embedded in a thick mucilage, the supporting hyphae thicker, interwoven, septate, with tapering, papillate pileocystidia, more or less reacting with SA.

Season: July–October.
Habitat: Broad-leaved trees, also conifers.
Distribution: Widespread. Quite common.

The most robust and offensively smelling species in this group.

---

p.204 *Russula subfoetens* W.G. Smith
Russulaceae, Russulales

Description: *Cap convex then expanded, somewhat rufous yellow. Smell of formalin.*
Cap 6–15 cm (2.4–6 in), globose then expanded to depressed, lubricious or slimy then dry, shiny, quite pale yellow-buff, then spotting somewhat russet. Margin sulcate tuberculate. *Gills* adnate, interveined at the base, whitish, later ochraceous cream or with rather rufous spots, edge sometimes rufous to brownish in places. Spore print pale cream. *Stipe* 4–8 × 1–3 cm (1.6–3.2 × 0.4–1.2 in), tapering, cavities soon developing, white or with somewhat rufous spots or tints below. *Flesh* rather fragile, dingy yellow to somewhat rufous in the stipe cavities. Smell fairly similar but less

strong than in *R. foetens*, like formalin. Taste acrid. *Chemical test* KOH golden yellow on the flesh. *Microscopic features* Sp 7–10.5 × 5.5–8 μm, shortly elliptic, with more or less strong warts partially joined by incomplete ridges, the SP poorly defined. Bas 45–60 × 7–13 μm. Cyst fusoid, the apex contracted rostrate, 60–120 × 7–12 μm, blackening with SA. CC hyphae 2–3.5 μm diameter, embedded in mucilage, the supporting hyphae thicker, interwoven, cylindric, septate, obtuse, 4–6 μm diameter; pileocystidia fusoid rostrate 80–120 × 4–8 μm, with refractive contents, only reacting slightly with SA.

Season: July–October.
Habitat: Broad-leaved trees, especially hornbeam and birch, usually on damp, heavy soils.
Distribution: Fairly widespread.

---

p.205 *Russula illota* Romagnesi
Russulaceae, Russulales

Description: *Cap brownish with vinaceous brown spots, edge dotted vinaceous brown.*
Cap 7–15 cm (2.8–6 in), globose then convex or depressed, extremely slimy when dry and shiny, often damaged in the centre, buff with violaceous grey to dirty crimson spots in the slime then on the surface. Margin distinctly sulcate tuberculate. *Gills* adnate, arched, quite close, whitish to somewhat pale rufous then vinaceous brown in patches, edge with crimson-brown dots. Spore print cream. *Stipe* 5–13 × 1.5–3.5 cm (2–5.1 × 0.6–1.4 in), cavities soon developing, vinaceous brown dots present under the gills, whitish then tinged brownish to vinaceous brown. *Flesh* white or pale, browning slightly where damaged and in the soft centre of the stipe. Smell a mixture of burnt horn and bitter almonds. Taste acrid. *Microscopic features* Sp 7.5–9 × 6.5–7.5 μm, subglobose, with incomplete spines and ridges, the SP not very amyloid. Bas 40–60 × 9–15 μm. Cyst cylindric, the apex fusoid rostrate, 60–120 × 8–12 μm, the contents blackish with SA. CC hyphae slender, embedded in

mucilage, the supporting hyphae with sometimes short, slightly inflated cells, 3–5 µm diameter; pileocystidia obtuse or slightly rostrate, 50–100 × 5–10 µm, reacting faintly with SA.

Season: July–October.
Habitat: Woods (especially with broad-leaved trees).
Distribution: Widespread. Sometimes very rare.

---

p.204 *Russula laurocerasi* Melzer
Russulaceae, Russulales

Description: *Cap yellow-buff. Smell a mixture of bitter almonds and burnt horn.*
Cap 5–8 cm (2–3.2 in), globose then expanded to depressed, slightly viscid, soon dry, smooth, yellow-buff to rather pale, somewhat rufous yellow. Margin sulcate tuberculate. *Gills* adnate, quite close, pale cream then spotted, edge white, later slightly rufous. Spore print pale cream. *Stipe* 5–6 × 1–1.5 cm (2–2.4 × 0.4–0.6 in), cavities soon developing, whitish to somewhat rufous below. *Flesh* white then spotted russet brown in the stipe cavities. Smell strong, of bitter almonds mixed with burnt horn. Taste acrid. *Chemical test* KOH no reaction. *Microscopic features* Sp 7.5–9.5 × 7–8.5 µm, subglobose, with ridges about 1 µm high and a few isolated spines, the SP not very amyloid. Bas 45–65 × 10–13 µm. Cyst cylindro-fusoid, rostrate, 60–100 × 8–10 µm, greying with SA. CC hyphae slender, in branching clusters, septate, 2–5 µm diameter; pileocystidia rare, cylindric obtuse, 5–6 µm diameter, scarcely reacting.

Season: July–September.
Habitat: Broad-leaved trees.
Distribution: Widespread. Not very common.

Also in this group are *R. fragrans* Romagnesi which has a pure smell of bitter almonds (flesh slightly paler, taste mild to nauseous) and *R. fragrantissima* Romagnesi which has a remarkable pure, heady smell of anise or marzipan (larger, with a scarcely sulcate margin and acrid taste).

p.203 *Russula amoenolens* Romagnesi
Russulaceae, Russulales

Description: *Medium-sized species, dull brown, with a very acrid taste and a smell of Jerusalem artichokes.*
Cap 5–8 cm (2–3.2 in), convex then depressed to deeply depressed, slightly viscid, soon dry, shiny, minutely wrinkled, dull brown, sometimes blackish in the centre or sometimes paler. Margin finely sulcate tuberculate. *Gills* adnate, not very close, white then pale greyish cream. Spore print pale cream. *Stipe* 3–5 × 1–1.5 cm (1.2–2 × 0.4–0.6 in), soon hollow, white, sepia or tinged brownish later. *Flesh* quite thin, whitish to greyish. Smell of Jerusalem artichokes (shellfish being cooked). Taste very acrid. *Chemical test* FeSO$_4$ pinkish.

Microscopic features Sp 7–8.5 × 5–6.5 µm, elliptic, with warts partially joined by slender connectives, the SP not very distinct. Bas 35–50 × 6.5–10 µm. Cyst fusiform, rostrate, 60–120 × 6–10 µm, black with SA. CC hyphae slender, 2–3.5 µm diameter, septate, branching slightly with a few short, conical or tapering to lageniform pileocystidia, 30–70 × 4–7 µm, more or less reacting with SA, the pigment intracellular, yellow.
Season: June–November.
Habitat: Woods, waysides, grassy clearings.
Distribution: Widespread. Rather common.

R. pectinatoides Peck, which is paler or rust spotted, and often has orange-buff spots at the stipe base, has a nauseous rubbery smell and a mild but unpleasant taste (woodland rides and clearings).

p.214 **Charcoal Burner**
Russula cyanoxantha (Sch.) Fr.
Russulaceae, Russulales

Description: *Cuticle with fine radial wrinkles. Gills white, greasy. Taste mild.*
Cap 5–15 cm (2–6 in), globose then convex to depressed, slightly viscid, with faint radial wrinkles, violaceous crimson but often

mixed with purple, lilac, pink, buff, yellow and green, (these colours may predominate). *Gills* adnate, sometimes subdecurrent, quite close, white, supple and greasy, sticking together when touched. Spore print white. *Stipe* 4–10 × 1–3.5 cm (1.6–4 × 0.4–1.4 in), solid or slightly stuffed, white, browning slightly below. *Flesh* quite thick, at least in the centre, firm, white. Taste mild. *Chemical test* $FeSO_4$ no reaction. *Microscopic features* Sp 7–9.5 × 5.5–6.5 µm, warts quite low, irregular or slightly drawn out, almost isolated, the SP fairly wide. Bas 35–55 × 8–12 µm. Cyst fusoid, rostrate, 55–120 × 5–9 µm, greying with SA. CC hyphae slender indistinct, 2–2.5 µm diameter, with a few rounded or rostrate pileocystidia, 40–100 × 2–4 µm, reacting slightly with SA. Pigment composed of small blackish granules.

Season: June–December.
Habitat: Woods; various places.
Distribution: Very common.

One of the best edible *Russula* spp. f. *peltereaui* Singer has a uniformly green cap (intermediate forms occur). *R. langei* M. Bon (on slightly more acid soils) seems more distinct on account of its very dark purplish colour, very firm flesh, pale lilac flush on the stipe and slow green reaction to $FeSO_4$.

---

p.207 *Russula cutefracta* Cooke
Russulaceae, Russulales

Description: *Cap dark green, very cracked. Gills greasy.*
Cap 5–10 cm (2–4 in), convex then flat, slightly viscid or soon dry, matt then cracked, at least at the edge, dark green, sometimes with dark violaceous tones, the white flesh showing through the cracks. *Gills* adnate, close, often forking at the stipe, white but staining greenish brown when touched, greasy. Spore print white. *Stipe* 4–7 × 1–3 cm (1.4–2.8 × 0.4–1.2 in), white but tinged or spotting brownish to somewhat rufous from the base.
Flesh hard, white, browning slightly

towards the base. Taste mild. *Chemical test* FeSO$_4$ no reaction. *Microscopic features* Sp 7–9.5 × 5.5–7 μm, warts irregular, not very numerous, sometimes partly joined, the SP warty in places. Bas, Cyst and CC as for *R. cyanoxantha*.

Season: July–November.
Habitat: Broad-leaved trees.
Distribution: France, UK.

Can be confused with dark forms of *R. virescens* (see p.710).

---

p.205 *Russula heterophylla* (Fr.:Fr.) Fr.
Russulaceae, Russulales

Description: *Cap a tender yellowish green. Gills anastomosing strongly near the stipe.*
Cap 4–10 cm (1.6–4 in), convex then depressed, shiny, often with fine radial wrinkles like *R. cyanoxantha*, tender green, yellowish green, green with a hint of lemon, sometimes with bluish, vinaceous, or yellow, olive or grey tones and then often darker in the centre. *Gills* adnate or subdecurrent, forking and anastomosing, even crinkled near the stipe, close, white, soft. Spore print white. *Stipe* 3–6 × 1–3 cm (1.2–2.4 × 0.4–1.2 in), slightly stuffed when old, white. *Flesh* of average thickness, quite firm, white to yellowish near the surface. *Chemical test* FeSO$_4$ strongly bright orange-pink. *Microscopic features* Sp 5.5–7.5 × 4–6 μm, warts sometimes arranged in incomplete or erratic lines, the SP rather small. Bas 30–50 × 6.5–9 μm. Cyst fusiform, 55–80 × 7–12 μm, rostrate, more or less reacting with SA. CC hyphae septate and branching, interwoven, the supporting cells often very short, the terminal cells tapering to a point. Very slender, long, hair-like, thick-walled, colourless hyphae present especially near the centre.

Season: June–October.
Habitat: Broad-leaved and coniferous trees.
Distribution: Widespread. Quite common.

A fairly good edible species.

p.209 **Bare-toothed Russula**
Russula vesca Fr.
Russulaceae, Russulales

Description: *Cap vinaceous pink, the cuticle not reaching to the extreme edge.*
Cap 5–10 cm (2–4 in), hemispherical then depressed, slightly viscid, with fine radial wrinkles, vinaceous pink, like ham but darker, but often with brown, greenish, olivaceous, hazel, grey, cream, etc., zones. The edge of the gills is exposed at the margin as the cuticle appears too small to cover the entire cap, or the edge often turns up. *Gills* close, white, slightly supple when young but quite soon brittle. Spore print white. *Stipe* 4–10 × 1–3 cm (1.6–4 × 0.4–1.2 in), white. *Flesh* hard when young, white or pinkish near the surface, sometimes yellowing or flushing rufous at the base. Taste of hazel nuts. *Chemical test* $FeSO_4$ strongly bright orange-pink. *Microscopic features* Sp 6.5–8.5 × 5–6.5 µm, subglobose, warts small and distant, isolated or faintly joined, the SP poorly defined. Bas 35–55 × 8–10 µm. Cyst cylindro-fusoid, obtuse or rostrate, 70–110 × 6–14 µm, reacting rather faintly with SA. CC hyphae septate, the cells rather long and cylindric, constricted, the terminal cell obtuse or tapering; pileocystidia rare and not well differentiated. Long filiform hair-like hyphae present, especially near the centre, the wall often more coloured, brownish.

Season: May–October.

Habitat: Broad-leaved trees, especially oak, often on acid soils; sometimes wood edges or clearings.

Distribution: Widespread. Becomes rarer northwards and in the mountains.

Possibly the best edible *Russula*, along with *R. virescens*.

---

p.206 *Russula virescens* (Sch.) Fr.
Russulaceae, Russulales

Description: *Cap pale green, very cracked. Spore print white. Taste mild.*

Cap 5–10 cm (2–4 in), hemispherical (young fruit body like a champagne cork) then convex, dry, velvety and cracked or even fissured, pale green, yellowish green then mid-green, sometimes rust spotted. *Gills* adnate, quite close, white, slightly greasy, quite soon brittle. Spore print white. *Stipe* 3–8 × 1.5–3 cm (1.2–3.2 × 0.6–1.2 in), fusiform, white, sometimes disrupting into somewhat rufous plates. *Flesh* thick, firm and even hard when young, white to greenish near the surface. Taste mild and pleasant, like hazel nut. *Chemical test* $FeSO_4$ bright orange-pink or orange. *Microscopic features* Sp 6–9 × 5–6.5 µm, the warts coarse and isolated or small and more or less joined by connectives sometimes forming an incomplete reticulum, the SP large, not amyloid. Bas 40–55 × 6–10 µm. Cyst 50–80 × 6–8 µm, cylindro-fusoid, obtuse or rostrate, scarcely reacting with SA. CC hyphae erect, short and thick, the supporting cells almost spherical, piled one on the other, the terminal cells tapering, conical or bottle-shaped.

Season: May–October.
Habitat: Broad-leaved trees. Mainly on acid soils.
Distribution: Common, throughout.

An excellent edible species, considered the best in the genus. Taste delicate and flesh nutty.

p.205 *Russula mustelina* Fr.
Russulaceae, Russulales

Description: *Cap ochraceous russet, finely grained. Spore print cream. Taste mild.*
Cap 5–15 cm (2–6 in), subglobose then expanded, lubricious or greasy, slightly shiny, finely wrinkled or grained, somewhat rufous to warm buff-brown, sometimes with paler patches, and with olivaceous or crimson tints. *Gills* adnate, close, greasy then brittle, white then pale ochraceous cream. Spore print cream. *Stipe* 3–10 × 1–3 cm (1.2–4 × 0.4–1.2 in), cylindro-

clavate or spindle-shaped, solid then slightly stuffed, white, later tinged or spotted yellowish to brownish below. *Flesh* often very firm, white or yellowish near the surface. Smell faint, of cheese. Taste mild. *Chemical test* $FeSO_4$ bright/dark orange-red to brownish red. *Microscopic features* Sp 7–11 × 6–8 μm, the warts often interconnected, sometimes subreticulate, the SP spotted or punctate. Bas 50–60 × 8–10 μm. Cyst fusiform, 80–110 × 8–15 μm, with a narrow appendage or many constrictions, scarcely reacting with SA. CC hyphae in clusters, septate, the supporting cells often wide or inflated, the terminal cells long tapering. Pileocystidia poorly defined, scarcely reacting with SA.

Season: July–October.
Habitat: Conifers, mainly spruce.
Distribution: Common, especially in mountain regions; very occasionally in lowland areas.

An excellent edible species, extremely variable macroscopically (cap colour) and microscopically (spore ornamentation particularly).

---

p.209 *Russula lepida* Fries
Russulaceae, Russulales

Description: *A pinkish red species, very hard. Stipe often pink. Spore print pale cream. Taste of menthol.*
Cap 5–12 cm (2–4.7 in), subglobose, expanding slowly, very matt, pruinose and often fissured or cracked, bright red to pinkish red, sometimes with paler, buff or cream marbling or blotching.
Gills adnate, close, white then pale cream, edge slightly brittle. Spore print pale cream. *Stipe* 3–8 × 1.5–3.5 cm (1.2–3.2 × 0.6–1.4 in), very hard, white or more often flushed pink, at least in part. *Flesh* thick, hard, white. Smell none or slightly fruity. Taste mild or of menthol. *Chemical test* $FeSO_4$ no reaction. *Microscopic features* Sp 7.5–9 × 6.5–8 μm, subglobose, with warts joined by strong ridges forming

an almost complete reticulum, the SP amyloid. Bas 40–55 × 9–13 μm. Cyst 50–120 × 7–15 μm, cylindric, obtuse to subcapitate, with refractive contents, not reacting with SA. CC hyphae slender, interwoven, septate and branching, 3–5 μm diameter, and cylindric tapering pileocystidia, 3–8 μm diameter, with the same contents as the hymenial cystidia. Rather numerous acid-resistant encrustations.

Season: July–October.
Habitat: Broad-leaved and coniferous trees, often under beech in western Europe.
Distribution: Very widespread. Common.

Has a wide colour range. The commonest satellite is probably var. *lactea* (Pers.) Möller & J. Schaeffer (= var. *ochroleucoides* ss auct europ), which is pinkish cream (*see* p.203).

---

p.209 *Russula aurora* Krombholz
Russulaceae, Russulales

Description: *Cap pink, the centre pinkish cream. Spore print white. Taste mild.*
Cap 4–10 cm (1.6–4 in), convex then depressed, often matt or slightly pruinose, pink, soon discolouring yellowish ochraceous cream in the centre, the intermediate zone salmon. *Gills* adnate, quite close, white then pale cream. Spore print whitish. *Stipe* 4–10 × 1–2 cm (1.6–4 × 0.4–1.6 in), irregularly clavate, the centre soon stuffed and supple, white. *Flesh* firm, soon soft, especially in the stipe, white. Smell slightly fruity. Taste mild. *Chemical test* $FeSO_4$ dingy pinkish. Sulphovanillin gives a redcurrant colour, especially bright on a dry fruit body.

Microscopic features Sp 6–8 × 5–6.5 μm, the warts irregular and often small, irregularly joined by incomplete connectives, the SP moderately amyloid. Bas 35–50 × 8–11 μm. Cyst fairly rare, fusiform, 50–80 × 7–12 μm, rostrate or mucronate, the contents greyish with SA. CC with a zone composed of short to subglobose cells, articulated like pieces of a jigsaw puzzle, supporting a few slender, erect, tapering hyphae, 2–2.5 µm diameter. Pileocystidia absent. Numerous acid-resistant encrustations present.

Season: July–November.
Habitat: Beech, hornbeam and lime.
Distribution: Widespread.

---

p.214 *Russula amoena* Quélet
Russulaceae, Russulales

Description: *Cap violaceous carmine, pruinose to finely cracked. Smell of Jerusalem artichokes.*
Cap 2–5 cm (0.8–2 in), convex then faintly depressed, velvety, pruinose to cracked towards the margin, violaceous carmine, sometimes more lilaceous red.
Gills adnate, quite close, white then cream, edge often violaceous reddish near the cap margin. Spore print cream.
Stipe 2–5 × 0.5–1 cm (0.8–2 × 0.2–0.4 in), slightly clavate fusoid, pruinose, often flushed lilaceous carmine.
Flesh firm, white to violaceous near the surface. Smell of Jerusalem artichokes (shellfish when cooking). Taste mild.
Chemical test $FeSO_4$ rather faint orange-pink. *Microscopic features* Sp 6.5–8.5 × 5.5–7 µm, subglobose, with low warts more or less joined by ridges forming an incomplete reticulum, SP not amyloid. Bas 40–55 × 10–15 µm. Cyst mainly marginal, with a swollen base and long tapering neck, 60–150 × 6–12 × 2–3 µm, hyaline or slightly brownish. CC as in *R. virescens* (see p.710) but the terminal cells are more elongated.

Season: July–October.
Habitat: Oak and sweet chestnut; also conifers. Tends to be continental and on acid soils.
Distribution: Widespread. Rather rare, especially in the north.

R. amoenicolor Romagnesi is taller and more variegated, violaceous with patches of lilac, green, yellow, red. The following species is in the same group.

p.205 *Russula violeipes* f. *citrina* Quélet
Russulaceae, Russulales

Description: *Cap matt, lemon yellow. Smell of Jerusalem artichokes.*
Cap 7–10 cm (2.8–4 in), subglobose then convex to depressed, velvety to scurfy, lemon yellow. *Gills* adnate, quite close, white or pale. Spore print cream. *Stipe* 4–7 × 1–2 cm (1.6–2.7 × 0.4–0.8 in), often fusoid, slowly becoming soft, white. *Flesh* firm, white or violaceous near the surface. Smell of Jerusalem artichokes. Taste mild. *Chemical test* Guaiac pale blue-green, rather slow. *Microscopic features* Sp 6.5–9 × 6–8 µm, subglobose, with ridges forming an incomplete reticulum, sometimes with a few warts, the SP not or faintly amyloid. Bas 50–65 × 10–12 µm. Cyst tapering or bottle-shaped, 80–130 × 10–20 µm, not reacting with SA. CC as in *R. amoena*.
Season: June–November.
Habitat: Broad-leaved and coniferous trees.
Distribution: Widespread. Sometimes quite common.

The type (*R. violeipes* Quélet), which typically has a carmine-purple cap often mixed with greenish yellow and purple colouring on the stipe, is rarer.

p.205 *Russula aeruginea* Lindblad
Russulaceae, Russulales

Description: *Cap more or less dark green. Spore print cream. Taste mild.*
Cap 4–10 cm (1.6–4 in), convex then depressed, slightly viscid or somewhat shiny when dry, with fine radial wrinkles, various shades of green, but often yellowish green to more or less dark sea-green. *Gills* adnate, quite close, white then ochraceous cream, sometimes with yellow tints at the base. Spore print cream. *Stipe* 5–8 × 1–2 cm (2–3.2 × 0.4–0.8 in), slightly tapering, white, yellowing or slightly rust below. *Flesh* quite firm except when old, white or yellowing slightly. Taste mild or slightly

hot in the young gills. *Chemical test* FeSO$_4$ rather faint pinkish grey. *Microscopic features* Sp 7–10 × 5–7 μm, elliptic, rather long for the genus, the warts elongated, partially joined by incomplete connectives, the SP slightly amyloid. Bas 35–45 × 6–8 μm. Cyst long cylindro-fusoid, 50–85 × 7–12 μm, rostrate or with a small apical head, greying with SA.

Season: July–November.
Habitat: Birch, sometimes conifers, especially spruce and pine.
Distribution: Common, especially northwards.

The species of the *Griseineae* section are particularly difficult to identify in the field (microscopic examination is essential).

---

p.207 *Russula parazurea* J Schaeffer
Russulaceae, Russulales

Description: *Cap violaceous blue-grey, the margin very pruinose.*
Cap 3–8 cm (1.2–3.2 in), convex then depressed, markedly pruinose, especially at the edge, even velvety in places, bluish grey, sometimes violaceous or brownish towards the centre. *Gills* adnate, quite close, whitish then pale cream. Spore print cream. *Stipe* 3–6 × 1–1.5 cm (1.2–2.4 × 0.4–0.6 in), tapering, white, sometimes slightly dingy below. *Flesh* white. Taste mild or slightly hot in the young gills. *Chemical test* FeSO$_4$ rather pale dirty orange-pink. *Microscopic features* Sp 6–8.5 × 5–6.5 μm, elliptic, with quite low warts joined by irregular ridges forming a slightly zebroid reticulum, the SP poorly defined. Bas 40–60 × 7–11 μm. Cyst fusoid, 50–90 × 7–12 μm, sometimes rostrate, more or less greying with SA. CC hyphae erect, septate, the supporting cells short and slightly inflated, the terminal one tapering, obtuse or conical. Pileocystidia cylindric obtuse or tapering, 60–90 × 4–10 μm, greyish with SA.

Season: June–November.
Habitat: Broad-leaved trees; rather disturbed places.
Distribution: Not fully known. Quite common.

Two similar but glabrous species are: *R. ionochlora* Romagnesi, which has a cap mottled with slightly lemony green and violaceous lilac, and is usually restricted to beech; and *R. grisea* (Pers.) Fr., which has a more or less dark violaceous cap, with violaceous flesh under the cuticle, and occurs under broad-leaved trees.

---

p.207 *Russula medullata* Romagnesi
Russulaceae, Russulales

Description: *Cap a mixture of greenish and olivaceous, stipe stuffed, spongy.*
Cap 4–10 cm (1.6–4 in), convex then depressed, viscid in damp weather then shiny, sometimes velvety at the edge, finely wrinkled when old, a mixture of bluish grey, greenish and lilac, the centre more buff to hazel. Margin slightly sulcate when old. *Gills* adnate, quite close, cream then rather dark yellow-buff. Spore print buff. *Stipe* 3–8 × 2–6 cm (1.2–3.2 × 0.8–2.4 in), soon stuffed, the centre spongy, whitish then flushed yellowish rust towards the base. *Flesh* white or pale. Taste mild, rarely hot in the young gills. *Chemical test* FeSO$_4$ orange pink (average intensity). *Microscopic features* Sp 6–8.5 × 5.5–6.5 µm, with quite low regular warts, often isolated or rarely joined, the SP amyloid. Bas 35–50 × 7–10 µm. Cyst cylindro-fusoid, 70–110 × 7–12 µm, rostrate, greying with SA. CC hyphae rather slender, bearing clusters of articulated cells, the supporting cells short and often inflated, 3–5 µm diameter, the terminal cells tapering, obtuse. Pileocystidia cylindro-clavate, 5–7 µm diameter, with refractive contents, greying with SA.

Season: June–October.
Habitat: Damp broad-leaved woods, grassy wood edges.
Distribution: Widespread but not fully known.

This species has the darkest spores in those *Griseineae* commonly encountered. Species in this section usually have cream spores but in this case the spores are distinctly buff.

p.212 *Russula puellaris* Fr.
Russulaceae, Russulales

Description: *Cap vinaceous brown in the centre. Fruit body yellowing strongly.*
Cap 2–6 cm (0.8–2.4 in), convex then depressed, slightly viscid to shiny, vinaceous crimson to vinaceous brown, darker in the centre. Margin striate to sulcate. *Gills* adnate, often not very close, pale cream then yellowish buff, with bright yellow patches when old. Spore print cream. *Stipe* 3–6 × 0.5–1 cm (1.2–2.4 × 0.2–0.4 in), slightly thickened towards the base, soon fragile, white, yellowing strongly or even orange-rust. *Flesh* yellowing strongly. Taste mild. *Microscopic features* Sp 6.5–9 × 5.5–7 µm, with rather tall conical warts, partly linked by sometimes zebroid ridges, the SP strongly amyloid. Bas 30–45 × 9–12 µm. Cyst cylindro-fusoid, rostrate, 50–85 × 8–14 µm, greying with SA. CC hyphae septate, the cells long and slender, 2–5 µm diameter, often obtuse; pileocystidia cylindro-clavate, sometimes with one or two (sometimes three) septa, obtuse, 5–12 µm diameter, more or less reacting with SA.
Season: July–November.
Habitat: Broad-leaved and coniferous trees.
Distribution: Widespread. Quite common.

There is another group of yellowing *Russula* spp., with a strong fruity smell.

---

p.212 *Russula unicolor* Romagnesi
Russulaceae, Russulales

Description: *Cap crimson, yellowing slightly. Spore print cream. Taste mild.*
Cap 3–5 cm (1.2–2 in), convex then slightly depressed, slightly viscid, shiny, smooth to grained half-way to the centre, vinaceous crimson or the centre sometimes darker, discolouring slightly from the edge. Margin finally slightly sulcate.
Gills adnate to nearly free, not very close, sometimes forking slightly, pale cream then

ochraceous cream. Spore print cream.
Stipe 3–6 × 0.5–1 cm (1.2–2.4 × 0.2–0.4 in), white, yellowing slightly from the base then with medium yellow spots.
Flesh white, unchanging or almost so. Smell fruity, very faint. Taste slightly acrid in the gills, mild elsewhere. *Microscopic features* Sp 6.5–8 × 5–6 µm, with wide, irregular and obtuse warts, joined by often slender connectives forming an almost complete reticulum, the SP distinctly amyloid. Bas 30–45 × 8–12 µm. Cyst cylindro-fusoid, 45–70 × 10–15 µm, papillate or rostrate, blackening with SA. CC hyphae rather short thick, obtuse, articulated, 2–5 µm diameter; pileocystidia with two or three septa, cylindro-clavate 4–9 µm diameter, greying with SA.

Season: June–October.
Habitat: Broad-leaved trees, usually birch.
Distribution: Probably widespread. Quite common, especially in temperate zones.

In the same group, *R. versicolor* J. Schaeffer, which has a more variegated cap, with vinaceous, lilac, cream or olivaceous colours, sometimes marbled with pink or grey, is mycorrhizal on birch. *R. pseudopuellaris* (M. Bon) M. Bon, a rather similar species also under birch (usually on sand, gravels, etc.), appears to yellow more strongly and often has a more shiny cap.

---

p.210 *Russula xerampelina* (Sch.) Fr.
Russulaceae, Russulales

Description: *Cap dark red to black in the centre. Spore print buff. Taste mild.*
Cap 6–15 cm (2.4–6 in), convex then depressed, shiny and slightly viscid then dry and matt at the edge, smooth, crimson, often dark or the centre almost black.
Gills adnate, quite close, pale cream then quite dark ochraceous cream, edge blackening slightly. Spore print dark buff.
Stipe 4–10 × 1.5–3 cm (1.4–4 × 0.6–1.2 in), pruinose, rose red at least in part, browning and finely wrinkled from

the base. *Flesh* browning with age. Smell soon unpleasant, like rotting shellfish. Taste mild. *Chemical test* $FeSO_4$ dark green. *Microscopic features* Sp 8–11 × 7–9 µm, with quite wide and irregular, spiny or obtuse warts, sometimes with the beginnings of slender incomplete connectives, the SP moderately amyloid. Bas 40–60 × 10–14 µm. Cyst rare, cylindro-fusoid, obtuse, 50–100 × 8–12 µm, more or less greyish with SA. CC hyphae densely interwoven, with tapering cylindric terminal cells, 3–6 µm diameter and a few pileocystidia not reacting with SA.

Season: September–December.
Habitat: Conifers, mainly pine.
Distribution: Widespread; especially coastal pine woods.

Best known of an extremely difficult group, the *Viridantes* (spore print typically buff, taste mild, smell of shellfish and a dark green reaction with $FeSO_4$). Numerous species have been described. This work gives just one example under conifers (*R. xerampelina*, see p.719) and another under broad-leaved trees (*R. subrubens*, see below). There are also representatives of this section in the dwarf shrub community of high mountain areas.

---

p.210 *Russula subrubens* (Lange) M. Bon
Russulaceae, Russulales

Description: *Cap coppery red. Spore print buff. Taste mild.*
Cap 7–12 cm (2.8–4.7 in), convex then slightly depressed, viscid then shiny to slightly viscid, brownish carmine, crimson, red, coppery to rust. *Gills* adnate, not very close, cream, somewhat rufous buff later, even spotted brownish, edge often spotted brownish. Spore print buff. *Stipe* 4–6 × 1–2 cm (1.6–2.4 × 0.4–0.8 in), slightly clavate, quite soon stuffed, the centre spongy, white then tinged or spotted yellow and somewhat rufous to brownish later. *Flesh* white then dingy rufous buff. Smell faint, fruity becoming unpleasant, like rotting shellfish. Taste mild. *Chemical*

test FeSO₄ dingy green. *Microscopic features* Sp 7.5–10 × 7–8.5 μm, with somewhat pointed but rather low warts, joined in small groups and interconnected by incomplete irregular low ridges, the SP rather small, amyloid. Bas 40–60 × 10–15 μm. Cyst cylindro-fusoid, 55–75 × 8–12 μm, obtuse or sometimes rostrate, greyish with SA. CC hyphae contorted, the ends hair-like, faintly articulated, cylindric, sometimes slightly inflated, wavy, 3–7 μm diameter; pileocystidia cylindric or fusoid, 50–80 × 4–8 μm, with refractive contents, reacting faintly with SA.

Season: July–November.

Habitat: Damp or marshy places, often near willow, alder, sometimes ash.

Distribution: Not fully understood. Rare.

---

p.210 *Russula velenovskyi* Melzer & Zvara
Russulaceae, Russulales

Description: *Cap zoned, bright red with a coppery orange centre. Cap* 2–8 cm (0.8–3.2 in), soon depressed or faintly umbonate, slightly viscid then shiny, smooth to finely wrinkled, zoned, brick-red to carmine towards the margin and orange or coppery in the centre, these shades sometimes paler or discolouring. *Gills* adnate to nearly free, becoming more distant with age, sometimes forking near the stipe, pale cream then pale buff, edge very often reddish or red near the cap margin. Spore print buff. *Stipe* 3–8 × 0.5–1.5 cm (1.2–3.2 × 0.2–0.6 in), the centre soon stuffed and spongy, white and very often with a pinkish or reddish patch on one side, near the base. *Flesh* white or cream, pinkish under the cuticle. Taste mild. *Microscopic features* Sp 7–9 × 5.5–7 μm, shortly elliptic, with rather slender warts, interconnected by occasional ridges or a few very slender connectives, the SP moderately amyloid. Bas 40–55 × 8–11 μm. Cyst cylindro-fusoid, rostrate, 55–75 × 8–12 μm, greying with SA. CC hyphae various, erect, more or less branching and clamped, the terminal cell

2.5–3.5 μm diameter, tapering to an obtuse apex. Pileocystidia cylindric, 50–120 × 3–7 μm, septate, tapering, with refractive contents, faintly greyish with SA and with rather slender acid-resistant encrustations.

Season: June–November.
Habitat: In woods or wood edges on acid soils, often under birch, conifers (pine).
Distribution: Widespread; rather northern.

---

p.211 *Russula nitida* (Pers.:Fr.) Fr.
Russulaceae, Russulales

Description: *Cap crimson, the centre discolouring greenish buff. Margin sulcate.*
Cap 2–6 cm (0.8–2.4 in), soon depressed, slightly viscid, shiny when dry, slightly grained, vinaceous crimson, sometimes more brown or rust and typically discolouring greenish or olivaceous in the centre. Margin sulcate when old. *Gills* adnate to nearly free, quite distant, pale cream then rather pale buff, edge sometimes slightly reddish towards the cap margin. Spore print buff. *Stipe* 3–8 × 0.5–1.5 cm (1.2–3.2 × 0.2–0.6 in), slightly irregular, sometimes pinkish below, the centre spongy or yellowing slightly. *Flesh* fragile, white, yellowing slightly. Taste mild. *Chemical test* $FeSO_4$ intense pinkish brown. *Microscopic features* Sp 8–12 × 6–9 μm, with spiny warts, often isolated or with occasional very slender connectives, the SP amyloid. Bas 40–55 × 1–13 μm. Cyst fusoid, 50–100 × 9–15 μm, rostrate, reacting faintly with SA. CC hyphae slender, the cells elongate, cylindric, slightly branching and with few clamps, 2–4 μm diameter; pileocystidia cylindro-clavate or long club-shaped, often slightly septate, 60–110 × 6–10 μm, faintly greyish with SA.

Season: July–November.
Habitat: Damp woodland, under birch, sometimes under conifers. Tends to be on acid soils.
Distribution: Widespread and mostly northern.

R. sphagnophila Kaufmann has a shiny violaceous pink cap with an olivaceous

p.211 *Russula decolorans* Maire
Russulaceae, Russulales

Description: *Cap pale brick-red. Blackening. Spore print pale buff. Taste mild.*
Cap 5–12 cm (2–4.7 in), globose then truncate or depressed, slightly viscid then matt, often grained, coppery orange, dull, sometimes more brick red. Margin finally slightly sulcate. *Gills* adnate, rather distant, whitish then ochraceous cream, edge quite thick, blackening. Spore print rather pale buff. *Stipe* 5–10 × 1.5–2.5 cm (2–4 × 0.6–1 in), firm then compressible, white then grey from the base and finally blackening. *Flesh* mild, soon brittle, white but quickly marbled black. *Microscopic features* Sp 9–13 × 7–10 µm, very large, with strong spiny warts partially joined by slender connectives forming an incomplete reticulum, the SP amyloid. Bas 50–55 × 12–15 µm. Cyst 70–110 × 10–15 µm, spindle-shaped inflated towards the middle, tapering or pointed but scarcely rostrate, black with SA. CC hyphae erect contorted tapering, 3–5 µm diameter, with few clamps; pileocystidia numerous, 50–90 × 5–8 µm, cylindro-clavate, with one or two septa, blackish with SA.
Season: June–September.
Habitat: Conifers in the mountains, marshy places.
Distribution: Mountains and northern Europe.

There is a look-alike of this species in warm southern districts: *R. seperina* Dupain (rare) which has a violaceous carmine cap, with a more olivaceous buff centre and flesh that reddens before blackening.

p.210 *Russula aurea* Pers.
Russulaceae, Russulales

Description: *Cap red but gills and superficial flesh bright yellow. Spore print pale yellow.*

Cap 4–8 cm (1.6–3.2 in), convex then depressed, slightly viscid then shiny, smooth or finely wrinkled, rather bright orange-red or coppery red with bright yellow zones or patches especially at the margin. *Gills* adnate, quite close, yellowish then ochraceous yellow, with orange tones, edge bright yellow. Spore print rather pale yellow. *Stipe* 3–8 × 1–2 cm (1.2–3.2 × 0.4–0.8 in), tuberculate to wrinkled, white to lemon yellow or with golden yellow patches. *Flesh* whitish but golden yellow under the cuticle. Taste mild. *Chemical test* $FeSO_4$ rather faint dingy pinkish.

Microscopic features Sp 7–10 × 6–8.5 µm, with warts joined by rather strong ridges, forming a more or less complete reticulum, the SP amyloid. Bas 40–65 × 10–15 µm. Cyst cylindro-fusoid, obtuse at the apex or slightly rostrate, 50–100 × 9–15 µm. CC hyphae erect, with many septa. No reactions with SA, acid-resistant encrustations absent. Pigment red, intracellular.

Season: July–October.
Habitat: Mainly broad-leaved woods in warm places; may prefer calcareous soils.
Distribution: Rather southern. Quite rare.

---

p.206 *Russula claroflava* Grove
Russulaceae, Russulales

Description: *Blackening; cap yellow. Spore print buff. Taste mild.*

Cap 5–15 cm (2–6 in), convex then expanded, finally irregularly sulcate, slightly viscid then shiny, more rarely matt when old, smooth or grained, rather bright yellow, sometimes dull or paler as the fruit body ages. *Gills* adnate, quite close, yellowish ivory then rather bright yellowish ochraceous, edge slightly thick, greying to blackening. Spore print buff. *Stipe* 4–9 × 1.5–3 cm (1.9–3.5 × 0.6–1.2 in), the centre spongy, finely wrinkled, white but greying or blackening from the base. *Flesh* white but greying then blackening in patches. Taste mild. *Microscopic features* Sp 8.5–10 × 6.5–8 µm, with more or less developed, dense warts, partially joined by fine, zebroid ridges, forming an incomplete reticulum, the SP rather faintly amyloid. Bas 40–50 × 10–12 µm. Cyst fusoid, 50–85 × 8–12 µm, rostrate, more or less blackish with SA. CC hyphae weakly articulated, 2–4 µm diameter, including numerous long primordial hyphae, 4–7 µm diameter, often with patches of very large, rather spectacular acid-resistant encrustations. No pileocystidia.

Season: July–November.
Habitat: Sphagnum or marshy places with birch but also other damp-loving broad-leaved trees.
Distribution: Rather northern or in submontane or montane zones.

---

p.211 *Russula laeta* J. Schaeffer
Russulaceae, Russulales

Description: *Cap orange-red with a paler centre. Spore print yellow. Taste mild.*
Cap 5–8 cm (2–3.2 in), convex then depressed, slightly viscid then shiny, finely grained, pinkish red to orange-red, sometimes coppery or spotted rust or olivaceous cream in the paler centre. Margin slightly sulcate tuberculate when old. *Gills* adnate, not very close, brittle, cream then mid-yellow. Spore print yellow. *Stipe* 4–6 × 1–1.5 cm (1.6–2.4 × 0.4–0.6 in), solid then stuffed, finely wrinkled, white or faintly greyish below *Flesh* becoming soft when old, whitish. Tastes a little like menthol, like *R. lepida*. *Chemical test* $FeSO_4$ pinkish. *Microscopic features* Sp 8–10 × 7–8.5 µm, with spiny, isolated or minutely crested warts, the SP not distinctive, amyloid. Bas 40–50 × 10–12 µm. Cyst fusoid or cylindric, slightly ventricose, slightly rostrate, greying with SA. CC hyphae misshapen, contorted or slightly inflated or with many constrictions, 2–4 µm diameter; pileocystidia relatively short and thick, cylindric or clavate, 50–90 × 5–9 µm, one to three septate, with refractive contents, greying with SA.

Season: July–October.
Habitat: Broad-leaved trees, especially beech.
Distribution: Not yet fully understood. Rare.

> *R. rubroalba* (Singer) Romagnesi, which is bright red to crimson, sometimes discolouring or with mixed colours, has stiff, very brittle gills and a faint smell of honey. In the same group but more variegated, *R. romellii* Maire, which has a violaceous to vinaceous cap with a greenish buff or olivaceous centre, also has very fragile gills that shatter when touched; it has a very faint smell of fruit.

---

p.212 *Russula melitodes* Romagnesi
Russulaceae, Russulales

Description: *Cap brownish crimson to olivaceous vinaceous. Spore print yellow.*
Cap 5–9 cm (2–3.6 in), convex then depressed, somewhat matt to velvety, at least towards the edge, sometimes with fine radial wrinkles half-way to the centre, rosewood to brownish crimson, sometimes mixed with vinaceous, rust-brown, with the centre discolouring dark buff or olivaceous. Margin slightly sulcate tuberculate when old. *Gills* adnate, not very close, cream then rather dark yellow. Spore print yellow. *Stipe* 4–7 × 1–2 cm (1.6–2.8 × 0.4–0.8 in), finely wrinkled, white to pale ochraceous cream towards the base. *Flesh* white or pale crimson under the cap cuticle. Smell slightly fruity. Taste mild. *Chemical test* $FeSO_4$ dull pinkish. *Microscopic features* Sp 8.5–10.5 × 6–9.5 µm, with rather tall spiny warts, finely crested in part or joined by short slender connectives, the SP distinctly amyloid. Bas 45–65 × 11–15 µm. Cyst cylindro-fusoid, 60–100 × 8–15 µm, obtuse or rarely rostrate, more or less grey with SA. CC hyphae slender filamentous septate, 2–3.5 µm diameter, the terminal cell long, tapering; pileocystidia septate, cylindric, 60–100 × 4–10 µm, with refractive

contents, greying with SA. Numerous acid-resistant encrustations present as large granules or sheathing the primordial hyphae and pileocystidia.

Season: July–October.
Habitat: Broad-leaved trees. Usually on calcareous soils.
Distribution: Mainly in lowland areas. Rare.

---

p.212 *Russula integra* (L.) Fr.
Russulaceae, Russulales

Description: *Cap brown to rather crimson or marbled and mottled with other colours. Spore print yellow.* *Cap* 5–12(15) cm (2–4.7(6) in), convex then faintly depressed, slightly viscid then shiny, dull when old and becoming grained to finely wrinkled, brown to crimson-brown, very often with a mixture of colours or discolouring or tinged with patches of buff, yellowish, olive, rosewood, etc. Margin slightly sulcate, often rather short and revealing the gills. *Gills* adnate, obtuse near the cap margin, somewhat brittle, becoming more distant when old, milky white then dark yellow, edge thick, appearing paler than the faces when mature. Spore print yellow. *Stipe* 4–8 × 2–6 cm (1.6–3.2 × 0.8–2.4 in), finely wrinkled, white then spotted or flushed brownish rust. *Flesh* white, slightly crimson under the cap cuticle. Taste of hazelnuts if chewed. *Chemical test* $FeSO_4$ dingy pinkish.

Microscopic features Sp 8–11 × 6–9.5 µm, conspicuously spiny, with isolated pointed warts, the SP wide, amyloid. Bas 45–65 × 10–15 µm. Cyst cylindric to fusoid, 70–150 × 7–13 µm, obtuse, often not rostrate, blackening in part with SA. CC hyphae slender, the ends hair-like, erect, tapering or pointed, 1–3 µm diameter; pileocystidia variable, 50–100 × 4–8 µm, more or less septate, with refractive contents, blackening with SA. Acid-resistant encrustations present on certain hyphae and the pileocystidia.

Season: June–October.

Habitat: Conifers, especially fir and spruce.
Distribution: Very widespread, in submontane and montane areas, extending to the subalpine zone. Unusual in lowland areas.

One of the best known edible species in the mountains.

---

p.213 *Russula cessans* Pearson
Russulaceae, Russulales

Description: *Cap shiny, vinaceous or with mixed colours. Spore print yellow. Taste mild.*
Cap 2–8 cm (0.8–3.2 in), often irregular, slightly viscid, shiny when dry, vinaceous but frequently mixed with brownish, buff, yellowish, coppery. *Gills* adnate to nearly free, not very close, cream then dark yellow when mature. Spore print dark yellow. *Stipe* 3–6 × 1–2 cm (1.2–2.4 × 0.4–0.8 in), fragile, white, the base becoming slightly dingy ochraceous cream. *Flesh* rather fragile, white or pale. Taste mild or slightly unpleasant but not acrid. *Microscopic features* Sp 8.5–11 × 7–8.5 µm, with irregular warts joined by slender connectives or rather pronounced ridges, forming an incomplete reticulum, the SP large, amyloid. Bas 30–50 × 10–12 µm. Cyst cylindro-fusoid, 50–90 × 6–10 µm, faintly greyish with SA. CC hyphae slender, interwoven, the ends hair-like, erect, cylindric or slightly clavate, septate, 3–4 µm diameter. Pileocystidia 50–120 × 6–10 µm, often with two to four septa, cylindric or the terminal cell slightly inflated, refractive, faintly reacting with SA.

Season: August–December.
Habitat: Pine, Mainly on sandy soils.
Distribution: Widespread, rather Atlantic (also in more continental sites).

The other two well-known species in this group have a striate to sulcate margin (and spores with more or less spiny, isolated warts). These are: *R. nauseosa* (Pers.) Fr., which occurs under spruce and *R. laricina* Vel. (with slightly greying flesh) which occurs mainly under larch, in the mountains.

p.211 *Russula risigallina* (Batsch) Saccardo
Russulaceae, Russulales

Description: *Cap yellow, often matt. Spore print yellow. Taste mild.*
Cap 1–5 cm (0.4–2 in), convex, soon depressed, fragile, dry to velvety, matt, smooth, yellow. Margin smooth then slightly sulcate to pectinate. *Gills* adnate or nearly free, not very close, cream then yellow, with bright or dark orange tints. Spore print yellow. *Stipe* $2–5 \times 0.5–1$ cm ($0.8–2 \times 0.2–0.4$ in), cylindric or irregular, fragile, white, faintly yellowing or cream towards the base. *Flesh* soon soft, white or pale. Smell of plums and roses, at least when over-mature. Taste mild. *Microscopic features* Sp $7–9 \times 6–7$ μm, elliptic, with rather slender, tall, spiny or obtuse warts, isolated or with occasional slender connectives, the SP rather small, moderately amyloid. Bas $30–45 \times 8–11$ μm. Cyst cylindro-fusoid, $45–70 \times 6–10$ μm, obtuse or slightly pointed but scarcely rostrate, blackening with SA. CC hyphae septate, the ends hair-like, erect, with rather short or inflated supporting cells, and sometimes subcapitate terminal cells; pileocystidia absent but acid-resistant encrustations numerous, more or less coarse or confluent, sometimes sheathing granules.

Season: June–November.
Habitat: Woods, mainly broad-leaved.
Distribution: Widespread.

The colours are rather variable: f. *luteorosella* (Britz.) M. Bon, with pink and yellow zones and f. *chamaeleontina* (Fr.) M. Bon, with a mixture of red and yellow, with some green or greenish have been described. *R. acetolens* S. Rauschert (= *R. vitellina*) has a sticky viscid cap, at least in the centre, even in dry weather (check with the kiss test), of a pure egg-yellow, sometimes paler at the margin. In addition, the flesh has a smell of vinegar, at least when over-ripe.

p.211 *Russula turci* Bresadola
Russulaceae, Russulales

Description: *Cap violaceous, zoned, viscid in the centre, the margin velvety.*
Cap 3–10 cm (1.2–4 in), quite soon depressed, viscid to slimy in the centre in both dry and damp weather, matt to velvety at the edge, zoned, violaceous or carmine to more blackish crimson in the centre, possibly also with yellow, orange, green colours. *Gills* adnate, not very close, cream then buff. Spore print dark buff to yellow. *Stipe* 3–7 × 1–2 cm (1.2–2.8 × 0.4–0.8 in), slightly clavate but often tapering at the extreme base, white, becoming slightly dingy. *Flesh* white, rather fragile. Smell slightly fruity under the gills but a strong smell of iodine at the stipe base. Taste mild.

Microscopic features Sp 8–9.5 × 6–7.5 μm, with warts joined by quite thick ridges forming an almost complete reticulum, the SP amyloid. Bas 35–60 × 9–12 μm. Cyst cylindro-fusoid, 60–100 × 7–12 μm, obtuse or rostrate, scarcely greying with SA. CC hyphae interwoven, the ends hair-like, contorted, 2–5 μm diameter, sometimes slightly clavate, mixed with numerous septate hyphae 4–9 μm diameter, the septa more or less close, strongly and even heavily encrusted (acid-resistant encrustations).
Season: August–December.
Habitat: Conifers; mainly pine, sometimes spruce.
Distribution: Widespread, often near the Atlantic coast.

R. amethystina Quélet, a very similar species with gills more yellow from the outset and a more fragile stipe, occurs mostly under spruce and fir in continental or submontane areas.

---

p.215 *Russula amara* Kucera
Russulaceae, Russulales

Description: *Cap purple, umbonate. Spore print yellow. Taste slightly or slowly bitter.*
Cap 4–10 cm (1.6–4 in), conico-convex, sometimes pointed then expanded,

retaining an obtuse umbo, slightly viscid and shiny, purple to blackish crimson, sometimes discolouring. *Gills* adnate, close, ivory then yellowish ochraceous cream. Spore print yellow. *Stipe* 4–8 × 1–2 cm (1.6–3.2 × 0.4–0.8 in), tapering below, pure white. *Flesh* white, scarcely browning or darkening. Smell faintly fruity. Taste mild or slightly to slowly bitter in the cap cuticle. *Microscopic features* Sp 8–9.5 × 7–8 µm, with slightly spiny rather dense warts, joined by fine dots or more or less complete connectives, forming an irregular reticulum, the SP amyloid. Bas 35–55 × 9–13 µm. Cyst cylindric or tapering, obtuse to papillate, 60–120 × 8–12 µm, faintly greying with SA. CC hyphae dense, the ends hair-like, articulated, tapering or slightly clavate, 2–4 µm diameter, the primordial hyphae cylindric, septate, 3–5 µm diameter, with rather large coarse acid-resistant encrustations. Pigment red in intracellular droplets.

Season: July–November.
Habitat: Pine. Usually on acid soils.
Distribution: Widespread. Common.

---

p.214 *Russula olivacea* (Sch.) Pers.
Russulaceae, Russulales

Description: *Cap with a mixture of colours, concentrically cracked at the edge. Stipe pinkish.*
Cap 8–20 cm (3.2–8 in), subglobose then convex or depressed, dry and matt, pruinose or velvety, with fine concentric wrinkles towards the edge, vinaceous red often mixed with greenish, buff, etc.
Gills adnate, more distant when old, cream then ochraceous yellow, edge sometimes slightly red towards the cap margin. Spore print yellow. *Stipe* 6–12 × 2–4 cm (2.4–4.7 × 0.8–1.6 in), brittle, finely wrinkled or grained, lilac pink, at least above, yellowing slightly from the base. *Flesh* white then tinged brownish yellow. Smell slightly fruity. Taste mild.

Microscopic features Sp 8–11.6 × 7–9.5 µm, with rather robust spiny, or often obtuse, warts, sometimes in pairs or irregular, with a few very slender and incomplete connectives, the SP amyloid. Bas 40–60 × 10–14 µm. Cyst fusoid, tapering and often pointed, 70–100 × 8–15 µm, only slightly blackening with SA. CC hyphae dense, the ends hair-like, erect, articulated, 3–7 µm diameter, certain supporting cells sometimes flask-shaped, up to 14 µm diameter. Pileocystidia and primordial hyphae absent.

Season: July–November.
Habitat: Mainly beech but sometimes spruce in the mountains. Often on acid soils.
Distribution: Widespread.

Although all the mild *Russula* spp. are considered edible, this species appears to be the exception since a few cases of slight poisoning in Italy have been attributed to it.

---

p.206 *Russula ochroleuca* (Hall.) Pers.
Russulaceae, Russulales

Description: *Cap uniformly ochraceous yellow, contrasting with the white gills.*
Cap 4–10 cm (1.6–4 in), very slightly viscid then slightly shiny, finely wrinkled to grained, fairly clear ochraceous yellow, sometimes bright and rather uniform.
Gills adnate, close, white and remaining so or sometimes greying when old. Spore print white. *Stipe* 4–8 × 2–6 cm (1.6–3.2 × 0.8–2.4 in), pruinose then finely wrinkled to wrinkled longitudinally, white, greying below, especially the wrinkles, or yellowish to dingy buff when old. *Flesh* white to yellowish buff below, greying slightly. Taste rather acrid.

Microscopic features Sp 8–11 × 6.5–8.5 µm, with unequal or irregular spiny warts, joined by slender connectives forming an often incomplete reticulum, the SP amyloid. Bas 40–55 × 10–13 µm. Cyst cylindro-fusoid, often not very prominent, 60–110 × 6–12 µm, blackening with SA.

CC hyphae slender, the ends hair-like, narrow, occasionally septate, with an obtuse apex, the cell wall slightly refractive and yellow with acid-resistant epimembranal pigment. Pileocystidia extremely rare.

Season: June–December.
Habitat: Broad-leaved and coniferous woods.
Distribution: Widespread, except perhaps high in the mountains.

Specimens occurring under conifers usually have more intense greenish tones.

---

p.206 *Russula fellea* (Fr.:Fr.) Fr.
Russulaceae, Russulales

Description: *Cap tawny orange with a pale margin. Gills cream. Smell of stewed apple.*
Cap 4–10 cm (1.6–4 in), soon expanded depressed, slightly viscid then dry, smooth or faintly grained, zoned, orange to tawny buff in the centre, cream to yellowish buff, paler and dull at the edge. Margin smooth then slightly sulcate. *Gills* adnate, quite close, pale cream then buff, concolorous with the cap margin. Spore print white. *Stipe* 4–6 × 1–2 cm (1.6–2.4 × 0.4–0.8 in), white then ochraceous cream from the base. *Flesh* white, becoming buff later in the stipe. Smell strong of stewed apple. Taste very acrid. *Microscopic features* Sp 7.5–9.5 × 6–7.5 µm, with conical, rather blunt warts, joined by slender connectives forming a more or less complete reticulum, the SP amyloid. Bas 40–50 × 8–11 µm. Cyst cylindro-clavate, 50–120 × 5–10 µm, often rostrate, black with SA. CC hyphae wavy, the ends hair-like, obtuse cylindric, scarcely septate; pileocystidia cylindric, 4–8 µm diameter, rather more densely septate, with yellowish contents greying with SA, the underlying hyphae with brownish yellow epimembranal pigment.

Season: July–November.
Habitat: Beech, sometimes in mixed woods.
Distribution: Widespread. Very common.

p.208 *Russula luteotacta* Rea
Russulaceae, Russulales

Description: *Cap red, discolouring. Gills arched, distant.*
Cap 3–8 cm (1.2–3.2 in), slightly viscid then a little shiny, often grained, the cuticle not peeling, rather bright red but discolouring, finally a mixture of red, pink or even cream to white. Margin scalloped, slightly sulcate, often spotted yellow when old. *Gills* decurrent, not very close, interveined at the base, white or greyish cream, spotting yellow where damaged. Spore print white. *Stipe* 2–6 × 1–1.5 cm (0.8–2.4 × 0.4–0.6 in), sometimes eccentric, firm, white, often flushed pink, at least towards the base, spotting yellow where damaged. *Flesh* white, yellowing. Taste acrid. *Microscopic features* Sp 7–9 × 6–7.5 µm, with regular or slightly elongated warts, rather dense and partially joined by occasional, fairly slender connectives, the SP amyloid. Bas 35–55 × 8–11 µm. Cyst cylindro-fusoid, 65–100 × 5–10 µm, obtuse or shortly rostrate, more or less blackish with SA. CC hyphae dense, slightly gelatinized, the ends hair-like, septae constricted, 2–3 µm diameter; pileocystidia cylindric, 60–110 × 3–10 µm, often slightly clavate, not very septate, greying with SA.

Season: June–October.
Habitat: Damp broad-leaved woodland, especially with hornbeam, on heavy clay soils.
Distribution: Widespread. Quite common.

R. persicina Krombholz, which has a drier cap that discolours less, more horizontal gills and a mid-cream spore print, yellows very faintly.

---

p.208 *Russula emetica* (Sch.:Fr.) Pers.
Russulaceae, Russulales

Description: *Cap red. Gills white. Spore print white. Taste acrid.*
Cap 5–10 cm (2–4 in), convex then expanded depressed, slightly viscid to shiny when dry, smooth, the cuticle almost

entirely peeling, pure bright red. Margin slightly sulcate when old. *Gills* adnate, not very close, white and remaining so, sometimes with pale yellowish tints when young. Spore print white. *Stipe* 5–8 × 1–2 cm (2–3.2 × 0.4–0.8 in), unchanging or faintly yellowing from the base. *Flesh* soon fragile, white and yellowing slightly towards the stipe base but pinkish under the cap cuticle. Smell slightly fruity. Taste very acrid. *Microscopic features* Sp 8–11 × 7–8.5 µm, with numerous, large conical, often pointed warts, joined by slender connectives, forming an almost continuous reticulum, the SP amyloid. Bas 40–50 × 10–12 µm. Cyst cylindro-fusoid, 50–100 × 8–13 µm, obtuse or rarely rostrate or umbonate, greying with SA. CC hyphae interwoven, the ends hair-like, slender, cylindro-clavate, 2–4 µm diameter; pileocystidia very numerous, cylindric or slightly clavate, with many septa, sometimes constricted at the septa, 5–13 µm diameter, greying with SA.

Season: June–September.
Habitat: Damp montane, mainly coniferous woods, with sphagnum or bilberry, sometimes under beech. Usually on acid soils.
Distribution: Mountains in Europe.

R. emetica var. *silvestris* Singer, is more slender and rather fragile species, with a fairly bright red cap which soon fades or discolours to pinkish cream patches in the centre. The cuticle peels fully, exposing flesh that is typically white or faintly pinkish near the surface and often has a distinct smell of coconut. It occurs mainly under broad-leaved trees on acid soils in lowland areas. Species in the *R. emetica* group, which are violently acrid, are among those responsible for causing stomach upsets.

---

p.208 *Russula betularum* Hora
Russulaceae, Russulales

Description: *Fragile, pinkish and discolouring. Spore print white. Taste very acrid.*

Cap 1–5 cm (0.4–2 in), often depressed, slightly viscid then shiny, the cuticle completely peeling, reddish pink, discolouring strongly and often finally almost white. Margin dentate and sulcate when old. *Gills* adnate, rather distant, white and remaining so, edge entire or slightly denticulate. Spore print white. *Stipe* 3–6 × 0.5–1 cm (1.2–2.4 × 0.2–0.4 in), very fragile, finely wrinkled, white but greying slightly when water-soaked. *Flesh* thin and very fragile, white. Smell of apple or coconut fainter than in *R. emetica*. Taste very acrid. *Microscopic features* Sp 8.5–12 × 7.5–8.5 µm, with strong dense, spiny warts, quite strongly joined by connectives forming an almost complete reticulum, the SP amyloid. Bas rarely 2-spored, 40–55 × 10–15 µm. Cyst pointed spindle-shaped, 50–100 × 7–13 µm, often awl-shaped, greying with SA. CC hyphae with hair-like, slender ends, 2–3 µm diameter; numerous pileocystidia as in *R. emetica*, 5–10 µm diameter.

Season: June–November.
Habitat: Birch. Usually in damp places.
Distribution: Widespread. Common.

---

p.208 *Russula fageticola* (Melzer) Lundell
Russulaceae, Russulales

Description: *Cap red. Underlying flesh pinkish. Gills with glaucous tints. Spore print white.*
Cap 3–10 cm (1.2–4 in), often depressed, soon dry, fairly shiny but sometimes more matt at the edge, the cuticle half peeling, fairly bright red, uniform or slightly paler towards the margin but not discolouring buff or yellowish. *Gills* adnate, quite close, white with a glaucous to bluish tint at the base. Spore print white. *Stipe* 4–8 × 1–1.5 cm (1.6–3.2 × 0.4–0.6 in), quite firm then fragile, sometimes yellowing slightly below. *Flesh* firm then becoming soft, white but pinkish under the cap cuticle, yellowing slightly when old. Smell fruity or of coconut. Taste very acrid. *Microscopic features* Sp 7–9 × 5–7 µm, with often quite low, irregularly

arranged warts, joined by connectives forming lines or a more or less complete reticulum, the SP amyloid. Bas 40–50 × 8–10 µm. Cyst long cylindro-fusoid, 50–100 × 6–11 µm, obtuse or rostrate, blackening with SA. CC hyphae with hair-like, slender or thicker, obtuse or clavate ends, up to 3–5 µm diameter, with rather short cells; pileocystidia not very septate, 50–120 × 6–10 µm, cylindric, greying with SA.

Season: July–November.
Habitat: Beech. Usually on neutral–acid soils.
Distribution: Western or temperate Europe.

The commonest species in a group differing from the *R. emetica* complex in its less separable cap cuticle, pinkish underlying flesh, relative firmness, smaller spores and the microscopic characters of the cuticle (wider terminal hyphae and less septate pileocystidia). *R. nobilis* Velenovsky, which has a convex, fleshy, bright red cap, short thick stipe and quite firm flesh, occurs in beech woods, usually on warm calcareous soils.

---

p.208 *Russula nana* Killerman
Russulaceae, Russulales

Description: *Dwarf species. Cap bright red or pink. Spore print white. Taste very acrid.*
Cap 1–4 cm (0.4–1.6 in), subglobose then rarely depressed, slightly viscid then shiny when dry, the cuticle half peeling, bright red to pinkish, sometimes more solidly coloured and almost garnet in the centre, discolouring in irregular ochraceous cream patches. Margin finally slightly sulcate. *Gills* adnate, rather close, whitish to greyish. Spore print white. *Stipe* 1–3 × 0.5–1 cm (0.4–1.2 × 0.2–0.4 in), fragile, short, white to greying slightly. *Flesh* very acrid, white to greyish, slightly reddish under the cuticle. *Microscopic features* Sp 6–9 × 5–7 µm, with quite low but dense warts, joined by an almost complete reticulum of slender and irregular connectives, the SP amyloid. Bas 40–55 ×

10–12 μm. Cyst fusoid, 50–80 × 7–10 μm, often tapering to a point, greying with SA. CC hyphae with rather slender, often obtuse hair-like ends, 2–4 μm diameter; pileocystidia not very septate, 5–8 μm diameter.

Season: June–August.
Habitat: Low alpine shrub community, dwarf willows.
Distribution: European alpine zone and at lower altitudes in Scandinavia.

Var. *alpina* (Blytt) M. Bon is almost mild.

---

p.215 *Russula fragilis* (Pers.:Fr.) Fr.
Russulaceae, Russulales

Description: *Cap purple with an olivaceous or greenish centre. Gills finely saw-edged.*
Cap 1–6 cm (0.4–2.4 in), soon depressed, slightly viscid then shiny, finely grained, the cuticle peeling fully, typically violaceous red or violaceous lilac, the centre darker then discolouring brownish or greenish. Margin with rather long striations or furrows.
Gills adnate, not very close or even distant, white, edge denticulate to serrulate. Spore print white. *Stipe* 2–5 × 0.5–1 cm (0.8–2 × 0.2–0.4 in), very fragile, white then slightly brownish yellow from the base. *Flesh* thin and soon very fragile, white. Smell strong and distinctive, like coconut with a hint of geranium, rather difficult to describe. Taste very acrid. *Microscopic features* Sp 7–10 × 6–8.5 μm, with quite low warts strongly joined by ridges into a reticulum, the SP faintly amyloid. Bas 30–50 × 9–12 μm. Cyst cylindro-fusoid or clavate, 60–95 × 10–14 μm, tapering or papillate, greying strongly with SA. CC hyphae slender, the ends hair-like, slender, contorted, 2–3 μm diameter, obtuse or clavate at the apex; pileocystidia, 50–100 × 6–10 μm, cylindric or clavate, not very septate, greying with SA. Pigment intracellular, pink, abundant.
Season: July–November.
Habitat: Mostly under broad-leaved trees.
Distribution: Widespread. Very common.

Numerous forms and varieties have been described, even satellite species. *R. fragilis* f. *fallax* (Fr.) Massee (broad-leaved trees) with dark olive and lilac zonation on the cap, is similar to *R. turci* (see p.730). *R. knauthii* (Singer) Hora (broad-leaved trees and conifers), is more robust and has a dark red to blackish centre with a lilac pink margin.

p.215 *Russula krombholzii* R. Shaffer
(= *R. atropurpurea*)
Russulaceae, Russulales

Description: *Cap red and black, sometimes violaceous and discolouring. Spore print white.*
Cap 3–10 cm (1.2–4 in), convex or slightly depressed, slightly viscid then shiny, smooth to finely wrinkled, the cuticle scarcely peeling, bright red at the edge and crimson to almost black in the centre, but very often more vinaceous, discolouring in buff to brownish patches. *Gills* adnate, quite close, white to pale cream. Spore print white. *Stipe* 3–5 × 1–2 cm (1.2–2 × 0.4–0.8 in), quite hard then softening, pruinose, often becoming finely wrinkled towards the base, white, typically greying on the wrinkles and sometimes yellow rust below. *Flesh* white, greying slowly. Smell faint, of raw apples. Taste slightly acrid. *Microscopic features* Sp 6.5–9 × 6–7 µm, with quite wide or elongated warts, joined by ridges and connectives forming a complete reticulum, the SP slightly amyloid. Bas 35–55 × 8–11 µm. Cyst fusoid clavate, 55–100 × 6–12 µm, rostrate, blackening with SA. CC hyphae slender, the ends hair-like, irregularly tapering, 1–3 µm diameter; pileocystidia cylindro-clavate, 55–90 × 4–10 µm, not very septate, the terminal cell often constricted or capitate, greying with SA.
Season: June–December.
Habitat: Broad-leaved and coniferous woods.
Distribution: Widespread. Very common.

F. *depallens* (Pers.:Fr.) M. Bon, which seems the most distinctive of the satellites described, has a strongly discolouring cap and more grey in the stipe.

---

p.215 *Russula norvegica* Reid
Russulaceae, Russulales

Description: *Small species. Cap dark vinaceous crimson. Spore print white. Taste acrid.*
Cap 1–4 cm (0.4–1.6 in), convex then depressed, pruinose then slightly viscid to shiny, smooth to finely wrinkled, the cuticle partially peeling, often crimson-black to vinaceous or crimson, sometimes discolouring in lilac or even whitish patches when old. Margin scarcely expanded, striate or sulcate when old.
Gills adnate, quite close, white. Spore print white. *Stipe* 1–3 × 0.5–1 cm (0.4–1.2 × 0.2–0.4 in), often short, smooth to wrinkled below, white or pale ochraceous cream at the base. *Flesh* quite thick, white to crimson under the cuticle. Smell sharp or like apple, rather faint. Taste acrid.
Microscopic features Sp 7–9 × 6–7.5 µm, with rather small warts, joined by often incomplete connectives, the SP only very faintly amyloid. Bas sometimes 2-spored, 35–50 × 8–11 µm. Cyst long fusoid, tapering to rostrate, 60–90 × 6–12 µm, greying with SA. CC hyphae with narrow, cylindric hair-like ends, 2–3 µm diameter; pileocystidia numerous, cylindric or variable, not very septate, 5–10 µm diameter, greying with SA.
Season: July–September.
Habitat: Low alpine shrubs, with dwarf willows.
Distribution: Alpine zone.

In the *R. krombholzii* group, this species is homologous to *R. nana* in the *R. emetica* group.

---

p.216 *Russula alnetorum* Romagnesi
Russulaceae, Russulales

Description: *Cap mauve. Spore print white. Taste slightly acrid.*

Cap 1–4 cm (0.4–1.6 in), soon expanded to depressed, slightly viscid then slightly shiny, the cuticle peeling, mauve or lilac to purple, often with brownish or buff patches towards the centre. Margin slightly sulcate. *Gills* adnate, rather distant, white to very pale greyish cream. Spore print white. *Stipe* 2–4 × 0.5–1 cm (0.8–1.6 × 0.2–0.4 in), soon hollow, finely wrinkled, white to yellowish when water-soaked, greying when old. *Flesh* white to yellowish then greyish. Taste not very acrid or even almost mild. *Microscopic features* Sp 7–8.5 × 5.5–7 µm, with irregular warts, joined by ridges or connectives forming a complete reticulum, the SP amyloid. Bas 45–55 × 1–13 µm. Cyst fusiform, rostrate, 70–90 × 8–12 µm, black with SA. CC hyphae filamentous, the ends hair-like, slender, contorted or elongate, 3–4 µm diameter; pileocystidia numerous, septate, cylindric to clavate or the terminal cell constricted, greying with SA.

Season: July–September.
Habitat: Green alder.
Distribution: Subalpine zone.

R. pumila Rouzeau & Massart, a related species, occurs in lowland sites, mainly with alder. It has a more violaceous cap with a very sulcate margin, a markedly wrinkled stipe and greys very strongly.

---

p.216 *Russula pelargonia* Niolle
Russulaceae, Russulales

Description: *Cap lilac to greenish grey. Margin sulcate. Spore print cream.*
Cap 2–5 cm (0.8–2 in), fragile, soon depressed, slightly viscid then shiny, the cuticle peeling, lilac, purplish red, soon fading to greenish grey in the centre. Margin undulating or sometimes lobed, striate to sulcate. *Gills* adnate to nearly free, rather distant, cream. Spore print cream. *Stipe* 2–6 × 0.5–1 cm (0.8–2.4 × 0.2–0.4 in), soon hollow, slightly wrinkled, white but greying slightly. *Flesh* very

fragile, white but greying somewhat. Smell very strong, of geranium mixed with stewed apple or fruit. Taste acrid.
Microscopic features Sp 7–9.5 × 6.5–9 µm, with spiny warts joined by ridges in lines forming an often complete reticulum, the SP amyloid. Bas 35–50 × 10–12 µm. Cyst cylindro-clavate, obtuse or rostrate, 50–75 × 7–13 µm, strongly greying with SA. CC hyphae slender, articulated, cylindric or clavate, 2–4 µm diameter; pileocystidia cylindric or inflated at the apex, not very septate, 6–10 µm diameter, greying with SA.

Season: July–October.
Habitat: Broad-leaved trees, especially poplar and aspen on damp ground.
Distribution: Not fully understood. Rather rare.

R. clariana Heim ex Kuyper & van Vuure, a more robust and less fragile species which has a scarcely sulcate or smooth margin and becomes more strongly grey, is usually restricted to poplars. *R. violacea* Quélet, another very similar species, has a cap that is clearly viscid in the centre and more distinct colours, usually violaceous with a green centre and yellowing flesh. It occurs under broad-leaved trees.

---

p.214 *Russula cavipes* Britzlemayer
Russulaceae, Russulales

Description: *Cap violaceous to greenish or pearly grey at the edge. Spore print cream.*
Cap 3–7 cm (1.2–2.8 in), quite soon depressed, sometimes irregular, viscid to shiny, grained, the cuticle peeling, violaceous but discolouring strongly greenish from the centre then a pretty pearl grey at the margin. The yellowing of the flesh also shows through near the margin which is finally striate to sulcate.
Gills adnate or nearly free, distant, interveined at the base, white then pale yellowish. Spore print pale cream.
Stipe 3–7 × 0.5–2 cm (1.2–2.8 × 0.2–0.8 in), brittle, white, yellowing

strongly (chrome yellow). *Flesh* brittle, white, yellowing strongly. Smell strong and complex, fruity, like *Pelargonium*. Taste very acrid. *Chemical test* ammonia rose-red. *Microscopic features* Sp 8.5–10.5 × 7–8.5 µm, with rather dense spiny warts, partially joined by slender connectives forming an incomplete or partial reticulum, the SP amyloid. Bas 50–65 × 7–11 µm. Cyst 70–120 × 8–15 µm, fusoid, often pointed, blackening with SA. CC hyphae with hair-like slender, elongate, cylindric or finely capitate ends, 1–3 µm diameter; pileocystidia 70–150 × 8–12 µm, not very septate, obtuse or constricted capitate, greying with SA.

Season: July–September.
Habitat: Conifers, especially spruce, in marshy places.
Distribution: Montane districts; very rare in lowland areas. Often rather rare.

---

p.207 *Russula gracillima* J. Schaeffer
Russulaceae, Russulales

Description: *Cap violaceous or violaceous pink. Stipe pinkish. Spore print cream.*
Cap 2–5 cm (0.8–2 in), sometimes depressed, viscid or greasy, shiny, smooth to finely wrinkled, pastel coloured, greenish in the centre and with a violaceous zone halfway to the margin, or a mixture of these colours, the margin more pinkish. *Gills* adnate or nearly free, quite close, white then pale cream. Spore print cream. *Stipe* 4–7 × 0.5–1 cm (1.6–2.8 × 0.2–0.4 in), soon brittle, fusoid or slightly clavate, white but flushed carmine-pink for most of its length, yellowing slightly from the base. *Flesh* slightly acrid, fragile, white to pinkish near the surface. Smell faint, of raw apple. *Chemical test* $FeSO_4$ pale pinkish.

Microscopic features Sp 7.5–9 × 5.5–7 µm, with irregular or minutely crested warts, often isolated, the SP amyloid. Bas 35–45 × 8–9 µm. Cyst prominent, 70–100 × 10–15 µm, cylindro-fusoid, rostrate or umbonate, even with many constrictions. CC hyphae with hair-like, slender, cylindric, obtuse or irregular ends, 2–4 µm diameter; pileocystidia thick, 70–130 × 5–12 µm, sometimes slightly septate, greying with SA.

Season: July–November.
Habitat: Birch, marshy places.
Distribution: Rather northern or submontane. Quite rare.

---

p.207 *Russula exalbicans* (Pers.) Melzer & Zvara
Russulaceae, Russulales

Description: *Cap violaceous pink, becoming very pale and dingy. Gills greyish.*
Cap 5–10 cm (2–4 in), fleshy, slightly viscid then shiny, smooth to finely wrinkled, rather dingy violaceous pink, often with some greenish tones in the centre and soon discolouring greyish or whitish, sometimes entirely white, the extreme edge remaining pinkish or even reddish. *Gills* adnate, quite close, greyish then dingy buff. Spore print buff.
Stipe 3–8 × 2–6 cm (1.2–3.2 × 0.8–2.4 in), pruinose then finely wrinkled, white or greyish, pinkish in places, greying from the base. *Flesh* white to dingy pinkish near the surface, slightly yellowish below, greying distinctly. Smell faint, slightly fruity. Taste slightly acrid. *Chemical test* $FeSO_4$ dingy pinkish. *Microscopic features* Sp 7–9 × 6–7 µm, with finely crested, irregular warts, joined by connectives forming a fine incomplete reticulum, the SP amyloid. Bas 35–50 × 8–12 µm. Cyst cylindric or fusoid, 60–150 × 8–13 µm, tapering, rostrate or obtuse, blackening with SA. CC hyphae dense, the ends hair-like, wavy, cylindric or irregular, 2–3 µm diameter; pileocystidia 70–150 × 5–10 µm, cylindric or clavate, irregular, sometimes septate, sometimes constricted to subcapitate at the apex, greying with SA.

Season: July–November.
Habitat: Birch (sometimes other broad-leaved trees) often in damp places. Usually on calcareous soils.
Distribution: Widespread. Quite common.

Completely white (but always somewhat dingy) forms can occur, causing confusion, and sometimes forms which discolour less.

---

p.215 *Russula artesiana* M. Bon
Russulaceae, Russulales

Description: *Cap vinaceous or with mixed colours. Stipe becoming yellow to russet.*
Cap 8–20 cm (3.2–8 in), hemispherical then depressed, remaining slightly greasy for a long time, even in dry weather, smooth or faintly wrinkled near the centre, vinaceous, but soon discolouring or with buff, violaceous, brownish, etc., patches. Margin obtuse, smooth then striate to sulcate. *Gills* adnate to subdecurrent, quite close, cream then slightly brownish when handled. Spore print cream.
Stipe 6–13 × 2–5 cm (2.4–5.1 × 0.8–2 in), robust, hard, finally compressible in the centre, white to yellowish, flushing rufous from the base. *Flesh* white but tinged rufous, especially in the base. Taste slightly acrid. *Chemical test* $FeSO_4$ dingy pinkish.
Microscopic features Sp 9–10.5 × 8–9.5 μm, with indistinct spiny warts, joined by sometimes ridge-like connectives forming an incomplete reticulum, the SP faintly amyloid. Bas 40–60 × 8–12 μm. Cyst quite rare, 50–80 × 10–15 μm, cylindro-fusoid, obtuse, drawn out or pointed to rostrate, greying with SA. CC hyphae dense, slightly gelatinized, the ends hair-like, narrow, contorted or constricted, 3–4 μm diameter; pileocystidia very long, septate, 80–150 × 6–8 μm, obtuse or constricted at the apex, greying with SA.
Season: July–November.
Habitat: Broad-leaved trees, especially on clay, in lowland areas.
Distribution: Not fully understood. Mainly in the sub-Atlantic lowlands of western Europe.

R. viscida Kudrna, which is distinctly continental in distribution, has a dark red cap with crimson, yellowish and olivaceous patches, a sharp fruity smell and occurs under conifers, especially spruce. *R. artesiana* is easily confused with *R. krombholzii* (see p.739) because it is frequent and occurs in similar habitats. The latter is a smaller species, has a more fruity smell and the stipe becomes grey rather than brown.

---

p.213 *Russula queletii* Fr.
Russulaceae, Russulales

Description: *Cap reddish crimson, the centre often greenish. Stipe bright pink.*
Cap 3–7 cm (1.2–2.8 in), not very fleshy, convex or depressed, viscid then shiny, finely wrinkled or grained, reddish crimson, the centre usually discolouring or with greenish to greyish buff tints. Margin faintly sulcate. *Gills* adnate, not very close or even distant, whitish then cream. Spore print cream. *Stipe* 2–7 × 1–2 cm (0.8–2.8 × 0.4–0.8 in), firm then stuffed, pruinose, flushed or tinted radish pink, often with redder, longitudinal wrinkles. *Flesh* quite firm then brittle, white to reddish near the surface. Smells strongly of apple. Taste fairly acrid. *Microscopic features* Sp 8–10 × 7–9 µm, with isolated, quite tall, pointed, spiny warts, the SP amyloid. Bas 50–60 × 8–11 µm. Cyst fusiform, protruding, 60–120 × 8–15 µm, tapering or capitulate, blackening with SA. CC hyphae close, gelatinized, the ends hair-like, cylindric, slightly lageniform or constricted, 2–5 µm diameter; pileocystidia 70–110 × 5–10 µm, often not very septate, greying with SA.
Season: July–October.
Habitat: Spruce. Mainly on neutral to calcareous soils.
Distribution: Submontane to montane. Very rare in lowland areas and in acid sites.

---

p.209 *Russula sanguinaria* (Pers.) Rauschert
Russulaceae, Russulales

Description: *Cap and stipe red. Gills arched decurrent. Spore print pale buff.*

Cap 4–10 cm (1.6–4 in), quite quickly expanded, depressed, regular, slightly viscid then matt, finely wrinkled or even tuberculate when old, bright reddish pink, fading slightly. *Gills* decurrent, arched, not very close, white then yellowish buff. *Stipe* 4–7 × 1–3 cm (1.6–2.8 × 0.4–1.2 in), quite firm, often tapering, reddish pink, but the colour less solid than the cap, sometimes spotting yellowish towards the base. *Flesh* firm then brittle, white but distinctly pink near the surface. Smell fruity, rather faint. Taste fairly acrid. *Microscopic features* Sp 7.5–9.5 × 6.5–8.5 μm, with quite tall but obtuse warts, often isolated but sometimes slightly ridged or with occasional, very slender connectives, the SP amyloid. Bas 40–55 × 9–12 μm. Cyst fusoid, 60–150 × 8–18 μm, tapering, or rostrate, blackening with SA. CC hyphae close, slightly gelatinized, the ends hair-like 3–4 μm diameter, contorted or with many constrictions; pileocystidia cylindric, 5–8 μm diameter, not very septate, the apex contracted, irregular or sometimes umbonate to papillate, greying with SA.

Season: July–November.
Habitat: Conifers, especially pine.
Distribution: Widespread but unequal.

Two other pure red species are much less common: *R. rhodopoda* Zvara (very rare) has a dark blood-red cap that shines as if lacquered, a lobed margin, pink stipe and faint fruity smell (under spruce, on marshy ground); and *R. helodes* Melzer (very rare) has a fairly bright red, sometimes discolouring cap that becomes matt more quickly, a wavy margin, white or pinkish, greying stipe and a faint smell (under conifers, in peat bogs with sphagnum).

---

p.213 *Russula drimeia* Cooke
Russulaceae, Russulales

Description: *Cap purple. Stipe violaceous. Gills lemon yellow. Taste acrid.*

Cap 4–10 cm (1.6–4 in), convex to subumbonate then expanded, slightly viscid then dry, purple, sometimes reddish or paler and spotting brownish or pale buff. *Gills* adnate, rather arched, close, a bright lemon yellow then ochraceous lemon when mature. Spore print dark cream. *Stipe* 3–8 × 1–2 cm (1.2–3.2 × 0.4–0.8 in), more or less tapering, pruinose, violaceous lilac, at least in the middle, rather yellowish or buff towards the base. *Flesh* white or slightly lemon to violaceous near the surface. Smell fruity, rather faint. Taste very acrid. *Chemical test* reddish pink with ammonia (gills and flesh turn this colour in a few minutes if exposed to the fumes). *Microscopic features* Sp 7–9.5 × 6–7.5 µm, with dot-like warts arranged more or less in lines or irregularly, sometimes rather dense, joined by variable connectives, sometimes forming ridges or bands, the SP amyloid. Bas 40–60 × 8–11 µm. Cyst abundant, fusiform to long fusiform-clavate, 60–150 × 8–12 µm, tapering or rostrate, blackening with SA. CC hyphae slender, slightly gelatinised, the ends hair-like 2–4 µm diameter, contorted or clavate, variable; pileocystidia long and slender, 3–5 µm diameter, cylindric and not very septate but the terminal cell with many constrictions or umbonate, greying with SA.

Season: June–November.
Habitat: Under pine.
Distribution: Widespread. Especially in lowland areas and in submontane areas.

Like all species in this group, it may occur in atypical, green or yellow colour forms.

---

p.212 *Russula torulosa* Bresadola
Russulaceae, Russulales

Description: *Cap violaceous. Stipe short, bluish lilac. Gills pale by contrast.*
Cap 4–10 cm (1.6–4 in), flattened or depressed, regular, slightly viscid or shiny, more or less finely wrinkled, rather dark violaceous crimson but quite often marbled or discolouring, sometimes with olivaceous

tones. *Gills* adnate, not very close, very pale in contrast to the cap and stipe colour. Spore print buff. *Stipe* 2–6 × 1–3 cm (0.8–2.4 × 0.4–1.2 in), short and thick, violaceous with bluish tones, sometimes whitish or yellowish at the extreme base. *Flesh* white to pale violaceous near the surface. Smells strongly of raw apple. Taste slightly acrid. *Microscopic features* Sp 7–9 × 6–7.5 µm, with finely crested warts, joined by slender connectives forming an almost complete reticulum, the SP amyloid. Bas 40–50 × 8–10 µm. Cyst fusiform-clavate, 60–150 × 7–12 µm, rostrate, browning or blackening with SA. CC hyphae slender, gelatinized, the ends hair-like 2–3 µm diameter, articulated and filiform or irregular; pileocystidia cylindric, 8–11 µm diameter, the apex with many constrictions or irregularly torulose, greying with SA.

Season: August–December.
Habitat: Under pine.
Distribution: Rather Atlantic.

A typical species of pine woods on sand on the western coasts of temperate and southern Europe.

---

p.213 *Russula fuscorubra* (Bresadola) Singer
Russulaceae, Russulales.

Description: *Cap blackish crimson. Stipe violaceous pink. Taste acrid.*
Cap 5–8 cm (2–3.2 in), quite soon depressed, regular, greasy, shiny, finely grained to finely wrinkled, dark violaceous crimson, almost black in the centre but sometimes more red at the edge, then marbled. *Gills* adnate, not very close, rather dark buff when mature, edge sometimes violaceous at least towards the margin. Spore print buff. *Stipe* 4–8 × 1–2 cm (1.6–3.2 × 0.4–0.8 in), slowly becoming soft, pruinose then wrinkled or punctate towards the base, violaceous pink. *Flesh* white or dingy, violaceous near the surface and often yellowish at the stipe base.

Taste very acrid. *Microscopic features* Sp 7.5–9 × 6.5–7.5 µm, with long warts, joined by thick connectives, forming ridges or sometimes lines making an almost complete reticulum, the SP amyloid. Bas 45–55 × 8–12 µm. Cyst fusoid-clavate, 70–120 × 8–15 µm, tapering or rostrate, blackening with SA. CC hyphae erect, gelatinized, the ends hair-like 3–4 µm diameter, cylindric, septate, fairly regular to subcapitate; pileocystidia short and thick, 5–12 µm diameter, not, or not very, septate, cylindric and regular, not constricted.

Season: August–October.
Habitat: Mainly spruce, sometimes pine. Mainly on neutral–calcareous soils.
Distribution: Hills and mountains.

---

p.213 *Russula badia* Quélet
Russulaceae, Russulales

Description: *Cap crimson, velvety. Stipe reddish pink. Spore print dark buff.*
Cap 6–10 cm (2.4–4 in), convex slightly depressed, soon dry and often velvety, sometimes even scurfy, quite bright red or crimson but variable, coppery, sometimes very dark or paler. *Gills* adnate, quite close, yellowish then rather pale yellow-buff, edge sometimes slightly brownish or red towards the cap margin. Spore print dark buff. *Stipe* 5–10 × 2–6 cm (2–4 × 0.8–2.4 in), whitish but pinkish or even reddish, browning slightly from the base. *Flesh* white to reddish near the surface. Smell of cedar wood, also like red wine corks. Taste slowly but extremely acrid.

Microscopic features Sp 8–11 × 6.5–8 µm, variable, with fine and rather dense, irregular warts, joined by slender and incomplete or dotted connectives, sometimes forming the beginnings of a reticulum, the SP amyloid. Bas 45–55 × 10–15 µm. Cyst long and slender, fusiform, 70–150 × 8–11 µm, rostrate, greying with SA. CC hyphae variable, the ends hair-like, narrow, tapering, 2–3 µm diameter; pileocystidia very large, 6–12 µm diameter,

often with many septa, cylindric or clavate at the apex, greying with SA.
Season: July–October.
Habitat: Woods, mainly coniferous, montane.
Distribution: Widespread but frequency uneven.

The appalling acridity of the flesh is not released immediately and the impatient sampler may take a second nibble at the cap before experiencing it distinctly. It then develops more rapidly, violently affecting the mouth and throat.

---

p.209 *Russula veternosa* Fr.
Russulaceae, Russulales

Description: *Cap reddish pink, ochraceous cream in the centre. Margin slightly sulcate.*
Cap 3–10 cm (1.2–4 in), not very fleshy, zoned, reddish pink at the edge, discolouring ochraceous cream in the centre. Margin shortly striate or sulcate.
Gills adnate to nearly free, quite close, cream then pale buff, with orange tones when mature. Spore print dark yellow.
Stipe 3–7 × 0.5–2 cm (1.3–2.4 × 0.2–0.8 in), finely wrinkled, white, becoming slightly dingy below. *Flesh* acrid, white or greyish. Smell of apple or stewed fruit then slightly of *Pelargonium* and finally of honey.
Microscopic features Sp 7–9.5 × 6.5–8 µm, with sharp spiny isolated warts, the SP amyloid. Bas 30–45 × 10–12 µm. Cyst cylindro-fusoid, 50–90 × 8–13 µm, fusoid-clavate or rostrate, greyish with SA. CC hyphae close, articulated, the ends hair-like 2–4 µm diameter; pileocystidia cylindric or clavate, 4–12 µm diameter, septate, the terminal cell(s) short and sometimes subglobose, greying with SA.
Season: June–December.
Habitat: Broad-leaved trees, especially beech.
Distribution: Widespread wherever beech occurs.

R. cuprea Krombholz is a rather common, related species, with a remarkable range of colours. There are bright red, orange, violaceous, pinkish vinaceous and greenish grey forms; the margin or the centre may be a different colour. *R. urens* Romell is well characterized by its olivaceous green cap, slightly greying stipe and habitat in hornbeam woods on calcareous clays. Other members of this group occur in the mountains.

---

p.210 *Russula mesospora* Singer
Russulaceae, Russulales

Description: *Cap coppery or orange-red. Spore print dark yellow. Taste acrid.*
Cap 6–12 cm (2.4–4.7 in), globose, slow to expand, slightly viscid then somewhat shiny, sometimes matt, orange-red or coppery, often bicoloured, with the centre orange-yellow. Margin faintly sulcate when old.
Gills adnate or nearly free, quite close then more distant, cream then yellow-buff with orange tints. Spore print yellow. *Stipe* 5–8 × 1.5–2.5 cm (2–3.2 × 0.6–1 in), white, almost unchanging. *Flesh* thick, firm and rather brittle. Smell of apple or slightly fruity. Taste distinctly acrid. *Chemical test* $FeSO_4$ medium orange-pink. *Microscopic features* Sp 7–9 × 6–7 µm, small, with more or less spiny or obtuse warts joined by connectives forming lines or an incomplete reticulum, the SP amyloid. Bas 35–50 × 9–12 µm. Cyst fusiform or clavate, 70–90 × 8–13 µm, rostrate, greying with SA. CC hyphae slender, the ends hair-like 2–3 µm diameter, cylindric or tapering, contorted; pileocystidia not very septate, cylindric, 4–7 µm diameter, greying with SA.
Season: August–October.
Habitat: Broad-leaved or coniferous trees, often in grassy places.
Distribution: Uneven, not fully assessed.

Mainly confused with *R. lundellii* Singer, the best known in this group of robust, more or less red or orange species, with a yellow spore print and acrid taste. The latter is taller (cap up to 18 cm (7 in)), bright coppery red and smells fruity or acid (spores with more isolated, spiny warts).

p.214 *Russula amarissima* Romagnesi & Gilbert
Russulaceae, Russulales

Description: *Cap carmine, matt. Gills with a reddish edge. Taste bitter.*
Cap 6–15 cm (2.4–6 in), very hard, long remaining convex, very dry, cracked or fissured, carmine, sometimes darker or, alternatively, discolouring in irregular buff or yellowish patches. *Gills* adnate, rather close, pale cream then dingy or spotting rusty buff, edge red, at least towards the cap margin. Spore print cream. *Stipe* 5–10 × 1.5–3.5 cm (2–4 × 0.6–1.4 in), very hard, slight hollows developing when old, pruinose or mealy, sometimes cracked, flushed carmine-pink and yellowing from the base. *Flesh* very hard, more brittle when old. Taste bitter. *Chemical test* $FeSO_4$ orange-pink. *Microscopic features* Sp 7.5–9 × 6.5–7.5 µm, subglobose, with finely crested warts, joined by irregular connectives forming a more or less complete reticulum, the SP rather faintly amyloid. Bas 50–60 × 6–8 µm. Cyst abundant, fusiform, 50–150 × 8–15 µm, obtuse, not reacting with SA. CC hyphae dense, interwoven, the ends hair-like 3–4 µm diameter, articulated, cylindric or tapering, septate; pileocystidia rather short, 4–8 µm diameter, not very septate, cylindric, not reacting with SA, with fine rather faint acid-resistant encrustations, all along the hyphae.

Season: July–October.
Habitat: Broad-leaved trees, especially in cool or damp places.
Distribution: Uneven. Rare.

The only really bitter *Russula*. A variant, *R. linnaei* (Fr.) Fr. has red-edged gills, but is not bitter.

---

p.216 *Lactarius vellereus* (Fr.:Fr.) Fr.
Russulaceae, Russulales

Description: *Cap deeply depressed, white, velvety. Gills distant.*
Cap 10–25 cm (4–10 in), soon deeply depressed, velvety or tomentose, white

but often mottled pale ochraceous cream when old. *Gills* decurrent, narrow, very distant, sometimes forking slightly near the stipe, whitish to ochraceous cream. *Stipe* 4–8 × 1.5–5 cm (1.6–3.2 × 0.6–2 in), firm, pruinose, white then a dingy pale ochraceous cream. *Flesh* hard, white to pale yellowish cream. Milk white, often abundant (flushing pink in var. *hometii* Boudier), very acrid.

Microscopic features Sp 9–11 × 8.5–10 µm, the ornament very low (a few striations or interconnected ridges). Cyst cylindro-clavate, obtuse. CC with filamentous hyphae, lacking sphaerocysts in the superficial layers.

Season: August–November.
Habitat: Broad-leaved trees, rarely under conifers.
Distribution: Very widespread. Common.

The milk of *L.. bertillonii* (Neuhoff ex Z. Schaefer) M. Bon turns yellow when mixed with KOH.

---

p.216 *Lactarius piperatus* (Scop.:Fr.) Pers.
Russulaceae, Russulales

Description: *Cap depressed, white, velvety. Gills decurrent, extremely close.*
Cap 5–12 cm (2–4.7 in), soon umbilicate funnel-shaped, smooth to velvety or indented in the centre, sometimes deeply cracked at the edge, white. *Gills* very decurrent, narrow, thin, extremely close, sometimes forking, whitish to pale flesh cream. *Stipe* 5–12 × 1–2 cm (2–4.7 × 0.4–0.8 in), pruinose, white. *Flesh* hard, white to yellowish cream. Milk white, often abundant, very acrid. *Chemical test* $FeSO_4$ more or less pinkish. *Microscopic features* Sp 7–9 × 5.5–7 µm, quite long, the ornament very low (almost smooth under an optical microscope), forming an incomplete reticulum. Cyst cylindro-fusoid, obtuse or capitulate. CC subcellular.

Season: June–August.
Habitat: Broad-leaved trees, especially beech and hornbeam.

Distribution: Quite widespread. Rather common.

L.. pergamenus (Sw.:Fr.) Fr., which is rather similar, has milk that turns bright orange in KOH and dries as greenish spots on the gills. *L.. glaucescens* Crossland is smaller, with milk reacting equally strongly in KOH but drying as dark olive green spots.

---

p.216 *Lactarius controversus* (Pers.:Fr.) Fr.
Russulaceae, Russulales

Description: *Cap depressed, whitish then with lilaceous or vinaceous spots.*
Cap 8–20(30) cm (3.2–8(12) in), depressed, sometimes irregular, slightly viscid then dry and even finely pubescent at the margin, whitish or pale, soon spotted vinaceous pink, and with a narrow zone of lilaceous pink or wine-red at the edge. *Gills* decurrent, quite narrow, close, forking near the stipe, pinkish cream to quite bright pink. *Stipe* 2–7 × 1–4 cm (0.8–2.8 × 0.4–1.6 in), hard, conical when short, white then spotted pink, often slightly yellowish below. *Flesh* firm, pinkish near the surface, rather mild without the milk. Milk white, acrid. *Chemical test* FeSO$_4$ faintly greenish. *Microscopic features* Sp 6–7.5 × 5–6 µm, with rather low, finely crested warts, joined by connectives into an almost complete reticulum. Cyst cylindro-clavate, obtuse. CC hyphae gelatinized, 2–4 µm diameter, interwoven, without superficial sphaerocysts.
Season: August–December.
Habitat: Poplars, more rarely willows.
Distribution: Uneven. This species has become rare in northern and central Europe.

---

p.217 *Lactarius torminosus* (Sch.:Fr.) S.F. Gray
Russulaceae, Russulales

Description: *Cap brick-red, zoned, woolly and bearded at the edge. Margin inrolled.*
Cap 6–12 cm (2.4–4.7 in), depressed then funnel-shaped, brick-red zoned with paler reddish and pinkish rings, woolly and

bearded, slightly viscid in the centre.
Gills adnate to subdecurrent, close, whitish to pinkish. *Stipe* 3–6 × 1–3 cm (1.2–2.4 × 0.4–1.2 in), quite firm, pruinose or sometimes scrobiculate, whitish then pinkish to beige with flesh-coloured tones.
Flesh white to pinkish near the surface, acrid. Milk white, often rather scanty, acrid.
Chemical test KOH turns milk orange.
Microscopic features Sp 8–9.5 × 5.5–6.5 µm, rather elongate, with low warts, joined by ridges and lines into a complete reticulum. Cyst fusiform, 60–80 × 8–10 µm, appendiculate. CC filamentous, with very long hyphae towards the margin, more like an ixocutis in the centre.

Season: June–December.
Habitat: Mycorrhizal with birch.
Distribution: Very widespread. Common.

L.. pubescens (Schrad.) Fr., which is very common under birch in urban grassland, has a smaller, less bearded cap, white or with faint pale pink zones, buff when old.
L.. resimus (Fr.:Fr.) Fr., with yellowing milk, has a shortly bearded margin, white, yellowing cap, white then pinkish cream gills and occurs mainly under pine.

---

p.217 *Lactarius necator* (Bull.:Fr.) P. Karsten
Russulaceae, Russulales

Description: *Cap dirty, olivaceous bronze. Gills greenish. Milk white.*
Cap 5–15 cm (2–6 in), convex then depressed, viscid in the centre then dry, with faint radial fibrils, putty, olivaceous brown, even blackish in the centre, sometimes paler, greenish yellow. Margin finely pubescent and narrowly zoned on the extreme edge.
Gills subdecurrent, close, white then with greenish, olivaceous or yellowish tints, finally spotted rufous or olivaceous brown.
Stipe 3–7 × 1–3 cm (1.2–2.8 × 0.4–1.2 in), slightly viscid and a little shiny when dry, smooth, rarely with a few scrobicules, almost concolorous or paler. *Flesh* white or dingy, especially near the surface, browning slightly

when exposed. Milk acrid, white, forming greenish yellow drops on the gill edge. *Chemical test* Ammonia turns milk purple. *Microscopic features* Sp 7.5–9 × 6–7 μm, with fine low warts, joined by lines and forming an incomplete reticulum. Cyst fusiform, 50–80 × 5–7 μm, attenuate or appendiculate. CC with hyphae 2–5 μm diameter forming a subtrichodermial ixocutis.

Season: July–December.
Habitat: Mixed woodland, especially with birch, sometimes also with pine.
Distribution: Widespread. Very common.

---

p.217 *Lactarius zonarius* (Bull.) Fr.
Russulaceae, Russulales

Description: *Cap glabrous, with yellow and rather rufous zones. Smell sharp.*
Cap 6–12 cm (2.4–4.7 in), convex then expanded, more or less deeply depressed, slightly viscid then shiny, glabrous, zoned, especially at the edge, with rather pale, dull yellow and rather rufous buff bands. *Gills* subdecurrent, rather close, whitish then cream, often with faint flesh tones. *Stipe* 3–5 × 1–3 cm (1.2–2 × 0.4–1.2 in), slightly conical when short, white or pale to brownish from the base. *Flesh* white, flushing slightly pink when cut then greying. Milk white, slightly or slowly acrid. Smell a little sharp or faintly fruity. *Microscopic features* Sp 7–8.5 × 6–6.5 μm, with low warts joined by ridges into an almost complete reticulum. Cyst fusiform, 40–60 × 6–8 μm, constricted or more or less capitate. CC an ixocutis, with slender, cylindric or clavate hyphae near the surface.

Season: July–November.
Habitat: Broad-leaved trees, especially in coppice and mixed woods.
Distribution: Widespread; usually in lowland areas.

L.. acerrimus Britzlemayr is remarkable on account of its irregular or lobed shape, gills which anastomose or crinkle near the stipe, and are often eccentric or twisted. *L. evosmus* Kühner is another pale, more regular species, with a strong smell of apple. *L.. porninsis* Rolland is transitional to the following species because of its much brighter colours (orange to yellowish tawny at the edge). It has mild or slightly to slowly bitter milk, a fruity smell and occurs under larch, in the mountains.

---

p.225 *Lactarius bresadolanus* Singer
Russulaceae, Russulales

Description: *Cap with rufous and rather bright orange zones. Flesh flushing slightly pink.*
Cap 5–10 cm (2–4 in), very soon depressed, the margin irregular, slightly viscid then shiny, pruinose towards the edge or with slight radial wrinkles, and with somewhat rufous and bright orange zones. *Gills* subdecurrent, close, cream then pale pinkish ochre. *Stipe* 3–5 × 1–2 cm (1.2–2 × 0.4–0.8 in), finally developing cavities, smooth or faintly scrobiculate, whitish to rather rufous buff from the base. *Flesh* white, slowly flushing pink, greying slightly when old. Milk white, very acrid. Smell fruity.
Microscopic features Sp 8–10.5 × 6.5–8.5 µm, the warts not very prominent, joined by transverse ridges and lines. Cyst fusiform, 45–65 × 5–7 µm, attenuate or with many constrictions, especially in the cheilocystidia, clearly greying in SA. CC an ixocutis with wavy or clavate elements near the surface.
Season: July–October.
Habitat: Mainly conifers, sometimes with broad-leaved trees.
Distribution: Mountains in western and southern Europe.

The bright colours and zoned cap are reminiscent of the *L.. deliciosus* (see p.786) group, but the milk is white. A related and rather rare species, *L.. zonarioides* K. & R., has darker zones and dingy greenish spots (under spruce, in the mountains).

p.217 *Lactarius circellatus* Fr.
Russulaceae, Russulales

Description: *Cap greyish beige, zoned. Gills ochraceous cream. Milk acrid.*
Cap 4–10 cm (1.6–4 in), convex, slowly becoming depressed, often regular, slightly viscid then dry, a little shiny but frosted, often with fine radial wrinkles, and with beige and greyish zones, sometimes tinged violaceous or olivaceous. *Gills* subdecurrent, rather close, cream then ochraceous yellow, with orange tones when old. *Stipe* 3–6 × 0.5–2 cm (1.2–2.4 × 0.2–0.8 in), corticate, concolorous with the cap or paler. *Flesh* whitish or greyish near the surface. Milk white, acrid. *Microscopic features* Sp 6–8 × 5–6 µm, with low warts, joined by ridges and lines. Cyst fusiform, 45–60 × 7–10 µm, attenuate or appendiculate. CC an ixocutis, moderately gelatinized. Pigment extra-cellular.
Season: June–November.
Habitat: Broad-leaved trees, especially hornbeam.
Distribution: Lowland and submontane areas.

---

p.217 *Lactarius pyrogalus* (Bull.:Fr.) Fr.
Russulaceae, Russulales

Description: *Cap not zoned, greyish beige. Gills very distant with orange tones. Milk very acrid.*
Cap 3–10 cm (1.2–4 in), quite soon depressed, irregularly funnel-shaped, slightly viscid then dry and shiny, sometimes frosted, often with fine radial wrinkles, greyish beige to buff, sometimes with greenish tints. *Gills* subdecurrent, very distant, ochraceous cream then bright orange-ochre. *Stipe* 3–7 × 0.5–2 cm (1.2–2.8 × 0.2–0.8 in), often finely wrinkled to compressed, slightly shiny, concolorous or paler. *Flesh* pale cream. Milk white, drying as olivaceous cream drops on the gills, very acrid. *Chemical test* KOH turns the milk golden yellow. *Microscopic features* Sp 6–8.5 × 5–6.5 µm, with low warts, joined by ridges and lines. Cyst abundant, fusiform, 50–80 × 6–9 µm, appendiculate or attenuate, greying in SA.

CC an ixocutis. Pigment extracellular, granular.

Season: July–November.
Habitat: Mainly hazel, sometimes hornbeam.
Distribution: Widespread. Rather common.

---

p.219 *Lactarius nanus* Favre
Russulaceae, Russulales

Description: *Cap violaceous brown then chestnut brown. Gills somewhat rufous. Milk slightly or slowly acrid.*
Cap 1–3 cm (0.4–1.2 in), depressed, with a persistent umbo, slightly viscid or shiny, finely grained, violaceous brown then chestnut brown, rather greyish when old and dry. *Gills* adnate to subdecurrent, not very close, flesh somewhat rufous then rather tawny, quite dark. *Stipe* 0.5–3 × 0.4–0.8 cm (0.2–1.2 × 0.16–0.32 in), sometimes twisted to subeccentric, pruinose, whitish then almost concolorous or paler, flesh somewhat tawny to brownish. *Flesh* fragile, cream to rather pale rufous beige, slightly violaceous near the surface. Milk not very abundant, whitish, slightly or slowly acrid. *Microscopic features* Sp 7–9.5 × 6–8 µm, with low warts, joined by incomplete ridges into a partial reticulum. Cyst fusiform, 45–60 × 7–10 µm, obtuse. CC an ixocutis. Pigment extracellular, granular.

Season: July–August.
Habitat: Dwarf alpine shrub community with dwarf willows. Often on acid soils.
Distribution: Alpine; western Europe and Scandinavia.

---

p.219 *Lactarius trivialis* (Fr.:Fr.) Fr.
Russulaceae, Russulales

Description: *Cap viscid, violaceous brown to pinkish cream. Stipe hollow. Milk acrid.*
Cap 8–15(20) cm (3.2–6(8) in), globulose then expanded or funnel-shaped, viscid then shiny, sometimes frosted at the edge, very variable and discolouring, violaceous then vinaceous brown, reddish beige to lilaceous buff, finally pinkish cream or whitish. *Gills* subdecurrent, rather close, pale cream, then

pale buff. *Stipe* 5–15 × 1.5–4 cm (2–6 × 0.6–1.6 in), soon hollow and then very watery, slightly viscid then shiny when dry, whitish or pale ochraceous cream. *Flesh* firm then fragile, whitish to almost concolorous near the surface. Milk white, often very abundant, drying as greenish beige droplets on the gills, acrid. *Microscopic features* Sp 9–11 × 7.5–9.5 µm, the warts joined by connectives into a rather wide-meshed reticulum. Cyst fusiform, not very numerous. CC an ixocutis, the hyphae 2–4 µm diameter, with close septa, the terminal cells obtuse or clavate.

Season: July–October.
Habitat: Conifers, often with birch; in marshy ground.
Distribution: Widespread; submontane.

---

p.218 *Lactarius vietus* (Fr.:Fr.) Fr.
Russulaceae, Russulales

Description: *Cap greyish lilaceous. Milk white, drying as grey-green droplets.*
Cap 3–8 cm (1.2–3.2 in), flattened or depressed, often with a central umbo, slightly viscid then shiny, rather frosted at the edge, often grained, lilaceous pinkish on a greyish, beige or brownish ground. *Gills* sub-decurrent, rather close, white then pale ochraceous cream. *Stipe* 3–7 × 0.5–2 cm (1.2–2.8 × 0.2–0.8 in), dented or irregular, shiny, almost concolorous or paler.
Flesh fragile, whitish to pinkish beige near the surface. Milk white, slightly or slowly acrid, drying as grey-green droplets on the gills. *Microscopic features* Sp 8–9 × 6–7.5 µm, elliptic, with low warts, clearly joined by slender connectives into an almost complete reticulum. Cyst fusiform, 50–80 × 10–18 µm, attenuate or appendiculate. CC an ixocutis. Pigment arranged in extracellular granules and patches.

Season: July–November.
Habitat: Birch, sometimes mixed with pine in damp places.
Distribution: Very widespread. Less frequent in southern areas.

p.218 *Lactarius blennius* (Fr.:Fr.) Fr.
Russulaceae, Russulales

Description: *Cap viscid, vinaceous brown. Gills white. Milk drying as olivaceous droplets.*
Cap 4–10 cm (1.6–4 in), depressed, sometimes funnel-shaped, extremely slippery in damp weather, shiny when dry, sometimes faintly zoned, reddish brown to brownish, sometimes with greenish tones (f. *virescens* J.E. Lange), often marked near the edge with concentric darker spots or small pits. *Gills* adnate to decurrent, very close, pure white then very pale cream. *Stipe* 3–7 × 2–6 cm (1.2–2.8 × 0.8–2.4 in), corticate, slightly viscid, sometimes scrobiculate, almost concolorous or paler. *Flesh* pale to almost concolorous near the surface. Milk abundant, white, acrid, drying as olivaceous grey droplets on the gills.

Microscopic features Sp 6–8 × 5–7 µm, with low warts, joined by irregular ridges forming an incomplete reticulum and a few lines. Cyst fusiform, not very obvious. CC strongly gelatinized, with slender wavy hyphae. Pigment in extra-cellular patches.

Season: July–December.
Habitat: Beech only.
Distribution: Widespread. Very common to common.

L.. fluens Boudier is similar (cap less viscid, brownish with greenish brown bands; gills rather pale ochraceous cream; milk extremely abundant, drying as rather reddish grey droplets on the gills) and is found under broad-leaved trees, especially beech and hornbeam.

p.221 *Lactarius hysginus* (Fr.:Fr.)
Russulaceae, Russulales

Description: *Cap greasy, coppery brown to yellowish brown. Gills lemon cream.*

Cap 5–10 cm (2–4 in), long remaining convex then depressed, the margin long continuing to be obtuse, viscid or greasy to waxy, shiny, with fine radial wrinkles, faintly zoned, red-brown to yellow-brown, sometimes with crimson brown tones or paler. *Gills* subdecurrent, close, yellowish cream then lemon ochre, the colour of box wood. *Stipe* 2–5 × 2–6 cm (0.8–2 × 0.8–2.4 in), short and thick, rather conical when short, slightly viscid, smooth or scrobiculate, concolorous with the gills. *Flesh* whitish to brownish cream. Milk white, often fairly abundant, acrid. Smell strong like forest-bugs. *Microscopic features* Sp 6–8.5 × 5.5–6.5 µm, with low warts, joined by ridges into a complete reticulum. Cyst fusiform, 50–70 × 8–12 µm, obtuse or appendiculate. CC hyphae gelatinized, slender, 3–5 µm diameter. Pigment intracellular.

Season: August–October.
Habitat: Broad-leaved trees, often birch, in damp places or marshy ground.
Distribution: Mainly northern.

---

p.224 *Lactarius rufus* (Scop.:Fr.) Fr.
Russulaceae, Russulales

Description: *Cap bright or dark rufous, velvety, umbonate. Milk white, very acrid.*
Cap 4–10 cm (1.6–4 in), with a prominent or even sharp umbo, the surrounding area soon depressed, dry, as if felted, bright or darker rufous brown, sometimes brick-red. *Gills* subdecurrent, rather close, whitish then pale cream. *Stipe* 5–10 × 1–2 cm (2–4 × 0.4–0.8 in), dry, concolorous or paler. *Flesh* whitish to almost concolorous near the surface, very acrid. Milk white, often not very abundant, very acrid. *Microscopic features* Sp 7–10 × 5.5–7 µm, rather elongate, with low warts, joined by incomplete connectives into a partial reticulum. Cyst fusiform, 50–70 × 8–10 µm, obtuse or appendiculate. CC with short erect

filaments (trichoderm) 5–6 μm diameter, the apex often clavate.
Season: August–November.
Habitat: Pine, sometimes other conifers but often with birch; damp places or preferring acid soils.
Distribution: Widespread, especially in the north.

The most acrid of the milk caps.

---

p.221 *Lactarius fuscus* Rolland
Russulaceae, Russulales

Description: *Cap very dry, violaceous brown. Smell of coconut. Cap* 4–10 cm (1.6–4 in), convex subumbonate then depressed, dry, fleecy to felted, sometimes pruinose at the edge, rather dark brown with violaceous or reddish tones. *Gills* subdecurrent, rather close, buff then salmon or with a hint of orange. *Stipe* 3–6 × 0.5–1.5 cm (1.2–2.4 × 0.2–0.6 in), dry to pruinose, pale cream to buff or almost concolorous. *Flesh* whitish to pale buff or almost concolorous near the surface. Milk often not very abundant, white or opalescent, slightly or slowly acrid. Smells strongly of coconut. *Microscopic features* Sp 7–8.5 × 5–6.5 μm, with rather low warts, joined by ridges and lines into an almost complete reticulum. Cyst not very numerous, fusoid or clavate attenuate, 30–60 × 6–10 μm, often obtuse. CC hyphae wavy, 3–6 μm diameter, in a short trichoderm.

Season: August–October.
Habitat: Conifers, sometimes mixed with birch. Often on acid soils.
Distribution: Usually in lowland areas, continental to submontane, in the north.

L.. glyciosmus (Fr.:Fr.) Fr., which is more widespread (much more frequent under birch, usually in damp acid places), is smaller and more delicate, with a silvery grey to pinkish lilaceous beige cap, yellowish cream gills, almost no milk and an identical smell of coconut.

p.220 *Lactarius lilacinus* (Lasch:Fr.) Fr.
Russulaceae, Russulales

Description: *Cap dry, crimson-pink. Gills distant, pale orange-ochre.*
Cap 3–8 cm (1.2–3.2 in), soon umbilicate then broadly depressed, sometimes lobed, felted subsquamulose or even squamulose in places, quite bright crimson-pink, sometimes more lilaceous vinaceous. Margin sometimes sulcate. *Gills* subdecurrent, distant, whitish then yellowish cream, with orange-ochre tones at maturity.
Stipe 4–6 × 0.5–1 cm (1.6–2.4 × 0.2–0.4 in), pruinose to wrinkled, pale cream or slightly vinaceous pinkish ochre.
Flesh fragile, whitish to almost concolorous near the surface. Milk not very abundant, watery, drying as pale greenish droplets, mild, slightly or slowly acrid. Smell developing on drying and reminiscent of chicory. *Microscopic features* Sp 7–9.5 × 6–7.5 µm, with low warts, joined by uneven ridges forming an incomplete reticulum, the apiculus prominent. Cyst (especially Cheilo) cylindric to fusoid, sometimes irregular, 40–60 × 5–8 µm, obtuse or appendiculate. CC hyphae 5–8 µm diameter, in a short trichoderm, the terminal elements more or less clavate. Pigment parietal, sometimes encrusting.
Season: August–December.
Habitat: Alders, among tall vegetation.
Distribution: Widespread but uneven.

---

p.220 *Lactarius spinosulus* Quélet
Russulaceae, Russulales

Description: *Cap with pointed scales at the margin, carmine-pink.*
Cap 2–5 cm (0.8–2 in), slightly depressed, with a small umbo, carmine-pink to violaceous lilac, sometimes zoned at the edge, slightly viscid then dry, velvety then finely scaly, especially at the margin which may be stiffly spinulose and sulcate.
Gills subdecurrent, not very close, pale ochraceous cream then with lilaceous

pinkish tones at the base. *Stipe* 2–6 × 0.5–1 cm (0.8–2.4 × 0.2–0.4 in), soon hollow, slightly lubricious then pruinose, often finely wrinkled towards the base, concolorous or paler. *Flesh* fragile, whitish to almost concolorous near the surface. Milk not very abundant, white or opalescent, sometimes drying as pale brownish green droplets, almost mild. *Microscopic features* Sp 7–9 × 6–7 μm, with low warts, joined by rather strong ridges or lines. Cyst fusoid or cylindric, 40–60 × 8–10 μm, obtuse. CC hyphae 4–8 μm diameter, a trichoderm or ixo-trichoderm, especially evident towards the margin.

Season: July–September.
Habitat: Birch, sometimes mixed with conifers on damp soils.
Distribution: Widespread. Rarer in the south.

Easy to identify when the hyphae of the cap are really erect and gathered into fine marginal spines. This character may be less obvious in old material.

---

p.221 *Lactarius cimicarius* (Batsch) Gillet
Russulaceae, Russulales

Description: *Cap reddish brown, dry, finely wrinkled. Milk cloudy, mild.*
Cap 3–7 cm (1.2–2.8 in), flat-topped, rarely umbilicate, soon dry, often finely wrinkled to slightly dented, medium red-brown to warm orange-ochre. Margin sinuate to crenulate. *Gills* subdecurrent, rather close, pale cream with pinkish ochre tones when old. *Stipe* 3–6 × 0.5–1 cm (1.2–2.4 × 0.2–0.4 in), brittle, dry, sometimes wrinkled below, almost concolorous or slightly paler, often yellowish above. *Flesh* ochraceous cream to somewhat rufous. Milk abundant but watery or weakly opalescent, unchanging, mild. Smell rather strong, of bugs. *Microscopic features* Sp 6.5–8 × 5.5–7 μm, subglobose or shortly elliptic, with rather low warts, joined by ridges of medium height into a more or less complete reticulum. Cyst absent, but MC misshapen or subfusiform. CC hyphae

short, clavate or wavy but wide, forming an irregular palisade.
Season: July–November.
Habitat: Broad-leaved trees, in damp places.
Distribution: Widespread. Relatively common.

L. serifluus (De Cand.:Fr.) Fr. is paler with close gills and cloudy or clear mild milk, has a weaker smell and occurs in mixed broad-leaved woods, in less damp places. *L. subumbonatus* Lindgren, which has a very dark to blackish cap, very roughened to crinkled, distant gills, less abundant watery milk, has a very strong smell and occurs in damper broad-leaved woods, ruts, etc.

---

p.223 *Lactarius atlanticus* M. Bon
Russulaceae, Russulales

Description: *Cap orange-russet. Gills reddish ochre. Milk opalescent, yellowing slightly.*
Cap 4–8 cm (1.6–3.2 in), convex then depressed, the centre flat or umbonate, slightly greasy then often matt, smooth or cracking slightly or minutely roughened, orange-russet to reddish brick, more buff when dry. *Gills* adnate, rather close, yellowish ochre then reddish ochre. *Stipe* 5–7 × 0.5–1 cm (2–2.8 × 0.2–0.4 in), dry, often with pale or rufous, strigose filaments at the base, almost concolorous or the base more reddish. *Flesh* whitish to buff or somewhat rufous. Milk not very abundant, watery or opalescent, yellowing slightly on the handkerchief, mild to slightly or slowly bitter. Smell rather similar to that of *L. quietus* (see p.768), of chicory when dry. *Microscopic features* Sp 7–9 × 6.5–8.5 µm, subglobose or shortly elliptic, the warts sometimes isolated, but most joined by low ridges forming an incomplete reticulum. Cyst absent but MC misshapen to capitate or lobed. CC hyphae very short, subglobose or club-shaped.
Season: September–December.
Habitat: Evergreen oaks, sometimes in mixed woodland.
Distribution: Mediterranean–Atlantic.

Russulaceae

p.222 *Lactarius camphoratus* (Bull.:Fr.) Fr.
Russulaceae, Russulales

Description: *Cap dark red-brown. Gills dark. Stipe darker below.*
Cap 2–6 cm (0.8–2.4 in), flattened or depressed, later deeply depressed, smooth to finely wrinkled in the centre, rather dark red-brown, sometimes with vinaceous tones. *Gills* subdecurrent, rather close, brownish ochre then darker, brownish with vinaceous tints. *Stipe* 1.5–5 × 0.5–1 cm (0.6–2 × 0.2–0.4 in), dry, sometimes finely wrinkled or veined towards the base and under the gills, brown and darker, vinaceous to blackish, below. *Flesh* brownish to vinaceous brown below. Milk fairly abundant, whitish or cloudy. Smell of L. quietus when fresh, then very strongly of chicory when dry. *Microscopic features* Sp 7.5–8.5 × 6–7.5 µm, with medium spiny warts, joined by connectives into a sometimes almost complete reticulum. Cyst fusoid, 50–70 × 7–10 µm, attenuate or appendiculate. CC hyphae erect, obtuse or clavate, 4–8 µm diameter.
Season: July–October.
Habitat: Broad-leaved trees, often in moist, mossy places.
Distribution: Widespread. Rather common.

---

p.222 *Lactarius quietus* (Fr.:Fr.) Fr.
Russulaceae, Russulales

Description: *Cap micaceous, somewhat rufous brown. Smell of bugs.*
Cap 5–12 cm (2–4.7 in), convex, then with a central depression, somewhat rufous brown or reddish grey, quite variable, often rather dull and zoned or with a ring of darker spots at the edge, frosted to micaceous (like dry snail trail). *Gills* adnate to decurrent, close, cream then somewhat rufous. *Stipe* 3–7 × 0.5–2 cm (1.2–2.8 × 0.2–0.8 in), almost concolorous but more vinaceous brown from the base. *Flesh* whitish then somewhat rufous beige, darker at the base. Milk abundant, milky cream, unchanging, mild.

Smell strong like bugs (like boiling laundry).
Microscopic features Sp 8–9.5 × 6–7.5 µm, with low warts joined by incomplete connectives into a partial reticulum. Cyst abundant, especially at the margin, fusiform, 50–80 × 6–10 µm, attenuate or appendiculate. CC hyphae arranged in a trichoderm, cylindric or fusoid cylindric.

Season: July–December.
Habitat: Strictly mycorrhizal with oak.
Distribution: Very common.

---

p.222 *Lactarius subdulcis* (Pers.:Fr.) S.F. Gray
Russulaceae, Russulales

Description: *Cap matt, beige chamois. Stipe apex very pale.*
Cap 2–6 cm (0.8–2.4 in), expanded to depressed, sometimes slightly sulcate or lobed, smooth or slightly rugose, very matt, brownish beige to pinkish chamois, sometimes darker or paler. Looks rubbery. *Gills* adnate to decurrent, rather close, whitish then pale cream. *Stipe* 3–6 × 0.5–1 cm (1.2–2.4 × 0.2–0.4 in), very matt, cream or beige with yellowish or pinkish tones, particularly pale under the gills, slightly more orange below. *Flesh* pale cream. Milk white, fairly abundant, unchanging, almost mild. Smell strong, like rubber. *Microscopic features* Sp 7–10 × 6–8 µm, with low warts joined by more or less well-developed connectives, forming an incomplete reticulum. Cyst mainly marginal, 50–85 × 6–10 µm, fusiform, appendiculate or obtuse. CC with short, erect, sometimes septate hyphae, more or less arranged in a palisade.

Season: July–December.
Habitat: Beech, sometimes other broad-leaved trees or conifers.
Distribution: Widespread. Rather common.

---

p.223 *Lactarius mitissimus* (Fr.:Fr.) Fr.
Russulaceae, Russulales

Description: *Cap orange or rather bright rufous. Milk white, absolutely mild.*

Cap 2–5 cm (0.8–2 in), expanded, with a broad central depression, rather matt, smooth or minutely roughened, orange or slightly tawny. *Gills* adnate to decurrent, close, cream to somewhat pale tawny-beige. *Stipe* 3–6 × 0.5–1 cm (1.2–2.4 × 0.2–0.4 in), concolorous. *Flesh* whitish to somewhat rufous cream near the surface. Milk white, sometimes fairly abundant, unchanging, absolutely mild. *Microscopic features* Sp 7.5–9.5 × 6–7.5 μm, with rather low warts, partly joined by short connectives, sometimes almost isolated. Cyst fairly abundant, fusiform, 50–90 × 6–10 μm, appendiculate or obtuse. CC with filamentous hyphae 5–8 μm diameter, rather short and thick, cylindric or wavy, lying parallel to the surface or slightly erect, partly gelatinized.

Season: July–October.
Habitat: Mixed broad-leaved woods on rather moist soils.
Distribution: Not well known. Rather rare.

L. aurantiofulvus Blum ex M. Bon, which is a beautiful uniform bright orange, with slightly acrid milk and slightly rubbery smell (spores more reticulate), occurs under conifers or sometimes in mixed broad-leaved woodland. *L. aurantiacus* (Vahl:Fr.) S.F. Gray, found in mixed broad-leaved woodland, is paler or yellowish orange, the stipe reddish towards the base; the smell is similar to that of *L. quietus* (*see* p.768) and the taste is acrid or slightly or slowly bitter (cap hyphae slender).

---

p.224 *Lactarius fulvissimus* Romagnesi
Russulaceae, Russulales

Description: *Cap reddish tawny. Milk unchanging. Smell rubbery, slightly sharp.*
Cap 4–8 cm (1.6–3.2 in), quite soon depressed or lobed, sometimes with a misshapen hump in the centre, dry or with a slightly greasy sheen in damp weather, smooth, sometimes cracking up in dry weather and when old, a uniform reddish

tawny or with orange or brownish zones. *Gills* adnate to decurrent, not very close, yellowish cream then somewhat rufous. *Stipe* 3–6 × 0.5–1.5 cm (1.2–2.4 × 0.2–0.6 in), more or less tapering, dry, often grooved just under the gills, almost concolorous or paler but frequently reddish or red-brown towards the base. *Flesh* cream to almost concolorous near the surface. Milk white, sometimes fairly abundant, absolutely unchanging, even on a white cloth, mild or slightly to slowly acrid. Smell rubbery or slightly sharp. *Microscopic features* Sp 7–8.5 × 6.5–8 μm, shortly elliptic to subglobose, with low warts, joined by connectives into an often complete reticulum or by raised ridges, some forming wings, the SP somewhat indistinct. Cyst fusiform, 50–80 × 6–10 μm, appendiculate or obtuse. CC hyphae slender, 3–7 μm diameter, more or less erect or lying parallel to the surface, not gelatinized.

Season: July–November.
Habitat: Broad-leaved woodland, sometimes with conifers. Prefers calcareous soils.
Distribution: Uneven. Rare.

L. subsericatus Kühner & Romagnesi ex M. Bon, a more common species, has a slightly darker cap, milk that yellows on white cloth in a few minutes and an almost mild or slightly to slowly bitter taste. An intermediate form (var. *pseudofulvissimus* M. Bon), also with yellowing milk, has a brighter orange cap with a rather pale margin and a more distinctly sharp rubbery smell (occurs in hornbeam coppice on calcareous clays).

---

p.223 *Lactarius lacunarum* (Romagnesi) ex Hora
Russulaceae, Russulales

Description: *Cap rufous to brick red. Stipe red-brown below. Cap* 4–8 cm (1.6–3.2 in), depressed, dry, sometimes scurfy or cracking when old or in dry weather, rufous or orange brick, fading slightly when dry. Margin quite often finely scalloped. *Gills* decurrent,

rather close, cream to rather pale orange-beige then spotted, somewhat rufous when old. *Stipe* 3–5 × 0.5–1 cm (1.2–2 × 0.2–0.4 in), sometimes irregular to curved, dry, finely wrinkled under the gills, almost concolorous or paler, sometimes orange above and often red-brown below.
Flesh whitish to somewhat pale rufous near the surface. Milk fairly abundant, white, yellowing quite slowly on white cloth, mild or slightly to slowly acrid. *Microscopic features* Sp 7–9 × 5–6 μm, with rather low warts, joined by connectives forming an incomplete reticulum, the apiculus prominent. Cyst fusiform, 60–80 × 5–10 μm, appendiculate. CC hyphae slender, wavy, above a layer of short or diverticulate cells.

Season: July–October.
Habitat: Damp places, pond edges.
Distribution: Widespread in western Europe.

---

p.222 *Lactarius tabidus* Fr.
Russulaceae, Russulales

Description: *Cap ochraceous beige, wrinkled in the centre. Milk yellowing on a white handkerchief.*
Cap 2–6 cm (0.8–2.4 in), often wrinkled or dented in the centre, sometimes with a sharp umbo, ochraceous beige to pinkish ochre, fading as it dries. *Gills* subdecurrent, rather close, pale cream then pinkish beige.
Stipe 2–5 × 0.3–0.8 cm (0.8–2 × 0.12–0.3 in), quite often irregular, dry, often finely wrinkled, almost concolorous or paler. *Flesh* fragile, whitish cream to pale beige near the surface. Milk sometimes not very abundant, white, yellowing rapidly (less than 15 s) on a white tissue, taste mild. *Microscopic features* Sp 8.5–10.5 × 6–8 μm, with low warts, joined by connectives in an almost complete reticulum. Cyst fusiform, 45–70 × 5–10 μm, appendiculate. CC hyphae short, arranged in an unequal palisade, the terminal cells slightly elongate or lageniform.

Season: July–December.
Habitat: Mainly under beech.
Distribution: Widespread. Common.

L. theiogalus (Bull.:Fr.) S.F. Gray, has sometimes been distinguished. It is more slender and fragile, with distinct long striations on the margin, has more watery milk, less reticulate spores and occurs in damp sites (bogs and marshy ground in woods).

---

p.222 *Lactarius hepaticus* Plowright
Russulaceae, Russulales

Description: *Cap reddish brown. Milk yellowing slowly on the gills.*
Cap 3–6 cm (1.2–2.4 in), convex or with a broad central depression, sometimes umbonate at the bottom, slightly greasy then dry, smooth to finely wrinkled, sometimes slightly cracking when dry, red-brown with a greenish tint, like liver, fading as it dries. Margin sometimes crenulate. *Gills* subdecurrent, rather close, dull yellowish buff, with rather reddish pink spots. *Stipe* 3–6 × 0.5–1 cm (1.2–2.4 × 0.2–0.4 in), rather irregular, dry, smooth or finely wrinkled, almost concolorous. *Flesh* cream to pinkish beige, yellowing slightly when cut. Milk fairly abundant, white, yellowing on the gills, and also on a white tissue, acrid or slightly to slowly bitter. *Microscopic features* Sp 7.5–9.5 × 6–7 µm, with rather low warts, joined by connectives forming an incomplete reticulum. Cyst fusiform, 50–80 × 5–10 µm, appendiculate. CC hyphae short, forming an irregular palisade, the terminal cells elongate or wavy, sometimes like long filaments.
Season: July–November.
Habitat: Pine.
Distribution: Widespread but uneven.

---

p.223 *Lactarius brunneohepaticus* Moser
Russulaceae, Russulales

Description: *Cap reddish brown with olivaceous tones. Milk yellowing on a handkerchief.*
Cap 1–4 cm (0.4–1.6 in), convex then depressed, dry to slightly scaly, rather dark

brown, with olivaceous tones, fading to reddish brown, at least at the margin.
Gills adnate, not very close, cream then pinkish buff or orange-beige. *Stipe* 1–2.5 × 0.3–0.5 cm (0.4–1 × 0.12–0.2 in), slightly pruinose, brownish with olivaceous tones above, more orange towards the base.
Flesh brownish to olivaceous grey. Milk whitish, rather rapidly lemon yellow on a handkerchief (1 min), mild. *Microscopic features* Sp 9–10 × 6–8 μm, with rather low spiny warts, joined by fine, sometimes incomplete connectives. Cyst more abundant near the gill edge, fusiform, 35–70 × 5–8 μm, appendiculate or constricted to capitate at the apex. CC with short, thick, irregular cells forming a subcellular palisade, the terminal cells often differentiated as long hairs.

Season: July–September.
Habitat: Green alder (*Alnus viridis*), in subalpine zones.
Distribution: Subalpine zones and Scandinavia. Rare.

Lactarius badiosanguineus Kühner & Romagnesi
Russulaceae, Russulales

Description: *Cap dark red-brown. Milk yellowing slightly on a handkerchief.*
Cap 3–6 cm (1.2–2.4 in), plano-convex or depressed, often umbonate, slightly greasy then more dry, smooth to finely wrinkled, dark red-brown to blackish crimson, fading when dry. *Gills* subdecurrent, rather close, rather pale reddish ochre, with orange-pink tones when old. *Stipe* 3–6 × 0.5–1 cm (1.2–2.4 × 0.2–0.4 in), pruinose to minutely roughened, almost concolorous but paler above or slightly more orange.
Flesh somewhat rufous cream to yellowish when cut, acrid or slightly to slowly bitter. Milk white or slightly watery or with whitish clouds, becoming rather quickly pale yellow on a white tissue, slightly to slowly acrid. Smell faint, sometimes slightly fruity. *Microscopic features* Sp 6.5–8.5 × 6–7 μm, shortly elliptic, with

rather low warts, joined by incomplete ridges making broken lines, sometimes almost reticulate. Cyst mostly marginal, fusiform or tapering to a neck at the apex, 3–50 × 5–10 µm, obtuse. CC hyphae short, more or less arranged in a palisade, mixed with longer cells, sometimes drawn out into slender hairs.

Season: July–September.
Habitat: Grassy edges of conifer woods, in subalpine zones.
Distribution: Quite widespread but limited to submontane areas. Rather rare.

---

p.218 *Lactarius volemus* (Fr.:Fr.) Fr.
Russulaceae, Russulales

Description: *Cap very matt, bright orange rufous. Milk white, very abundant.*
Cap 6–12 cm (2.4–4.7 in), subglobose then convex to depressed, very dry, pruinose or velvety, cracking when old and in dry weather, bright orange to somewhat rufous yellow, sometimes more reddish or paler, especially at the margin.
Gills subdecurrent, rather close, pale cream then pale orange-yellow, bruising brown.
Stipe 5–10 × 1–3 cm (2–4 × 0.4–1.2 in), hard, pruinose, smooth to finely wrinkled or grooved, especially at the apex, concolorous or paler. *Flesh* hard, white to somewhat rufous cream near the surface, browning slightly where exposed. Milk very abundant, flowing freely, white, browning or reddening slightly when exposed, mild. Smell strong, of Jerusalem artichokes (cooking shellfish).
Chemical test $FeSO_4$ dark green. *Microscopic features* Sp 8–10.5 × 7–9 µm, subglobose, with rather low warts, joined by connectives into an incomplete reticulum. Cyst very abundant, tapering and pointed, 50–100 × 8–12 µm, thick-walled and sometimes brownish. CC cellular, with numerous awl-shaped brownish hairs.

Season: July–October.
Habitat: Various broad-leaved trees, sometimes conifers.
Distribution: Uneven. Decreasing.

L. rugatus Kühner & Romagnesi, which is very similar, has a very wrinkled to strongly concentrically cracked cap, is darker and does not turn green with $FeSO_4$. The smell is also less strong.

---

p.223 *Lactarius decipiens* Quélet
Russulaceae, Russulales

Description: *Cap pinkish ochre or flesh coloured. Milk yellowing on the gills.*
Cap 3–7 cm (1.2–2.8 in), almost umbonate then depressed, slightly viscid then dry, smooth or faintly grained, pinkish ochre or flesh-beige, sometimes more reddish or paler. Margin sometimes sulcate.
Gills adnate to decurrent, rather close, pale pinkish cream. *Stipe* 3–7 × 0.5–1 cm (1.2–2.8 × 0.2–0.4 in), dry, smooth to finely wrinkled, almost concolorous.
Flesh whitish to almost concolorous near the surface, somewhat more rufous towards the stipe base. Milk often not very abundant, white or milky white, yellowing quite rapidly (1 min) on the gills and the cut flesh, almost mild. Smell strong, of *Pelargonium*. *Microscopic features* Sp 6.5–8 × 6–7 µm, with low warts joined by connectives, into an almost complete reticulum. Cyst fusiform, 50–70 × 6–9 µm, appendiculate. CC with filamentous hyphae, sometimes short or entangled, certain terminal cells tapering or fusoid.
Season: August–November.
Habitat: Mainly hornbeam and oak, more rarely conifers.
Distribution: Very uneven. Absent from more northerly regions.

---

p.218 *Lactarius chrysorrheus* Fr.
Russulaceae, Russulales

Description: *Cap orange-beige, zoned to scrobiculate. Milk turning golden yellow.*
Cap 4–8 cm (1.6–3.2 in), subglobose then slightly depressed, slightly viscid then micaceous with fine radial wrinkles and

concentrically zoned with irregular bands or pits, orange-beige or rather pale apricot yellow. *Gills* subdecurrent, close, pinkish cream to pale-orange, edge sometimes yellow. *Stipe* 3–7 × 0.5–2 cm (1.2–2.8 × 0.2–0.8 in), slightly viscid then dry, smooth to finely wrinkled, whitish then pinkish cream or almost concolorous. *Flesh* white or almost concolorous near the surface, soon bright yellow when cut. Milk abundant, white then rapidly golden yellow on the gills and on the flesh, acrid. *Microscopic features* Sp 8–9 × 7–7.5 µm, with low warts joined partially by connectives into an incomplete reticulum with uneven ridges. Cyst fusiform, 40–80 × 6–10 µm, appendiculate. CC hyphae lying parallel to the surface or more or less erect, fine.

Season: August–November.
Habitat: Broad-leaved trees, mainly oak.
Distribution: Widespread but uneven.

---

p.226 *Lactarius acris* (Bolt.:Fr.) S.F. Gray
Russulaceae, Russulales

Description: *Cap very pale, umbonate, viscid. Milk white, turning bright pink.*
Cap 6–10 cm (2.4–4 in), umbonate, sometimes slightly depressed, viscid then shiny when dry, very pale whitish, sometimes mottled the colour of milk coffee or pale greyish buff.
Gills subdecurrent, rather close, rather pale yellowish buff, pinkish when old.
Stipe 4–8 × 1–2 cm (1.6–3.2 × 0.4–0.8 in), quite hard, slightly viscid, pruinose or later finely wrinkled, white then flushed pinkish buff. *Flesh* white, soon pink when cut, buff on drying. Milk abundant, white, rapidly bright rose-red, even when isolated (on a finger nail), acrid. *Microscopic features* Sp 8–9 × 7–8 µm, subglobose or elliptic, with low warts, joined by short connectives forming an incomplete reticulum, sometimes with stronger spiny warts or more prominent, almost ridged connectives. Cyst mainly marginal, fusiform or developing a neck, 40–80 ×

5–10 μm, appendiculate or obtuse. CC hyphae gelatinized, arranged in a palisade, rather narrow or cylindric, sometimes with an inflated base.

Season: July–October.
Habitat: Broad-leaved trees, mainly beech. Often on medium to warm, calcareous soils.
Distribution: Widespread. Rather rare.

This is the first of a group of *Lactarius* with pink milk, or at least flushing pink when exposed. *L. acris*, which always has a pale cap, is the only one with milk that turns pink when isolated from the gills and flesh (pink is reminiscent of some toothpastes).

---

p.226 *Lactarius pterosporus* Romagnesi
Russulaceae, Russulales

Description: *Cap buff, strongly wrinkled or crisped in the centre. Gills close.*
Cap 3–8 cm (1.2–3.2 in), convex then slightly depressed, slightly viscid then dry or velvety, wrinkled with brain-like folds in the centre, buff milk-coffee colour, sometimes mottled yellowish. Margin sometimes faintly scalloped. *Gills* decurrent, often very close, ochraceous cream then salmon. *Stipe* 3–6 × 0.5–1 cm (1.2–2.4 × 0.2–0.4 in), often tapering, like satin or finely wrinkled, beige to whitish, sometimes pinkish in places when old. *Flesh* whitish, bright pink in a few minutes when cut, then brick red on drying. Milk abundant, creamy white, unchanging when isolated, becoming somewhat rufous pink on the flesh and gills, acrid and bitter. Smell faintly fruity. *Microscopic features*

Sp 7–8 × 6–7 μm, the warts joined by spectacular ridges, with wings 2–4 μm high, often independent and not reticulate. Cyst fusiform, 50–80 × 5–10 μm, attenuate or obtuse. CC hyphae septate, arranged in a palisade, the terminal cells tapering into long hairs up to 80 μm long.

Season: July–November.
Habitat: Broad-leaved trees, mainly hornbeam, sometimes beech; on calcareous clays.
Distribution: Quite widespread.

L. fuliginosus (Fr.:Fr.) Fr., although its spores are not, or not very, winged, is similar (and common). It has a smooth to finely wrinkled, brownish to greyish milk-coffee colour cap, an almost concolorous stipe, and milk that flushes pink or reddens rather slowly. It also occurs in coppice woodland on calcareous clays with hornbeam. Var. *albipes* (Lange) M. Bon, which has a very slightly paler cap with a white margin and, in particular, a wholly white stipe, occurs mainly under beech.

---

p.226 *Lactarius ruginosus* Romagnesi
Russulaceae, Russulales

Description: *Cap brownish with a scalloped margin. Gills distant.*
Cap 5–10 cm (2–4 in), convex then slightly depressed, slightly viscid then dry or velvety, smooth or slightly dented towards the centre, brownish buff, sometimes with slightly yellowish patches. Margin sulcate, with wide scallops. *Gills* subdecurrent, distant, pale buff. *Stipe* 3–6 × 1–2 cm (1.2–2.4 × 0.4–0.8 in), tapering or dented below, dry, pruinose or finely wrinkled, almost concolorous or slightly paler. *Flesh* whitish, developing carroty rufous dots when cut. Milk abundant, white, unchanging when isolated, turning rufous pink on the gills and flesh, acrid. *Microscopic features* Sp 6–9 μm diameter, subglobose, with medium warts, joined by rather spectacular ridges about 2–3 μm tall, and forming a complete reticulum. Cyst fusiform or cylindric, 50–80 × 6–10 μm, obtuse or appendiculate. CC hyphae septate, arranged in a palisade, the terminal cells tapering into obtuse hairs.

Season: July–November.
Habitat: Broad-leaved trees, mainly beech.
Distribution: Widespread but uneven.

---

p.226 *Lactarius romagnesii* M. Bon
Russulaceae, Russulales

Description: *Cap dull dark brown. Gills not very close. Stipe almost concolorous.*

Cap 4–10 cm (1.6–4 in), globose then flat or depressed, dry, matt to velvety, smooth or minutely roughened in the centre, rather dull dark brown, sometimes mottled with paler, milk coffee or light buff patches. Margin smooth or slightly sulcate.
Gills subdecurrent, not very close, ochraceous cream to rather dark ochre.
Stipe 4–10 × 2–6 cm (1.6–4 × 0.8–2.4 in), sometimes dented below, dry, pruinose, almost concolorous or slightly paler at the apex. *Flesh* white to ochraceous cream near the surface, gradually becoming carroty pink when cut, and rusty vinaceous on drying. Milk abundant, white, unchanging when isolated, slowly turning carroty rufous on the gills and flesh, slightly acrid.
Microscopic features Sp 7.5–9.5 μm diameter, subglobose or spherical, with medium warts, joined by irregular ridged connectives, up to 1.5 μm tall, forming an incomplete reticulum. Cyst fusiform, 40–70 × 5–9 μm, appendiculate or obtuse. CC hyphae septate, arranged in a palisade, with rather short and rather inflated club-shaped elements.

Season: August–October.
Habitat: Broad-leaved trees, mainly beech and hornbeam.
Distribution: In lowland areas, mainly continental.

This is the lowland, deciduous woodland equivalent of the following, typically montane, coniferous species.

---

p.226 *Lactarius picinus* Fr.
Russulaceae, Russulales

Description: *Cap blackish brown. Gills close. Stipe concolorous.*
Cap 6–12 cm (2.4–4.7 in), convex then flat or slightly depressed, dry, matt to velvety, smooth to finely wrinkled in the centre, blackish brown, sometimes mottled with greyish or dull beige patches. Margin smooth or slightly scalloped.
Gills subdecurrent, rather close, ochraceous cream, remaining pale. *Stipe* 3–6 ×

1.5–3 cm (1.2–2.4 × 0.6–1.2 in), often short and thick, pruinose, concolorous or slightly paler at the apex. *Flesh* firm then slightly more brittle, whitish, and gradually very pinkish rufous when cut. Milk fairly abundant, white, unchanging when isolated, but slowly turning somewhat rufous pink on the flesh and gills, acrid. *Microscopic features* Sp 8.5–10 μm diameter, subglobose to elliptic, with medium warts, joined by rather coarse thick ridges 1–1.5 μm tall, forming an incomplete reticulum. Cyst not very abundant, fusiform, 50–75 × 6–10 μm, appendiculate or obtuse. CC hyphae septate, in a palisade, the terminal cells tapering or cylindric.

Season: July–October.
Habitat: Conifers in the mountains.
Distribution: Continental and montane.

p.226 *Lactarius lignyotus* Fr.
Russulaceae, Russulales

Description: *Cap dark sooty brown. Gills decurrent as a thread. Stipe wrinkled at the apex.*
Cap 4–7 cm (1.6–2.8 in), convex umbonate, slightly viscid or matt to velvety, wrinkled to grained around the umbo, dark sooty brown. Margin sulcate to lobed. *Gills* decurrent with a tooth down the stipe, quite distant, white or very pale. *Stipe* 5–8 × 0.5–1.5 cm (2–3.2 × 0.2–0.6 in), often watery, sometimes irregular, pruinose, deeply grooved at the apex, almost concolorous right to the apex. *Flesh* fragile, white or very pale, scarcely changing when cut. Milk fairly abundant, whitish or almost watery, unchanging when isolated, slowly turning reddish on the flesh and gills, almost mild. *Microscopic features* Sp 9–10.5 μm diameter, subglobose, with rather tall indistinct warts, joined by connectives like ridges or even like wings, up to 2 μm tall, and forming a complete reticulum. Cyst often abundant, fusiform or cylindric, 40–70 × 6–10 μm, appendiculate or obtuse. CC hyphae septate, arranged in a

palisade, the terminal cells cylindric or club-shaped.
Season: July–October.
Habitat: Peaty bogs in the mountains.
Distribution: Mountains and northern Europe. Rather rare to rare.

p.219 *Lactarius uvidus* (Fr.:Fr.) Fr.
Russulaceae, Russulales

Description: *Cap beige-grey, later spotted lilaceous. Milk white, turning purple.*
Cap 4–10 cm (1.6–4 in), convex then more or less depressed, viscid, shiny when dry, smooth, not zoned, a rather pale lilaceous grey, sometimes flushed pinkish beige to purplish. *Gills* subdecurrent, rather close, cream to pale buff, edge sometimes spotted violaceous. *Stipe* 4–10 × 0.5–2 cm (1.6–4 × 0.2–0.8 in), broadly stuffed, slightly viscid then drier but shiny, whitish to almost concolorous, sometimes pale buff or slightly yellow below. *Flesh* white to yellowish cream, the taste unpleasant, slightly acrid or bitter. Milk white, unchanging when isolated but turning purple or wine-red on the flesh and gills, mild. *Microscopic features* Sp 9–11 × 7.5–8 µm, with rather low warts, joined by unequal connectives, sometimes with short ridges, forming an incomplete or irregular reticulum. Cyst fusiform, 50–80 × 6–12 µm, appendiculate or pointed. CC an ixocutis, the hyphae slender, lying parallel to the surface. Pigment rather intracellular.
Season: August–November.
Habitat: Broad-leaved trees, sometimes conifers in mixed woods. Tends to be on damp acid soils.
Distribution: Widespread. Rare in places.

The foremost species in the *Lactarius* group with violet milk or milk that becomes violet when exposed on the flesh and gills. Var. *pallidus* Bres., with a whitish cap has been described. *L. flavidus* Boudier has a rather bright yellow to yellowish buff cap and the milk turns vinaceous purple then violaceous chocolate-brown.

p.220 *Lactarius uvidus* var. *candidulus* Neuhoff
Russulaceae, Russulales

Description: *Cap whitish or very pale. Milk turning lilaceous purple.*
Cap 3–5 cm (1.2–2 in), convex then depressed, viscid to slightly viscid, shiny when dry, smooth, whitish to tinged buff or yellowish beige. *Gills* subdecurrent, rather close, cream to pale buff. *Stipe* 3–5 × 0.5–1 cm (1.2–2 × 0.2–0.4 in), fragile and watery in the centre, concolorous.
Flesh rather watery and fragile, whitish to pale beige, spotted dingy lilac when cut, the taste not very pleasant. Milk fairly abundant, rather watery or cloudy, unchanging when isolated but turning lilaceous purple on the flesh and gills, mild. *Microscopic features* Identical with *L. uvidus*.
Season: July–October.
Habitat: Broad-leaved trees, mainly willows; very damp places.
Distribution: Less widespread than *L. uvidus*.

This pale form is not always easy to distinguish from similar and rarer taxa, which would probably have a pubescent margin. Other pale species also occur in the mountains (*see* the comments under *L. robertianus*, p.784).

---

p.220 *Lactarius luridus* S.F. Gray
Russulaceae, Russulales

Description: *Cap somewhat rufous brown to brownish grey. Milk turning purplish chocolate.*
Cap 4–10 cm (1.6–4 in), convex then faintly depressed, viscid, shiny when dry, smooth with fine radial wrinkles, dull, faintly zoned at the edge, somewhat rufous brown to brownish grey, sometimes paler. *Gills* subdecurrent, rather close, creamy white then spotted purple. *Stipe* 4–8 × 1–2 cm (1.6–3.2 × 0.4–0.8 in), slightly viscid, lumpy below, often paler than the cap, whitish to dingy yellowish cream.
Flesh white or pale, more buff near the surface, chocolate-brown when cut. Milk

white, unchanging when isolated, turning deep purple or dark purplish chocolate on the flesh and gills, slightly to slowly acrid. Smell somewhat fruity. *Microscopic features* Sp 8–10.5 × 7–8.5 μm, elliptic, with rather low warts, joined by connectives into an almost complete reticulum. Cyst fusiform, 50–70 × 6–10 μm, appendiculate or obtuse. CC an ixocutis, with slender hyphae lying parallel to the surface.

Season: July–November.
Habitat: Broad-leaved trees; mainly in lowland areas.
Distribution: Widespread but uneven.

L. violascens (Otto:Fr.) Fr., a violaceous brown species of warmer areas, zoned or spotted at the cap margin, has milk that becomes dark violaceous and is slightly or slowly bitter.

---

p.219 *Lactarius robertianus* M. Bon
Russulaceae, Russulales

Description: *Cap dark crimson brown, umbonate. Gills and stipe whitish.*
Cap 2–5 cm (0.8–2 in), conico-convex then expanded, umbonate, viscid then shiny, violaceous brown to blackish crimson, the umbo remaining darker, sometimes lilaceous grey at the edge when old. Margin obtuse. *Gills* adnate, not very close, white. *Stipe* 2–4 × 0.5–1 cm (0.8–1.6 × 0.2–0.4 in), pruinose at the apex, fibrillose silky below, white, spotting violaceous when handled. *Flesh* white then violaceous. Milk white, turning purple only on the flesh. Smell of cedar wood (pencils). *Microscopic features* Sp 9–12 × 6.5–8.5 μm, with rather low warts, the ridges making lines or a more or less complete reticulum. Cyst (especially Pleuro) fusiform or tapering at the apex or with many constrictions, 50–90 × 6–12 μm, almost no reaction in SA. CC an ixocutis with hyphae 2–3 μm diameter, sometimes slightly clavate. Pigment intracellular.
Season: July–September.
Habitat: Dwarf willow (*Salix herbacea*), in snow gullies and dwarf alpine shrub communities.

Distribution: The Alps. Rather common within its range.

L. dryadophilus Kühner has a convex umbilicate then depressed cap, viscid but dry at the bearded, finally glabrous edge, cream to ivory-yellow or pale brownish yellow, yellowish; it occurs mainly among *Dryas,* on calcareous soils. *L. salicis-herbaceae* Kühner which has a yellow to ochraceous yellow cap, sometimes tinged brownish, pale brownish yellow colours at the stipe base, occurs in dwarf alpine shrub communities with dwarf willows, on damp acid soils.

---

p.224 *Lactarius sanguifluus* (Paulet:Fr.) Fr.
Russulaceae, Russulales

Description: *Cap reddish. Gills pinkish vinaceous. Stipe scrobiculate. Milk vinaceous red.*
Cap 4–12 cm (1.6–4.7 in), flattened, then depressed, slightly lubricious, shiny in dry weather, micaceous or frosted, zoned, reddish orange, with pinkish vinaceous or somewhat crimson-buff zones. Greening absent or faint. *Gills* subdecurrent, close, vinaceous pink, spotting green when damaged. *Stipe* 3–7 × 2–6 cm (1.2–2.8 × 0.8–2.4 in), scrobiculate, rather violaceous reddish orange in places, with darker, vinaceous red pits. *Flesh* thick, vinaceous red. Milk vinaceous red, turning brown, slightly or slowly bitter. *Microscopic features* Sp 7.5–9.5 × 6–8 µm, with medium warts, joined by connectives in an irregular reticulum. Cyst not very abundant, fusiform, 40–60 × 4–10 µm, refractive. CC a slightly gelatinized cutis, with slender filamentous hyphae.
Season: June–October.
Habitat: Pine. Often on warm sites.
Distribution: Southern, but occurring, albeit more rarely, in certain northern districts.

An excellent, much sought-after edible species. *L. vinosus* Quélet, which is very similar and occurs in the same habitats, has a very pruinose cap margin, a stipe whitened by the abundant pruina, violaceous brown scrobicules and only slightly acrid milk.

---

p.224 *Lactarius deliciosus* (L.:Fr.) S.F. Gray
Russulaceae, Russulales

Description: *Cap orange, zoned, greening slightly. Stipe scrobiculate. Milk carrot-colour.*
Cap 5–15 cm (2–6 in), gradually becoming depressed, slightly viscid or shiny when dry, sometimes radially grained, zoned, orange ochre, sometimes more brownish red. Moderately greening. *Gills* subdecurrent, rather close, bright orange, spotting green when old. *Stipe* 3–7 × 2–6 cm (1.2–2.8 × 0.8–2.4 in), concolorous or paler, with bright orange scrobicules. *Flesh* pale to concolorous near the surface, mild. Milk carrot-colour, unchanging. Smell faint like carrot or slightly fruity.
Microscopic features Sp 7–9 × 5.5–6.5 µm, with rather low warts, joined by ridges or fine connectives, into a dense, contorted or incomplete reticulum. Cyst fusiform, 40–70 × 6–10 µm, appendiculate or obtuse. CC a slightly gelatinized cutis, with slender hyphae.
Season: August–December.
Habitat: Pine.
Distribution: Widespread. Common.

Edible but less good than the preceding species.

---

p.224 *Lactarius quieticolor* Romagnesi
Russulaceae, Russulales

Description: *Cap orange-brown, with greyish zones. Stipe not very scrobiculate.*
Cap 4–10 cm (1.6–4 in), soon flattened, then depressed, a little shiny or slightly micaceous, a dull orange-brown and quite dark, often with greyish or violaceous grey zones. *Gills* subdecurrent, close, orange then brownish, spotting dingy green. *Stipe* 3–4.5 × 1–2 cm (1.2–1.8 × 0.4–0.8 in), concolorous or tinged greenish

or greyish, often frosted, not very scrobiculate. *Flesh* pale then somewhat rufous near the surface, greening slowly. Milk orange, slightly or slowly bitter, greening. Smell rather light, fruity.
Microscopic features Sp 7–9 × 6–7.5 μm, with rather tall warts, joined by connectives sometimes in the shape of raised ridges, into an often complete reticulum. Cyst fusiform, 40–70 × 5–10 μm, appendiculate or obtuse. CC a slightly gelatinized cutis, with slender filamentous hyphae.

Season: July–December.
Habitat: Pine.
Distribution: Uneven.

Slightly less good to eat than the preceding species. *L. semisanguifluus* Heim & Leclair always occurs under pine (mainly Scots pine, on calcareous soils); a strongly greening species, it has a zoned, dull orange to vinaceous brown cap, with a frosted margin and an almost concolorous, strongly greening, not very scrobiculate stipe. The flesh is pale, turning blood red in a few minutes before becoming vinaceous then green. It is a good edible species but rarer than the preceding ones.

---

p.225 *Lactarius deterrimus* Gröger
Russulaceae, Russulales

Description: *Cap orange, not very zoned, greening. Stipe not scrobiculate.*
Cap 4–12 cm (1.6–4.7 in), soon flattened then depressed, slightly viscid, somewhat shiny when dry, with fine radial wrinkles, not very zoned, rather bright orange but soon dingy or dark green. *Gills* sub-decurrent, rather close, orange ochre or slightly flesh coloured at the base, rapidly dark green in places and on the edge.
Stipe 4–8 × 2–2.5 cm (1.6–3.2 × 0.8–1 in) concolorous but soon becoming green and dingy, not scrobiculate, with a pruinose white ring at the apex. *Flesh* pale or orange in the cortex, reddish then greenish if exposed. Milk orange, mild then slightly or

slowly bitter. Smell often distinctly carrot-like. *Microscopic features* Sp 8.5–10.5 × 7–8.5 μm, with rather low warts, often isolated but also partly joined by medium ridges. Cyst fusiform, 60–80 × 6–8 μm, long and slender, the apex tapering or constricted. CC a slightly gelatinized cutis, with slender filamentous hyphae.

Season: August–October.
Habitat: Spruce.
Distribution: Widespread. Common.

Edibility very average.

---

p.225 *Lactarius salmonicolor* Heim & Leclair
Russulaceae, Russulales

Description: *Cap bright salmon orange. Stipe slightly scrobiculate. Milk salmon orange.*
Cap 6–15 cm (2.4–6 in), convex then faintly depressed, slightly viscid or dry, a little shiny, generally smooth, finely zoned, especially towards the edge, uniformly salmon orange, sometimes with orange-brown spots but only very slightly greening. *Gills* subdecurrent, close, a beautiful salmon orange, edge almost unchanging *Stipe* 4–10 × 1.5–3 cm (1.6–4 × 0.6–1.2 in), almost concolorous, slightly scrobiculate. *Flesh* pale but more orange near the surface. Milk salmon orange, almost unchanging or slowly turning slightly brown to faintly vinaceous. Taste slightly or slowly bitter. *Microscopic features* Sp 9–12 × 6–7.5 μm, quite long, with rather low warts, joined by connectives or rather tall ridges into an often incomplete reticulum. Cyst fusiform, 40–70 × 5–10 μm, attenuate or appendiculate. CC a slightly gelatinized cutis, with slender filamentous hyphae.

Season: August–November.
Habitat: Firs.
Distribution: Widespread. Rather rare.

STROPHARIACEAE

The *Strophariaceae* is a relatively heterogeneous family whose defining characters are: fleshy or slender, with a fibrous texture, saprotrophic; spore print brown or violaceous brown to purplish black; spores usually with a germ pore; gills notched or more or less decurrent.

Although all species whose spores are more or less violaceous in the mass are placed here, the family also includes typically brown-spored species. Therefore, the family consists of two tribes, the *Stropharieae* and the *Pholioteae*. There is a relatively large number of genera.

The nature of the partial veil and the chrysocystidia play an important part in the classification. Constant characters are: margin inrolled then curved and finally expanded, thin, often appendiculate or slightly veiled and then overhanging shortly; gills not very close; gill edge finely frosted, whitish; stipe cylindric; smell and taste insignificant; spores smooth, the wall slightly thick, the endosporium orthochromatic (bright red) in saturated Congo Red–ammonium hydroxide solution, the germ pore axial; basidia 4-spored, clavate, with clamps; trama subregular; pigment parietal, in encrusting bands, proportional in intensity to the cap colour; clamps present.

---

p.150 *Kuehneromyces mutabilis* (Scop.:Fr.) A.H. Smith & Singer (= *Pholiota m.*) *Strophariaceae*, Cortinariales

Description: *Cap hygrophanous, date-brown to pale honey-ochre. Stipe with a fleecy sheath.*
Cap 1.5–8 cm (0.6–3.2 in), subglobose then almost flat, hygrophanous, smooth, greasy or shiny when fresh, zoned, rather light date-brown (moist peripheral zones) to yellow-ochre or yellowish honey (dry, central zones). Margin slightly appendiculate with a rather short, fragile, brown membrane. *Gills* adnate to subdecurrent, close, rather pale buff then rust-brown. Spore print brown. *Flesh* pale to almost concolorous. Taste fungal, ordinary. *Stipe* 3–10 × 0.2–1 cm (1.2–4 ×

0.08–0.4 in), often curved, pale ochraceous yellow at the apex, brownish or rust-brown elsewhere. Veil forming a membranous but fragile ring, at the top of a fleecy or fibrillose to finely scaly sheath. In clusters. *Microscopic features* Sp 6–7.5 × 3.5–4.5 µm, elliptic or slightly almond-shaped, smooth (under an optical microscope), the GP quite wide. Bas 20–25 × 6–8 µm. Cheilo 20–30 × 4–8 µm, fusiform lageniform with a wavy, long cylindric obtuse neck. CC an ixocutis, the hyphae 2–7 µm diameter. Pigment parietal, smooth.

Season: June–November.
Habitat: Stumps and fallen trunks of broad-leaved trees, more rarely on conifers.
Distribution: Widespread. Common.

K. mutabilis is a very good edible species often occurring in large numbers. However, *Galerina marginata* and similar closely related species (*see* p.526), which may have fatal effects (they include the same toxins as *Amanita phalloides*) are real look-alikes in the field.

p.151 *Pholiota squarrosa* (Müll.:Fr.) Kummer
Strophariaceae, Cortinariales

Description: *Cap with rufous brown scales. Stipe with a scaly sheath.*
Cap 3–12 cm (1.2–4.7 in), subglobose then convex, not umbonate, dry, with dark brown, rust-brown or orange-brown, upturned triangular scales, on a paler, yellowish to yellowish buff ground.
Gills subdecurrent, close, yellowish then rust-brown. Spore print brown.
Flesh yellowish. *Stipe* 5–15 × 2–6 cm (2–6 × 0.8–2.4 in), curved, concolorous or darker to quite bright rusty at the base. Sheath scaly like the cap, ending in a wide, membranous then fibrillose ring, finally disintegrating. In clusters.
Microscopic features Sp 6–9 × 3.5–5 µm, elliptic or ovoid, the GP quite wide. Bas 20–25 × 5–7 µm. Cheilo 25–50 ×

7–15 μm, cylindric-lageniform or faintly clavate (mixed with a few chrysocystidia). Pleuro (chrysocystidia) 25–60 × 10–15 μm, clavate, mucronate or appendiculate, with refractive contents, yellow in ammonia. CC a trichoderm, the hyphae septate 8–30 μm diameter, thick-walled, brown. Pigment parietal, smooth or slightly encrusting.

Season: August–November.
Habitat: Stumps and base of standing broad-leaved trees.
Distribution: Widespread. Rather common.

Easily recognizable by the dry cap with contrasting, regular, upturned scales. *P. jahnii* (*see* below) is similar but can be distinguished by the viscid cuticle.

---

p.151 *Pholiota jahnii* Tjallingii-Beukers & Bas
Strophariaceae, Cortinariales

Description: *Cap viscid, with black-brown scales on a yellow-ochre ground. Sheath scaly.*
Cap 2–10 cm (0.8–4 in), very similar to that of *P. squarrosa* (*see* p.790) but with a viscid surface, the scales sometimes disappearing, brown with a black point. *Gills* as in *P. squarrosa*. Spore print brown. *Stipe* 5–14 × 0.5–1.5 cm (2–5.5 × 0.2–0.6 in), as in *P. squarrosa* but with a lubricious cuticle. In clusters or troops. *Microscopic features* Sp 5–7 × 3–4 μm, elliptic to cylindro-elliptic or almost bean-shaped, the GP indistinct. Cheilo 25–40 × 5–10 μm, cylindro-clavate (mixed with a few chrysocystidia). Pleuro (chrysocystidia) 25–50 × 6–12 μm, clavate, mucronate or appendiculate, with refractive contents, yellow in ammonia. CC an ixotrichoderm, the hyphae septate 2–5 μm diameter, brown.

Season: August–November.
Habitat: Stumps and base of standing broad-leaved trees; rarer on conifers.
Distribution: Uneven. Rather rare.

See comment under *P. squarrosa*, p.790.

p.151 *Pholiota flammans* (Batsch:Fr.) Kummer
Strophariaceae, Cortinariales

Description: *Cap with rufous orange fleecy scales on a bright yellow ground.*
Cap 2–8 cm (0.8–3.2 in), subglobose then convex, bright yellow to golden yellow, with golden yellow then orange or very bright orange-rust, appressed then more or less upturned fleece or scales; dry to lubricious. *Gills* adnate, close, sulphur yellow, sometimes with olivaceous tones then rust-brown. Spore print brown. *Flesh* slightly or slowly bitter, sulphur yellow to orange-brown. Smell slight, of radish. *Stipe* 3–10 × 0.3–0.7 cm (1.2–4 × 0.12–0.28 in), concolorous or more yellow. Sheath scaly, ending in a fragile, fugitive ring. *Microscopic features* Sp 3.5–5 × 2–2.5 µm, small, elliptic, the GP indistinct. Bas 15–25 × 3.5–5 µm, cylindro-clavate. Cheilo 20–30 × 5–10 µm, cylindro-clavate (mixed with a few chrysocystidia). Pleuro (chrysocystidia) 20–40 × 5–12 µm, cylindric or faintly lageniform, with refractive contents, often misshapen, yellow in ammonia. CC an ixocutis with trichodermial zones, the hyphae septate, 2–10 µm diameter. Pigment mainly intracellular.
Season: August–November.
Habitat: Stumps and rotten branches of conifers; rare on broad-leaved trees.
Distribution: Continental or northern.

---

p.147 *Pholiota tuberculosa* (Sch.:Fr.) Kummer
Strophariaceae, Cortinariales

Description: *Cap fibrillose to finely scaly, yellowish rust. Stipe bright red-brown at the base.*
Cap 1–5 cm (0.4–2 in), hemispherical to broadly convex, finely fibrillose scaly to glabrous, dry or slightly viscid, rather bright yellow, soon orange or rufous orange, finally rufous to rufous brown, the margin often denticulate, remaining lighter for a long time. *Gills* adnate, rather close, yellow then rust with yellow or brownish

tones, rather pale. Spore print brown.
Flesh sulphur yellow to bright orange at the base. Taste slightly or slowly bitter.
Stipe 1–6 × 0.2–0.7 cm (0.4–2.4 × 0.08–0.28in), usually curved, often swollen or bulbous, pruinose at the apex, fibrillose-scaly towards the base, concolorous or more yellow but bright orange-red to red-brown at the base, when bruised. Veil leaving a slight ring or ring zone, rather short-lived.
Microscopic features Sp 6.5–10 × 4–6 µm, cylindro-elliptic or with a depressed ventral face, bean-shaped, GP absent. Bas 20–35 × 5–7 µm, cylindro-clavate. Cheilo 25–70 × 3–15 µm, clavate or with a rounded head. Chrysocystidia absent. CC subtrichodermial, the hyphae septate 3–12 µm diameter, the terminal elements sometimes fusiform or clavate. Pigment mainly intracellular.

Season: July–December.
Habitat: Branches and fallen trunks of broad-leaved trees.
Distribution: Widespread.

---

p.142 *Pholiota graminis* (Quélet) Singer
Strophariaceae, Cortinariales

Description: *Cap rust yellow, almost smooth. Stipe orange rust at the base.*
Cap 1–4 cm (0.4–1.6 in), convex or almost flat, almost smooth, slightly viscid to lubricious, rather bright yellow or more dull then rust from the centre.
Gills adnate, close, pale yellowish buff then rusty beige, pale. Spore print brown.
Flesh slightly or slowly bitter, sulphur yellow to rust at the base. Smell slightly herbaceous. *Stipe* 2–5 × 0.2–0.5 cm (0.8–2 × 0.08–0.2 in), concolorous or more orange to rust at the base. Cortina rather short-lived. *Microscopic features* Sp 5.5–8 × 2.5–4.5 µm, ovoid or the ventral face slightly depressed, the GP small. Bas 25–30 × 5–6 µm, cylindro-clavate. Cheilo 20–30 × 5–8 µm, lageniform, with a wavy neck, sometimes subcapitate (mixed with a few chrysocystidia). Pleuro (chrysocystidia)

30–45 × 10–15 µm, ventricose or conical umbonate at the apex, with refractive contents, yellow in ammonia. CC a cutis, the hyphae 2–10 µm diameter.

Season: September–December.
Habitat: Grasses or herbaceous plants, in damp places.
Distribution: Rather rare or not understood.

---

p.152 *Pholiota alnicola* (Fr.:Fr.) Singer
Strophariaceae, Cortinariales

Description: *Cap bright yellow to somewhat rufous. Stipe with a ring zone, becoming rufous at the base.*
Cap 2–6 cm (0.8–2.4 in), subglobose then convex, glabrous or with a few fleecy fibrils towards the margin when young, lubricious then drier, bright yellow, sometimes with greenish or lemon tones, more tawny or somewhat rufous from the centre when old. *Gills* adnate, rather close, sulphur yellow then rather pale rusty beige. Spore print brown. *Flesh* pale lemon-yellow in the cap, rust-brown to cinnamon at the stipe base. Smell aromatic, complex and strong.
Stipe 5–10 × 0.5–0.8 cm (2–4 × 0.2–0.32 in), quite often curved, fibrillose, pale yellow, becoming progressively warm rust-brown towards the base. Cortina rather short-lived. In clusters or trooping.
Microscopic features Sp 7–10 × 4–5 µm, almond-shaped or ovoid, the GP extremely faint. Bas 25–35 × 5–8 µm, cylindro-clavate. Cheilo 20–60 × 3–8 µm, cylindric or wavy, faintly ventricose at the base or slightly inflated at the apex. Chrysocystidia absent. CC an ixocutis, the hyphae 4–8 µm diameter. Pigment faint.
Season: September–December.
Habitat: Stumps and base of alder and sometimes willow, in damp places.
Distribution: Not well known. Rather rare.

P. salicicola (Fr.) Arnolds has no smell and occurs mainly on willow. *P. apicrea* (Fr.) Moser ex Singer is slightly more robust and has a strong smell of cinnamon at the stipe base (scratch before smelling).

p.136 *Pholiota lenta* (Pers.:Fr.) Singer
Strophariaceae, Cortinariales

Description: *Cap very slimy, pale ochraceous beige, with whitish fleecy scales floating in the mucilage.*
Cap 3–10 cm (1.2–4 in), hemispherical then almost flat, slimy in damp weather, shiny when dry, smooth or with innate radial brownish fibrils, and with small white, short-lived flecks of veil embedded in the slime, ochraceous beige with pinkish or yellowish tints, paler towards the edge.
Gills adnate, rather close, pale yellowish then slightly more rust. Spore print brown.
Flesh whitish or brownish at the base.
Stipe 3–10 × 0.5–1 cm (1.2–4 × 0.2–0.4 in), cylindro-clavate, white or whitish, beige to pale brownish at the base. Cortina finally disappearing, above a fibrillose-fleecy sheath.
Microscopic features Sp 5–8 × 3–4.5 µm, elliptic or almost bean-shaped, the GP indistinct. Bas 20–35 × 6–9 µm, cylindro-clavate. Cheilo 35–75 × 10–15 µm, fusiform-lageniform, with cloudy yellowish contents, sometimes slightly thick-walled or minutely coated with small crystals. Pleuro with larger and yellower contents. CC a very well-developed ixocutis, the hyphae 2–8 µm diameter. Pigment rather faint.

Season: August–November.
Habitat: Often in groups, among dead leaves and plant remains.
Distribution: Uneven. Rather common.

This species often confuses the beginner. Admittedly, it is not easy to identify the genus in the field, since *Pholiota* spp. are usually more highly coloured and often lignicolous.

---

p.150 *Hemipholiota populnea* (Pers.:Fr.) M. Bon
(= *Pholiota destruens*)
Strophariaceae, Cortinariales

Description: *Cap strongly scaly, whitish to brownish. Stipe scaly.*
Cap 5–20 cm (2–8 in), subglobose or remaining conico-convex for a long time,

the centre slightly raised to umbonate, ochraceous beige to brownish tawny or rufous in the centre, with large fibrous white or whitish scales more or less overlapping in irregular rings, glabrous from the centre, viscid in damp weather then dry. Margin strongly appendiculate. *Gills* adnate, close, light beige to brownish then chocolate brown. Spore print brownish. *Flesh* slightly or slowly bitter, white or very pale, slightly more strongly coloured at the stipe base. Smell aromatic, complex. *Stipe* 5–20 × 1–4 cm (2–8 × 0.4–1.6 in), curved and often swollen, fibrillose-scaly like the cap, concolorous. Sheath thick, ending in a rather wide fleecy, but soon torn, ring. In clusters or trooping. *Microscopic features* Sp 6.5–10 × 4.5–6 µm, elliptic, the GP narrow. Bas 20–30 × 6–10 µm, cylindro-clavate. Cheilo 20–40 × 3–10 µm, clavate or cylindric. Chrysocystidia absent. CC an (ixo)cutis with trichodermial zones, the hyphae septate, 5–12 µm diameter.

Season: September–November.
Habitat: Stumps, trunks of dead or standing poplars.
Distribution: Widespread. Rather rare to rare.

H. heteroclita (Fr.:Fr.) M. Bon, a slightly smaller species with a mild taste, is more yellowish with more fibrillose brown scales and occurs on birch and alder.

---

p.161 **Verdigris Agaric**
Stropharia caerulea Kreisel (= *S. cyanea pp*)
Strophariaceae, Cortinariales

Description: *Cap greenish blue, yellowing. Gills chamois. Stipe with a short-lived ring.*
Cap 2–7 cm (0.8–2.8 in), convex or almost flat, fibrillose-fleecy, smooth, viscid then dry and slightly shiny, rather pale or tender greenish blue, discolouring in yellow or ochraceous yellow patches. Margin appendiculate, with short-lived white fibrils or flecks. *Gills* adnate, rather close, very pale then pinkish beige to pale crimson chamois, edge concolorous. Spore

print lilaceous brown. *Flesh* pale, blue-green at the base. *Stipe* 4–10 × 0.3–1 cm (1.6–4 × 0.12–0.4 in), with white rhizoids, concolorous or a more tender blue, gradually becoming dull or dingy. Sheath fibrillose-fleecy, topped by a fleecy, rather short-lived ring-like cortina. *Microscopic features* Sp 7–9.5 × 4.5–6 µm, elliptic or slightly almond-shaped in side view, the GP absent. Bas 25–30 × 6–8 µm, cylindro-clavate. Cheilo 25–55 × 10–18 µm, ventricose, the upper part narrowing into an obtuse, more or less elongated or sometimes mucronate neck, with rounded contents, bright yellow in ammonia (chrysocystidia). Pleuro identical or sometimes longer fusoid. CC an ixocutis, with hyphae 2–12 µm diameter.

Season: July–December.
Habitat: Disturbed sites or damp humus.
Distribution: Widespread but uneven.

S. aeruginosa (Curt.:Fr.) Quélet, is darker (dark blue-green), the margin retains its white fleecy scales for longer, the ring is larger but fragile and the rhizoids less abundant; it mainly occurs in woods. The gills darken more rapidly, becoming dark blackish crimson at maturity and the gill edge has a distinct whitish frosting, contrasting clearly with the faces. *S. pseudocyanea* (Desm.:Fr.) Morgan, which is taller and more slender, a very tender greenish blue, soon fading or discolouring yellowish white, is rarer and occurs on grasses.

---

p.161 *Stropharia ochrocyanea* M. Bon
Strophariaceae, Cortinariales

Description: *Cap greenish blue, soon ochraceous yellow. Gills crimson-beige.*
Cap 1–4 cm (0.4–1.6 in), hemispherical then expanded, smooth, lubricious then slightly shiny, tender greenish blue, discolouring from the centre to yellow or rather bright ochraceous yellow.
Gills notched, quite distant, pale then pinkish beige to pale lilaceous brown. Spore

print lilaceous brown. *Flesh* pale to bluish in the cap and yellow at the base.
Stipe 1–2 × 0.3–0.6 cm (0.4–0.8 × 0.12–0.24 in), pale greenish at the base, white or yellowish elsewhere, yellowing when old. Cortina short-lived. *Microscopic features* Sp 7–9 × 4.5–6 µm, elliptic or slightly almond-shaped in side view, the GP absent. Bas 25–30 × 6–8 µm, cylindro-clavate. Cheilo 30–65 × 2–5 × 8–10 µm, elongated, the base cylindric and the apex inflated, clavate or ventricose fusoid, sometimes capitate. Pleuro (chrysocystidia) rare, 20–40 × 8–12 µm, fusoid, the apex often papillate or mucronate. CC an ixocutis, the hyphae 2–8 µm diameter.

Season: September–January.
Habitat: Grassland, especially on sandy substrates.
Distribution: Coastal or Atlantic, sometimes also continental. Rare.

---

p.143 *Stropharia coronilla* (Bull.:Fr.) Quélet
Strophariaceae, Cortinariales

Description: *Cap bright yellow-ochre. Gills crimson-grey. Ring striate.*
Cap 2–5 cm (0.8–2 in), subglobose, flattening late, smooth to minutely roughened, somewhat viscid or slightly greasy in damp weather, then dry, rather bright ochraceous yellow, sometimes more dull or pale. *Gills* notched, not very close, pale greyish beige then crimson-grey or blackish crimson. Spore print dark crimson-brown. *Flesh* white. Smell of radish. *Stipe* 2–5 × 0.4–1 cm (0.8–2 × 0.16–0.4 in), white or with a slight tinge of ochraceous sepia. Ring rather thin, white, striate above, rather fragile. *Microscopic features* Sp 7.5–9 × 4–5 µm, elliptic or slightly almond-shaped in side view, the GP absent. Bas 20–25 × 5–8 µm, cylindro-clavate. Cheilo 25–45 × 5–12 µm, fusiform-clavate the apex papillate or mucronate, with more or less rounded, refractive contents, bright yellow in ammonia (chrysocystidia). Pleuro identical. CC an (ixo)cutis, the hyphae 2–12 µm diameter.

Season: May–December.
Habitat: Lawns and grassy places.
Distribution: Uneven.

Often quite frequent in urban lawns.

---

p.149 *Hypholoma fasciculare* (Huds.:Fr.) Kummer
Strophariaceae, Cortinariales

Description: *Cap orange-yellow. Gills greenish yellow then crimson-black. Stipe yellow.*
Cap 1–6 cm (0.4–2.4 in), hemispherical then almost flat, glabrous or slightly dotted with yellowish fleece at the margin, yellow, orange-yellow or orange-apricot, soon becoming yellow-ochre then somewhat tawny rufous from the centre. *Gills* adnate, close, pale yellow then greenish, finally dark crimson-brown or blackish crimson. Spore print violaceous brown. *Flesh* yellow. Taste very bitter. *Stipe* 2–10 × 0.2–1 cm (0.8–4 × 0.08–0.4 in), often curved, bright yellow then darkening slightly from the base. In tufts. Ring zone rather narrow, becoming dark violaceous from the spores. *Microscopic features* Sp 6–8 × 3.5–4.5 µm, elliptic or the dorsal face slightly flatter, the GP quite distinct but narrow. Bas 15–20 × 3–6 µm, cylindro-clavate. Cheilo 15–40 × 4–8 µm, wavy cylindric to fusiform-clavate or clavate (the contents sometimes yellow slightly in ammonia). Pleuro 25–45 × 7–13 µm, fusoid to fusiform-clavate or clavate, the apex mucronate or papillate, the contents irregular, refractive, bright yellow in ammonia (chrysocystidia). CC a cutis, the hyphae 2–8 µm diameter. Pigment mostly intracellular.
Season: Practically all year round.
Habitat: Stumps and dead wood.
Distribution: Very widespread. Very common.

F. *pusillum* (Lange) has a cap not more than 1 cm (0.4 in) diameter and a very slender stipe. *H. capnoides* (Fr.:Fr.) Kummer is more dull, with white then greyish to violaceous brown gills and a mild taste. It occurs mainly on conifers, in continental sites.

p.149 *Hypholoma sublateritium* (Fr.) Quélet
Strophariaceae, Cortinariales

Description: *Cap red-brown in the centre, the margin with appressed fleecy veil remains.*
Cap 3–10 cm (1.2–4 in), hemispherical then broadly convex, often with yellowish or yellow then dingy fleece at the margin; dry or slightly micaceous, brick brown, fading to rufous orange, the margin sometimes becoming yellowish buff to pale yellow. *Gills* whitish then grey and darkening, finally becoming dark crimson-grey. Spore print violaceous brown. *Flesh* bitter, white or yellowish. *Stipe* 5–12 × 0.4–1.5 cm (2–4.7 × 0.16–0.6 in), often curved, whitish then yellowish to buff or brownish from the base. In clusters. Cortinal zone quite narrow and short-lived, becoming dark violaceous from the spores. *Microscopic features* Sp 5.5–8 × 3–5 μm, elliptic, the GP distinct (about 0.5–1 μm). Bas 15–20 × 5–8 μm, cylindro-clavate. Cheilo 20–35 × 5–10 μm, cylindric lageniform or fusoid ventricose. Pleuro 25–40 × 7–10 μm (chrysocystidia as in *H. fasciculare*). CC an (ixo)cutis, the hyphae 3–10 μm diameter, over a subcellular layer. Pigment mostly intracellular.

Season: August–December.
Habitat: Stumps and dead wood of broad-leaved trees.
Distribution: Widespread but uneven.

Should not be confused with robust, highly coloured forms of *H. fasciculare* (see p.799).

---

p.129 *Hypholoma elongatum* (Pers.) Ricken
Strophariaceae, Cortinariales

Description: *Cap yellowish honey. Margin striate. Gills distant. Stipe tall and slender.*
Cap 0.5–2 cm (0.2–0.8 in), convex, glabrous or slightly fibrillose at the margin, slightly viscid or lubricious, translucent striate, yellowish honey, sometimes rufous orange in the centre. *Gills* adnate, rather distant, pale yellow then lilaceous beige to

crimson-brown. Spore print rather pale violaceous brown. *Stipe* 4–12 × 0.1–0.2 cm (1.6–4.7 × 0.04–0.08 in), yellowish white then honey to buff from the base, with silky white fibrils. Cortina fine, short-lived. *Microscopic features* Sp 9–12 × 5.5–7 µm, elliptic, the GP indistinct or very narrow. Bas 25–30 × 8–10 µm, cylindro-clavate. Cheilo 15–40 × 5–9 µm, cylindro-lageniform to lageniform with a wide obtuse neck. Pleuro 25–50 × 10–15 µm, fusoid to ventricose, the apex tapering and obtuse or sometimes elongated into a poorly defined neck, with irregular refractive contents, bright yellow in ammonia (chrysocystidia). CC an (ixo)cutis, the hyphae 2–8 µm diameter, over a subcellular layer.

Season: Practically all year round.
Habitat: Tall damp moss in bogs.
Distribution: Widespread. Rather common.

The commonest of the *Hypholoma* spp. occurring in moss.

---

p.120 *Hypholoma ericaeum* (Pers.:Fr.) Kühner
Strophariaceae, Cortinariales

Description: *Cap reddish brown, fading. Gills olivaceous brown to blackish violaceous.*
Cap 1–5 cm (0.4–2 in), conico-convex then expanded, hygrophanous, slightly fibrillose fleecy at the margin when young, lubricious, shiny when dry, brown, with warm, reddish brown or orange-brown tones, fading from the margin to somewhat tawny rufous, then reddish buff to yellowish honey. *Gills* adnate, not very close, pale olivaceous beige then olivaceous brown to dark violaceous brown, edge frosted, white. Spore print violaceous brown. *Flesh* dingy yellowish to brownish or dark reddish brown at the stipe base. Smell often of iodine. *Stipe* 3–10 × 0.1–4 cm (1.2–4 × 0.04–1.6 in), wavy, with silky whitish fibrils, yellowish white at the apex, then turning brown from the base. Cortina short-lived. *Microscopic features* Sp

12–15 × 7–8 µm, elliptic, the GP quite distinct. Cheilo 30–40 × 4–6 µm, cylindric or lageniform with a wavy neck. Pleuro often rare, 25–45 × 7–10 µm, fusoid to fusoid-ventricose, the apex elongated into a more or less distinct, obtuse neck, the contents more or less yellow in ammonia (chrysocystidia). CC an ixocutis, the hyphae 2–5 µm diameter, over a subcellular layer.

Season: June–December.
Habitat: Damp places, pond edges.
Distribution: Not well known. Rather rare.

In this group, *H. ericaeoides* Orton, which has distinctly yellow gills when young and *H. udum* (Pers.:Fr.) Kühner, which has a dark red-brown cap, seem fairly easy to identify.

---

p.130 *Psilocybe squamosa* (Pers.:Fr.) Orton
Strophariaceae, Cortinariales

Description: *Cap yellow to orange, with white flecks. Stipe stiff with a thin ring.*
Cap 1.5–6 cm (0.6–2.4 in), conico-convex then broadly umbonate, with short-lived, fibrillose fleecy, white scales, lubricious or viscid in damp weather, yellow, yellow-ochre, yellow-brown, orange-brown, often paler at the margin. *Gills* adnate, rather close, grey then violaceous brown to blackish, edge frosted, white. Spore print violaceous brown. *Flesh* pale cream to yellowish or brownish in the stipe.
Stipe 6–12 × 0.2–5 cm (2.4–4.7 × 0.08–2 in), quite stiff, hollow, whitish to pale ochraceous cream at the apex, almost concolorous or more brown under the ring. Sheath fibrillose-fleecy with upturned scales, topped by a rather wide, short-lived ring. *Microscopic features* Sp 10–16 × 5.5–8.5 µm, elliptic, the GP wide. Bas 25–30 × 8–12 µm, constricted cylindric.

Cheilo 45–100 × 5–10 µm, narrow cylindric or cylindro-clavate, sometimes fusiform or more or less long lageniform. Chrysocystidia absent. CC an ixocutis, the hyphae 2–7 µm diameter, over a layer of septate elliptic hyphae.

Season: September–November.
Habitat: Undergrowth, often in troops.
Distribution: Widespread. Rare.

P. thrausta (Schulz.) M. Bon is considered to be just an orange form.

---

p.134 *Psilocybe pratensis* Orton
Strophariaceae, Cortinariales

Description: *Cap brown, with a separable gelatinized layer. Gills broadly adnate.*
Cap 0.5–2 cm (0.2–0.8 in), hemispherical then almost flat, hygrophanous, smooth, lubricious, covered with a separable gelatinous layer (when fresh), dark date-brown to rusty chocolate-brown, sometimes with crimson or more buff tones when dry. Margin finely striate. *Gills* subdecurrent, not very close, ochraceous beige then crimson-brown. Spore print violaceous brown. *Stipe* 5–20 × 1–3 mm (0.2–0.8 × 0.04–0.12 in), pale at the apex, ochraceous beige, becoming brown from the base, with faint whitish fibrils. *Microscopic features* Sp 9–12.5 × 6–7.5 × 5–7 µm, slightly lens-shaped, elliptic, the dorsal face slightly flatter, the GP fairly distinct. Bas 25–35 × 7–12 µm, cylindro-clavate. Cheilo 25–30 × 6–10 µm, lageniform, with a wavy cylindric neck. CC an ixocutis, the hyphae 2–5 µm diameter.
Season: July–December.
Habitat: Lawns and grassy places.
Distribution: Uneven; mainly coastal dunes.

The small *Psilocybe* spp. with broadly adnate or even decurrent gills are very difficult to identify.

---

p.134 *Psilocybe chionophila* Lamoure
Strophariaceae, Cortinariales

Description: *Cap shallowly convex, hygrophanous, red-brown. Gills chocolate brown.*
Cap 0.5–2 cm (0.2–0.8 in), hemispherical then almost flat, hygrophanous, glabrous,

smooth, lubricious, without a detachable layer, dark red-brown, becoming dingy buff when dry. Margin finely striate. *Gills* subdecurrent, not very close, pale ochraceous beige then dark chocolate brown. Spore print violaceous brown. *Stipe* 1–3 × 0.1–0.3 cm (0.4–1.2 × 0.04–0.12 in), brownish to dark brown below. Bands of rather dense whitish fibrils, all along the stipe. *Microscopic features* Sp 6.5–9 × 4–5 × 4.5–6 μm, slightly lens-shaped, wider on the ventral face, elliptic in side view, the GP quite distinct. Bas 20–25 × 5–9 μm, cylindro-clavate. Cheilo 15–30 × 5–8 μm, lageniform, with a more or less distinct, wavy cylindric neck. CC an ixocutis, the hyphae 2–5 μm diameter, over a lower layer with wider hyphae. Pigment mixed. Clamps occasional.

Season: June–September.

Habitat: Snow gullies and dwarf alpine shrub communities, on acid soils; often attached to moss.

Distribution: Alps and mountains in northern Europe. Rare.

Attacks moss, which it parasitizes and destroys. Can be seen from some distance as it leaves brown circles, like burn marks, on moss patches.

---

p.86 *Melanotus phillipsii* (Berk. & Br.) Singer
Strophariaceae, Cortinariales

Description: *Cap beige to somewhat tawny grey. Gills not very close. Stipe rudimentary.*
Cap 0.3–1 cm (0.12–0.4 in), subglobose then almost flat, circular then kidney-shaped, hygrophanous, finely velvety to glabrous, sometimes slightly wrinkled radially, not lubricious, somewhat tawny beige to somewhat tawny grey, often with flesh tones, paler when dry. Margin striate. *Gills* converging towards the point of attachment, quite distant, light beige to pinkish beige then slightly darker. Spore print light violaceous brown. *Stipe* 1–3 × 0.5–1 mm (0.04–0.12 × 0.02–0.04 in), rudimentary, central then eccentric,

recurved, reddish beige to brownish, darker than the cap. *Microscopic features* Sp 5–7 × 2.5–3.5 µm, long elliptic, the GP fairly distinct. Bas 15–20 × 3–6 µm, cylindro-clavate. Cheilo 15–40 × 3–7 µm, lageniform, with a wavy, more or less elongated neck. CC a trichoderm, the hyphae 4–5 µm diameter.

Season: July–November.
Habitat: Marshy and damp ground; attached to dead grasses, more rarely on Cyperaceae or Juncaceae and exceptionally on other herbaceous plants.
Distribution: Not fully assessed.

M. caricicola (Orton) Guzman can be distinguished by its larger spores but mainly (in the field) by its quite clearly gelatinized cuticle (ixotrichoderm) and specific habitat (nearly always on *Cyperaceae*). *M. horizontalis* (Berk.) Singer, a rather similar but darker species, occurs on willow bark, always in damp sites.

TRICHOLOMATACEAE

The *Tricholomataceae* is a heterogeneous family which may be tentatively accorded the following characters: fleshy or slender to very reduced species with a fibrous texture; gills decurrent or notched, never free; spore print white or pale, cream to pinkish; cap cuticle not distinctive, a cutis, sometimes a trichoderm; silhouette tricholomatoid but also collybioid, clitocyboid, omphalinoid or pleurotoid.

The scale of the family is such that it has to be divided into several subfamilies:
- *Lyophylloideae* have siderophilous basidia which fix metals, such as iron, when treated with acetocarmine and then appear punctate: *Lyophyllum* species are tricholomatoid, *Tephrocybe* are mycenoid–collybioid and *Calocybe* and *Rugosomyces* (with a cuticle transitional to the Dermolomataceae) are rather collybioid, sometimes tricholomatoid.
- *Leucopaxilloideae*, whose species have amyloid spores, is composed of three tribes: Leucopaxilleae (spore ornament

amyloid, more or less fleeting), Porpolomateae (smooth spores, partial veil absent) and Biannularieae (smooth spores, partial veil present).
- *Clitocyboideae*, whose species typically have an omphalinoid–clitocyboid silhouette, comprises several tribes.
- *Tricholomatoideae* comprises only one tribe, *Tricholomateae* (*Tricholoma*, *Tricholomopsis* and *Callistosporium*).

Species are presented in the following order: *Clitocyboideae*, *Tricholomatoideae*, *Leucopaxilloideae* and *Lyophylloideae*.

There are very few constant characters: gills thin, edge thin, entire, concolorous; flesh white; taste and smell insignificant; spores hyaline, smooth, non-amyloid, thin-walled, germ pore absent; basidia 4-spored, with clamps; pleurocystidia absent; terrestrial.

---

p.84 *Resupinatus applicatus* (Batsch:Fr.) S.F. Gray
Tricholomataceae, Tricholomatales

Description: *Tiny, pleurotoid, attached by the top of the cap. Grey.*
Cap 0.2–0.5(1) cm (0.08–0.2(0.4) in), globose then hemispherical, finally flattened, attached by the top of the cap then rather lateral, velvety to bristly near the point of attachment, grey-brown.
Gills converging towards the centre or eccentric, edge rather thick. *Microscopic features* Sp 4–6 µm diameter, globose, hyaline, smooth. Cheilo 10–25 × 1–7 µm, clavate but with one or several apical filiform outgrowths. CC a trichoderm, with hyphae 2–5 µm diameter. Pigment parietal.
Season: Practically all year round.
Habitat: Underside of fallen wood of broad-leaved trees, slightly damp places.
Distribution: Widespread but uneven.

R. striatulus (Pers.) Murrill is very similar but has a striate to sulcate margin. *R. trichotis* (Pers.) Singer, is rarer, with very stiff black hair, especially abundant near the

p.86 *Hohenbuehelia algida* (Fr.:Fr.) Singer
Tricholomataceae, Tricholomatales

Description: *Cap ochraceous beige, with a gelatinized layer. Gills pale yellow.*
Cap 1–6 cm (0.4–2.4 in), hemispherical then shell- to kidney-shaped, tomentose-pubescent near the point of attachment, finely pubescent at the edge, ochraceous beige to greyish, often rather pale.
Gills confluent near the point of attachment, quite close, whitish, soon yellowish or butter yellow. *Flesh* with a dark gelatinous layer beneath the superficial tomentum. *Microscopic features* Sp 6.5–10 × 3–4 µm, cylindric to allantoid. Cyst (Pleuro and Cheilo) metuloid, 50–100 × 10–20 µm, dextrinoid, fusoid, pedicellate, pointed, thick-walled, with a thick conical sheath of refractive crystals. Cheilo mixed with numerous MC with a clavate base and slender neck, sometimes bifid or diverticulate, the apex often narrowly capitulate, sometimes crowned with a refractive globule. CC a trichoderm, with hyphae 1–5 µm diameter, over a thick layer of finer gelatinized hyphae. Pigment parietal, in irregular lines
Clamps present.
Season: August–December.
Habitat: On broad-leaved trees; gregarious or growing singly.
Distribution: Quite widespread. Rather rare to rare.

H. atrocaerulea (Fr.:Fr.) Singer is darker and bluish, with whitish cream gills.

p.82 *Faerberia carbonaria* (A. & S.) Pouzar
Tricholomataceae, Tricholomatales

Description: *Cap grey-brown, depressed. Gills ridge-like, forked, grey.*
Cap 1–5 cm (0.4–2 in), soon depressed or funnel-shaped, radially fibrillose, sometimes slightly disrupted in the centre, grey-

brown, sometimes reddish or blackish grey. Margin thin, fibrillose and splitting when mature. *Gills* very decurrent, like folds, forked or anastomosing, narrow, whitish to pale grey. *Stipe* 1–6 × 0.2–0.8 cm (0.4–2.4 × 0.08–0.32 in), sometimes eccentric, greyish to grey or reddish brown. *Flesh* thin but rather leathery, at least in the stipe, pale. *Microscopic features* Sp 7.5–10 × 4–5.5 µm, cylindro-elliptic or more or less curved. Bas quite long. Cyst 70–200 × 10–18 µm, dextrinoid, long cylindric-fusoid and pedicellate, pointed, thick-walled, encrusted with refractive crystals near the middle. CC a cutis, the hyphae 1–3 µm diameter. Pigment parietal. Clamps present.

Season: August–October.
Habitat: Bonfire sites and burnt ground, in woods.
Distribution: Quite widespread. Rather rare to very rare.

Although the ridge-like gills are reminiscent of the chanterelles (especially the group around *Cantharellus cinereus*, see p.361), the distinctive cystidia suggest affinities with the genus *Hohenbuehelia*.

---

p.86 *Arrhenia acerosa* (Fr.:Fr.) Kühner
Tricholomataceae, Tricholomatales

Description: *Pleurotoid. Cap grey, felted, lobed.*
Cap 0.5–2(3) cm (0.2–0.8(1.2) in), spathulate or reniform, hygrophanous, wavy or lobed, thinly felted to pubescent, especially near the stipe, sometimes zoned, grey to greyish beige. Margin finely striate. *Gills* subdecurrent, convergent, rather thick, sometimes slightly forked, rather distant, grey. *Stipe* eccentric then lateral, short, 2–10 × 2–5 mm (0.08–0.4 × 0.08–0.2 in), whitish, pubescent. *Flesh* very thin, concolorous. *Microscopic features* Sp 6–8.5 × 3–5 µm, elliptic, almost tear-shaped. CC a cutis, the hyphae 3–10 µm diameter. Pigment parietal. Clamps present.

Season: July–December.
Habitat: Bare ground or plant remains; gregarious,

sometimes confluent.
Distribution: Widespread. Easily overlooked.

---

p.86 *Arrhenia lobata* (Pers.:Fr.) Kühner & Lamoure ex Redhead
Tricholomataceae, Tricholomatales

Description: *Pleurotoid. Cap lobed, brown, shiny.*
Cap 1–4 cm (0.4–1.6 in), spathulate to fan-shaped, hygrophanous, wavy to strongly lobed, shiny, moist, smooth to finely wrinkled, reddish brown to brown.
Gills replaced by wavy folds, especially towards the margin, distant and forked, subconcolorous, shiny. *Stipe* absent.
Flesh slightly gelatinous or elastic, sometimes almost translucent. *Microscopic features* Sp 6–10.5 × 4.5–8.5 µm, tear-shaped. CC a cutis, the hyphae 3–12 µm diameter. Pigment parietal. Clamps present.
Season: June–September.
Habitat: Wet mossy places in well-lit marshy areas.
Distribution: Occurs as high as mountain tops, where it is less rare than in lowland districts.

---

p.85 *Arrhenia spathulata* (Fr.:Fr.) Redhead
Tricholomataceae, Tricholomatales

Description: *Pleurotoid. Cap lobed, zoned, brownish grey.*
Cap 0.5–2.5 cm (0.2–1 in), spathulate to semicircular or fan-shaped, hygrophanous, wavy to lobed, pruinose, zoned, smooth to finely wrinkled, sometimes more felted towards the point of attachment, greyish brown to pale greyish when dry.
Gills replaced by distant, forked, sometimes very faint folds, subconcolorous or paler. *Stipe* absent or lateral, rudimentary. *Flesh* thin, fragile, concolorous. *Microscopic features* Sp 5.5–9 × 4–6 µm, elliptic to tear-shaped. CC a cutis, the hyphae 5–10 µm diameter. Pigment parietal. Clamps present.
Season: June–January.
Habitat: Moss in short turf; dunes.
Distribution: Mainly Atlantic or coastal.

p.86 *Arrhenia auriscalpium* (Fr.:Fr.) Fr.
Tricholomataceae, Tricholomatales

Description: *Cap shaped like a hood, with a lateral stipe, brown. Hymenophore with a few folds.*
Cap 0.5–1 cm (0.2–0.4 in), elliptic hood-shaped or faintly spathulate to cup-like, hygrophanous, sometimes wavy, smooth to finely wrinkled, greyish brown or reddish, pale greyish when dry. *Gills* replaced by a few distant, wavy or forked folds, sometimes not very developed, subconcolorous or paler. *Stipe* lateral, rather thin, 4–10 × 0.5–1.5 mm (0.16–0.4 × 0.02–0.06 in), concolorous. *Flesh* thin, fragile, concolorous. *Microscopic features* Sp 7–10.5 × 4.5–6 µm, elliptic to almost tear-shaped. CC a cutis, the hyphae 5–15 µm diameter. Pigment parietal. Clamps present.
Season: June–August.
Habitat: Moss, damp places in alpine areas.
Distribution: Alpine. Rare to very rare.

---

p.94 *Phaeotellus rickenii* (Singer ex Hora) M. Bon
(= *Omphalina r.*)
Tricholomataceae, Tricholomatales

Description: *Cap beige-grey, striate. Gills very distant, forked.*
Cap 0.5–2 cm (0.2–0.8 in), depressed or umbilicate, hygrophanous, sometimes lobed, smooth or slightly wrinkled, brownish grey to rather pale beige-grey, sometimes with violaceous tones, paler when dry. Margin striate. *Gills* decurrent, very distant, becoming faint towards the margin, slightly thick, forked and more or less anastomosing, concolorous.
Stipe 0.5–1.5 × 0.1–0.2 cm (0.2–0.6 × 0.04–0.08 in), concolorous or paler to whitish below. *Microscopic features* Sp 6–9.5 × 3–5 µm, cylindro-elliptic. Bas 4-spored. CC a cutis, the hyphae 2–5 µm diameter. Pigment rather faint.
Season: August–December.
Habitat: With moss in short turf, sometimes even on old walls, often in warm places.

Distribution: Mainly coastal. Rare.

Morphologically, *P. rickenii* forms a link between *Arrhenia* and *Omphalina*.

---

p.97 *Phaeotellus griseopallidus* (Desm.) Kühner & Lamoure ex Courtecuisse
Tricholomataceae, Tricholomatales

Description: *Cap dark grey-brown, hygrophanous, striate. Gills distant.*
Cap 0.5–3 cm (0.2–1.2 in), soon depressed or even umbilicate, hygrophanous, sometimes lobed, smooth or with fine radial wrinkles, sometimes faintly zoned, dark grey-brown, the margin often with a whitish edge, fading to beige-grey when dry. Margin striate. *Gills* decurrent, quite distant, sometimes forked, grey-brown. *Stipe* 0.5–2.5 × 0.1–0.2 cm (0.2–1 × 0.04–0.08 in), thinly pruinose, concolorous to whitish below. *Flesh* concolorous. Smells slightly of *Pelargonium*. Taste unpleasant. *Microscopic features* Sp 8.5–11.5 × 4.5–6.5 μm, cylindro-elliptic. Bas 2-spored. CC a cutis, the hyphae 2–4 μm diameter, sometimes with clavate tips. Pigment parietal, encrusting. Clamps present.
Season: June–December.
Habitat: Grassy or mossy places. Usually on neutral to acid soils.
Distribution: Not fully assessed. Rare.

---

p.96 *Omphalina galericolor* (Romagnesi) Bon
Tricholomataceae, Tricholomatales

Description: *Cap honey buff. Gills and stipe concolorous.*
Cap 0.5–2.5 cm (0.2–1 in), depressed to umbilicate, hygrophanous, smooth, rather pale honey buff, fading in small radial streaks to pale cream or pinkish. Margin slightly striate. *Gills* very decurrent, distant, subconcolorous or paler. *Stipe* 1–3 × 0.1–0.25 cm (0.4–1.2 × 0.04–0.09 in), glabrous, concolorous with the gills. *Microscopic features* Sp 6.5–8.5 ×

5.5–7 µm, shortly elliptic to subglobose. Cheilo scattered. CC a cutis, slightly interwoven, the hyphae 4–8 µm diameter. Pigment parietal, rather faint. Clamps present.

Season: September–December.
Habitat: On moss in short grass, on sand.
Distribution: Coastal. Rather rare to very rare.

---

p.96 *Omphalina pyxidata* (Bull.:Fr.) Quélet
Tricholomataceae, Tricholomatales

Description: *Cap tawny brown to rather bright rufous, striate. Margin often crenulate.*
Cap 1–3 cm (0.4–1.2 in), depressed or funnel-shaped, not very hygrophanous, glabrous and mat, fulvous-brown or rather bright rufous, fading faintly when dry. Margin often finely scalloped, with rather long striations. *Gills* very decurrent, distant, cream to pale beige. *Stipe* 2–6 cm × 0.15–0.3 cm (0.8–2.4 × 0.06–0.12 in), glabrous, almost concolorous or paler. *Microscopic features* Sp 7–9.5 × 3.5–4.5 µm, elliptic or almond-shaped. Cheilo rather difficult to find, cylindric or diverticulately branched to irregular. CC a trichoderm, the hyphae interwoven, 3–5 µm diameter. Pigment parietal in irregular lines. Clamps present.

Season: August–December.
Habitat: On mossy grass or herbaceous plants.
Distribution: Widespread but uneven.

Probably the commonest *Omphalina*.

---

p.96 *Omphalina hepatica* (Fr.→Gillet) Orton
Tricholomataceae, Tricholomatales

Description: *Cap dark red-brown, sometimes with olivaceous tones, hygrophanous.*
Cap 1–3 cm (0.4–1.2 in), almost flat then funnel-shaped, hygrophanous, glabrous or thinly scurfy in the centre, matt, dark red-brown, sometimes with olivaceous tones, fading in pale buff or even whitish cream spots when dry. Margin slightly crenulate,

scarcely striate. *Gills* very decurrent, distant, whitish then pale pinkish beige. *Stipe* 2–6 cm × 0.1–0.3 cm (0.8–2.4 × 0.04–0.12 in), with a white pruina on a ground almost concolorous with the cap. *Microscopic features* Sp 6–7.5 × 4–5.5 µm, elliptic. Cheilo very few. CC a trichoderm, the hyphae interwoven, 5–10 µm diameter. Pigment parietal, in irregular lines. Clamps present.

Season: July–November.
Habitat: Terrestrial, mossy or grassy, often damp, places.
Distribution: Widespread. Rather rare to rare.

Among the dark *Omphalina* spp. occurring in their usual habitat (mosses and grass), *O. obscurata* (Kühner) ex Reid is rather similar, but dark brown, with no red tones.

---

p.96 *Omphalina rivulicola* (Favre) Lamoure
Tricholomataceae, Tricholomatales

Description: *Cap striate, reddish brown.*
Cap 0.5–2 cm (0.2–0.8 in), soon umbilicate, often finely scalloped, hygrophanous, matt to felted, glabrous, red-brown, lacking olivaceous, but sometimes with rather crimson, tones, buff when dry. Margin with long striations. *Gills* very decurrent, sometimes forked, distant, rather rufous beige, gill edge pale. *Stipe* 1–3 cm × 0.15–0.3 cm (0.4–1.2 × 0.06–0.12 in), pruinose or fibrillose to felted below, glabrous, almost concolorous or paler. *Microscopic features* Sp 8–10.5 × 6–7.5 µm, elliptic, the apex obtuse. CC a cutis or a trichoderm towards the centre, the hyphae 2–7 µm diameter with a tapered apex. Pigment parietal. Clamps present.

Season: June–September.
Habitat: Very damp and wet places, especially in alpine zones; rarer in submontane areas.
Distribution: The Alps and mountains in Scandinavia. Globally rare.

The numerous Alpine species of *Omphalina* are rather difficult to identify.

p.96 *Omphalina pseudomuralis* Lamoure
Tricholomataceae, Tricholomatales

Description: *Cap silky, dark crimson-brown. Stipe slightly fibrillose.*
Cap 0.5–2 cm (0.2–0.8 in), soon umbilicate or funnel-shaped, hygrophanous, silky fibrillose, soon glabrous, dark crimson-brown then reddish brown, paler and dull when dry. Margin with rather long striations, often finely scalloped.
Gills decurrent, distant, rather thick, brownish to rather rufous beige.
Stipe 1–3 cm × 0.1–0.3 cm (0.4–1.2 × 0.04–0.12 in), silky fibrillose then glabrous, almost concolorous or paler. *Microscopic features* Sp 7–8.5 × 4.5–5.5 µm, elliptic. CC a cutis, the hyphae 2–6 µm diameter. Pigment parietal. Clamps present.
Season: June–October
Habitat: Old walls, grassy or mossy places.
Distribution: Not very well known. Rather rare.

---

p.95 *Omphalina barbularum* (Romagnesi) Bon
Tricholomataceae, Tricholomatales

Description: *Cap very hygrophanous, brown then whitish beige from the centre.*
Cap 1–3 cm (0.4–1.2 in), barely convex then slightly depressed, very hygrophanous, glabrous, very slightly gelatinous and shiny, often changing colour and becoming zoned when dry, rather dull dark brown then greyish beige or almost whitish. Margin striate, opaque when dry. *Gills* not very decurrent, quite close, whitish to pale greyish beige. Spore print white.
Stipe 1–3.5 × 0.15–0.4 cm (0.4–1.4 × 0.06–0.16 in), soon glabrous or with a pruinose ring at the apex, grey-brown (like horn), paler when dry. *Microscopic features* Sp 5.5–7.5 × 3–4.5 µm, elliptic to cylindric. Bas 4-spored. CC an (ixo)cutis, the hyphae 3–8 µm diameter. Pigment mixed. Clamps present.
Season: October–January.
Habitat: Moss in grassland on dunes.
Distribution: Uneven. Rather rare.

This species is intermediate between *Clitocybe* and *Omphalina*.

---

p.97 *Omphalina oniscus* (Fr.:Fr.) Quélet
Tricholomataceae, Tricholomatales

Description: *Cap striate, fibrillose, dark grey-brown. Gills grey.*
Cap 1–4 cm (0.4–1.6 in), almost flat then umbilicate, hygrophanous, fibrillose or slightly felted in the centre, black-brown, dark grey-brown then a dull dark brown, fading slightly when dry. Margin with long striations. *Gills* very decurrent, distant and sometimes forked, grey. *Stipe* 2–6 × 0.2–0.5 cm (0.8–2.4 × 0.08–0.2 in), glabrous and often shiny, dark grey-brown, whitish from mycelium at the base. *Microscopic features* Sp 6.5–9.5 × 4–5 µm, elliptic to almost tear-shaped. CC a cutis or a trichoderm towards the centre, the hyphae 3–8 µm diameter. Pigment parietal. Clamps present.
Season: August–October.
Habitat: Damp moss in marshy ground.
Distribution: Mainly continental or montane. Rather rare to very rare.

Other dark species also occur in the mountains.

---

p.97 *Phytoconis ericetorum* (Pers.:Fr.) Redhead & Kuyper (= *Omphalina umbellifera*)
Tricholomataceae, Tricholomatales

Description: *Cap yellow-brown or buff, striate. Gills pale yellow. Stipe slightly violaceous.*
Cap 0.5–2 cm (0.2–0.8 in), barely convex then umbilicate, hygrophanous, yellow-brown to yellow-buff or yellowish, paler when dry. Margin with long striations, often crenulate. *Gills* very decurrent, distant, pale yellow. *Stipe* 1–3 × 0.1–0.2 cm (0.4–1.2 × 0.04–0.08 in), glabrous or slightly pruinose, pale yellow or yellowish buff but often lilaceous at the apex. *Microscopic features* Sp 8–10.5 ×

6–7.5 μm, elliptic. CC a sometimes subtrichodermial cutis, the hyphae 2–6 μm diameter. Pigment mixed. Clamps absent.
Season: April–December.
Habitat: On soil or peat, sometimes on plant remains. Often in damp places.
Distribution: Mainly continental. Rather rare.

A lichenized species: sometimes, the small dark green tubercles of the algal partner can be found at the base of the stipe.

---

p.95 *Rickenella fibula* (Bull.:Fr.) Raith.
Tricholomataceae, Tricholomatales

Description: *Cap orange. Gills white. Stipe pale orange.*
Cap 3–10 mm (0.12–0.4 in), hemispherical then narrowly umbilicate, flattened when old, bright orange, especially in the centre, fading and sometimes whitish at the edge. Margin with long striations. *Gills* very decurrent, distant, white, orange at the base. *Stipe* 1–5 × 0.1–0.2 cm (0.4–2 × 0.04–0.08 in), thinly velvety, concolorous or paler below *Microscopic features* Sp 4.5–6.5 × 2–2.5 μm, cylindro-elliptic. Cyst 30–60 × 5–10 μm, fusoid lageniform, the neck sometimes slightly clavate. CC with pileocystidia comparable to the hymenial cystidia. Pigment mainly intracellular. Clamps present.
Season: July–December.
Habitat: Moss, woods, lawns, etc.
Distribution: Widespread. Very common or common.

R. mellea (Singer & Clémençon) Lamoure, which occurs on damp moss in the alpine zone, is yellowish buff to honey.

---

p.94 *Rickenella swartzii* (Fr.:Fr.) Kuyper
Tricholomataceae, Tricholomatales

Description: *Cap beige with a violaceous brown centre. Gills white.*
Cap 4–15 mm (0.16–0.6 in), barely convex, umbilicate, soon flattened, brownish beige, sometimes greyish or

whitish at the edge but violaceous brown or blackish in the centre. Margin with long striations. *Gills* very decurrent, distant, white. *Stipe* 2–7 × 0.1–0.2 cm (0.8–2.8 × 0.04–0.08 in), finely velvety, pale beige, the apex violet to dark violaceous brown. *Microscopic features* Sp 4–5 × 2–3 μm, cylindro-elliptic. Cyst 40–60 × 5–12 μm, fusiform or tapering, the neck sometimes slightly subcapitate. CC a cutis, with numerous pileocystidia. Pigment mainly intracellular. Clamps present.

Season: July–December.
Habitat: Moss. Often in sunny places.
Distribution: Widespread but uneven.

---

p.98 *Gerronema chrysophyllum* (Fr.:Fr.) Singer
Tricholomataceae, Tricholomatales

Description: *Cap yellow-brown. Gills bright golden yellow.*
Cap 2–6 cm (0.8–2.4 in), convex, depressed to umbilicate, smooth to finely scaly in the centre, with slight radial streaking, rather warm yellow-brown, darker in the centre. *Gills* very decurrent, not very close, golden yellow to orange-buff. Spore print yellowish. *Stipe* 1–4 × 0.2–0.5 cm (0.4–1.6 × 0.08–0.2 in), hollow when mature, glabrous, almost concolorous or slightly paler. *Microscopic features* Sp 9.5–12 × 5–6 μm, elliptic or faintly almond-shaped. CC a cutis, the hyphae 1–5 μm diameter. Pigment mainly intracellular. Clamps absent.

Season: August–October.
Habitat: Rotten wood (broad-leaved and coniferous).
Distribution: Continental to submontane. Rare.

---

p.102 *Cantharellula umbonata* (Gmel.:Fr.) Singer
Tricholomataceae, Tricholomatales

Description: *Cap umbonate, brown. Gills white.*
Cap 1–3 cm (0.4–1.2 in), depressed or umbilicate, with a central umbo, fibrillose, sometimes finely wrinkled or minutely scaly in the centre, more or less dark brown and often with reddish or violaceous tones,

sometimes also more olivaceous. Margin finely striate. *Gills* decurrent, quite close, white then spotting reddish brown. *Stipe* 3–8 × 0.2–0.4 cm (1.2–3.2 × 0.08–0.16 in), with whitish fibrils then glabrous, on a brownish to almost concolorous ground. *Flesh* white but sometimes reddening faintly when cut. *Microscopic features* Sp 8.5–11 × 2.5–4 µm, elliptic to cylindric-fusoid, amyloid. CC a cutis, the hyphae 2–6 µm diameter. Pigment encrusting, parietal. Clamps present.

Season: September–December.
Habitat: Damp places, moss on acid soils.
Distribution: Continental to submontane. Rather rare.

---

p.99 *Clitocybe clavipes* (Pers.:Fr.) Kummer
Tricholomataceae, Tricholomatales

Description: *Cap compressible in the centre. Gills pale yellowish. Stipe clavate.*
Cap 3–8 cm (1.2–3.2 in), flat to depressed, sometimes umbonate, soft and compressible in the centre, matt, sometimes slightly greasy, yellowish beige but sometimes more greyish or darker, even a rather dark brown. *Gills* strongly decurrent, pale yellowish to butter yellow. *Stipe* 4–8 × 0.5–2 cm (1.6–3.2 × 0.2–0.8 in), compressible and soft, clavate, greyish beige, often more dull than the cap and contrasting with the gills. *Flesh* pale and soft. Smell slightly sharp. *Microscopic features* Sp 7.5–10 × 5–6 µm, elliptic tear-shaped or slightly fusoid. Bas long, 35–50 × 6–8 µm. CC (sub)trichodermial, the hyphae 3–7 µm diameter. Pigment mainly intracellular.

Season: September–November.
Habitat: Broad-leaved woodland, sometimes mixed conifer woods.
Distribution: Widespread but uneven.

A long basidia is a primitive character, common to the earliest *Clitocybe* spp. It represents an echo of the *Hygrophoraceae* from which the *Clitocybe* spp. probably evolved.

p.92 *Clitocybe geotropa* (Bull.:Fr.) Quélet
Tricholomataceae, Tricholomatales

Description: *Cap umbonate in a depression, beige. Stipe subclavate.*
Cap 5–20 cm (2–8 in), conico-convex then expanded and funnel-shaped, almost always with a central umbo, felted to almost smooth, matt, ochraceous beige, sometimes with pinkish or paler, almost whitish tones. *Gills* strongly decurrent, close, white or pale cream. *Stipe* 6–18 × 1–3 cm (2.4–7 × 0.4–1.2 in), firm, often slightly clavate, faintly felted when young, concolorous. *Flesh* thick, pale. Smell pleasant. *Microscopic features* Sp 5–7.5 × 4.5–6 µm, short, tear-shaped or subglobose. Bas long, 35–45 × 6–8.5 µm. CC a cutis, the hyphae 5–7 µm diameter. Pigment mainly intracellular.
Season: August–November.
Habitat: Undergrowth, wood edges, coppice woodland. Often on calcareous soils.
Distribution: Uneven.

A good edible species but care should be taken with *Clitocybe* spp., some of which (especially the small ones) are highly poisonous.

p.98 *Clitocybe gibba* (Pers.:Fr.) Kummer
Tricholomataceae, Tricholomatales

Description: *Cap funnel-shaped, often with a pimple at the bottom, pinkish buff to pale beige.*
Cap 2–7 cm (0.8–2.8 in), button- then funnel-shaped, often with a small conical umbo, glabrous or faintly velvety, smooth, matt, yellowish buff, pinkish buff or ochraceous beige, sometimes marbled with watery zones that are darker in damp weather or very pale when dry. Margin thin and often wavy. *Gills* strongly decurrent, close. *Stipe* 2–5 × 0.3–1 cm (0.8–2 × 0.12–0.4 in), soon becoming soft, slightly velvety below, concolorous or

paler. *Flesh* thin. Smell pleasant.
Microscopic features Sp 6–8 × 3.5–5.5 μm, tear-shaped to long ovoid. CC a subtrichodermial cutis, the hyphae 3–10 μm diameter. Pigment mainly parietal in irregular lines.

Season: July–December.
Habitat: Undergrowth, various.
Distribution: Rather common everywhere.

Not poisonous (but confusion is possible with related species containing muscarine). *C. costata* Kühner & Romagnesi is similar but darker and has a distinctly ribbed margin; *C. squammulosa* (Pers.:Fr.) Kummer is pinkish brown and has a finely scaly cuticle.

---

p.102 *Clitocybe lateritia* Favre
Tricholomataceae, Tricholomatales

Description: *Cap depressed, dark red-brown. Gills cream to pinkish.*
Cap 1.5–5 cm (0.6–2 in), hemispherical then flat to convex or depressed, frosted, soon felted to scurfy, matt, dark red-brown (sometimes with an olivaceous tone) to brownish brick, often with darker patches or spots towards the margin which is inrolled, often wavy and slightly sulcate, sometimes frosted. *Gills* decurrent, not very close to distant. *Stipe* 2–4 × 0.2–0.5 cm (0.8–1.6 × 0.08–0.2 in), compressible, slightly clavate, streaked with fibrils, almost concolorous.
Flesh thin, concolorous near the surface.
Microscopic features Sp 6.5–9.5 × 4–5.5 μm, elliptic or ovoid. CC a trichoderm, the hyphae 5–10 μm diameter. Pigment parietal in irregular lines.

Season: July–September.
Habitat: With dwarf willow among dwarf shrubs or with *Dryas*, in alpine zones.
Distribution: The Alps and Scandinavian mountains.

C. sinopica (Fr.:Fr.) Kummer, a bright orange-brown species with a strong mealy smell, usually occurs in spring.

p.100 **Clouded Agaric**
Clitocybe nebularis (Batsch:Fr.) Kummer
Tricholomataceae, Tricholomatales

Description: *Cap greyish beige, slightly fibrillose. Gills subdecurrent. Stipe robust.*
Cap 6–15 cm (2.4–6 in), barely convex then plano-convex to faintly depressed, often with slight radial lines, matt, often with patches of tomentum on old specimens, greyish beige or rather grey, quite pale especially at the sometimes slightly sulcate margin. *Gills* slightly decurrent, close, white to pale cream. *Stipe* 5–12 × 1–5 cm (2–4.7 × 0.4–2 in), slightly clavate, pruinose or fibrillose, whitish then light beige-grey. *Flesh* thick. Smells rather strongly of polypores or slightly aromatic, sometimes very strong and then unpleasant. *Microscopic features* Sp 6–8 × 3–4 µm, elliptic or slightly tear-shaped. CC a cutis, the hyphae 2–5 µm diameter. Pigment parietal, in irregular lines.
Season: September–December.
Habitat: Broad-leaved and coniferous woods, quite often in groups or even in rings.
Distribution: Widespread but uneven.

Considered a good edible species, sometimes causing ill effects, with a strong, possibly repellent flavour. Often confused with the poisonous *Entoloma lividum* (see p.535). They can be clearly separated by cap and gill colour, spore print, smell and microscopic details.

---

p.92 *Clitocybe inornata* (Sow.:Fr.) Gillet
Tricholomataceae, Tricholomatales

Description: *Cap ochraceous beige. Gills not very decurrent. Smell of fish.*
Cap 5–10 cm (2–4 in), weakly convex then faintly depressed, obscurely umbonate, finely pruinose or frosted, velvety to glabrous, ochraceous beige, sometimes brownish or greyish. Margin slightly sulcate. *Gills* adnate, wavy, quite close,

greyish then dingy brownish grey.
Stipe 4–8 × 1–1.5 cm (1.6–3.2 × 0.4–0.6 in), sometimes tapering, whitish to almost concolorous. Smell unpleasant, of fish when mature. *Microscopic features* Sp 7–11 × 3.5–4.5 µm, elliptic to fusiform. CC a cutis, the hyphae 2–7 µm diameter. Pigment mixed.

Season: August–November.
Habitat: Conifers, sometimes mixed with broad-leaved trees. Often on calcareous soils.
Distribution: Uneven. Rather rare to very rare.

---

p.99 *Clitocybe odora* (Bull.:Fr.) Kummer
Tricholomataceae, Tricholomatales

Description: *Cap green, fading. Gills not very decurrent. Smells very strongly of aniseed.*
Cap 2–7 cm (0.8–2.8 in), weakly convex then plano-convex or depressed, pruinose or frosted, glabrous, smooth or with fine radial wrinkles, green, the intensity varying, pale green, dark green, blue-green, etc., often fading. *Gills* slightly decurrent, close, white or pale, sometimes greenish.
Stipe 3–6 × 0.4–0.8 cm (1.2–2.4 × 0.16–0.32 in), pruinose to fibrillose, almost concolorous. *Flesh* medium. Smell sweet, of aniseed. *Microscopic features* Sp 5.5–8 × 3.5–5 µm, elliptic. CC a cutis, the hyphae 1.5–5 µm diameter. Pigment mainly intracellular.

Season: August–December.
Habitat: Broad-leaved and coniferous woods. Often on calcareous soils.
Distribution: Widespread but uneven.

Discoloured or pale forms may be confused with *C. phyllophila* var. *ornamentalis* (Vel.) Raith. (*see* p.823), which has a glabrous frosted cuticle, becoming marbled or all one colour, rather ochraceous beige or pinkish, and the same smell, but slightly less strong. *C. fragrans* (With.:Fr.) Kummer also has an aniseed smell (cap distinctly hygrophanous, brownish, paler at the edge, margin very striate) and occurs mainly under broad-leaved trees. In the same habitat, *C. obsoleta*

(Batsch:Fr.) Quélet is more pinkish brown, pale to whitish when dry and the margin not striate. *C. acicola* Singer (cap dark pinkish brown, scarcely fading, margin not striate) occurs under conifers.

---

p.89 *Clitocybe dealbata* (Sow.:Fr.) Kummer
(poisonous)
Tricholomataceae, Tricholomatales

Description: *Cap frosted then mottled pinkish buff. Gills not very decurrent.*
Cap 1–4 cm (0.4–1.6 in), plano-convex or flat, strongly frosted white or pale greyish, glabrous (when rubbed or handled), often with more coloured patches or zones, white then marbled pinkish beige to dingy buff. *Gills* adnate, quite close, whitish then dingy beige. *Stipe* 2–3 × 0.2–0.5 cm (0.8 –1.2 × 0.08–0.2 in), rather fragile, pruinose, concolorous. Smell slightly mealy or spermatic. *Microscopic features* Sp 4–5 × 3–3.5 μm, elliptic. CC a cutis, the hyphae branched or contorted and strongly interwoven (frosting) 1.5–3 μm diameter.
Season: June–December.
Habitat: Lawns and grassy places, sometimes in undergrowth.
Distribution: Widespread but uneven.

Representative of a large group of small white, very poisonous *Clitocybe* spp. (muscarine syndrome, *see* p.17) mainly characterized by colour and the not very decurrent gills. The frosting on *C. dealbata* is patchy and fleeting, while *C. candicans* (Pers.:Fr.) Kummer, a closely related species, has a frosty, pure white and unchanging cap.

---

p.89 *Clitocybe phyllophila* (Pers.:Fr.) Kummer
(poisonous)
Tricholomataceae, Tricholomatales

Description: *Cap white or mottled when old. Gills not very decurrent.*
Cap 3–10 cm (1.2–4 in), barely convex or slightly depressed, with white or pale

greyish frosting, glabrous (when rubbed or handled), somewhat mottled, white to pale ochraceous beige. Margin sometimes slightly sulcate or lobed. *Gills* not very decurrent or adnate, quite close, white or very pale. *Stipe* 4–10 × 0.5–1.5 cm (1.6–4 × 0.2–0.6 in), often with abundant cottony mycelium at the base, white. *Flesh* medium. Smell mealy to spermatic or aromatic. *Microscopic features* Sp 4–5.5 × 3–4 µm, elliptic. CC irregular, the hyphae more or less erect and diverticulate, interwoven, 4–15 µm diameter.

Season: August–December.
Habitat: Undergrowth in broad-leaved woodland.
Distribution: Not fully assessed.

C. cerussata (Fr.:Fr.) Kummer is sometimes slightly more umbonate and with a faintly pinkish spore print occurs on acid soils. All these species are highly poisonous!

---

p.95 *Clitocybe graminicola* M. Bon
Tricholomataceae, Tricholomatales

Description: *Cap pale creamy buff. Gills subdecurrent.*
Cap 1–4 cm (0.4–1.6 in), barely convex to more or less depressed, slightly hygrophanous, not striate, glabrous, with a greasy sheen, whitish or pale yellowish buff, paler when dry. *Gills* subdecurrent, not very close, whitish. *Stipe* 2–4 × 0.2–0.5 cm (0.8–1.6 × 0.08–0.2 in), fibrillose-pruinose, whitish or cream to slightly dingy buff below. *Flesh* thin, pale. Smell of coumarin (new-mown hay) or faintly earthy. *Microscopic features* Sp 4–6 × 2.5–4 µm, elliptic to almost tear-shaped. CC a cutis, sometimes slightly gelatinized, the hyphae 2–5 µm diameter.

Season: September–December.
Habitat: Lawns and grassy places.
Distribution: Uneven. Mainly Atlantic.

The more or less hygrophanous species of *Clitocybe* are often poisonous to a greater or lesser degree and should be strictly avoided.

p.99 *Clitocybe phaeophthalma* (Pers.) Kuyper
(= *C. hydrogramma*)
Tricholomataceae, Tricholomatales

Description: *Cap depressed, ochraceous beige, hygrophanous. Smell of hen houses.*
Cap 1.5–6 cm (0.6–2.4 in), flat or depressed, hygrophanous, striate, glabrous, slightly moist then dry, ochraceous beige to pale brownish, whitish when dry.
Gills decurrent, rather close, whitish.
Stipe 2.5–5 × 0.2–0.6 cm (1–2 × 0.08–0.24 in), fibrillose-pruinose, sometimes tufted with mycelium at the base, whitish or dingy cream below. *Flesh* pale, hygrophanous. Smell rather strong like a hen house or earthy, sometimes also slightly honeyish. *Microscopic features* Sp 5–7 × 2.5–4.5 µm, elliptic almost tear-shaped. CC an (ixo)cutis, the hyphae 2–5 µm diameter with inflated elements with refractive contents scattered among them (elliptic or bladder-like). Pigment intracellular.
Season: August–October.
Habitat: Broad-leaved woods; rarer under conifers.
Distribution: Widespread but uneven.

---

p.97 *Clitocybe ditopa* (Fr.:Fr.) Gillet
Tricholomataceae, Tricholomatales

Description: *Cap grey, very pruinose, not striate, hygrophanous. Gills grey.*
Cap 2–6 cm (0.8–2.4 in), weakly convex then depressed, hygrophanous, not striate, strongly pruinose, sometimes cracking or more or less zoned as the pruina rubs off, rather dark grey, paler to whitish when dry.
Gills not very decurrent, quite close, grey.
Stipe 3–6 × 0.3–0.8 cm (1.2–2.4 × 0.12–0.32 in), pruinose or faintly striate, grey, fading. *Flesh* greyish. Smell of cucumber. *Microscopic features* Sp 3–5 × 2.5–4 µm, elliptic to subglobose. CC a cutis, the hyphae 1–5 µm diameter, interwoven near the surface (pruina).
Season: August–November.
Habitat: Spruce or other conifers, sometimes broad-leaved trees (especially alder), in damp places.

Distribution: Widespread but uneven.

C. vibecina (Fr.:Fr.) Quélet is similar but paler and has a finely striate margin. There are numerous other similar species.

---

p.99 *Clitocybe umbilicata* (Sch.) Kummer
Tricholomataceae, Tricholomatales

Description: *Cap umbilicate, tan. White ring at the apex.*
Cap 4–7 cm (1.6–2.8 in), rather narrowly umbilicate, hygrophanous, not striate, glabrous and smooth, rather warm or dark tan, sometimes more greyish or rather olivaceous, pale greyish beige when dry. *Gills* decurrent, quite close, pale beige-grey. *Stipe* 4–8 × 0.4–1 cm (1.6–3.2 × 0.16–0.4 in), silky fibrillose but with a white ring at the apex, almost concolorous or more greyish towards the base.
Microscopic features Sp 5.5–7.5 × 3–4.5 µm, elliptic. CC a cutis, the hyphae 2–5 µm diameter. Pigment mainly intracellular.
Season: September–November.
Habitat: Conifers. Usually in fairly warm places.
Distribution: Mainly continental.

C. subspadicea (Lange) M. Bon & Chevassut, which also has a white pruinose ring at the stipe apex, is very similar but occurs under broad-leaved trees.

---

p.98 *Clitocybe decembris* Singer
Tricholomataceae, Tricholomatales

Description: *Cap depressed, very pale when dry apart from a brown 'eye'. Gills greyish.*
Cap 2–6 cm (0.8–2.4 in), barely convex then depressed, hygrophanous, not very striate, glabrous and smooth, brownish to ochraceous beige, cream or whitish when dry, apart from a persistent brown spot in the centre. *Gills* decurrent, close, rather pale greyish cream. *Stipe* 3–6 × 0.2–0.7 cm (1.2–2.4 × 0.08–0.28 in), slightly silky fibrillose towards the base, greying then browning or dingy from the base.

Flesh thin. Smell faintly earthy.
Microscopic features Sp 5.5–7.5 × 3–4.5 µm, elliptic. CC a cutis, the hyphae 2–5 µm diameter.
Season: September–December.
Habitat: Various broad-leaved trees.
Distribution: Usually in lowland districts and in the Atlantic zone.

Probably the commonest of the hygrophanous *Clitocybe* spp. in lowland broad-leaved woodland.

---

p.103 *Armillaria tabescens* (Scop.) Emeland
Tricholomataceae, Tricholomatales

Description: *In clusters. Cap finely scaly, brown. Stipe bare.*
Cap 2–8 cm (0.8–3.2 in), convex then depressed, finely fibrillose-scaly towards the centre, brownish with strong yellowish, buff or pinkish tones, more dull and dark when old. Margin striate. *Gills* not very decurrent, rather distant, pale then pinkish buff or with rust spots. *Stipe* 5–12 × 0.5–1 cm (2–4.7 × 0.2–0.4 in), often tapering, fibrillose-fleecy or subglabrous, without a ring, concolorous or paler, sometimes greying to yellowish grey below. In clusters.
Microscopic features Sp 8.5–11 × 5–6.5 µm, elliptic, slightly thick-walled. Cheilo scattered, 35–55 × 5–12 µm, clavate. CC a trichoderm, the hyphae septate, with elements 30–70 × 10–18 µm. Pigment parietal, smooth. Clamps absent.
Season: September–November.
Habitat: Stumps or roots of various broad-leaved trees, especially oak.
Distribution: Uneven. Quite rare or rare.

---

p.104 **Honey Fungus**
Armillaria mellea (Vahl.:Fr.) Kummer
Tricholomataceae, Tricholomatales

Description: *Cap olive-brown to yellow, speckled. Stipe with a ring.*

Cap 2–15 cm (0.8–6 in), convex then depressed, finely scaly in the centre, glabrous, yellow-buff with olive tones, sometimes olive-brown or occasionally quite bright yellow. Margin striate when mature. *Gills* decurrent, quite close, white or cream to pale beige then spotting rusty. *Stipe* 8–20(25) × 0.5–2 cm (3.2–8(10) × 0.2–0.8 in), cylindric or clavate, quite stiff, fibrous, glabrous, soon almost smooth, whitish then ochraceous yellow to rusty honey from the base. Ring well-formed, membranous, white or yellow. Often in dense clusters.
Flesh sometimes flushing pink or rufous. *Microscopic features* Sp 7–9 × 5.5–6.5 µm, elliptic, slightly thick-walled, sometimes irregular. Bas without clamps. Cheilo scattered, 25–35 × 5–10 µm, more or less clavate, utriform or extended into a narrow wavy neck. CC a trichoderm, the hyphae septate, with elements 30–60 × 13–25 µm. Pigment mixed.

Season: August–December.
Habitat: Trunks and stumps of various broad-leaved trees.
Distribution: Widespread. Common.

Armillaria spp. with a ring, which were for a long time grouped together under the collective name of *Armillaria mellea*, have been divided into five species on the basis of their morphology and tests carried out on laboratory cultures. They are formidable parasites, with black anastomosing rhizomorphs as tough as leather, which force their way between bark and trunk. These can be seen on the rotting wood, once the tree has fallen. Considered to be a good edible species, although rather mediocre, honey fungus should be picked young and eaten as soon as possible, as some ill effects or more or less serious cases of poisoning may occur, possibly caused by the later action of micro-organisms. It would appear that *A. ostoyae* (*see* comments on *A. gallica* below) is the species most usually associated with digestive upsets.

p.103 *Armillaria gallica* Marxmüller & Romagnesi
Tricholomataceae, Tricholomatales

Description: *Cap with fine yellow-brown scales. Ring with yellow fleece.*
Cap 5–12 cm (2–4.7 in), long remaining convex, brownish to buff brown, with small yellow-brown to dingy yellowish grey scales. Margin slightly floccose. *Gills* subdecurrent, quite close, white or brownish cream, often spotting brownish when old. *Stipe* 5–12 × 0.5–2.5(3) cm (2–4.7 × 0.2–1(1.2) in), clavate, fibrillose-striate with a few irregular fleecy bands, pale at the apex, darker, yellow-brown to dark grey-brown at the base. Ring soon fleecy or torn, fragile, yellow to yellowish grey. *Flesh* thin, brownish below.
Microscopic features Sp 6.5–10.5 × 5–6.5 µm, elliptic or cylindro-elliptic. Bas with clamps. Cheilo 20–40 × 5–12 µm, sometimes septate or branched. CC a trichoderm, the hyphae septate, with elements 20–120 × 12–30 µm, constricted at the septa. Pigment parietal.
Season: August–December.
Habitat: Stumps and buried wood of broad-leaved trees; often solitary or in loose groups.
Distribution: Widespread. Rather common.

A. ostoyae (Romagnesi) Herink (= *A. obscura*) differs in its dark brown cap with small erect brownish but rather fleeting scales, fleecy margin, cylindric, fleecy to fleecy banded stipe and rather persistent membranous whitish ring with dingy brown fleece on the underside (found on broad-leaved and coniferous trees). *See also* comment under *A. mellea* (p.827).

p.145 **Amethyst Deceiver**
Laccaria amethystina (Huds.) Cooke
Hydnangiaceae, Tricholomatales.

Description: *Wholly dark violet, fading when dry. Gills thick and distant.*
Cap 1–7 cm (0.4–2.8 in), soon expanded, weakly convex, hygrophanous, often

finely scalloped or scalloped, wavy, glabrous, soon scurfy, even almost minutely scaly, deep or bright violet, fading to lilaceous cream. *Gills* adnate, thick and distant, concolorous then dusted with white (spores). *Stipe* 3–10 × 0.2–0.8 cm (1.2–4 × 0.08–0.32 in), quite tough then fragile and hollow, often wavy, fibrillose to striate, sometimes twisted, concolorous or paler. *Flesh* quite thin, concolorous. *Microscopic features* Sp 7.5–10 µm, subglobose to globose, with sharp spines about 0.5–1 µm tall. CC a trichoderm, the hyphae 4–10 µm diameter. Clamps present.

Season: June–December.
Habitat: Various, mainly in woodland.
Distribution: Very widespread. Common.

An excellent edible species despite its small size and somewhat tough stipe (remove before cooking). However, one should be aware that this species is one of the most efficient accumulators of radioactive elements (should be eaten in moderation). It can be confused with two poisonous species: *Mycena pura* (*see* p.643, gills white, thin, and a smell of radish) and *Inocybe geophylla* var. *lilacina* (*see* p.508, cap conical, gills brown when mature, thin, and smell spermatic).

p.133 **Deceiver**
Laccaria laccata (Scop.:Fr.) Cooke
Hydnangiaceae, Tricholomatales

Description: *Cap pinkish beige to pinkish brown. Gills thick and distant, pinkish beige.*
Cap 1–4 cm (0.4–1.6 in), hemispherical then almost flat, hygrophanous, finely scalloped, wavy, striate, often smooth, rather pinkish brown to pinkish buff then pinkish cream. *Gills* broadly adnate, thick and distant, concolorous or paler, pinkish beige. Spore print white. *Stipe* 2–8 × 0.2–0.5 cm (0.8–3.2 × 0.08–0.2 in), quite tough, fibrillose to striate, sometimes wavy or twisted, concolorous or paler, sometimes

more rufous to orange. *Microscopic features* Sp 7.5–10.5 × 6–7.5 μm, broadly elliptic, with spines about 0.5–1 μm high, sharp. CC an almost hymenidermial trichoderm, the hyphae clavate 5–12 μm diameter. Clamps present.

Season: June–December.
Habitat: Damp or cool places in undergrowth. Often on acid soils.
Distribution: Widespread. Common.

The genus includes a number of somewhat pinkish brown species that can only be separated by their spore characters (round or elliptic spores and height of the spines on the spores). Like the preceding species, *L. laccata* (and its allies) is edible but care must be taken to avoid the confusion which is always possible (and sometimes dangerous) with small more or less brownish species.

---

p.133 *Laccaria tortilis* (Bolt.) Cooke
Hydnangiaceae, Tricholomatales

Description: *Very small species of muddy places. Cap like a parachute, brownish pink.*
Cap 0.3–1.5 cm (0.12–0.6 in), subglobose then shaped like a parachute, flat when old or even depressed in the centre, very finely scalloped or lobed, often wavy, with long striations, smooth, a pretty brownish pink or pinkish beige, fading to whitish pink from the margin (the striations retaining their colour). *Gills* broadly adnate, thick and very distant, often crisped, a tender pink. Spore print white. *Stipe* 4–10 × 1 mm (0.16–0.8 × 0.04 in), often very slender and short, wavy, concolorous or more rufous. *Microscopic features* Sp 12–16 μm, globose, very large and strongly spiny (spines about 1.5–2.5 μm high, sharp). Bas 2-spored. CC a cutis or a trichoderm, the hyphae 3–10 μm diameter. Clamps present.

Season: August–November.
Habitat: Muddy or very damp places.
Distribution: Uneven.

p.93 *Ripartites tricholoma* (Alb. & Schw.:Fr.) P. Karsten
Tricholomataceae, Tricholomatales

Description: *Cap margin bearded bristly. Gills subdecurrent, dingy buff.*
Cap 2–5 cm (0.8–2 in), weakly convex then depressed, sometimes umbonate, lubricious in the centre in damp weather, white to ochraceous cream where rubbed. Margin bearded bristly then fibrillose. *Gills* more or less decurrent, quite close, dingy beige to greyish buff. Spore print brownish. *Stipe* 2–6 × 0.2–0.7 cm (0.8–2.4 × 0.08–0.28 in), whitish then somewhat rufous buff to almost concolorous. *Flesh* pale. Smell weak or slightly mealy. *Microscopic features* Sp 4.5–6 × 3.5–5.5 µm, elliptic to subglobose, finely spiny (spines about 0.5–1 µm high, blunt). CC a cutis or a trichoderm towards the margin, the hyphae 4–6 µm diameter. Clamps present.
Season: August–November.
Habitat: Conifers, sometimes broad-leaved trees, rather varied.
Distribution: Widespread but uneven.

R. strigiceps (Fr.:Fr.) P. Karsten has hairs arranged in concentric zones, especially at the edge and the stipe is strigose at the base (under conifers). Other species are glabrous: *R. metrodii* Huijsman (cap white or very pale, glabrous and gills pinkish beige) found especially under broad-leaved trees and *R. helomorphus* (Fr.) P. Karsten, which is distinctly umbonate and usually found under conifers. Other species are rarer.

---

p.99 **Tawny Funnel Cap**
Lepista inversa (Scop.) Patouillard
Tricholomataceae, Tricholomatales

Description: *Cap brownish tan or rather tawny, smooth. Gills pale.*
Cap 4–10 cm (1.6–4 in), soon funnel-shaped, slightly hygrophanous, tan, somewhat rufous tawny to somewhat

rufous beige, paler when dry, quite uniform. Margin not striate, sometimes finally sulcate. *Gills* very decurrent, close, rather pale rufous beige to almost concolorous. *Stipe* 2–5 × 0.4–0.8 cm (0.8–2 × 0.16–0.32 in), almost concolorous or paler, often tomentose below and attached to needles and leaf litter. *Flesh* slightly elastic. *Microscopic features* Sp 3.5–5 × 3–4.5 µm, elliptic to subglobose, finely spiny or finely warted. CC a cutis, the hyphae 2–6 µm diameter. Pigment mainly intracellular.

Season: August–December.
Habitat: Conifers, sometimes in mixed woods.
Distribution: Very widespread but uneven.

L. flaccida (Sow.:Fr.) Patouillard is more elastic or soft and occurs under broad-leaved trees (the two species are perhaps synonymous). *L. gilva* (Pers.:Fr.) Roze has a creamy buff cap, with concentric brownish spots towards the edge and a slightly more ribbed margin.

---

p.93 *Lepista martiorum* (Favre) M. Bon
Tricholomataceae, Tricholomatales

Description: *Cap frosted in places, brownish buff. Gills buff.*
Cap 3–8 cm (1.2–3.2 in), soon expanded, flat or slightly depressed, initially frosted, glabrous in irregular patches, mottled, dingy whitish or pinkish beige-grey then with brownish or somewhat rufous patches where rubbed. Margin wavy, not striate. *Gills* not very decurrent, close, separable from the flesh, rather dingy rufous beige to almost concolorous. Spore print dirty pinkish. *Stipe* 4–7 × 1–2 cm (1.6–2.8 × 0.4–0.8 in), soon hollow, sometimes lumpy, felted then glabrous, concolorous or darker at the base. *Flesh* pale. Smell mealy or spermatic. Taste slightly or slowly bitter. *Microscopic features* Sp 4.5–5.5 × 2.5–3.5 µm, elliptic to almost tear-shaped, smooth under the optical microscope. CC subtrichodermial, the hyphae slightly diverticulate.

Season: August–October.
Habitat: Conifers, sometimes mixed with broad-leaved trees.
Distribution: Not known.

Rather strong resemblance to the darker *Clitocybe* spp. in the *C. phyllophila* group (*see* p.823).

---

p.165 *Lepista panaeola* (Fr.) P. Karsten (= *L. luscina*)
Tricholomataceae, Tricholomatales

Description: *Cap brownish beige, with dark concentric spots near the margin.*
Cap 4–12 cm (1.6–4.7 in), hemispherical then expanded or slightly depressed, wavy, not striate, pruinose or fibrillose, greyish beige to brownish, darker in the centre, with small darker concentric pits towards the margin. *Gills* adnate, close, separable from the flesh, pale beige to pinkish beige. Spore print brownish pink. *Stipe* 4–7 × 1–2 cm (1.6–2.8 × 0.4–0.8 in), pruinose to felted, concolorous or darker at the base. *Flesh* pale. Smell fungal, sometimes very (too) strong when old, like certain vitamin complexes. *Microscopic features* Sp 4.5–6.5 × 3–4 µm, elliptic, finely spiny. CC subtrichodermial, the hyphae interwoven. Pigment parietal. Clamps present.
Season: August–November.
Habitat: Meadows, grassy coppice woodland; quite often in rings.
Distribution: Uneven.

Often said to be a good edible species but the taste is sometimes too strong. Care must be taken, as with all grassland species, not to collect the small white *Clitocybe* spp. whose rings may intersect with those of edible taxa.

---

p.163 *Lepista glaucocana* (Bres.) Singer
Tricholomataceae, Tricholomatales

Description: *Cap whitish with bluish or pale lilaceous tones.*
Cap 5–12 cm (2–4.7 in), convex then flat or slightly depressed, pruinose, almost whitish

but with bluish or lilaceous tones.
Gills notched, quite close, separable from the flesh, pinkish cream. Spore print pinkish beige. *Stipe* 5–8 × 1–2 cm (2–3.2 × 0.4–0.8 in), very pruinose, concolorous. *Flesh* pale. Smell aromatic, with a herbaceous or floral component, weak in very old specimens. *Microscopic features* Sp 5.5–8 × 3–4.5 µm, elliptic, very finely spiny to almost smooth. CC subtrichodermial, the hyphae 2–5 µm diameter. Clamps present.

Season: September–November.
Habitat: Mixed woodland, wood edges, waysides.
Distribution: Widespread. Rather rare generally.

---

p.174 **Wood Blewit**
Lepista nuda (Bull.:Fr.) Cooke
Tricholomataceae, Tricholomatales

Description: *Cap lilac-beige to violaceous brown, dull or pale when old.*
Cap 5–15 cm (2–6 in), convex then expanded, almost flat or broadly umbonate, glabrous or slightly pruinose towards the margin, violet or lilac, more or less beige or brown, often rather pale and dull (sometimes brownish) when old. *Gills* notched, quite close, separable from the flesh, violet to violaceous beige or pinkish beige. Spore print quite deep rosy buff. *Stipe* 5–10 × 2–6 cm (2–4 × 0.8–2.4 in), with a white pruina on a concolorous ground. *Flesh* more or less violet, quite firm then soft. Smell often strongly aromatic, fruity or mealy to spermatic. *Microscopic features* Sp 6.5–9 × 4–5 µm, elliptic, finely spiny. CC a cutis, the hyphae 2–6 µm diameter. Clamps present.

Season: September–December.
Habitat: Broad-leaved or coniferous trees.
Distribution: Very widespread. Common.

L. nuda often appears late in the season and is considered a good edible species, although the taste can be too strong and it is occasionally indigestible. *L. personata* (Fr.:Fr.) W.G. Smith (= *L. saeva*), with a greyish beige cap and purplish colours on the fibrillose stipe only, prefers more sunny situations.

---

p.144 *Lepista sordida* (Fr.:Fr.) Singer
Tricholomataceae, Tricholomatales

Description: *Cap lilac to violaceous brown, hygrophanous.*
Cap 2–10 cm (0.8–4 in), convex then almost flat or even depressed, sometimes umbonate, hygrophanous, glabrous, distinctly lilaceous or violet or more dull, beige or brown, with pinkish or lilac tones, often with a paler zone when dry. Margin finely striate.
Gills notched, not very close, separable from the flesh, cream to pinkish. Spore print rosy buff. *Stipe* 2–5 × 0.4–0.8 cm (0.8–2 × 0.16–0.32 in), striate to fibrillose or slightly pruinose, almost concolorous. *Flesh* almost translucent towards the margin of the cap (against the light). Smell weak. *Microscopic features* Sp 6–7.5 × 3.5–4.5 µm, elliptic, finely spiny. CC a cutis, the hyphae 2–8 µm diameter. Clamps present.
Season: July–December.
Habitat: Occurs in a range of habitats, sometimes in disturbed sites.
Distribution: Widespread but uneven.

There are a number of varieties based on colour or shape.

---

p.162 *Tricholoma album* (Sch.:Fr.) Kummer
Tricholomataceae, Tricholomatales

Description: *Cap white, becoming dingy. Margin not sulcate. Gills uneven.*
Cap 2–5 cm (0.8–2 in), convex, then almost flat or flat, dry, smooth to minutely roughened or felted, white, remaining so or sometimes becoming dingy or ochraceous cream or with rather rufous beige patches. Margin not sulcate. *Gills* notched, of uneven width, not very close, white and remaining so. *Stipe* 5–6 × 0.4–0.7 cm (2–2.4 × 0.16–0.28 in), often tall and slender and sometimes wavy, smooth or fibrillose, white, sometimes becoming a little dingy.

Flesh quite thin. Smell rather weak, of flour mixed with an earthy, slightly unpleasant component. Taste acrid and bitter. *Microscopic features* Sp 5–6.5 × 3.5–4 µm, elliptic or slightly ovoid. Bas 25–30 × 5–7 µm, clavate. CC a cutis, the hyphae interwoven. Pigment almost absent. Clamps rare.

Season: September–November.
Habitat: Broad-leaved trees, usually on acid soils.
Distribution: Widespread.

T. lascivum (Fr.:Fr.) Gillet, which occurs in similar habitats, is medium-sized, pale beige then uniformly pale buff, with a more fruity or aromatic smell and weak taste. *T. inamoenum* (Fr.:Fr.) Gillet, which occurs under conifers, is more continental, medium-sized, whitish to ochraceous cream, with distant gills and a smell of insecticide or gas. *T. sulfurescens* Bresadola, which occurs on warm calcareous soils, has a larger, strongly yellowing cap, sickening fruity smell and slightly acrid taste.

---

p.162 *Tricholoma pseudoalbum* M. Bon
Tricholomataceae, Tricholomatales

Description: *Cap white. Margin sulcate. Gills regular. Smell strong.*
Cap 5–14 cm (2–5.5 in), convex, often irregular, sometimes umbonate, smooth to felted, sometimes finely wrinkled, white, and almost unchanging or slightly spotted when old. Margin sulcate (vary the way the light falls on the cap). *Gills* notched, regular, quite close, white, bruising slightly to a rather rufous cream where rubbed. *Stipe* 5–15 × 1–2 cm (2–5 × 0.4–0.8 in), often quite robust, smooth to fibrous, white, becoming slightly dingy like the cap. *Flesh* relatively thick, firm then fibrous. Smell strong and unpleasant, of insecticide. Taste acrid and bitter. *Microscopic features* Sp 5.5–6.5 × 3.5–4 µm, elliptic to ovoid. Bas 25–30 × 5–7 µm, clavate. CC a cutis, sometimes very slightly gelatinized, the hyphae slender, interwoven. Pigment almost absent. Clamps rare.

Season: September–October.
Habitat: Broad-leaved or coppice woodland and disturbed places, usually on calcareous or neutral soils.
Distribution: Not fully assessed.

See comments under *T. album* (p.836).

---

p.162 *Tricholoma columbetta* (Fr.:Fr.) Kummer
Tricholomataceae, Tricholomatales

Description: *Cap white, silky. Smell mealy. Bluish or glaucous spots.*
Cap 6–10 cm (2.4–4 in), conico-convex then expanded, sometimes domed, slightly viscid or lubricious in the centre, soon dry, finely innately silky, slightly shiny, white, sometimes somewhat buff in the centre and often spotted bluish or greenish, even pinkish. Margin splitting. *Gills* notched, quite close, white, often pinkish (sometimes a rather vivid pink). *Stipe* 5–10 × 2–6 cm (2–4 × 0.8–2.4 in), fibrous, white, often tinged or spotted blue, green or violaceous pink below. *Flesh* mild, white. Smell mealy. *Microscopic features* Sp 5.5–7 × 3.5–5 µm, elliptic to almost tear-shaped. Bas 30–35 × 6–8 µm, clavate. CC a cutis, sometimes slightly gelatinized in the centre, the hyphae slender. Pigment almost absent. Clamps absent.
Season: September–October.
Habitat: Broad-leaved woods, mainly on acid or neutral soils.
Distribution: Widespread but uneven.

Although the preceding species are inedible (their taste is unpleasant), this species is edible (but avoid confusion with species in other groups, especially the deadly white *Amanita* spp.).

---

p.162 *Tricholoma acerbum* (Bull.:Fr.) Quélet
Tricholomataceae, Tricholomatales

Description: *Cap yellowish cream to somewhat rufous. Margin inrolled, sulcate.*

Cap 5–15 cm (2–6 in), hemispherical or slightly convex, sometimes irregular or lobed, dry, smooth to felted, sometimes wrinkled towards the centre, yellowish cream, with the centre sometimes more rufous. Margin inrolled, sulcate.
Gills notched, close, whitish to pale cream then spotting pale rust, edge notched.
Stipe 5–9 × 1–3 cm (2–3.5 × 0.4–1.2 in), often pruinose or even fleecy at the apex, whitish to concolorous with the cap.
Flesh firm, fibrous in the stipe. Taste unpleasant, styptic. *Microscopic features* Sp 4.5–5.5 × 3.5–4.5 µm, elliptic or subglobose. Bas 25–35 × 6–7 µm, clavate, edge with wavy marginal cells. CC somewhat trichodermial, the hyphae more or less erect 3–6 µm diameter. Pigment mixed, intracellular, faint, near the apex of the superficial hyphae, more parietal, sometimes with sheathing encrustations on hyphae in lower layers. Clamps absent.

Season: September–November.
Habitat: Mainly broad-leaved trees.
Distribution: Uneven. Appears to be threatened.

---

p.170 *Tricholoma virgatum* (Fr.:Fr.) Kummer
Tricholomataceae, Tricholomatales

Description: *Cap conical or umbonate, silvery grey, fibrillose. Taste acrid.*
Cap 4–7 cm (1.6–2.8 in), conical then remaining umbonate, often lobed or irregular, slightly viscid or almost dry, radially fibrillose, more silky on the umbo, silvery grey. *Gills* notched, quite close, waxy grey with yellowish or bluish tones. *Stipe* 5–10 × 0.5–1.5 cm (2–4 × 0.2–0.6 in), the base almost bulbous, sometimes with fine scales, whitish or very pale.
Flesh medium, fibrous. Smell earthy, weak. Taste acrid. *Microscopic features* Sp 6–7.5 × 5–5.5 µm, elliptic. Bas 30–45 × 5–10 µm. Cheilo scattered or in groups, 20–25 × 5–15 µm, clavate, sometimes misshapen, not or scarcely coloured. CC rather like an ixocutis, the hyphae 3–6 µm diameter, radial. Pigment parietal, slightly

encrusting, especially in lower layers, mixed near the surface with intracellular, grey pigment. Clamps absent.

Season: August–October.
Habitat: Broad-leaved trees, conifers.
Distribution: Continental to submontane. Rarer in Atlantic or southern zones.

T. sciodes (Pers.) C. Martin, has a cap which is slightly less umbonate or expands more widely. It often has lilaceous tints, radial fibrils or minute scales, black dots on the gill edge and a bitter taste (under mixed broad-leaves). *T. bresadolanum* Clémençon, on the other hand, is not umbonate, slate coloured, with small dark concentric scales, gills with an irregular black-dotted edge, and dark scales on the lower part of the stipe. It has a bitter taste and occurs under broad-leaved trees.

---

p.171 *Tricholoma portentosum* (Fr.:Fr.) Quélet
Tricholomataceae, Tricholomatales

Description: *Cap grey, innately streaky, the underlying flesh yellow. Gills yellowish.*
Cap 5–12 cm (2–4.7 in), conico-convex, expanded, more or less umbonate, slightly viscid or soon dry, smooth and shiny, with fine innate radial streaks, rather dark slate-grey, showing yellow under the cuticle. *Gills* notched, quite close, white to greyish at the base, often yellowish to pale yellow in places. *Stipe* 6–12 × 1–2 cm (2.4–4.7 × 0.4–0.8 in), white or tinged yellow to yellowish, especially at the apex. *Flesh* firm, white but pale yellow under the cuticles. Smell and taste mealy. *Microscopic features* Sp 5–7.5 × 4–5.5 μm, elliptic to ovoid. Bas 25–35 × 6–10 μm, clavate. CC somewhat like an ixocutis towards the centre, the hyphae slender. Pigment intracellular or mixed, mainly parietal in lower layers. Clamps absent.
Season: September–November.
Habitat: Various conifers, more rarely broad-leaved trees.
Distribution: Widespread but uneven.

p.172 *Tricholoma atrosquamosum* (Chev.) Saccardo
Tricholomataceae, Tricholomatales

Description: *Cap blackish grey, finely scaly. Smell peppery.*
Cap 5–8 cm (2–3.2 in), convex with a broad and obtuse umbo, dry, tomentose then finely scaly, blackish grey on a greyish ground. Margin fibrillose. *Gills* notched, whitish, edge black dotted. *Stipe* 4–8 × 0.5–1.5 cm (1.6–3.2 × 0.2–0.6 in), soon fragile, fibrillose or more or less finely scaly, pale greyish, the scales, if present, sometimes concolorous with the cap. *Flesh* not very thick, rather fragile, almost concolorous near the surface. Smell of pepper. Taste mild. *Microscopic features* Sp 6.5–8.5 × 4.5–6 µm, elliptic. Bas 25–30 × 6–7 µm. Cheilo scattered, or in clusters, 25–40 × 4–8 µm, clavate or irregular, with grey pigment. CC a trichoderm, the hyphae with short and rather thick elements 8–12 µm diameter. Pigment parietal, encrusting and in irregular lines, blackish grey. Clamps absent.

Season: August–December.

Habitat: Mainly broad-leaved trees. Often on calcareous soils.

Distribution: Mainly western Europe; in lowland districts.

T. squarrulosum Bresadola, which is very closely related but occurs in warmer areas, can be distinguished by the concolorous stipe which is scaly like the cap.

---

p.171 *Tricholoma terreum* (Sch.:Fr.) Kummer
Tricholomataceae, Tricholomatales

Description: *Cap fibrillose or tomentose, grey. Gills grey.*
Cap 4–8 cm (1.6–3.2 in), conical then plano-convex or broadly umbonate, dry, radially fibrillose to innately streaky, sometimes slightly tomentose on the umbo, blackish grey. Margin fibrillose. *Gills* notched, not very close, pale greyish. *Stipe* 3–6 × 0.5–1.5 cm (1.2–2.4 × 0.2–0.6 in), rather brittle, whitish or very pale. *Flesh* medium, rather fragile, pale or almost concolorous. *Microscopic features* Sp

6–7.5 × 4.5–5.5 µm, elliptic. Bas 25–30 × 5–8 µm, edge sometimes with scattered deformed basidioles. CC a cutis, sometimes with bundles of hyphae angled slightly upwards, 5–12 µm diameter, the elements sometimes short; under the suprapellis, the hyphae have isodiametric elements. Pigment parietal, encrusting, in irregular lines, dark grey-brown. Clamps absent.

Season: September–December.
Habitat: Conifers. Often on calcareous soils.
Distribution: Widespread but uneven.

An excellent edible species, very popular in certain regions. *T. myomyces* (Pers.:Fr.) Lange, which has a rather dark mouse-grey felted cap and whitish gills, has a more or less definite ring zone on the stipe. *T. gausapatum* (Fr.:Fr.) Quélet, which has a dark brownish grey shaggy cap, very deep uneven pale greyish gills and a white stipe, has a slightly more distinct ring zone. Two other species turn red: *T. orirubens* Quélet, which has a fleecy fibrillose blackish grey to brownish cap, reddens at the cap margin and gill edge and a blue-green or yellowish mycelium visible at the stipe base, occurs mainly in broad-leaved, coppice woodland on calcareous soils; *T. basirubens* (Bon) Riva & Bon reddens distinctly at the stipe base, which is never blue-green.

---

p.171 *Tricholoma scalpturatum* (Fr.) Quélet
Tricholomataceae, Tricholomatales

Description: *Cap beige-grey, finely scaly. Yellowing distinctly when old.*
Cap 3–6 cm (1.2–2.4 in), conical then plano-convex, or broadly umbonate, dry, the outer half finely scaly, the centre felted or tomentose, various shades of beige-grey on a pale ground. Margin fibrillose.
Gills notched, rather close, white or slightly dingy, spotting yellow when bruised or old. *Stipe* 3–6 × 0.5–1 cm (1.2–2.4 × 0.2–0.4 in), soon brittle, almost smooth to fibrillose with a rather faint fleeting ring zone, whitish or greyish,

sometimes almost concolorous.
Flesh fragile, white or almost concolorous near the surface, yellowing when old or rotting. Smell distinctly mealy, at least when bruised. *Microscopic features* Sp 5.5–6.5 × 2.5–3.5 µm, elliptic or long. Bas 25–30 × 5–7 µm. CC somewhat trichodermial, the hyphae parallel or in clusters 6–15 µm diameter. Pigment parietal, often almost smooth or smooth, more or less pale. Clamps absent.

Season: September–December.
Habitat: High forest and coppice woodland. Often on calcareous soils.
Distribution: Widespread. Sometimes rare.

T. argyraceum (Bull.:Fr.) Gillet has a pale grey or silvery cap; f. *alboconicum* (Lange) M. Bon, which is almost wholly white or very pale, also occurs.

---

p.171 *Tricholoma cingulatum* (Almf.) Jacobasch
Tricholomataceae, Tricholomatales

Description: *Cap brownish grey, slightly scaly. Ring cottony.*
Cap 4–6 cm (1.6–2.4 in), hemispherical or plano-convex, not umbonate, finely scaly but sometimes more felted in the centre and more fibrillose at the edge, brownish grey. *Gills* notched, not very close, white or pale cream, yellowing weakly from the edge. *Stipe* 4–6 × 0.5–1 cm (1.6–2.4 × 0.2–0.4 in), fibrous, with a very distinct cottony ring, whitish. *Flesh* almost concolorous near the surface, only yellowing slightly. Smell weakly mealy. *Microscopic features* Sp 5.5–7.5 × 2.5–3.5 µm, elliptic or rather long. Bas 20–30 × 5–7 µm. CC somewhat trichodermial, the hyphae slightly inflated at the free apices 5–15 µm diameter. Pigment parietal, smooth or faintly encrusted, brownish grey. Clamps absent.

Season: August–December.
Habitat: Willow coppice.
Distribution: From sea-level to the mountains.

T. ramentaceum (Bull.:Fr.) Ricken, which has a

grey stipe with several irregular bands under the ring, often has a more umbonate cap.

---

p.170 *Tricholoma pardinum* (Pers.) Quélet (poisonous)
Tricholomataceae, Tricholomatales

Description: *Cap with dull grey-brown scales on a pale ground. Gills glaucous.*
Cap 8–15 cm (3.2–6 in), hemispherical then convex or broadly umbonate, fleshy, finely scaly or scaly, especially half-way to the centre (however, there are purely fibrillose forms: var. *filamentosum* Alessio), the scales grey-brown, sometimes almost blackish, on a paler ground, often flushing slightly rufous with age. Margin obtuse, overhanging, sometimes irregular.
Gills notched, close, whitish with bluish or glaucous tones. *Stipe* 7–12 × 1–3 cm (2.8–4.7 × 0.4–1.2 in), firm, fibrous and fleshy, often swollen, whitish, browning slightly from the base. *Flesh* thick, firm. Smell of meal, slightly nauseous when old.
Microscopic features Sp 9–11.5 × 5.5–7.5 µm, elliptic. Bas 30–50 × 10–14 µm, with clamps. CC a radial cutis, the hyphae slender 3–6 µm diameter. Pigment parietal, smooth or slightly rough, sometimes in irregular lines in lower layers and also intracellular near the surface. Clamps abundant.
Season: August–October.
Habitat: Conifers sometimes mixed with broad-leaved trees. Often on calcareous soils.
Distribution: Submontane or montane. Much rarer in lowland districts or in the south.

The most toxic of the *Tricholoma* spp., responsible for most of the cases of poisoning in mountain regions.

---

p.161 *Tricholoma saponaceum* (Fr.:Fr.) Kummer
Tricholomataceae, Tricholomatales

Description: *Cap typically greenish grey. Stipe tapering and flushing pink below. Smell of soap.*

Cap 6–13 cm (2.4–5.1 in), hemispherical, then convex, sometimes with a broad umbo, sometimes wavy, slightly viscid or lubricious, then dry and smooth or cracking when old, typically with greenish tones on a grey or greyish beige ground, sometimes more yellowish, brownish, etc.
Gills notched, not very close, whitish with a greenish or yellowish tinge, often dingy beige or dingy buff, edge sometimes flushing pink. *Stipe* 5–15 × 0.5–1.5 cm (2–6 × 0.2–0.6 in), tapering or even fusiform, sometimes disrupted, whitish or dingy, coloured like the cap but paler, flushing pink below. *Flesh* soft in the cap. Smells strongly of soap. Taste dull or unpleasant. *Microscopic features* Sp 6–7.5 × 3.5–4.5 µm, narrowly elliptic or somewhat long tear-shaped. Bas 25–35 × 5–7 µm, clavate. CC an (ixo)cutis, the hyphae slender, interwoven. Pigment intracellular. Clamps abundant.

Season: August–December.
Habitat: Broad-leaved trees, especially oak.
Distribution: Widespread. Very common.

There are numerous varieties. Some have a smooth stipe like var. *atrovirens* (Pers.:Fr.) P. Karsten (cap dark green), var. *napipes* (Krombholz) M. Bon (cap yellowish grey, gills yellow), var. *sulfurinum* (Quélet) Rea (cap bright yellow, gills and stipe yellowish), f. *carneifolium* Bertault (cap white and gills pink). Others have a finely scaly stipe: var. *squamosum* (Cooke) Rea (cap greyish brown, more or less scaly, less so than the stipe; fairly easy to recognize when all the characters are present; linked to the type species by sometimes confusing intermediates), f. *ardosiacum* (Bresadola) M. Bon (very dark, blackish, with bluish or violaceous tones on the cap).

---

p.161 *Tricholoma sejunctum* (Sow.:Fr.) Quélet
Tricholomataceae, Tricholomatales

Description: *Cap greenish yellow with brownish fibrils. Taste slightly or slowly bitter.*

Cap 4–10 cm (1.6–4 in), conical then with a persistent and finally low umbo, irregular, lubricious then dry, often slightly shiny, radially fibrillose, sometimes rimose, greenish yellow, obscured by more or less brownish and more or less dense fibrils. Margin sometimes slightly contracted or inrolled. *Gills* coarsely notched, not very close, whitish or very pale, edge often irregular. *Stipe* 5–8 × 0.5–1.5 cm (2–3.2 × 0.2–0.6 in), almost lubricious then fibrillose, white or tinged yellowish, and rather rufous below. Smell weakly mealy. Taste slightly or slowly bitter. *Microscopic features* Sp 6–8 × 5–6 μm, elliptic or slightly tear-shaped. Bas 35–45 × 6–10 μm. A few sterile, deformed to contorted cylindric elements present near the gill edge. CC a cutis or ixocutis in the centre, sometimes somewhat trichodermial, the hyphae rather slender 3–6 μm diameter. Pigment intracellular, more or less yellow near the surface, more parietal, faintly encrusting at lower levels. Clamps absent.

Season: August–October.
Habitat: Broad-leaved trees. Often on acid soils.
Distribution: Widespread. Rarer in some areas.

---

p.161 *Tricholoma viridilutescens* Moser
Tricholomataceae, Tricholomatales

Description: *Cap olive-brown to bright greenish yellow at the edge, edge yellow.*
Cap 4–8 cm (1.6–3.2 in), conical then with an obtuse umbo, slightly viscid then dry, almost smooth to radially fibrillose, quite dark, dark olive-brown in the centre, fading towards the edge to greenish yellow. Margin slightly contracted to vertical.
Gills notched, very pale, edge irregular, yellow towards the margin. *Stipe* 4–10 × 0.5–1.5 cm (1.6–4 × 0.2–0.6 in), whitish to pale yellowish below. *Flesh* greenish near the surface. Smell and taste mealy.
Microscopic features Sp 6–8 × 4.5–6.5 μm, elliptic to subglobose. Bas 25–40 × 5–9 μm. CC a cutis or ixocutis towards the centre, sometimes somewhat trichodermial, the

Season:	August–October.
Habitat:	Broad-leaved trees. Usually on damp acid soils.
Distribution:	Widespread. Rather rare to rare.

hyphae slender 3–7 μm diameter. Pigment intracellular, olive yellow near the surface and developing with age, sometimes slightly parietal, encrusting, in lower layers.

T. fucatum (Fr.:Fr.) Kummer has a more conical, darker and more distinctly green cap. *T. viridifucatum* M. Bon is an intense greenish yellow, with broad, pale greenish gills and a rather pale greenish yellow stipe. *T. joachimii* M. Bon & Riva is finely scaly (cap and stipe) and has olive-yellow to dark olive-brown colours. It often occurs in clusters, especially under conifers.

---

p.167 *Tricholoma equestre* (L.:Fr.) Kummer
Tricholomataceae, Tricholomatales

Description: *Cap orange-yellow, slightly scaly in the centre. Stipe and flesh bright yellow.*
Cap 5–10 cm (2–4 in), conical then with an obtuse umbo, slightly viscid or soon dry, disrupting into small scales in the centre, smooth or slightly fibrillose elsewhere, orange-yellow and rather tan in the centre.
Gills notched, close, bright yellow.
Stipe 6–10 × 0.5–1 cm (2.4–4 × 0.2–0.4 in), often fusoid at the extreme base, sometimes with occasional small fleecy or fibrillose scales, bright yellow.
Flesh mild, bright yellow throughout.
Smell weak of meal or orange blossom.
Microscopic features Sp 6–8 × 3.5–5 μm, elliptic or slightly almond-shaped. Bas 30–35 × 6–8 μm. CC a cutis or ixocutis towards the centre, somewhat trichodermial near the small scales, the hyphae 3–6 μm diameter. Pigment mixed, intracellular, reddish brown, refractive, especially in dry areas, also parietal, concolorous, dotted near the superficial layers, not very gelatinized, more encrusting in lower layers. Clamps absent.
Season: September–November.

848 Tricholomataceae

Habitat: Broad-leaved and coniferous trees.
Distribution: Rather common.

An excellent edible species.

---

p.167 *Tricholoma auratum* (Fr.) Gillet
Tricholomataceae, Tricholomatales

Description: *Cap orange-yellow to brownish. Stipe pale yellow. Innermost flesh white.*
Cap 8–15 cm (3.2–6 in), hemispherical then convex to almost flat, not or not very umbonate, slightly viscid or lubricious, almost smooth or very slightly disrupted in the centre when old, orange-yellow or more brownish in the centre. *Gills* notched, not very close, yellow. *Stipe* 4–8 × 1–3 cm (1.6–3.2 × 0.4–1.2 in), short and thick, almost smooth, pale yellow. *Flesh* thick, yellowish to almost concolorous but innermost flesh whitish. *Microscopic features* Sp and hymenium like *T. equestre*. CC more gelatinized, less trichodermial and the pigment mixed, intracellular and parietal, smooth to almost smooth. Clamps absent.
Season: September–December.
Habitat: Conifer forests on sandy soils.
Distribution: Mainly Atlantic. Rarer than *T. equestre*.

This is also a very good edible species. *T. equestre* (see p.847) and *T. auratum* can be confused with potentially poisonous yellow species of *Cortinarius* (spore print rusty and stipe with a cortina).

---

p.167 *Tricholoma sulfureum* (Bull.:Fr.) Kummer
Tricholomataceae, Tricholomatales

Description: *Cap more or less dark yellow, dry. Gills concolorous, thick and distant.*
Cap 4–8 cm (1.6–3.2 in), hemispherical or domed then plano-convex, dry, almost velvety, more or less dark yellow.
Gills deeply notched, distant and rather thick, concolorous. *Stipe* 5–10 × 0.4–1 cm (2–4 × 0.16–0.4 in), often tapering, wavy, smooth to fibrillose, almost concolorous or

paler. *Flesh* rather brittle, concolorous. Smell strong and penetrating, like sulphur, said to be like coal gas. *Microscopic features* Sp 9–11 × 5.5–7 µm, almond-shaped, relatively thick-walled and with a prominent apiculus. Bas 40–55 × 6–12 µm, large for the genus. Cystidia absent. CC a trichoderm, the hyphae slender, erect, the apex sometimes slightly thickened or clavate. Pigment intracellular, cloudy, brownish yellow. Clamps on some slender hyphae.

Season: August–November.
Habitat: Mainly broad-leaved trees. Often on acid soils.
Distribution: Rather common.

Var. *coronarium* (Pers.) Gillet, which has a brown cap or has crimson tints in the centre and a floral then nauseous smell, occurs in warmer sites. *T. bufonium* (Pers.:Fr.) Gillet, a species with a wholly dark crimson-brown cap, a slightly less strong smell and habitat mainly under conifers, is rather rare and continental.

---

p.166 *Tricholoma imbricatum* (Fr.:Fr.) Kummer
Tricholomataceae, Tricholomatales

Description: *Cap umbonate, rufous brown, fibrillose to finely scaly. Stipe tapering.*
Cap 5–10 cm (2–4 in), conical or trapezoidal, then with an obtuse umbo, dry, velvety then fibrillose or with small slightly erect scales, especially towards the centre, dull red-brown, on a paler, sometimes yellowish ground. *Gills* notched, close, whitish to pale cream, sometimes pinkish and spotting reddish. *Stipe* 7–12 × 1–2 cm (2.8–4.7 × 0.4–0.8 in), slightly swollen in the lower one-third but tapering at the base, fibrillose, rather rufous. *Flesh* pale cream. Smell faintly mealy. Taste slightly or slowly bitter. *Microscopic features* Sp 6–7.5 × 4.5–5.5 µm, elliptic or oval. Bas 30–40 × 5–8 µm. CC a trichoderm, the hyphae cylindric or clavate 5–10 µm diameter. Pigment parietal, almost smooth

p.166 *Tricholoma vaccinum* var. *fulvosquamosum*
M. Bon
Tricholomataceae, Tricholomatales

Description: *Cap subumbonate, reddish orange-brown, with concentric scales.*
Cap 5–8 cm (2–3.2 in), hemispherical then umbonate, finally expanded, woolly, then with often well-defined, concentric scales, warm brown or distinctly rufous. Margin long remaining vertical or contracted, often woolly and appendiculate. *Gills* notched, pale cream to pale beige, spotting reddish when old. *Stipe* 3–6 × 1–1.5 cm (1.2–2.4 × 0.4–0.6 in), tapering slightly, pale at the apex, fibrillose or even with fine rufous scales below. *Flesh* pale cream, more rufous near the surface, flushing very faintly pink. *Microscopic features* Sp 6.5–8 × 4–5 µm, elliptic or ovoid. Bas 30–35 × 6–8 µm. CC a trichoderm, the hyphae septate, with short thick elements 10–15 µm diameter. Pigment parietal, almost smooth or slightly encrusting in lower layers, mixed near the surface, with pigment intracellular, diffused. Clamps absent.
Season: September–December.
Habitat: In and on the edges of conifer woods. Often in warm places.
Distribution: Mainly southern.

A variety of *T. vaccinum* (Pers.:Fr.) Kummer which is also a coniferous species but more tolerant and, in particular, less confined to warm sites, occurs as far north as Scandinavia. It can be distinguished by its dark, much more umbonate cap, with more untidy or shaggy to woolly scales and flesh which flushes deeper pink when cut.

p.169 *Tricholoma psammopus* (Kalchbr.) Quélet
Tricholomataceae, Tricholomatales

Description: *Cap brownish buff, granular. Veil distinct.*
Cap 4–8 cm (1.6–3.2 in), hemispherical then convex umbonate, very dry, granular, especially in the centre, brownish buff, pale and dull. *Gills* notched, pale cream to buff, sometimes somewhat rufous. *Stipe* 5–8 × 0.5–1.5 cm (2–3.2 × 0.2–0.6 in), granular to powdery, concolorous, beneath a high-placed ring zone, whitish and smooth above the veil. *Flesh* slightly or slowly bitter. *Microscopic features* Sp 6–7.5 × 4–5.5 µm, elliptic. Bas 25–30 × 5–8 µm. CC a trichoderm, the hyphae erect and septate, with frequently inflated or even capitate terminal elements 5–10 µm diameter. Pigment parietal, almost smooth or slightly punctate in lower layers, mixed with intracellular pigment, which is dominant near the surface. Clamps absent.

Season: August–November.
Habitat: Conifers, especially larch and pine.
Distribution: Widespread. Rather rare to very rare.

---

p.166 *Tricholoma fulvum* (Bull.:Fr.) Saccardo
Tricholomataceae, Tricholomatales

Description: *Cap tawny brown, the margin sulcate. Gills yellow. Flesh bright yellow.*
Cap 5–10 cm (2–4 in), conical then umbonate to plano-convex, slightly viscid then dry, somewhat shiny then matt and almost fibrillose or silky in dry weather and when old, rather dark tawny brown, especially in the centre, the margin paler, sometimes with underlying yellowish colours. Margin sulcate to finely ribbed. *Gills* notched, rather close, quite bright yellow, often dull or spotted brownish rust when old. *Stipe* 6–10 × 0.5–1 cm (2.4–4 × 0.2–0.4 in), smooth to fibrillose, brownish yellow, paler above and browning from the base. *Flesh* slightly or slowly bitter, yellow throughout and brightly so near the surface, slightly paler in the

middle. Smell mealy. *Microscopic features*
Sp 5–7 × 3.5–5 µm, elliptic, sometimes subglobose. Bas 25–35 × 5–8 µm. CC an ixocutis, subtrichodermial in the centre, the hyphae 3–5 µm diameter. Pigment parietal, almost smooth to encrusting in irregular lines in lower layers, mixed with a brownish intracellular pigment, which is dominant near the surface. Clamps absent.

Season: August–November.
Habitat: Broad-leaved trees, wood edges, often with birch.
Distribution: Widespread. Rather common.

T. pseudonictitans M. Bon, which occurs under conifers or evergreen broad-leaved trees in the Mediterranean, is very similar but the margin is not sulcate. It has whitish gills, a whitish or ochraceous cream stipe and the flesh is yellow only near the surface.

---

p.168 *Tricholoma pessundatum* (Fr.:Fr.) Quélet
Tricholomataceae, Tricholomatales

Description: *Cap often spotted at the edge, bronze brown. Smell and taste mealy.*
Cap 5–8 cm (2–3.2 in), convex then faintly umbonate, slightly viscid to viscid in damp weather, drier at the edge which may be slightly fibrillose or spotted with darker zones, brown, sometimes with bronze tones. Margin often slightly sulcate.
Gills notched, quite distant, white to somewhat reddish when old. *Stipe* 5–9 × 0.5–1.5 cm (2–3.5 × 0.2–0.6 in), pruinose at the apex, pale cream, browning below when old. *Flesh* pale or almost concolorous near the surface. Smell and taste very strong of meal or cucumber. *Microscopic features* Sp 4.5–6 × 2.5–3.5 µm, elliptic. Bas 20–30 × 5–7 µm. CC a thick ixotrichoderm, the hyphae 3–6 µm diameter, over a denser zone of hyphae lying flat. Pigment encrusting, parietal, in irregular lines, visible beneath the gelatinized layer. Clamps absent.

Season: September–December.
Habitat: Conifers (typically pine), on sand.
Distribution: Mainly Atlantic; rarer inland.

T. populinum Lange is taller, with a warm pinkish brown cap, short thick stipe, white flesh and a very strong smell of watermelon (open, often grassy undergrowth, especially under poplars, quite often in tufts). It is a good edible species, while *T. pessundatum* is more or less indigestible and not recommended.

---

p.168 *Tricholoma ustale* (Fr.:Fr.) Kummer
Tricholomataceae, Tricholomatales

Description: *Cap warm buff-brown, with a mild viscid coating. Stipe lacking a ring zone.*
Cap 5–10 cm (2–4 in), convex, almost flat, not umbonate, sometimes lobed, slightly viscid to viscid, somewhat fibrillose or almost innately streaky when dry, buff-brown, rather warm towards the margin but darker in the centre. Viscid coating mild. *Gills* notched, quite close, pale cream then with somewhat rufous to rust spots or dots when old. *Stipe* 4–8 × 0.5–1.5 cm (1.6–3.2 × 0.2–0.6 in), white then rather rufous from the base, with no distinct limit. *Flesh* white or reddish near the surface. Smell faintly mealy. Taste slightly or slowly bitter. *Microscopic features* Sp 6–7.5 × 4.5–5 μm, elliptic. Bas 25–35 × 5–8 μm. CC an ixotrichoderm, the hyphae erect, 4–8 μm diameter. Pigment fairly parietal, smooth to encrusting in irregular lines in lower layers, mixed with intracellular pigment which is dominant near the surface. Clamps absent.
Season: September–November.
Habitat: Broad-leaved trees. Often on calcareous soils.
Distribution: Widespread. Rather common.

---

p.168 *Tricholoma ustaloides* Romagnesi
Tricholomataceae, Tricholomatales

Description: *Cap viscid, tawny brown. viscid coating very bitter. Stipe with an annular line.*
Cap 5–12 cm (2–4.7 in), hemispherical

then almost flat, not umbonate, viscid in damp weather, the viscid coating very bitter if touched with the tongue, tawny brown. Margin often sulcate or ribbed. *Gills* notched, not very close, whitish to cream, with somewhat rufous spots when old. *Stipe* 5–10 × 1–2 cm (2–4 × 0.4–0.8 in), rather short and thick, cylindric or tapering, pruinose and white or very pale at the apex, fibrillose and somewhat rufous beneath a distinct annular line. *Flesh* bitter, firm. Smells strongly of cucumber or, reputedly, of corduroy.
Microscopic features Sp 5.5–6.5 × 4.5–5.5 µm, ovoid. Bas 25–35 × 5–8 µm. CC an ixotrichoderm, the hyphae erect, 3–5 µm diameter. Pigment parietal, often smooth, mixed with intracellular pigment which is dominant near the surface. Clamps absent.

Season: August–November.
Habitat: Broad-leaved trees on acid soils.
Distribution: Widespread. Rather rare.

---

p.170 *Tricholoma fracticum* (Britz.) Kreisel
Tricholomataceae, Tricholomatales

Description: *Cap reddish brown, slightly-viscid. Ring zone slightly raised.*
Cap 8–15 cm (3.2–6 in), hemispherical then convex or almost flat, slightly viscid, soon dry and radially fibrillose, more or less dark reddish brown. Margin long remaining inrolled. *Gills* notched, quite close, white then slightly spotted or dotted, rather rufous when old. *Stipe* 5–10 × 1.5–3 cm (2–4 × 0.6–1.2 in), compact, fibrous, tapering or suddenly contracted below, pruinose and white or very pale at the apex, with rather rufous chestnut or almost concolorous fibrils under a very distinct, slightly raised ring zone. *Flesh* thick, compact. Smell mealy. Taste bitter.
Microscopic features Sp 5–6 × 4–5 µm, elliptic, with a large apiculus. Bas 30–35 × 5–8 µm. CC an ixocutis, the hyphae erect, 4–8 µm diameter. Pigment parietal, almost smooth or slightly encrusting or in irregular lines in

lower layers, intracellular pigment very faint near the surface. Clamps absent.

Season: August–November.

Habitat: Conifers, sometimes mixed with broad-leaves, grassy wood edges. Often on warm calcarous soils.

Distribution: Mainly in southern Europe or warm regions.

T. colossus (Fr.) Quélet, another robust species with a ring zone, is taller, with a pale cap, whitish to pinkish gills and a swollen stipe with tiers of irregular bands under the ring zone. It occurs under conifers, especially in continental or submontane areas (rare to very rare).

---

p.166 *Tricholoma aurantium* (Sch.:Fr.) Ricken
Tricholomataceae, Tricholomatales

Description: *Cap rufous orange, speckled. Sheath distinct with irregular orange bands.*
Cap 5–10 cm (2–4 in), convex or flat, viscid and almost smooth, disrupting in dry weather into small, orange or quite bright rufous orange speckles, with olivaceous or bronze tones. *Gills* notched, rather close, white then slightly rufous, edge soon dotted orange. *Stipe* 5–10 × 1–2 cm (2–4 × 0.4–0.8 in), firm, the sheath with irregular orange bands beneath a slightly raised ring zone, white above. *Flesh* bitter, firm. Smell of cucumber. *Microscopic features* Sp 5–6 × 3–4 μm, rather variable to shortly elliptic. Bas 15–20 × 5–6 μm. CC an ixotrichoderm, the hyphae erect, 2–5 μm diameter. Pigment parietal, rather smooth in the lower layers, distinctly dominated in superficial layers by abundant yellow intracellular pigment. Clamps absent.

Season: August–November.

Habitat: Mainly spruce, sometimes broad-leaved trees or open places. Often on warm calcareous soils.

Distribution: Mainly southern. Rather rare.

Certain forms of *T. focale* (Fr.) Ricken are also sometimes quite brightly coloured. This latter, rather polymorphous, species has a rufous orange cap, innately streaked with red and is sheathed beneath the thick ring with irregular orange-red bands.

---

p.170 *Tricholoma caligatum* (Viv.) Ricken
Tricholomataceae, Tricholomatales

Description: *Cap with brown scales. Sheath concolorous.*
Cap 6–15 cm (2.4–6 in), convex or almost flat, felted then with broad appressed, more or less concentric brown scales on an ochraceous cream ground. Margin inrolled then slightly overhanging. *Gills* notched, not very close, yellowish white or pale. *Stipe* 8–12 × 1–3 cm (3.2–4.7 × 0.4–1.2 in), compact, with a concolorous scaly or irregularly disrupted sheath beneath an initially wide but fragile ring, smooth and white above. *Flesh* thick and compact. Smell aromatic, strong. Taste slightly or slowly bitter. *Microscopic features* Sp 6–7.5 × 5–6.5 µm, elliptic or subglobose. Bas 25–35 × 6–10 µm. CC a sometimes trichodermial cutis, the hyphae 5–15 µm diameter, cylindric. Pigment mainly parietal, smooth. Clamps absent.
Season: August–November.
Habitat: Mainly pine, sometimes fir or spruce but also evergreen broad-leaved trees.
Distribution: Southern. Rather uneven.

An edible species that should be protected on account of its rarity. Popular in Japan (matsu-také) and North America.

---

p.150 *Tricholomopsis decora* (Fr.) Singer
Tricholomataceae, Tricholomatales

Description: *Cap yellow, finely scaly.*
Cap 4–8 cm (1.6–3.2 in), convex, often depressed in the centre, radially fibrillose, with fine dark, brownish or brown scales at least in the centre, on a rather bright, yellow-buff ground. *Gills* adnate, quite close, golden yellow or covered, when mature, with a fine white pruina, edge eroded or fimbriate. *Stipe* 4–6 × 0.4–1 cm

(1.6–2.4 × 0.16–0.4 in), cylindric or slightly thickened below, often curved, fibrillose, concolorous or paler.
Flesh slightly or slowly bitter, yellowish to almost concolorous near the surface.
Microscopic features Sp 5.5–7.5 × 4–5.5 µm, elliptic. Bas 25–35 × 5–8 µm. Cheilo abundant, clavate or inflated, sometimes tapering to a more or less distinct or narrow neck, 30–70 × 10–18 µm. CC a trichoderm, the hyphae rather slender or clavate 7–15 µm diameter. Pigment intracellular, brown, abundant. Clamps abundant.

Season: September–November.
Habitat: Lignicolous, on conifers.
Distribution: Often montane or continental. Rather rare.

T. flammula Métrod (very rare), is also yellow but distinctly smaller and with small violaceous brown scales. Moreover, it has distinct pleurocystidia. *T. ornata* (Fr.) Singer (slightly more widespread but also northern or in the mountains) has a medium-sized, rather olivaceous yellow cap, with small tawny brown scales (found on rotten conifers). It can be confused with species of *Pholiota* or *Gymnopilus* which share the same habitat (the latter have a brown to rust spore print).

---

p.173 *Tricholomopsis rutilans* (Sch.:Fr.) Singer
Tricholomataceae, Tricholomatales

Description: *Yellow, but cap and stipe covered with small violet fibrillose scales.*
Cap 8–15 cm (3.2–6 in), convex, sometimes slightly depressed or with a low umbo, fibrillose, and with small radial violet-brown to lilaceous brown scales, on a rather bright yellow ground. Margin often slightly sulcate or ribbed. *Gills* adnate, quite close, bright yellow, edge eroded or fimbriate. *Stipe* 5–10 × 0.5–1.5 cm (2–4 × 0.2–0.6 in), sometimes curved, covered, especially from the base, with small scaly fibrils, concolorous with the cap or more yellow. *Flesh* bright yellow. Taste slightly or slowly bitter. *Microscopic features*

Sp 7–9 × 5.5–6.5 µm, elliptic. Bas 30–40 × 5–8 µm. Cheilo abundant, 50–120 × 15–30 µm, fusoid or clavate. CC a trichoderm, the hyphae in clusters, quite broadly clavate or fusiform, sometimes appendiculate. Pigment intracellular, violaceous brown. Clamps abundant.

Season: August–December.
Habitat: Conifer stumps.
Distribution: Very widespread. Rather common.

---

p.135 *Melanoleuca iris* Kühner
Tricholomataceae, Tricholomatales

Description: *Cap with a low umbo, grey, pruinose. Stipe pruinose.*
Cap 2–6 cm (0.8–2.4 in), conico-convex then slightly umbonate, pruinose, matt, dry, greyish beige, sometimes whitish from the pruina. *Gills* notched or with a decurrent tooth, close, pale yellowish cream. *Stipe* 3–6 × 0.4–1 cm (1.2–2.4 × 0.16–0.4 in), pruinose, especially at the apex, white or very pale, sometimes rather rufous beige from the base. *Flesh* with a strong smell of iris, or floral. *Microscopic features* Sp 8–10 × 5–6 µm, elliptic, with rather regular, medium amyloid warts, the SP smooth. Cyst (especially marginal) 35–55 × 3–7 µm, lageniform, the neck delimited by a septum, narrow, conical or almost pointed, crowned with small crystals. CC a cutis or a trichoderm.

Season: September–December.
Habitat: Lawns or grassy places.
Distribution: Not well known. Rather rare.

The genus, fairly easy to recognize from its silhouette, is very difficult at species level.

---

p.169 *Melanoleuca grammopodia* (Bull.:Fr.) Patouillard
Tricholomataceae, Tricholomatales

Description: *Cap yellow-brown, umbonate. Stipe striate, ochraceous beige.*
Cap 5–15 cm (2–6 in), broadly convex

with a prominent umbo, often wavy, glabrous, slightly shiny, rather warm yellow-brown or brownish beige, sometimes more dull, greyish brown, with a rather pale margin. *Gills* notched or with a decurrent tooth, close, white then pale beige. *Stipe* 5–12 × 1–2 cm (2–4.7 × 0.4–0.8 in), strongly grooved and sometimes twisted, pale sepia then becoming brown or warm buff. *Flesh* pale. Smell fungal. *Microscopic features* Sp 8–10.5 × 5–6.5 µm, elliptic or cylindro-elliptic, with amyloid, more or less encrusted warts, the SP smooth. Cyst (especially marginal) 20–30 × 1–2 µm, the base inflated and the neck (separated by a septum) very narrow, sharp, capped with small crystalline barbs. CC a cutis, the hyphae interwoven.

Season: August–November.
Habitat: Grassy places, clearings, wood edges.
Distribution: Widespread but uneven.

---

p.172 *Melanoleuca cinereifolia* var. *maritima* (Huijsm.) ex M. Bon
Tricholomataceae, Tricholomatales

Description: *Cap greyish brown or tawny. Stipe turning brown. Gills ash grey.*
Cap 2–8 cm (0.8–3.2 in), broadly convex, sometimes with a low umbo, glabrous, slightly shiny in damp weather, brownish, with rather tawny to tan tones, often rather dull, the margin sometimes sulcate, lighter or whitish. *Gills* notched, close, pure white but soon grey to brownish grey. *Stipe* 3–6 × 0.4–1.2 cm (1.2–2.4 × 0.16–0.47 in), fibrillose-striate, greyish beige then browning, cinnamon to dark brown. *Microscopic features* Sp 7–9.5 × 5–6.5 µm, elliptic, with weakly defined, amyloid, more or less finely encrusted warts, the SP smooth. Cyst (especially marginal) 35–60 × 10–20 µm, fusoid-lageniform, the neck not septate, more or less wavy or with many constrictions, capped by small crystals. CC a cutis or trichoderm, the hyphae slender,

Season: September–December.
Habitat: Maritime dunes, often bare sand.
Distribution: Atlantic and Mediterranean. Rare.

M. cinereifolia (M. Bon) M. Bon, which is more widespread, occurs in cooler regions. The cap is darker and the gills grey from the outset.

---

p.92 *Leucopaxillus paradoxus* (Costantin & Dufour) Boursier
Tricholomataceae, Tricholomatales

Description: *Cap whitish to buff. Gills pale cream. Stipe concolorous.*
Cap 5–8 cm (2–3.2 in), convex then almost flat or umbonate, felted to subglabrous, sometimes cracking when old, dry, white to yellowish cream. Margin long, remaining inrolled, often slightly sulcate.
Gills decurrent, not very close, white.
Stipe 5–7 × 1–2 cm (2–2.8 × 0.4–0.8 in), fibrillose to fleecy, even rough or coarsely felted, concolorous. *Flesh* white. Smell complex, like orange peel at the stipe apex but sometimes more unpleasant, like insecticide on top of the cap. Taste mild.
Microscopic features Sp 6.5–9 × 4.5–6 μm, elliptic to ovoid, with weakly defined, amyloid warts. CC a cutis or trichoderm.
Season: September–November.
Habitat: Open conifer woods, sometimes mixed with broad-leaves. Often on calcareous soils.
Distribution: Uneven. Rather rare to rare.

---

p.103 *Catathelasma imperiale* (Fr.→Quélet) Singer
Tricholomataceae, Tricholomatales

Description: *Cap golden brown, fibrillose. Gills decurrent. Ring double.*
Cap 8–20 cm (3.2–8 in), hemispherical then convex, fibrillose or slightly squamulose in the centre, golden brown, sometimes brownish buff or with quite dark brown spots. Margin inrolled.
Gills decurrent, quite close, white then

cream. *Stipe* 5–10 × 1–4 cm (2–4 × 0.4–1.6 in), short and thick, often conical, with rufous gold fibrils or fine scales below on a pale ground, white at the apex. Ring double, striate. *Flesh* thick, white. Smells strongly of cucumber. Taste slightly or slowly bitter. *Microscopic features* Sp 8–15 × 4.5–6.5 µm, elliptic to cylindro-fusoid, amyloid. A few cylindric or wavy MC. CC an (ixo)cutis, the hyphae fasciculate 2–5 µm diameter, with clamps. Pigment parietal.

Season: August–October.
Habitat: Spruce, in grassy places.
Distribution: Montane. Rare.

---

p.172 *Lyophyllum fumosum* (Pers.:Fr.) Kühn & Romagn. ex Orton
Tricholomataceae, Tricholomatales

Description: *In clusters. Cap innately streaky, yellowish brown to blackish sepia. Gills dingy.*
Cap 3–12 cm (1.2–4.7 in), convex then almost flat, radially fibrillose to innately streaky, sometimes lubricious, yellowish brown to blackish sepia, sometimes irregularly marbled. Margin wavy or irregular. *Gills* adnate, close, white then dingy cream to greyish or brownish in patches. *Stipe* 5–10 × 0.5–1.5 cm (2–4 × 0.2–0.6 in), sometimes branching below, often curved, pruinose to fibrillose, pale to almost concolorous at the base. In clusters. *Flesh* quite elastic or slightly cartilaginous, white. *Microscopic features* Sp 5.5–7 µm, globose, slightly thick-walled, not amyloid. Bas with siderophilous granulations. CC an (ixo)cutis, the hyphae 2–4 µm diameter. Pigment parietal. Clamps present.

Season: July–November.
Habitat: In grass or undergrowth, on buried wood.
Distribution: Widespread but uneven.

L. decastes (Fr.:Fr.) Singer (= *L. aggregatum*), is less innately streaky on the yellow-brown to brownish tawny cap. *L. loricatum* (Fr.) Kühner is distinguished by its very elastic

p.141 *Rugosomyces carneus* (Bull.:Fr.) M. Bon
Tricholomataceae, Tricholomatales

Description: *Cap pink to pinkish beige. Gills white. Stipe concolorous.*
Cap 1–4 cm (0.4–1.6 in), convex then almost flat, hygrophanous, glabrous and smooth, pinkish to pinkish beige, sometimes slightly more dull, fading when dry. *Gills* notched, not very close, white. *Stipe* 1–4 × 0.2–0.5 cm (0.4–1.6 × 0.08–0.2 in), often with a longitudinal groove, pruinose above to faintly fibrillose, concolorous and contrasting with the gills, sometimes darker below. *Flesh* white or pinkish near the surface. *Microscopic features* Sp 4–6 × 2–3.5 µm, cylindro-elliptic. Bas with siderophilous granulations. CC a disjunct trichoderm, the hyphae sometimes coralloid, interwoven. Pigment parietal. Clamps present.

Season: August–December.
Habitat: Lawns.
Distribution: Quite widespread but uneven.

p.144 *Rugosomyces ionides* (Bull.:Fr.) M. Bon
Tricholomataceae, Tricholomatales

Description: *Cap and stipe violaceous lilac. Gills white or yellowish cream.*
Cap 2–5 cm (0.8–2 in), slightly convex then flat, glabrous and smooth or slightly felted to fibrillose, violaceous lilac, sometimes more violaceous or bluish. *Gills* like *R. carneus* (see above), white to pale yellowish cream. *Stipe* 2–5 × 0.3–0.7 cm (0.8–2 × 0.12–0.28 in), fibrillose or striate, concolorous or darker. *Flesh* white. Smell mealy. *Microscopic features* Sp 4.5–6 × 2–3 µm, cylindro-elliptic or faintly fusiform. Bas with siderophilous granules. CC a subtrichodermial cutis, the superficial hyphae sometimes shorter and more erect, or diverticulate to branched. Pigment

mixed. Clamps present.
Season: August–November
Habitat: Broad-leaved or mixed woodland, on moist soils. Often in disturbed places.
Distribution: Widespread. Rare to very rare.

---

p.163 **Saint George's Mushroom**
Calocybe gambosa (Fr.:Fr.) Singer ex Donk
Tricholomataceae, Tricholomatales

Description: *Cap white or pale. Gills close. Strong smell of meal.*
Cap 3–12 cm (1.2–4.7 in), hemispherical then convex or even flat, glabrous and matt to felted or velvety, white or very pale, sometimes with patches of cream or dingy buff. *Gills* notched, close, white. *Stipe* 3–7 × 2–6 cm (1.2–2.8 × 0.8–2.4 in), pruinose-fibrillose, glabrous, white. *Flesh* thick, white. Smell strongly mealy. *Microscopic features* Sp 4–6 × 2.5–3.5 µm, elliptic. Bas with siderophilous granules, short, 15–25 × 3–5 µm. CC a cutis, the hyphae 2–5 µm diameter. Clamps present.
Season: Mainly April–June.
Habitat: Hedges, wood edges, meadows and undergrowth, often in rings.
Distribution: Uneven.

A highly regarded edible species, eagerly sought after in some regions. The season in which it appears is a safeguard against mistakes which could be dangerous (but look out for *Inocybe patouillardii* (p.494) which has a reddening fruit body and a spermatic smell).

---

p.122 *Tephrocybe palustris* (Peck) Donk
Tricholomataceae, Tricholomatales

Description: *Cap striate, brown to olivaceous buff. Stipe tall and slender.*
Cap 0.5–2 cm (0.2–0.8 in), convex then flat or subumbonate, hygrophanous, glabrous, slightly lubricious or shiny, more or less dark brown or with a rather olivaceous tinge in the centre, fading.

Margin with long striations.
Gills notched, quite close, white.
Stipe 5–12 × 0.1–0.25 cm (2–4.7 × 0.04–0.1 in), fragile, soon glabrous, almost concolorous or paler, with white mycelium below. *Flesh* white. Smell rancid.
Microscopic features Sp 6–8.5 × 3.5–5 μm, elliptic to almost tear-shaped, smooth. Bas with siderophilous granules. CC a cutis, the hyphae 3–10 μm diameter. Clamps present.

Season: July–November.
Habitat: Sphagnum and tall mosses in bogs and marshy places.
Distribution: Widespread. Rather rare.

---

p.133 *Tephrocybe rancida* (Fr.:Fr.) Donk
Tricholomataceae, Tricholomatales

Description: *Cap lead-grey to grey-brown. Gills grey. Stipe rooting.*
Cap 1–5 cm (0.4–2 in), conico-convex then campanulate or umbonate, not hygrophanous, glabrous to fibrillose or finely wrinkled, slightly shiny, lead-grey or sometimes more brown. *Gills* notched, quite close, grey to brownish grey.
Stipe 6–12 × 0.2–0.5 cm (2.4–4.7 × 0.08–0.2 in), thicker at ground level and with quite a long root, fibrillose to velvety or strigose towards the base, almost concolorous. *Flesh* grey. Smell very strongly rancid. *Microscopic features* Sp 6.5–8.5 × 3.5–4.5 μm, elliptic to fusoid, smooth. Bas with siderophilous granules. CC a cutis, the hyphae 2–8 μm diameter. Clamps quite rare.

Season: September–November.
Habitat: Broad-leaved and coniferous trees.
Distribution: Widespread. Rather rare.

---

p.117 *Nyctalis agaricoides* (Fr.:Fr.) M. Bon
(= *N. asterophora*)
Tricholomataceae, Tricholomatales

Description: *Cap white then covered with buff-brown powder. Gills aborted, glassy.*
Cap 1–3 cm (0.4–1.2 in), hemispherical or

convex, pruinose then very powdery to fleecy, white then brownish to rather pale buff-brown. *Gills* adnate, often aborted, with a slightly glassy appearance, greyish. *Stipe* 1–3 × 0.2–0.5 cm (0.4–1.2 × 0.08–0.2 in), pruinose, white to whitish. *Flesh* pale. Smell mealy. *Microscopic features* Sp 4–6.5 × 2.5–4 μm, elliptic. CC with a mass of elliptic chlamydospores, often biapiculate and irregularly lumpy, sometimes forked.

Season: September–November.
Habitat: On old blackening *Russula* spp.
Distribution: Widespread but uneven.

The genus *Nyctalis* has very unusual habitat requirements (old *Russulaceae*).

---

p.114 *Nyctalis parasitica* (Bull.:Fr.) Fr.
Tricholomataceae, Tricholomatales

Description: *Cap conical, white. Gills well formed.*
Cap 1–2 cm (0.4–0.8 in), conical then expanded umbonate, silky fibrillose, white to whitish, sometimes greyish when old. *Gills* notched, well developed, white to greyish or dingy beige. *Stipe* 1–4 × 0.2–0.3 cm (0.4–1.4 × 0.08–0.12 in), fibrillose, white. *Flesh* pale. Smell very mealy. *Microscopic features* Sp 4.5–6 × 3–4.5 μm, elliptic to subglobose. CC a cutis. Chlamydospores mainly on the gills, fusiform, smooth.

Season: September–November.
Habitat: On old blackening *Russula* spp.
Distribution: Widespread but uneven.

C. GASTEROMYCETIDEAE: 'STOMACH FUNGI'
This final subclass (commonly known as Gasteromycetes) is distinguished from other Homobasidiomycetes by the special form of its Frb, which is more or less globose or spherical. The fertile part, whether it is organized into a hymenium or not, is internal (at least initially). The principal orders are defined according to the structure of the fruit bodies, the way

they develop, the presence (or absence) of a well-ordered hymenium, etc. They have no direct relationship with the *Agaricomycetideae*.

The species in the photographs and descriptions are arranged in the following order: gasteroid, lycoperdoid, bird's nest fungi, etc. and phalloid. In order to underline the phenomena of convergent evolution leading to a hypogeous life style (a feature of some groups in the *Gasteromycetideae*), the succession of photographs is deliberately chosen so that they end with some Ascomycetes. Indeed, if truffles are highly evolved Ascomycetes, nothing (apart from body shape) connects them with certain equally hypogeous Basidiomycetes.

In this group, the constant characters are: Frb initially globose or spherical; spores spherical, brown, not forcibly discharged at maturity (statismospores), often with a persistent sterigmal appendage; basidia difficult to observe (gleba still young); hyphae of the gleba often brown, thick-walled.

---

p.239 **Common Earthball**
Scleroderma citrinum Pers.:Pers.
Sclerodermataceae, Sclerodermatales

Description: *Globose, scaly, yellow-buff. Peridium thick. Gleba purplish black.*
Frb 2–10 cm (0.8–4 in), sometimes flattened to misshapen or contracted into a pseudostipe, with large, variable, scales or warts, yellow-buff, sometimes with orange-yellow or, on the contrary, with paler patches. Peridium thick (2–5 mm (0.08–0.2 in)), splitting irregularly. Gleba violaceous then blackish violaceous. Smell strong, unpleasant (when cut). *Microscopic features* Sp 9–12 μm diameter, with crests (up to 1.5 μm) forming an incomplete reticulum. Bas 5–15 × 3–5 μm.
Capillitium 3–6 μm diameter, with clamps.
Season: June–December.
Habitat: Bare ground or moss in woods. Usually on

acid soils.
Distribution: Widespread. Very common or common.

Two other common species have a thinner peridium (about 1 mm (0.04 in)): *S. areolatum* Ehrenb., which is smaller with a very short pseudostipe and fine areolate scales; and *S. verrucosum* (Bull.: Pers.) Pers., which is medium-sized, often with a well developed pseudostipe with rhizoids, and a nearly smooth then finely scaly surface.

---

p.240 *Scleroderma polyrrhizon* Pers.:Pers.
(= *S. geaster*)
Sclerodermataceae, Sclerodermatales

Description: *Globose, semihypogeous, star-shaped at maturity. Peridium very thick.*
Frb 5–12 cm (2–4.7 in), globose or pyriform, hypogeous to semihypogeous, covered with earth, tomentose or coarsely wrinkled. Peridium very thick (4–6 mm (0.16–0.24 in)), tearing irregularly into a star with four or five rays, yellow-buff, sometimes slightly olivaceous. Gleba violaceous, marbled whitish then blackish. Smell unpleasant. *Microscopic features* Sp 7.5–11 μm diameter, with crests (up to 1 μm) forming an incomplete reticulum. Capillitium 2–5 μm diameter. Clamps present.
Season: October–January.
Habitat: Coppice woods and wood edges, especially on sandy soil. Usually in warm places.
Distribution: Mainly southern. Quite common in some areas but rare everywhere.

---

p.240 *Scleroderma bovista* Fr.
Sclerodermataceae, Sclerodermatales

Description: *Globose. Pseudostipe with rhizoids. Gleba yellow to brown.*
Frb 2–6 cm (0.8–2.4 in), sometimes slightly clavate, almost smooth then minutely scaly, yellowish with brown scales, and sometimes with reddish or violaceous tones near the base. Peridium quite thin (1 mm (0.04 in)),

tearing irregularly at maturity. Gleba whitish then brownish yellow, brown to black. Smell rather faint. *Microscopic features* Sp 12–17 μm diameter, with crests (up to 2.5 μm) forming a rather regular reticulum. Bas 8–20 × 5–7 μm. Capillitium 2–6 μm diameter. Clamps present.

Season: August–November.
Habitat: On sandy or gravelly soils.
Distribution: Uneven.

Rather difficult to distinguish from some forms of *S. verrucosum*.

---

p.241 *Tulostoma brumale* Pers.:Pers.
(= *T. mammosum*)
Tulostomataceae, Tulostomatales

Description: *Stalked. Head globose, pale, with a rufous peristome.*
Frb 2–5 cm (0.8–2 in) tall. Head globose or flattened (0.5–1 cm (0.2–0.4 in)), white to greyish cream, the peristome prominent, surrounded by quite a bright reddish brown ring. Peridium thin. Gleba powdery, pale rufous. *Stipe* cylindric, 2–3 mm (0.08–0.12 in) diameter, smooth to mottled fibrillose, white then somewhat red-brown to dark grey-brown below. *Microscopic features* Sp 3.5–4.5 μm diameter, with rather low, isolated warts. Capillitium 3–6 μm diameter.

Season: September–January.
Habitat: In moss or short grass, often on sandy or stony soil.
Distribution: Widespread but uneven.

Probably the commonest *Tulostoma* sp.

---

p.243 *Battarraea phalloides* (Dicks.:Pers.) Pers.
Battarraeaceae, Tulostomatales

Description: *Head conico-convex. Stipe very tall, scaly. Base volva-like.*
Frb 12–22 cm (4.7–8.7 in) tall. Head hemispherical or conico-convex to flattened (1–3 cm (0.4–1.2 in)), membranous and fragile, rufous brown. Gleba powdery, dark

brown. *Stipe* cylindric, 10–22 × 1–2 cm (4–8.7 × 0.4–0.8 in), scaly or finely fibrillosely tufted, somewhat rufous brown to rufous. Base volva-like, membranous but the inner part slightly gelatinous when fresh, pale. *Microscopic features* Sp 5–6.5 µm diameter, with fine dense warts. Hyphae (elaters) 60–80 × 3–6 µm, with thickened parts forming more or less complete spirals.

Season: September–November.
Habitat: Sandy soil, banks, wood edges.
Distribution: Very rare, usually Mediterranean or Atlantic or at least in very dry places.

B. stevenii (Liboschitz) Fr. is taller (up to 50 cm (20 in)), with a rather thick, dry volva. Even rarer, more warmth-loving species.

---

p.239 *Pisolithus arrhizus* (Pers.) Rauschert
Pisolithaceae, Sclerodermatales

Description: *Globose or irregular. Gleba divided into iridescent chambers.*
Frb 5–30 × 3–12 cm (2–11.8 × 1.2–4.7 in), subglobose, pyriform or clavate, sometimes elongated, semihypogeous then epigeous, buff to olivaceous brown or more yellow at the base. Peridium thin at maturity, the top tearing irregularly. Gleba powdery, forming small elliptic or pyriform chambers, iridescent then yellow-brown. *Microscopic features* Sp 8–12 µm diameter, warty to spiny. Hyphae hyaline, septate.

Season: September–December.
Habitat: Sandy or gravelly soil, dunes, road sides, spoil heaps, gardens. Usually in warm places.
Distribution: Widespread. Rare.

Extremely variable. May be overlooked.

---

p.241 *Astraeus hygrometricus* (Pers.:Pers.) Morgan
Astraeaceae, Sclerodermatales

Description: *Frb star-shaped, hygroscopic, with a dark grey sphere in the centre.*
Frb 3–10 cm (1.2–4 in), initially spherical.

Exoperidium, quite thin, membranous or hard, opening like a star with 5–20 hygroscopic rays (the rays spread wide in damp conditions and close when dry), brown, cracking. Endoperidium 1–3 cm (0.4–1.2 in), more or less spherical, membranous, the apex tearing irregularly, grey to grey-black. *Microscopic features* Sp 8–13 μm diameter, with prominent warts (about 1 mm (0.04 in)). Hyphae 3–7 μm diameter, somewhat thick-walled.

Season: August–December.
Habitat: Sandy or gravelly soils. Usually in warm places.
Distribution: Widespread. Quite rare to very rare.

---

p.242 *Sphaerobolus stellatus* Tode:Pers.
Sphaerobolaceae, Sclerodermatales

Description: *A minute species, globose then opening like a star.*
Frb globose, up to 3 mm (0.12 in), whitish to yellowish. Peridium opening like a star, orange-yellow or bright orange, enclosing a translucent, milky or yellowish to orange sphere (1–2 mm (0.04–0.08 in)). Very gregarious. *Microscopic features* Sp 7.5–10 × 4–5.5 μm, elliptic, hyaline, smooth, thick-walled. Bas elliptic, short.

Season: June–December.
Habitat: Rotten wood, sawdust, plant debris.
Distribution: Not fully assessed. Overlooked.

The firing of the central sphere to a considerable distance from the Frb (several metres) is very unusual in the Gasteromycetes. Many species use various external factors to achieve spore dispersal (wind, rain, dietary requirements of small animals, etc.).

---

p.241 **Earthstar**
Geastrum triplex Junghuhn
Geastraceae, Lycoperdales

Description: *Frb star-shaped. Peridium thick, splitting into three layers.*
Frb 4–15 cm (1.6–6 in). Initially

hypogeous and globose umbonate, onion or pear-shaped. The thick (5 mm (0.2 in)) and rubbery exoperidium opens like a star (five to seven rays) and splits into two or three layers, the innermost often forming a collar around the endoperidium. The latter (2–4 cm (0.8–1.6 in)) is spherical, thin-walled, opening at the apex with a prominent slightly ciliate peristome surrounded by a pale ring. Brownish.
Microscopic features Sp 4–6 µm diameter, with rather coarse warts (< 1 µ (0.04 in) high). Capillitium 1.5–10 µm diameter. Clamps present.

Season: August–December.
Habitat: Broad-leaved and coniferous woodland and coppice woodland.
Distribution: Widespread but uneven.

One of the more common earthstars.

---

p.240 *Geastrum fimbriatum* Fr. (= *G. sessile*)
Geastraceae, Lycoperdales

Description: *Frb star-shaped. Exoperidium medium. Peristome fimbriate.*
Frb 2–7 cm (0.8–2.8 in), initially ovoid. Exoperidium pale cream to pinkish, of average thickness (1–3 mm (0.04–0.12 in)), opening like a star (6–10 rays) and only rarely splitting into two layers. Endoperidium 0.5–3.5 cm (0.2–1.4 in), sessile, spherical, ochraceous beige, thin-walled, opening by a conical, fimbriate peristome, no surrounding ring.
Microscopic features Sp 3–4.5 µm diameter, with slender, sometimes confluent warts (< 0.5 mm (0.2 in) high). Capillitium 1–2.5 µm diameter. Clamps absent.

Season: August–November.
Habitat: Broad-leaved and coniferous woodland, parks, rides.
Distribution: Widespread but uneven.

Probably the commonest species in the genus in Europe.

p.241 *Geastrum morganii* Lloyd
Geastraceae, Lycoperdales

Description: *Frb star-shaped. Exoperidium medium. Peristome conical, furrowed.*
Frb 2–6 cm (0.8–2.4 in), initially very umbonate. Exoperidium beige grey, reddish brown to dark brown, of average thickness (2–3.5 mm (0.08–0.14 in)), opening like a star (six to eight rays). Endoperidium 0.5–2.5 cm (0.2–1 in), sessile, more or less spherical, brown to grey-brown, thin-walled, opening by a pore at the top of a raised, strongly furrowed cone. *Microscopic features* Sp 4–6 µm diameter, rather finely warted, sometimes confluent (about 0.8 mm (0.32 in) high). Capillitium 1.5–4 µm diameter. Clamps present.
Season: October–December.
Habitat: False acacia and pine, on sand.
Distribution: Atlantic coast. Rare.

---

p.241 *Myriostoma coliforme* (With.:Pers.) Corda
Geastraceae, Lycoperdales

Description: *Frb star-shaped. Endoperidium on several stalks, with numerous pores.*
Frb 6–15 cm (2.4–6 in), initially spherical. Exoperidium light beige then brownish, of medium thickness (1–2.5 mm (0.04–0.1 in)), opening like a star (5–12 unequal rays). Endoperidium 3–8 cm (1.2–3.2 in), raised on many stalks 2–5 mm (0.08–0.2 in) tall, spherical or flattened, whitish then brownish grey, thin-walled, opening by numerous pores (up to 50). *Microscopic features* Sp 3.5–5 µm diameter, the warts linked by their interwoven crests, up to 1.5 mm (0.06 in) high. Capillitium 1.5–3 µm diameter.
Season: July–December.
Habitat: False acacia, pines, in dunes.
Distribution: Coastal, especially in warm places near the Atlantic. Rare to very rare.

Recognizable at a glance.

p.238 *Calvatia excipuliformis* (Scop.: Pers.) Perdeck
Lycoperdaceae, Lycoperdales

Description: *Surface covered with small, short-lived spines and floccose granules. Brownish buff.*
Frb 8–20 cm (3.2–8 in), pyriform then developing into a stalk 2–5 cm (0.8–2 in) tall and clavate head 5–12 cm (2–4.7 in) diameter, with short-lived, slender spines and floccose granules. Peridium brownish buff, tearing irregularly or gradually disappearing. Gleba whitish then greenish yellow to brownish. *Stipe* sterile. *Microscopic features* Sp 4.5–6.5 µm diameter, with low, rather dense warts. Capillitium 2–4 µm diameter, with pores. Clamps present.

Season: August–December.
Habitat: Undergrowth, wood edges.
Distribution: Widespread. Rather common.

The stipe of old fruit bodies from the previous year may be found at the beginning of the fungus season.

---

p.238 *Lycoperdon piriforme* Sch.:Pers.
Lycoperdaceae, Lycoperdales

Description: *White then brownish, soon glabrous, in clusters.*
Frb 2–5 cm (0.8–2 in), pyriform (head 1–4 cm (0.4–1.6 in) diameter). Surface with short-lived, erect, conical warts, glabrous, soon smooth. Peridium white then beige to brownish, brown at maturity, opening by a rather wide apical pore. Base with branched rhizoids and often giving rise to a string of several fruit bodies. Gleba whitish then olivaceous to brown. *Microscopic features* Sp 3.5–5.5 µm diameter, smooth. Capillitium 2–6 µm, without pores. Clamps absent.

Season: July–December.
Habitat: In clusters, on stumps and rotten wood.
Distribution: Widespread but uneven.

The genus *Lycoperdon* is characterized by a rather small regular pore (apical ostiole).

p.238 *Lycoperdon mammiforme* Pers.:Pers.
Lycoperdaceae, Lycoperdales

Description: *Rather squat, white. Peridium with short-lived, floccose plates.*
Frb 4–10 cm (1.6–4 in), pyriform or rather squat (head 3–5 cm (1.2–2 in) diameter), initially covered with a creamy or floccose veil, breaking into very fleeting plates, persisting longer at the base; white. Gleba whitish then yellowish brown to brown. *Microscopic features* Sp 4.5–6.5 µm diameter, with short, finely spiny, rather distant warts. Capillitium 3–6 µm diameter, pores rare.
Season: August–October.
Habitat: Especially beech. Warm places on calcareous soils.
Distribution: Southern. Rare.

---

p.238 *Lycoperdon perlatum* Pers.:Pers.
Lycoperdaceae, Lycoperdales

Description: *White then brownish, with conical spines and smaller beads.*
Frb 4–10 cm (1.6–4 in), pyriform or sometimes with a rather narrow stipe, the head 3–6 cm (1.2–2.4 in) diameter, slightly conical at the apex, covered with conical warts (1–2 mm (0.04–0.08 in) high) and fine granules, all short-lived and leaving more or less polygonal scars; white then brownish to brown. Gleba whitish then olivaceous yellow to brown. *Microscopic features* Sp 3.5–4.5 µm diameter, with fine distant warts. Capillitium 2–7 µm diameter, pores rare.
Season: June–January.
Habitat: Quite varied.
Distribution: Very widespread but uneven.

One of the commonest puffballs. When the brown stage is reached, this species can be confused with *L. foetidum* Bonorden, which is brown from the outset and has a strong, unpleasant smell. Two other common species have persistent, finer and softer spines: *L. umbrinum* Pers.:Pers. is brown, with yellow-brown spores (the colour of the gleba) and occurs on acid soils, while *L. molle* Pers. Pers. is paler, has spines mixed with mealy granules, reddish chocolate-brown spores and usually occurs on neutral soils.

---

p.238 *Lycoperdon echinatum* Pers.:Pers.
Lycoperdaceae, Lycoperdales

Description: *Frb with erect, brown, convergent spines.*
Frb 4–7 cm (1.6–2.8 in), pyriform or with a rather short stipe, the head (3–6 cm (1.2–2.4 in) diameter), regular, covered with long slender spines (3–6 mm (0.12–0.24 in)), converging into more or less short-lived tufts and leaving reticulate scars; brown. Gleba whitish then olive to olivaceous brown. *Microscopic features* Sp 4–5.5 µm diameter, with rather fine warts. Capillitium 2–7 µm diameter, with rather numerous, wide pores.
Season: August–October.
Habitat: Mature beech woods, rarer under other broad-leaved trees, on calcareous soils. Often in warm places.
Distribution: Quite widespread. Rather rare.

---

p.238 *Vascellum pratense* (Pers.:Pers.) Kreisel
Lycoperdaceae, Lycoperdales

Description: *Warty then smooth. Fertile part separated by a distinct diaphragm.*
Frb 2–5 cm (0.8–2 in), subglobose then pyriform but often flattened or truncate (3–6 cm (1.2–2.4 in) diameter) and more or less pinched into a thick base, covered with short, fleeting, granular spines, white then buff brown. Gleba whitish then yellow to olive, finally brown, separated from the sterile base by a very distinct diaphragm. *Microscopic features* Sp 3.5–4.5 µm diameter, with fine warts, thick-walled. Capillitium 2–5 µm diameter, without pores but encrusted with amorphous matter.
Season: June–December.
Habitat: Lawns and meadows, sometimes in clearings.
Distribution: Widespread but uneven.

p.239 *Bovista limosa* Rostrup
Lycoperdaceae, Lycoperdales

Description: *Very small, subspherical, smooth, whitish then brownish.*
Frb very small, 5–15 mm (0.2–0.6 in), subglobose, nearly smooth. Peridium separating into two layers, the outer white then pale ochraceous beige, the inner reddish brown to dark brown. Gleba whitish then yellowish to brownish.
Microscopic features Sp 4–6 μm diameter, finely warted, the sterigmal remnant 5–10 μm long. Hyphae nonporoid.
Season: August–November.
Habitat: Sandy calcareous substrates.
Distribution: Rather northern. Rare to very rare, in scattered coastal areas, very rare inland.

B. plumbea Pers.:Pers., which has a white exoperidium lifting off like an egg-shell, and a membranous or papery, lead-grey endoperidium, is larger (frequent in lawns and grassy places).

---

p.240 *Mycenastrum corium* (Guern.) Desvaux
Mycenastraceae, Lycoperdales

Description: *Tearing coarsely into an irregular star. Gleba powdery, brown.*
Frb 5–13 cm (2–5.1 in), subglobose to pyriform, partly buried then epigeous, with well-developed mycelial cords. Peridium white then grey to grey-brown, separating into two parts, the outer breaking into short-lived scales, the inner membranous, quite thick, opening like an irregular star when mature (5–10 unequal rays). Gleba whitish then olivaceous to brown, powdery.
Microscopic features Sp 7.5–1 μm diameter, subspherical, reticulate. Hyphae 10–18 μm, sinuate, with straight or recurved spines, with a characteristic appearance of barbed wire.
Season: August–December.
Habitat: Meadows, lawns, disturbed coppice areas, in warm, very dry places.
Distribution: Rather southern. Rare.

p.242 **Common Bird's Nest**
Crucibulum laeve (Huds) Kambly
Nidulariaceae, Nidulariales

Description: *Frb cushion- then cup-shaped, yellow-buff then brown, filled with small cream-coloured 'eggs'.*
Frb globose or cushion-shaped, 5–10 × 4–7 mm (0.2–0.4 × 0.16–0.28 in), beige then brownish or brown, closed by an orange to light yellow membrane, which tears to reveal a chamber containing small cream-coloured 'eggs' (5–15 peridioles). Outer surface finely felty. Inner surface smooth. *Microscopic features* Sp 7–10 × 3–5 µm, elliptic, hyaline, smooth. Hyphae hyaline to brown, 1–4 µm diameter, with clamps, the outer surface of the peridium composed of thick-walled, contorted, brown hyphae ornamented with coarse spines.

Season: July–November.
Habitat: On wood or a range of plant debris.
Distribution: Widespread. Quite common.

The *Nidulariaceae* has spectacular fruit bodies looking like bird's nests (hence the common name). Often hard to find.

---

p.242 *Cyathus olla* (Batsch:Pers.) Pers.
Nidulariaceae, Nidulariales

Description: *Frb brown, cup-shaped. Inner surface lead grey, with large peridioles.*
Frb ovoid or cylindric to obconical, 5–15 mm (0.2–0.6 in) high, brownish or grey-brown, closed by a whitish membrane which tears from the centre to reveal 5–10 grey then silvery beige peridioles. Outer surface finely felty. Inner surface smooth and shiny, lead grey. *Microscopic features* Sp 9–12 × 5–7 µm, elliptic, hyaline, smooth, somewhat thick-walled. Hyphae 1–4 µm diameter, with clamps, somewhat thick-walled, the outer surface of the peridium composed of thicker hyphae, 5–15 µm diameter, clavate, thick-walled.

Season: July–December.
Habitat: Gregarious, on soil or plant remains.
Distribution: Widespread but uneven.

p.242 *Cyathus striatus* (Huds.:Pers.) Willdenow
Nidulariaceae, Nidulariales

Description: *Frb funnel-shaped, fluted inside, brown and strigose outside.*
Frb cylindric to obconical, 8–20 × 5–15 mm (0.32–0.8 × 0.2–0.6 in), dark chestnut-brown, closed by a white membrane, which breaks up to reveal 5–15 grey peridioles. Outer surface with stiff erect hairs. Inner surface glabrous, fluted to striate, grey-brown. *Microscopic features* Sp 15–20 × 6–9 µm, ovoid-elliptic, hyaline, smooth, thick-walled. Hyphae 2–4 µm diameter, with clamps, the outermost surface of the peridium composed of thick constricted hyphae, 5–15 µm diameter, the terminal elements elliptic, brown.
Season: August–November.
Habitat: Often gregarious, on wood (branches) or plant remains.
Distribution: Widespread but uneven.

Probably the commonest of the *Nidulariaceae*. *C. stercoreus* (Schw.) De Toni, which often occurs on dung or animal droppings, and sometimes on burnt ground, is intermediate to *C. olla* since the Frb is not striate and has less erect hairs.

p.242 *Nidularia farcta* (Willd.:Pers.) Fr. & Nordh.
Nidulariaceae, Nidulariales

Description: *Yellowish cream, the pale then golden brown peridioles embedded in mucilage.*
Frb globose or irregular, 5–12 mm (0.2–0.47 in) diameter, yellowish white then dingy buff, closed by a gelatinous white membrane, breaking up to reveal 10–25 yellowish white then golden chestnut peridioles, embedded in a hyaline mucilage. Outer surface with soft erect hairs. *Microscopic features* Sp 5–12 × 5–7 µm, ovoid elliptic, hyaline, smooth, thick-walled. Outer peridial hyphae thick, branching and diverticulate, 2–5 µm diameter.

Season: August–October.
Habitat: Rotten wood, in damp places.
Distribution: Widespread. Rare to very rare.

Old specimens can be seen, reduced to a soft membrane attached to the substrate and surrounded by the peridioles which have been dispersed (by rain drops for example).

---

p.242 **Stinkhorn**
Phallus impudicus L.:Pers.
Phallaceae, Phallales

Description: *Phallic-shaped when mature, the head ridged like a honeycomb.*
Frb spherical or ovoid (2–6 cm (0.8–2.4 in) diameter), white, with thick, well developed rhizoids. In section: outer layer gelatinous, hyaline to brownish (covered with a thin papery membrane); middle layer a greenish brown paste; central part conical, white, sometimes hollow. At maturity, the white part elongates, forming a fusiform, pitted and fragile, white stipe (up to 25 cm (10 in)), bearing the dark green, very evil-smelling gleba all over the conical, honeycomb-like head set at the apex of the stipe. The gelatinous part remains as a soft volva at the base. *Microscopic features* Sp $4–5 \times 1–2$ µm, cylindro-elliptic, brownish. Bas with 6–8 spores.
Season: July–December.
Habitat: Broad-leaved and coniferous woodland; rather varied.
Distribution: Widespread but uneven.

Can be located from a distance thanks to the very strong smell of carrion, which can be detected before the fruit bodies are seen. The white part of the young eggs is edible with a pleasant flavour of young raw peas.

---

p.243 *Phallus hadrianii* Vent.:Pers.
Phallaceae, Phallales

Description: *Like* P. impudicus *but with a lilaceous pink volva, smell faint.*

Same characters as *P. impudicus* (see p.879) but often slightly smaller, the volva lilaceous pink, the gleba less evil-smelling. The head (just poised on the stipe apex) is often more ovoid than in *P. impudicus* (margin pressed against the stipe). *Microscopic features* Sp 3–4 × 1–1.8 µm, cylindro-elliptic, yellowish, thick-walled. Bas 8-spored.

Season: September–December.
Habitat: Coastal dunes in warm sheltered places but also, still on sandy substrates, in continental to rather steppe-type areas.
Distribution: Widespread. Quite rare to very rare.

---

p.243 **Dog Stinkhorn**
Mutinus caninus (Huds.:Pers.) Fr.
Phallaceae, Phallales

Description: *Similar to the Stinkhorn but more slender, the head reddish, attached.*
Frb ovoid or cylindric (up to 4 × 2 cm (1.6 × 0.8 in)), white, with rhizoids. In section, it has the same structure as the Stinkhorn but the central part is more elongated and red or reddish. When mature, the stipe is slender, cylindric, soon fragile and often curved (up to 15 × 1 cm (6 × 0.4 in)), reddish to orange, fading from the base. It bears an ovoid elliptic head, attached to the stipe by its margin, covered with the dark olive, offensively smelling gleba. Volva basal, soft, elongated. *Microscopic features* Sp 4–5 × 1–2.5 µm, cylindro-elliptic, hyaline, smooth. Bas 6-spored.

Season: July–November.
Habitat: Broad-leaved, more rarely coniferous woods.
Distribution: Widespread. More common in warmer districts.

Closely related to this elegant fungus are a few, very rare species, often with more red colouring and a proportionately larger head. These exotic species, originating in America or the Tropics, sometimes occur in greenhouses.

p.243 *Clathrus archeri* (Berk.) Dring
Clathraceae, Phallales

Description: *Like the Stinkhorn but forming a red star when mature.*
Frb initially ovoid (up to 4×3 cm (1.6×1.22 in)), whitish, with pinkish or lilaceous rhizoids. In section, it has the same structure as the Stinkhorn but the central part is divided into several red sections. When mature, four to eight bright red star-like rays (attached to a common stipe) emerge (up to 15 (6 in) or even 20 cm (8 in) long), bearing the dark greenish brown, evil-smelling gleba, on the inner face of the rays. Volva basal, soft, elongated. *Microscopic features* Sp $5-6.5 \times 2-2.5$ μm, cylindro-elliptic, hyaline, smooth.

Season: August–October.
Habitat: Moist humus under broad-leaved trees, rarer under conifers, damp grassland.
Distribution: Very common in some districts but generally rather rare.

An Australian species, introduced initially into eastern France at the beginning of the twentieth century, from where it has spread. Now very common in some woodlands.

---

p.244 *Clathrus ruber* Pers.:Pers.
Clathraceae, Phallales

Description: *Like the Stinkhorn but the central part is a bright red, lattice-like sphere.*
Frb spherical (up to 3 cm (1.2 in)), whitish, with white rhizoids. In section, it has the same structure as the Stinkhorn but the central part is a fretwork of wavy red bars. When mature, a bright red, spherical, lattice-work cage is exposed (up to 12 cm (4.7 in) diameter), bearing the greenish brown, evil-smelling gleba on the inner surface. Volva basal, soft. *Microscopic features* Sp $5-6 \times 1.5-2$ μm, cylindro-elliptic, greenish, smooth.

Season: July–October.
Habitat: Often in gardens, parks, warm disturbed places.

Distribution: Rare to very rare, Mediterranean, very occasionally occurring in more northerly areas (Ireland) where it is always sporadic.

A spectacular species of warm sites, confined to particularly protected places in its more northerly locations.

Part III

Appendices

GLOSSARY

Adnate (hymenophore* especially gills or tubes) attached to the stipe at (approximately) a right angle.

Allantoid (spore) sausage-shaped (arching).

Amarescent describes flesh which is slightly bitter or becomes so when chewed.

Amyloid describes material which gives a blue-black reaction with Melzer's reagent.

Anamorph asexual reproductive state.

Apiculus a small appendage on Basidiomycetes* spores (*see* p.28), the remains of the point of attachment to the basidium*.

Apothecium (pl. apothecia) the disc- or cup-shaped sporophore* of the Pezizomycetideae (*see* p.256).

Appressed (e.g. a scale) lying very closely and in weak relief to the surface.

Areolate divided into block- or net-like areas.

Ascoma (pl. ascomata) Ascomycete sporophore* (*see* p.27).

Ascospore sexual reproductive spore in the Ascomycetes (*see* p.27).

Ascus fertile cell in the Ascomycetes (*see* p.27), often shaped like a long tube, enclosing the sexual reproductive spores.

Basidioma (pl. basidiomata) Basidiomycete sporophore (*see* p.27).

Basidiospore sexual reproductive spore in the Basidiomycetes (*see* p.28).

Basidium (pl. basidia) fertile, often club-shaped cell in the Basidiomycetes (*see* p.27), bearing the sexual reproductive spores on the tips of the sterigmata*.

Buff a pale brownish yellow or ochre colour; yellowish or very light tan.

Caespitose in clumps or clusters.

Callus a convex protrusion on the germ pore*.

Campanulate bell-shaped.

Capitate tip (e.g. of cystidia) shaped like a head.

Capitulate with a small apical swelling (term usually used in microscopy).

Cauline pertaining to the stipe*.

Cheilocystidia cystidia* situated on the edge of the hymenophore.

Chrysocystidia thin-walled cystidia* with contents which stain strongly yellow in aqueous alkali solutions.

Clamp a specialized form of hyphal outgrowth which makes a connection between two cells.

Clavate club-shaped.

Coenocytic (structure) consisting of numerous nuclei in adjacent cells which are not separated by septa* (siphon).

Collar a thin membrane surrounding the stipe apex and so creating a clear space around it, to which the gills are attached.

Concolorous all one colour; the same colour (as another part).

Conidiophore a hypha producing and bearing conidia.

Conidium (pl. conidia) asexual reproductive spore.

Coralloid coral-shaped, i.e. branching many times.

Corticate when the cortex (outer part) has a different texture from the innermost layers.

Cortina a filamentous veil remaining on the stipe as very fine threads after the cap has expanded.

Crenate (cap margin) with rounded teeth, scalloped.

Crenulate (cap margin) with very small rounded teeth.

Crested ornamented with low ridges.

Cuticle cap or stipe surface; the outermost layer of hyphae.

Cutis (cuticle) when the outer layer of hyphae lies parallel to the surface.

Cylindro-clavate almost cylindrical, but faintly clavate*.

Cystidioles small cystidia.

Cystidium (pl. cystidia) a sterile cell inserted between the basidia*.

Decurrent (hymenophore) extending down the stipe on which it is inserted.

Dendroid a branching arrangement with slender branched elements.

Denticulate with very fine teeth.

Dextrinoid describes material which gives a reddish brown reaction with Melzer's reagent.

Dikaryophase that phase of the reproductive cycle when the cells have two nuclei.

Dimitic (structure) composed of two different types of hypha.

Discoid more or less disc-shaped.

Dissepiment partition between the pores of a polypore.

Diverticulate	with numerous peg-like outgrowths.
Emarginate	(hymenophore*) with a small notch where it is inserted on the stipe.
Endoperidium	in the Gasteromycetes (*see* p.865), the inner envelope of the fruit body.
Endosporium	innermost layer of the spore wall.
Epigeous	forming sporophores above the surface of the soil.
Erumpent	arising below the bark and pushing through it to the open air.
Exoperidium	in the Gasteromycetes (*see* p.865), outer envelope of the fruitbody.
Fimbriate	very finely divided, eroded.
Fusiform	spindle-shaped.
Germ pore	often apical thin area or hollow in the spore wall through which the spore may germinate.
Gleba	fertile part of the Gasteromycetes (*see* p.866) and hypogeous* Ascomycetes (*see* p.257). Means many-angled.
Grandinioid	surface irregularly warty or rough, with small points, crests etc.
Guttule	small drop, droplet (often used for intracellular inclusions).
Hispid	with stiff, rather dense hairs.
Hygrophanous	changing colour as it dries.
Hymeniderm	a cuticle formed of hymenium-like units, i.e. one composed of a layer of single cells or hyphal tips.
Hymeniform	a term used in microscopy to describe a structure composed of clavate cells, arranged parallel to each other forming a palisade and resembling the structure of a hymenium*.
Hymenium	a single layer of fertile cells (asci or basidia, depending on the type of fungus).
Hymenophore	surface bearing a hymenium*.
Hypha	(pl. hyphae) name given to fungal cells.
Hyphidium	a slender, sterile, sometimes slightly branched cell, more or less projecting beyond the other cells in the hymenium (term usually applied to the Aphyllophorales).
Hypogeous	forming its sporophores below the soil surface.

Inferior (ring) ascending.
Intracellular (pigment) occurring within the cell.
Ixocutis cuticle composed of barely differentiated hyphae lying parallel to the surface (cutis), but gelatinized (ixo).
Ixotricho- type of (cap or stipe) covering made of a
dermium gelified (ixo) trichodermium (with long erected hairs).

Karyogamy reproductive phase when cell nuclei fuse.

Laciniate finely but deeply cut or torn.
Lactifer a specialized hypha in sporophores, carrying and secreting a latex (e.g. the 'milk' in *Lactarius* species).
Lageniform bottle-shaped.
Lamprocystidia in the Aphyllophorales, thick-walled cystidia encrusted at the apex with refractive crystals.
Lanceolate shaped like the point of a spear.
Latex term for the milk in *Lactarius*.
Lecithiform carafe-shaped, i.e. with a swollen base, narrow neck and spherical head at the apex (stopper of the carafe).
Lignicolous growing on or in wood (lignin); not on the bark.

Margin edge of the cap.
Marginate (bulb at stipe base) when there is a distinct margin or rim round the edge of the bulb.
Medallion a clamp* which has a distinct gap between
clamp the main hypha and the hook connecting the cells.
Metachromatic describes structures which turn reddish to violet in Cresyl Blue.
Metuloid (cystidia) thick-walled and encrusted with refractive crystals at the apex.
Micaceous looking like mica, i.e. rather shiny.
Mitriform (spore) reminding of the mitre (bishop hat), with two lateral spurs close to the apiculus (for example in *Coprinus micaceus*).
Monomitic (structure) composed of one kind of hyphae.
Mucro a small rather narrow point at the tip of a microscopical element; also used for certain small narrow pointed umbos* (on caps).
Mucronate having a mucro*.
Mycelium (pl. mycelia) mass of hyphae forming the nutritional range of a fungus.
Mycetism poisoning from eating the higher fungi.

Mycorrhizal (species) forming mycorrhizae (*see* p.24).
Mycotoxicosis poisoning from ingesting the toxins produced by lower fungi.

Ochraceous more or less buff or ochre.
Ostiole minute or small opening through which spores of Pyrenomycetes (*see* p.247) or Gasteromycetes (*see* p.865) are released.

Papillate having a small papilla, i.e. a small terminal outgrowth, on a hypha or some other body.
Paraphysis sterile cell inserted between the asci* in an Ascomycete hymenium.
Parasite living at the expense of the host.
Parietal (pigments) occurring on the wall.
Pectinate rather well-marked, comb-like lines on the cap margin.
Pedicellate borne on a slender stalk.
Peridiole 'egg' of a bird's nest fungus.
Peridium in Gasteromycetes, envelope protecting the gleba.
Peristome in Gasteromycetes, an edging round an opening.
Perithecium (pl. perithecia) elementary ascoma in the *Pyrenomycetideae*.
Pileocystidia cystidia occurring on the cap surface.
Plasmodium state of certain 'false fungi' (Myxomycetes, *see* p.26), resembling a giant amoeba and capable of moving across a substrate engulfing its prey.
Plasmogamy reproductive stage when cytoplasms fuse.
Pleurocystidia those situated on the sides of the hymenophore.
Pruina a fine substance resembling very light flour (e.g. on grapes).
Pruinose covered in pruina and, accordingly, looking rather matt.
Pubescent with fine hair, downy.
Pulvinate cushion-shaped.
Punctate with small spots or dots.
Pyriform pear-shaped.

Refractive shiny.
Reniform kidney-shaped.
Resupinate lying flat on the substrate, lacking a distinct stipe or cap.
Reticulum a net-like pattern.
Rhizoid a fine thread composed of a mass of mycelium and looking like rootlets (N.B. fungi do not have roots).

Rimose	radially fissured.
Ring	a partial veil of more or less membranous texture, remaining on the stipe of certain adult fungi.
Rivulose	concentrically fissured.
Rooting	when an extension of the stipe reaches down into the substrate.
Rostrate	beaked; with a narrow appendage.
Rufous	quite dark reddish brown.
Rugose	rough.
Rugulose	minutely roughened.
Saprotroph	living by feeding on dead organic matter.
Sclerotium	an oblong, often hard form in which a fungus may resist unfavourable conditions.
Scrobiculate	with scrobicules*.
Scrobicule	a small pit.
Scurfy	finely granular.
Septate	with a septum/septa.
Septum	(pl. septa) a cell wall or partition.
Serrulate	coarsely toothed.
Sessile	lacking a stalk or stipe.
Sheath	veil forming a kind of sock starting at the stipe base and reaching a more or less well-defined limit which may be like a ring or ring zone.
Siphon	very long element, composed of nonseptate, multinucleate cells.
Spathulate	shaped like a spatula.
Sphaerocyte	round, isodiametric cells occurring in the cap cuticle of certain genera and in *Amanita* veils.
Sphaero-pedunculate	(term used in microscopy) with a broadly rounded head on a narrow stalk.
Spore	reproductive cell of a fungus (and other organisms).
Sporophore	structure bearing the fertile cells which produce the spores.
Squamulose	with small scales.
Squarrose	extremely scaly, the scales upturned.
ss. auct. europ.	in the sense of European authors.
Sterigma	(pl. sterigmata) small, spore-bearing point growing at the apex of a basidium.
Stipe	stalk of a fungus.
Stipitate	with a stipe*.
Styptic	(taste) very acrid, attacking the throat.
Strigose	with large, very stiff hairs.
Stroma	structure bearing several ascomata.
Stuffed	(stipe) when the innermost part has a different consistency from the outermost.

Sub-	prefix meaning 'almost', which may be attached to many words.
Subcellular	(cuticle) composed of juxtaposed rounded cells.
Subiculum	layer of superficial hyphae covering the host and bearing sporophores*.
Sulcate	grooved.
Superior	(ring) hanging.
Suprahilar plage	flattened or depressed area on the spore wall near the hilar appendix (apiculus*).
Symbiont	a partner in a symbiotic relationship (see p.24).
Teleomorph	sexual reproductive state (opposite of anamorph).
Thallus	term used for the vegetative apparatus of the lower plants reproducing by spores (Cryptogams), which once included the fungi. Since these are now placed in a separate kingdom, the term no longer applies.
Tomentose	with fine dense, long and entangled hairs.
Torulose	looking like a necklace of beads.
Trichoderm	type of cap cuticle in which hair-like elements project from the surface.
Trichodermial	like a trichoderm*.
Trimitic	(structure) composed of three different types of hyphae.
Turbinate	top-shaped.
Umbilicate	depressed, like a navel.
Umbo	a central bump or raised area on the cap.
Umbonate	with an umbo.
Utriform	bladder-shaped.
Veil	layer of protective tissue on an agaric; either universal or partial (see pp.363–5).
Ventricose	(gills) broad in the middle and narrower at both ends.
Verrucose	warty.
Viscid	very sticky.
Zygospore	Zygomycetes spore (see p.27).

* Refer to this word in the Glossary

Picture Credits

t = top
c = centre
b = bottom
uc = upper centre
lc = lower centre (lcl = lower centre left)

All the photographs that appear in this guide are by Jacques Vast apart from the following:

André Bidaud: 103bl
Patrick Boisselet: 44br, 122tr, 191tl, 200b, 220c, 235tr, 236cl, 241cl, 243c
René Chalange: 102br, 138cl, 141tl, 153lc, 158ucl, 171b, 183tr, 193b, 199tr, 214tr, 229c, 231tr, 232cr, 233b, 234t, 234bl
Maxime Chiaffi: 58bl, 76br, 86tr, 89b, 132b, 156tr, 189cl, 199br, 208ucl, 234c, 239c
Sandy Clelland: 82t, 133cl, 137cl, 234br, 242tr
Jean-Pierre Cornu: 64tr, 70bl, 72tl, 92b, 101tr, 167lc, 176c
Gilles Corriol: 52b, 53br, 112tr, 120tl, 182b, 228tr, 229tl
Régis Courtecuisse: 121tr, 121lc, 150cl, 165bl, 178c, 179t, 180bl, 184b, 200cl, 225c
Guillaume Eyssartier: 187bl
Mathilde Guény: 144tl
Jacques Guinberteau: 226br
Pascal Hériveau: 40cl, 57cr, 64cr, 94tl, 99cr, 104tl, 110bl, 119t, 126tl, 128lc, 143c, bl; 144bl, 162tl, 169tr (both), 173c, 191c, 209bl, 212uc, 213tr, b; 215tr, 222bl, 235b, 241t
Robert Le Coz: 69lcl, 119br, 162c, 165cr, 181t
Gérard Martin: 39cr, 40tl, 40bl, 63bl, 72cr, 86tl, 111br, 157tr, cr; 216bl, 231c, 235cr
Pierre-Arthur Moreau: 63c, 64cl, 72cl, 80b, 81t, 86bl, 91cr, 93lcl & r, 98cl, 125br, 172tr, 181cr (both)
Natural Image: B. Candy: 47tr, 112bl, 180t, 202cl, 209br; S. Cleland: 105br, 111tl, 151t; G. Dickson: 38b, 42bl, 47cl,

49tr, 49cl, 52cr, 58t, 60b, 64tl, 68lc, 71ucr, 77cl, 78tr, 80c, 81b, 95tl, 99b, 111bl, 113t, cl; 114tr, 123br, 126br, 129cr, 139tr, 140tr, 141cl, 144tr, 145tl, 150cl, 151br, 160tl, 166cl, br; 167bl, 168bl, br; 171cr, 187br, 205t, 209c, 211cr, br; 217bl (both), 218cr, 219tl, 230tr, 232cl, 233tl, 236cr, 238cl, bl; N. Diserens: 54cr, 62bl, 114bl, 126c, 133cr, 184tl, 185t, 216br; Bob Gibbons: 39cl, 40tr, 42t, 43c, 48t, 48br, 50b, 59tl, 60tr, 67tl, 68bl, 70tr, c; 77bl, 78bl, 100cl, 112br, 152tr, 185bl, 232tl, 239b, 242br, 244t; N. Legon: 56cl, cr, b; 57tr, 98b, 100t, 142b, 171cl, 172tl, 178c, 226br; P.R. Perfect: 49br; G. Redeuilh: 42cr, 59b, 69uc, 80t, 91br (both), 96tr, 98cr, 110c, 124cr, 128tl, 145bl, 148br, 150tl, 158t, 159tr, 163tl, tr; 168tl, 169tl, 170cr, 173b, 174b, 189cr, 206tl; 215b, 217t, 221b, 224tl, 225t, 226bl, 229bl, 230b, 232tr, 233c, 235tl, 235cl, 236t; John Roberts: 40cr, 47tl, 47b, 48cr, 49tl, 49bl, 50tl, 54b, 58br, 59c, 60tl, 62t, 63br, 66tl, tr, cr; 67bl, 69t, 70br, 71ucl, lcl; 74br, 75b, 82c, 85b, 87tl, 88lcr, 89lcr, 91cl, 92cl, 94b, 103tl (both), 103cl (both), bl, br; 105tl, tr; 108b, 109t, 111tr, c; 114tl, 115cl, 118bl, 121tl, 122tl, 124b, 127t, 128b, 131bl, 132t, 133br, 134tr, 135tl, 136br, 137t, 138tr, 140c, 140bl, 144c, 146tl, 148tl, 149t, 149bl, 152tl, 154b, 161bl, 162bl, 165br, 166cr, bl; 168tr, 171t, 174tl, uc; 175cr, 177cl, 178tr, 179cr, 180br, 181br, 182tr, cl; 184tr, 185br, 188bl, 189t, 193c, 195c, br; 196b, 198b, 207uc, 208tr, 210c, 214tl, 214b, 215cr, 217br, 218c, cl; 224tr, bl (both); 228c, 230tl, 231bl, br; 233tr, 236br (both), 238tl, 238br, 239tr, 240br, 241cr, br; 242tl, cl, bl, 243t; K. Rowland: 43br, 48bl, 147lc, 176t, 242cr; A. Wason: 43bl, 57b, 58c, 68t, 76tl, 131cl, 143br, 149br, 190c; Peter Wilson: 238tr

Alan Outen: 104tr, 187t

Jean-Paul Priou: 230c

Guy Redeuilh: 54t, 75tr, 77t, 115tl, 115cr, 137cr, 186tr, 206b, 219b

Pierre Roux: 161br

Gérard Sulmont: 244br

INDEX

Numbers in **bold** refer to colour plates (pp.37–244). The circle preceding Latin names of mushrooms makes it easy for you to keep a record of the mushrooms you see.

A

○ *Abortiporus biennis* 67, **334**
  Agaric, Clouded 821
    Fly 386
    Verdigris 796
○ *Agaricus arvensis* 177, **373**
○   *bernardii* 177, **370**
○   *bisporus* 176, **367**
○   *bitorquis* 178, **369**
○   *bohusii* 180, **367**
○   *campestris* 176, **366**
○   *cupreobrunneus* 180, **366**
○   *devoniensis* 176, **369**
○   *essettei* 177, **374**
○   *haemorrhoidarius* **373**
○   *koelerionensis* 179, **372**
○   *langei* **373**
○   *littoralis* **370**
○   *maleolens* **370**
○   *menieri* 178, **377**
○   *nivescens* **374**
○   *porphyrocephalus* **367**
○   *porphyrrhizon* 179, **375**
○   *praeclaresquamosus* 178, **376**
○   *romagnesii* 178, **371**
○   *silvaticus* 180, **373**
○   *silvicola* **374**
○   *spissicaulis* 179, **371**
○   *subperonatus* 179, **368**
○   *vapourarius* **368**
○   *variegans* **372**
○   *xanthoderma* 177, **375**
○ *Agrocybe aegerita* 148, **396**
○   *cylindracea* **396**
○   *dura* **397**
○   *molesta* **397**
○   *pediades* 143, **398**
○   *praecox* 143, **397**
○   *semiorbicularis* **398**
○   *vervacti* 142, **398**
○ *Albatrellus confluens* **346**
○   *cristatus* **346**
○   *pes-caprae* 72, **346**
○ *Aleuria aurantia* 48, **283**
○   *bicucullata* 43, **283**
○ *Aleurodiscus amorphus* 50, **314**

○ *Alnicola badia* **485**
○   *melinoides* 120, **485**
○   *scolecina* 126, **485**
○ *Amanita arctica* **385**
○   *aspera* **389**
○   *asteropus* **395**
○   *badia* **384**
○   *battarae* **381**
○   *beckeri* **381**
○   *beillei* 189, **391**
○   *boudieri* **391**
○   *caesarea* 186, **385**
○   *ceciliae* 185, **380**
○   *citrina* 189, **395**
○   *crocea* 184, **383**
○   *dunensis* 187, **393**
○   *echinocephala* 187, **390**
○   *franchetii* 188, **389**
○   *friabilis* **382**
○   *fulva* **383**
○   *fuscoolivacea* **381**
○   *gemmata* **387**
○   *groendlandica* **381**
○   *inaurata* **380**
○   *lactea* **385**
○   *lividopallescens* **384**
○   *mairei* 183, **384**
○   *malleata* 184, **382**
○   *muscaria* 185, **386**
○   *muscaria* ssp. *americana* 185, **386**
○   *nivalis* 185, **384**
○   *ovoidea* 186, **392**
○   *pantherina* 186, **387**
○   *phalloides* 187, **393**, **394**
○   *porphyria* 189, **395**
○   *proxima* **392**
○   *pseudofriabilis* 184, **381**
○   *rubescens* 188, **388**
○   *singeri* 188, **391**
○   *spissa* 189, **388**
○   *strangulata* **380**
○   *strobiliformis* 188, **389**
○   *submembranacea* 184, **380**
○   *vaginata* 183, **383**
○   *verna* **394**
○   *virosa* 186, **394**
○ *Anellaria semiovata* 113, **407**

○ *Anthracobia maurilabra* 46, **276**
○ *Arachnopeziza aurata* 45, **283**
○ *Armillaria gallica* 103, **829**
○   *mellea* 104, **827**
○   *obscura* **829**
○   *ostoyae* **829**
○   *tabescens* 103, **827**
○ *Arrhenia acerosa* 86, **808**
○   *auriscalpium* 86, **810**
○   *lobata* 86, **809**
○   *spathulata* 85, **809**
○ *Artomyces pyxidatus* 78, **350**
○ *Ascobolus furfuraceus* 48, **273**
○ *Ascotremella faginea* **269**
○ *Astraeus hygrometricus* 241, **869**
○ *Auricularia auricula-judae* 59, **302**
○   *mesenterica* 59, **302**
○ *Auriscalpium vulgare* 64, **323**

B

○ *Balsamia vulgaris* 244, **300**
  Bare-toothed Russula 710
○ *Battarraea phalloides* 243, **868**
○   *stevenii* **869**
  Bird's Nest, Common 877
○ *Bisporella citrina* 42, **263**
○   *sulfurina* **263**
○ *Bjerkandera adusta* 66, **342**
  Black Bulgar 269
  Blewit, Wood 835
  Blusher, The 388
  Blushing Bracket 337
○ *Bolbitius tener* 113, **399**
○   *vitellinus* 128, **399**
  Bolete, Birch 431
    Devil's 428
    Larch 413

- ○ *Boletinus cavipes* 229, 412
- ○ *elegans* 413
- ○ *flavus* 413
- ○ *Boletopsis grisea* 347
- ○ *leucomelaena* 72, 347
- ○ *Boletus aereus* 232, 424
- ○ *aestivalis* 423
- ○ *albidus* 420
- ○ *appendiculatus* 232, 421
- ○ *calopus* 233, 421
- ○ *depilatus* 232, 424
- ○ *dupainii* 234, 426
- ○ *edulis* 232, 422
- ○ *erythropus* 234, 425
- ○ *impolitus* 425
- ○ *luridus* 234, 427
- ○ *luteocupreus* 233, 427
- ○ *mamorensis* 423
- ○ *persoonii* 423
- ○ *pinophilus* 232, 423
- ○ *pseudoregius* 422
- ○ *pulverulentus* 229, 420
- ○ *queletii* 235, 425
- ○ *radicans* 233, 420
- ○ *regius* 234, 422
- ○ *reticulatus* 423
- ○ *satanas* 233, 428
- ○ *subappendiculatus* 422
- ○ *venturii* 423
- ○ *Bondarzewia mesenterica* 72, 346
- ○ *Bovista limosa* 239, 876
- ○ *plumbea* 876
- Brown Roll-rim 681
- ○ *Bryoglossum gracile* 49, 271
- ○ *Bulbillomyces farinosus* 56, 309
- ○ *Bulgaria inquinans* 43, 269
- Butter Cap 622

C

- ○ *Callorina fusarioides* 42, 261
- ○ *Calocera cornea* 74, 305, 306
- ○ *viscosa* 75, 306
- ○ *Calocybe gambosa* 163, 863
- ○ *Calvatia excipuliformis* 238, 873
- ○ *Calyptella campanula* 640
- ○ *capula* 48, 640

- ○ *Camarophyllopsis atropuncta* 676
- ○ *foetens* 94, 675
- ○ *Campanella caesia* 84, 630
- ○ *inquilina* 84, 630
- Candle Snuff 252
- ○ *Cantharellula umbonata* 102, 817
- ○ *Cantharellus cibarius* 82, 359
- ○ *cinereus* 82, 361
- ○ *friesii* 360
- ○ *ianthinoxanthus* 362
- ○ *lutescens* 80, 360
- ○ *melanoxeros* 81, 361
- ○ *tubiformis* 81, 360, 361
- ○ *Capitotricha bicolor* 260
- ○ *bicolor* var. *rubi* 45, 260
- ○ *Catathelasma imperiale* 103, 860
- Cep 422
- ○ *Cerocorticium confluens* 57, 312
- ○ *Chalciporus piperatus* 415
- ○ *Chamaemyces fracidus* 191, 678
- Chanterelle 359
- False 680
- Charcoal Burner 707
- ○ *Cheilymenia stercorea* 46, 277
- Chicken of the Woods 339
- ○ *Chlorociboria aeruginascens* 50, 265
- ○ *aeruginosum* 265
- ○ *Chondrostereum purpureum* 58, 316
- ○ *Chroogomphus fulmineus* 562
- ○ *rutilus* 104, 561
- ○ *Ciboria batschiana* 43, 265
- ○ *conformata* 267
- ○ *Clathrus archeri* 243, 881
- ○ *ruber* 244, 881
- ○ *Clavaria falcata* 74, 352
- ○ *zollingeri* 76, 352
- ○ *Clavariadelphus pistillaris* 75, 354
- ○ *truncatus* 355
- ○ *Claviceps purpurea* 40, 254
- ○ *Clavulina cinerea* 351
- ○ *cristata* 76, 351
- ○ *rugosa* 351

- ○ *Clavulinopsis corniculata* 75, 352
- ○ *fusiformis* 75, 353
- ○ *Clitocybe acicola* 823
- ○ *candicans* 823
- ○ *cerussata* 824
- ○ *clavipes* 99, 818
- ○ *costata* 820
- ○ *dealbata* 89, 823
- ○ *decembris* 98, 826
- ○ *ditopa* 97, 825
- ○ *fragrans* 822
- ○ *geotropa* 92, 819
- ○ *gibba* 88, 819
- ○ *graminicola* 95, 824
- ○ *hydrogramma* 825
- ○ *inornata* 92, 821
- ○ *lateritia* 102, 820
- ○ *nebularis* 100, 535, 821
- ○ *obsoleta* 822
- ○ *odora* 99, 822
- ○ *phaeophthalma* 99, 825
- ○ *phyllophila* 89, 823
- ○ *phyllophila* var. *ornamentalis* 822
- ○ *sinopica* 820
- ○ *squamulosa* 820
- ○ *subspadicea* 826
- ○ *umbilicata* 99, 826
- ○ *vibecina* 826
- ○ *Clitopilus hobsonii* 556
- ○ *hobsonii* var. *daamsii* 85, 556
- ○ *prunulus* 91, 555
- ○ *Collybia butyracea* 137, 622
- ○ *confluens* 114, 626
- ○ *cookei* 114, 621
- ○ *distorta* 137, 623
- ○ *dryophila* 136, 625
- ○ *dryophila* var. *funicularis* 626
- ○ *fusipes* 148, 623
- ○ *kuehneriana* 626
- ○ *luteifolia* 626
- ○ *maculata* 162, 624
- ○ *ocior* 626
- ○ *peronata* 117, 627
- ○ *platyphylla* 625
- ○ *racemosa* 114, 626
- ○ *radicata* 672
- ○ *succinea* 626
- ○ *velutipes* 674
- ○ *Coltricia perennis* 72, 327
- ○ *Conocybe dunensis* 127, 401
- ○ *inocyboeides* 128, 401
- ○ *tenera* 128, 400

Index

Coprinus acuminatus 442
  ammophilae 125, 443
  atramentarius 125, 442
  auricomus 123, 435
  cinereus 444
  comatus 125, 441
  disseminatus 124, 437
  kuehneri 123, 436
  lagopides 125, 443
  lagopus 443
  leiocephalus 436
  macrocephalus 444
  micaceus 124, 438
  narcoticus 124, 437
  niveus 113, 440
  picaceus 125, 444
  plicatilis 436
  radicans 438
  romagnesianus 442
  saccharinus 124, 439
  truncorum 439
  xanthothrix 124, 440
Coprobia granulata 46, 277

Coral Spot 253

Cordyceps microcephala 255
  militaris 40, 255
  sinensis 256

Cortinarius acutus 126, 473
  alboviolaceus 174, 461
  anomalus 160, 463
  arcuatorum 164, 478
  armillatus 140, 468
  barbatus 482
  bataillei 467
  betulinus 483
  bicolor 145, 469
  bivelus 164, 473
  bolaris 144, 461
  bulliardii 469
  caerulescens 478
  caligatus 175, 476
  calochrous 479
  camphoratus 174, 462
  caninus 169, 463
  cavipes 139, 471
  cephalixus 475
  cinnamomeoluteus 467
  cinnamomeus 141, 466
  claricolor 474
  collinitus 483
  cotoneus 458
  croceocaeruleus 145, 481
  delibutus 143, 482
  elatior 485

  elegantissimus 167, 479
  epsomiensis var. *alpicola* 138, 464
  evernius 469
  favrei 139, 483
  fragrantior 139, 471
  gentilis 460
  herculeus 165, 474
  hinnuleus 140, 470
  humicola 142, 460
  infractus 477
  malicorius 467
  melanotus 142, 458
  mucosus 483
  ochroleucus 482
  olidus 475
  olivaceofuscus 468
  orellanus 138, 459
  paleaceus 472
  paleifer 145, 472
  pholideus 138, 460, 465
  polaris 138, 467
  praestans 175, 475
  puniceus 144, 465
  purpurascens 175, 478
  rubicundulus 461
  rufoolivaceus 175, 481
  safranopes 471
  salor 483
  sanguineus 466
  saniosus 142, 459
  speciosissimus 459
  sphagneti 141, 468
  splendens 167, 480
  stillatitius 168, 484
  subtortus 169, 476
  terpsichores 175, 477
  torvus 174, 470
  traganus 174, 462
  triumphans 166, 475
  trivialis 168, 483
  uliginosus 138, 466
  variiformis 476
  vibratilis 482
  violaceus 175, 457, 458
  vitellinus 480
Cotylidia pannosa 78, 348

Cramp Balls 251

Craterellus cornucopioides 80, 358
  konradii 358
Creolophus cirrhatus 324
Crepidotus calolepis 522
  cesatii 523
  mollis 85, 522
  variabilis 85, 522

Crinipellis stipitarius 111, 631
Crucibulum laeve 242, 877

Crumble Cap, Trooping 437

Cudoniella acicularis 48, 264
  clavus 265
Cuphophyllus berkeleyi 90, 564
  borealis 90, 564
  cereopallidus 90, 566
  colemannianus 91, 567
  grossulus 98, 563
  niveus 565
  ochraceopallidus 564
  pratensis 95, 563
  radiatus 567
  russocoriaceus 90, 566
  subradiatus 567
  virgineus 90, 565
Cyathus olla 242, 877
  stercoreus 878
  striatus 242, 878
Cylindrocolla urticae 262
Cystoderma amianthinum 143, 676
  carcharias 117, 676
  fallax 677
  intermedium 677
  terreyi 144, 677
  tuomikoskii 677
Cystolepiota adulterina 194, 593
  bucknallii 201, 595
  hetieri 195, 594
  seminuda 194, 594
  sistrata 595
Cytidia salicina 314

D

Daedalea quercina 71, 335
Daedaleopsis confragosa 70, 337
  septentrionalis 338
  tricolor 71, 338
Daldinia concentrica 40, 251
  vernicosa 252
Dasyscyphella nivea 261

Dead Man's Fingers 253

Death Cap 393
  False 395

Deceiver 830
  Amethyst 829

○ *Delicatula integrella* 88, 669
○ *Dendropolyporus umbellatus* 70, 332
○ *Dermoloma atrocinereum* 140, 674
○ *cuneifolium* 675
Destroying Angel 394
○ *Diatrype bullata* 38
○ *Disciotis venosa* 52, 296
Dryad's Saddle 344

E

Earthball, Common 866
Earth Fan 348
Tongue 272
Earthstar 870
○ *Echinoderma asperum* 593
○ *echinaceum* 197, 592
○ *perplexum* 197, 593
○ *Entoloma alpicola* 136, 539
○ *ameides* 136, 542
○ *aranaeosum* 132, 552
○ *bloxamii* 536
○ *cephalotrichum* 550
○ *cetratum* 120, 551
○ *chalybaeum* 146, 545
○ *clypeatum* 141, 536
○ *corvinum* 543
○ *dysthales* 553
○ *euchroum* 146, 543
○ *excentricum* 135, 541
○ *griseoluridum* 541
○ *hebes* 122, 551
○ *hirtipes* 552
○ *icterinum* 128, 549
○ *incanum* 129, 549
○ *inopiliforme* 173, 536
○ *jubatum* 538
○ *lampropus* 146, 547
○ *lazulinum* 146, 544
○ *lividoalbum* 173, 541
○ *lividum* 163, 535
○ *mougeotii* 147, 545
○ *olorinum* 551
○ *pernitrosum* 553
○ *phaeocyathus* 555
○ *poliopus* 147, 547
○ *politum* 93, 553
○ *porphyrophaeum* 141, 537
○ *prunuloides* 536
○ *querquedula* 147, 546
○ *rhodocylix* 93, 553
○ *rhodopolium* 541
○ *rhodopolium* f. *nidorosum* 115, 540

○ *roseum* 130, 548
○ *rusticoides* 94, 555
○ *sericellum* 89, 550
○ *sericeum* 136, 538
○ *serrulatum* 547
○ *sordidulum* 137, 539
○ *subradiatum* 541
○ *tjallingiorum* 146, 545
○ *undatum* 97, 554
○ *Epichloe typhina* 39, 254
Ergot 254
○ *Eriopeziza caesia* 45, 259
○ *Exidia glandulosa* 60, 304
○ *truncata* 269

F

○ *Faerberia carbonaria* 82, 807
Fairy Ring Champignon 639
○ *Favolus europaeus* 344
○ *Fistulina hepatica* 68, 331
○ *Flammulaster carpophilus* 524
○ *carpophilus* var. *autochtonoides* 121, 523
○ *ferrugineus* 121, 524
○ *velutipes* 151, 674
○ *Fomes fomentarius* 68, 335
○ *Fomitopsis pinicola* 68, 336
Foxy Spot 624
Fungus, Artist's 330
Beefsteak 331
Brain 351
Cauliflower 351
Ear-pick 323
Hedgehog 325
Honey 827
Hoof 335
Oyster 687
Poached Egg 673
Porcelain 673
Scarlet Caterpillar 255
Silver-leaf 316
Tinder 335
Tripe 302
Yellow Brain 303
Funnel Cap, Tawny 832

G

○ *Galerina autumnalis* 527

○ *heterocystis* 127, 524
○ *laevis* 127, 525
○ *marginata* 150, 526
○ *paludosa* 126, 526
○ *subfusispora* 121, 527
○ *Ganoderma australe* 330
○ *carnosum* 331
○ *lipsiense* 70, 330
○ *lucidum* 69, 330
○ *Geastrum fimbriatum* 871
○ *morganii* 241, 872
○ *sessile* 240, 871
○ *triplex* 241, 870
○ *Geoglossum cookeianum* 74, 272
○ *Geopora arenicola* 47, 279
○ *clausa* 244, 257, 280
○ *sumneriana* 47, 280
○ *Gerronema chrysophyllum* 98, 817
○ *Gloeophyllum abietinum* 339
○ *saepiarium* 71, 338
○ *Gomphidius glutinosus* 561
○ *maculatus* 561
○ *roseus* 103, 561
○ *viscidus* 561
○ *Gomphus clavatus* 81, 362
○ *Grandinia granulosa* 311
○ *Grifola frondosa* 70, 332
Grisette 383
○ *Gymnopilus hybridus* 529
○ *penetrans* 152, 529
○ *sapineus* 529
○ *spectabilis* 151, 528
○ *Gyrodon lividus* 228, 411
○ *Gyromitra esculenta* 54, 292
○ *Gyroporus castaneus* 228, 410
○ *cyanescens* 228, 410

H

○ *Hapalopilus rutilans* 67, 334
Hare's Ear 284
○ *Hebeloma crustuliniforme* 163, 487
○ *edurum* 163, 488
○ *kuehneri* 139, 487
○ *mesophaeum* 139, 486
○ *radicosum* 164, 489
○ *sinapizans* 164, 488

Helvella acetabulum 295
  atra 294
  corium 295
  crispa 53, 293
  elastica 53, 294
  lactea 293
  lacunosa 294
  leucomelaena 53, 296
  macropus 53, 295
  sulcata 53, 294
  unicolor 53, 295
Hemimycena candida 670
  cephalotricha 667
  cucullata 667
  delectabilis 671
  lactea 112, 667
  mairei 670
  ochrogaleata 93, 670
  tortuosa 112, 666
Hemipholiota heteroclita 796
  populnea 150, 795
Hericium clathroides 63, 323, 324
  coralloides 324
  erinaceum 62, 324
  ramosum 323
Heterobasidion annosum 336
Hexagonia nitida 343
Heyderia cucullata 49, 270
Hirneola auricula 302
Hohenbuehelia algida 86, 807
Horn of Plenty 358
Humaria hemisphaerica 47, 278
Hydnellum concrescens 63, 321
  ferrugineum 322
  peckii 63, 321
Hydnum repandum 64, 325
  rufescens 64, 325
Hydropus marginellus 98, 671
  scabripes 132, 671
Hygrocybe aurantiolutescens 581
  aurantiosplendens 107, 569
  aurantiosplendens f. *luteosplendens* 568
  calyptriformis 108, 580
  cantharellus 105, 572
  ceracea var. *vitellinoides* 105, 570

  chlorophana 107
  cinereifolia 109, 578
  coccinea 106, 569
  coccineacrenata 107, 573
  conica 109, 577
  conicoides 109, 578
  fornicata 576
  fornicata var. *streptopus* 108, 576
  insipida 107, 571
  intermedia 581
  konradii 108, 580
  laeta 105, 574
  lepida 572
  luteolaeta 574
  miniata 574
  olivaceonigra 109, 579
  perplexa 574
  phaeococcinea 570
  psittacina 108, 574
  punicea 106, 567
  quieta 108, 570
  reai 571
  reidii 106, 572
  sciophana 574
  spadicea 110, 579
  splendidissima 106, 568
  subminutula 105, 571
  tristis 110, 577
  turunda 573
  unguinosa 109, 575
Hygrophoropsis aurantiaca 105, 680
  fuscosquamulosa 680
Hygrophorus agathosmus 92, 585
  arbustivus 101, 582
  atramentosus 585
  calophyllus 585
  camarophyllus 102, 584
  chrysaspis 586
  chrysodon 92, 590
  cossus 586
  discoideus 587
  discoxanthus 586
  eburneus 91, 586
  fagi 91, 582
  fuscoalbus 589
  hypothejus 101, 591
  latitabundus 101, 590
  leucophaeo-ilicis 583
  leucophaeus 100, 583, 587
  limacinus 590
  lindtneri 100, 587

  marzuolus 172, 585
  olivaceoalbus 102, 588
  penarius 91, 581
  penarius var. *barbatulus* 582
  persoonii 101, 589
  piceae 582
  pudorinus 101, 587
  russula 102, 583
Hymenochaete cruenta 57, 313
  mougeotii 313
  rubiginosa 317
  tabacina 58, 316
Hymenoscyphus umbilicatus 263
Hyphoderma radulum 62, 310
  roseocremeum 56, 310
Hyphodontia aspera 57, 311
Hypholoma capnoides 799
  elongatum 129, 800
  ericaeoides 802
  ericaeum 120, 801
  fasciculare 149, 799
  sublateritium 149, 800
  udum 802
Hypocrea pulvinata 341
Hypocreopsis lichenoides 39, 251
Hypoxylon fragiforme 39
  mammatum 251
  multiforme 38

I

Incrucipulum ciliare 45, 260
Ink Cap, Glistening 438
  Shaggy 441
Inocybe acuta 516
  acutella 156, 515
  agardhii 158, 490
  arenicola 158, 497
  asterospora 157, 520
  atripes 159, 514
  bongardii 160, 501
  caesariata 490
  calamistrata 152, 500
  calospora 152, 518
  cervicolor 159, 501
  cincinnata 506
  cookei 154, 499

- *corydalina* **160**, 503
- *curreyi* **156**, 498
- *curvipes* **156**, 516
- *dulcamara* **154**, 493
- *erinaceomorpha* 503
- *fastigiata* **157**, 496
- *fastigiella* **157**, 498
- *flocculosa* **153**, 510
- *friesii* 510
- *geophylla* **113**, 507
- *geophylla* var. *lilacina* **145**, 508
- *haemacta* **156**, 504
- *heimii* **155**, 490
- *hirtella* **154**, 515
- *hystrix* **152**, 504
- *incarnata* 503
- *jurana* **158**, 495
- *kuehneri* **117**, 512
- *kuthanii* 500
- *lacera* **154**, 507
- *lanuginosa* **160**, 517
- *maculata* **158**, 497
- *margaritispora* 520
- *mixtilis* 520
- *nitidiuscula* **153**, 510
- *obscura* **153**, 506
- *paludosa* **155**, 492
- *patouillardii* **159**, 494
- *petiginosa* **120**, 518
- *pholiotinoides* **154**, 513
- *piriodora* **159**, 502
- *pisciodora* 502
- *praetervisa* **157**, 519, 520
- *pruinosa* **160**, 509
- *psammophila* **158**, 508
- *quietiodor* 500
- *rhodiola* **152**
- *salicis* **156**, 520
- *splendens* 512
- *splendentoides* **159**, 511
- *squamata* 499
- *squarrosa* **153**, 505
- *squarrosoannulata* **155**, 492
- *tenebrosa* 514
- *terrigena* **155**, 491
- *umbrina* 516
- *umbrinofusca* **155**, 494
- *vulpinella* **153**, 512
- ○ *Inonotus dryadeus* **68**, 329
- *hispidus* **68**, 328
- *radiatus* 329
- ○ *Iodophanus carneus* **44**, 273

J

- Jelly Babies 269
- ○ *Junghuhnia nitida* **66**, 320, 326

K

- King Alfred's Cakes 251
- ○ *Kuehneromyces mutabilis* **150**, 789

L

- ○ *Laccaria amethystina* **145**, 644, 829
- *laccata* **133**, 830
- *tortilis* **133**, 831
- *Lachnellula subtilissima* **45**, 259
- ○ *Lachnum virgineum* **45**, 261
- ○ *Lacrymaria velutina* 444
- ○ *Lactarius acerrimus* 757
- *acris* **226**, 777, 778
- *atlanticus* **223**, 767
- *aurantiacus* 770
- *aurantiofulvus* 770
- *badiosanguineus* **221**, 774
- *bertillonii* 754
- *blennius* **218**, 762
- *bresadolanus* **225**, 758
- *brunneohepaticus* **223**, 773
- *camphoratus* **222**, 768
- *chrysorrheus* **218**, 776
- *cimicarius* **221**, 766
- *circellatus* **217**, 759
- *controversus* **216**, 755
- *curtus* 762
- *decipiens* **223**, 776
- *deliciosus* **224**, 786
- *deterrimus* **225**, 787
- *dryadophilus* 785
- *evosmus* 758
- *flavidus* 782
- *fluens* 762
- *fuliginosus* 779
- *fulvissimus* **224**, 770
- *fuscus* **221**, 764
- *glaucescens* 755
- *glyciosmus* 764
- *hepaticus* **222**, 773
- *hysginus* 221
- *lacunarum* **223**, 771
- *lignyotus* **226**, 781
- *lilacinus* **220**, 765
- *luridus* **220**, 783
- *mitissimus* **223**, 769
- *nanus* **219**, 760
- *necator* **217**, 756
- *pergamenus* 755
- *picinus* **226**, 780
- *piperatus* **216**, 754
- *porninsis* 758
- *pterosporus* **226**, 778
- *pubescens* 756
- *pyrogalus* **217**, 759
- *quieticolor* **224**, 786
- *quietus* **222**, 768
- *robertianus* **219**, 784
- *romagnesii* **226**, 779
- *rufus* **224**, 763
- *rugatus* 776
- *ruginosus* **226**, 779
- *salicis-herbaceae* 785
- *salmonicolor* **225**, 788
- *sanguifluus* **224**, 785
- *semisanguifluus* 787
- *serifluus* 767
- *spinosulus* **220**, 765
- *subdulcis* **222**, 769
- *subsericatus* 771
- *subumbonatus* 767
- *tabidus* **222**, 772
- *theiogalus* 773
- *torminosus* **217**, 755
- *trivialis* **219**, 760
- *uvidus* **219**, 782
- *uvidus* var. *candidulus* **220**, 783
- *vellereus* **216**, 753
- *vietus* **218**, 761
- *vinosus* 785
- *violascens* 784
- *volemus* **218**, 775
- *zonarioides* 758
- *zonarius* **217**, 757
- ○ *Laetiporus sulfureus* **69**, 339
- ○ *Lanzia luteovirescens* **44**, 266
- ○ *Lasiobolus ciliatus* 277
- *cuniculi* **46**, 276
- ○ *Lasiosphaeria ovina* 38
- Lawyer's Wig 441
- ○ *Leccinum aurantiacum* **236**, 429
- *brunneogriseolum* **236**, 432
- *carpini* **236**, 433
- *crocipodium* **235**, 433, 434
- *duriusculum* **235**, 430
- *lepidum* **235**, 434
- *piceinum* 430

- *pulchrum* 431
- *quercinum* 430
- *scabrum* 235, 431
- *variicolor* 236, 432
- *versipelle* 430
- *vulpinum* 430
- *Lentinellus bisus* 685
- *cochleatus* 86, 684
- *micheneri* 87, 684
- *omphalodes* 684
- *tigrinus* 686
- *Lentinus tigrinus* 88, 685
- *Lenzites betulinus* 71, 336
- *warnieri* 337
- *Leotia atrovirens* 270
- *lubrica* 49, 269
- *Lepiota acerina* 200, 601
- *alba* 597
- *audreae* 198, 600
- *boudieri* 200, 600
- *brunneoincarnata* 201, 604
- *brunneolilacea* 200, 605
- *castanea* 602
- *clypeolaria* 196, 597
- *cristata* 198, 599
- *felina* 196, 603, 607
- *fulvella* 601
- *fuscovinacea* 201, 608
- *grangei* 199, 603
- *helveola* 604
- *ignipes* 199, 601
- *ignivolvata* 194, 596
- *josserandii* 198, 604
- *laevigata* 193, 596
- *ochraceodisca* 192, 597
- *ochraceofulva* 197, 608
- *ochraceosulfurescens* 598
- *pseudofelina* 196, 602
- *pseudohelveola* 606
- *pseudohelveola* var. *sabulosa* 199, 606
- *rhodorrhiza* 199, 606
- *saponella* 599
- *sublaevigata* 597
- *ventriosospora* 196, 599
- *ventriosospora* var. *fulva* 598
- *Lepista flaccida* 833
- *gilva* 833
- *glaucocana* 163, 834
- *inversa* 99, 832
- *luscina* 834
- *martiorum* 93, 833
- *nuda* 174, 835
- *panaeola* 165, 834
- *personata* 835
- *saeva* 165, 835
- *sordida* 144, 836
- *Leptosphaeria acuta* 38, 248
- *Leucoagaricus badhamii* 193, 614
- *carneifolius* 615
- *cinerascens* 615
- *croceovelutinus* 200, 613
- *gauguei* 199, 611
- *holosericeus* 615
- *leucothites* 192, 615
- *littoralis* 192, 612
- *macrorrhizus* 191, 610
- *naucinus* 615
- *pinguipes* 191, 611
- *coagaricus purpureorimosus* 201, 613
- *sericatellus* 609
- *subcretaceus* 615
- *sublittoralis* 612
- *wychanskyi* 612
- *Leucocoprinus brebissonii* 195, 616
- *Leucocortinarius bulbiger* 164, 456
- *Leucopaxillus paradoxus* 92, 860
- *Limacella guttata* 190, 379
- *illinita* 191, 379
- *ochraceolutea* 380
- *Lycoperdon echinatum* 238, 875
- *foetidum* 874
- *mammiforme* 238, 874
- *molle* 875
- *perlatum* 238, 874
- *piriforme* 238, 873
- *umbrinum* 874
- *Lyomyces sambuci* 56, 308
- *Lyophyllum aggregatum* 861
- *decastes* 861
- *fumosum* 172, 861
- *loricatum* 861

M

- *Macrocystidia cucumis* 127, 560
- *Macrolepiota fuliginosa* 190, 616
- *mastoidea* 618
- *procera* 617
- *rhacodes* 190, 618
- *rickenii* 190, 617
- *Macrotyphula fistulosa* 75, 354
- Magpie, The 444
- *Marasmiellus candidus* 112, 628
- *omphaliformis* 95, 629
- *ramealis* 116, 627
- *trabutii* 630
- *vaillantii* 628
- *Marasmius alliaceus* 137, 635
- *androsaceus* 111, 632
- *anomalus* 110, 637
- *bulliardii* 634
- *cohaerens* 116, 637
- *curreyi* 110, 634
- *epiphylloides* 111, 633
- *epiphyllus* 633
- *globularis* 638
- *limosus* 634
- *oreades* 116, 639
- *rotula* 110, 634
- *scorodonius* 636
- *torquescens* 638
- *undatus* 126, 636
- *urens* 627
- *wynneae* 114, 638
- *Marcelleina benkertii* 52, 274
- *Megacollybia platyphylla* 165, 625
- *Melanoleuca cinereifolia* 860
- *cinereifolia* var. *maritima* 172, 859
- *grammopodia* 169, 858
- *iris* 135, 858
- *Melanophyllum eyrei* 195, 620
- *haematospermum* 195, 619
- *Melanotus caricicola* 805
- *horizontalis* 805
- *phillipsii* 86, 804
- *Melasmia acerina* 258
- *Melastiza chateri* 49, 282
- *Meripilus giganteus* 70, 333
- *Meruliopsis corium* 58, 317
- *Merulius papyrinus* 317
- *tremellosus* 58, 318
- *Microglossum fuscorubens* 272
- *viride* 74, 271
- *Micromphale perforans*

633
Miller, The 555
○ *Mitrophora semilibera* 54, 297
○ *Mitrula paludosa* 49, 270
○ *Mollisia cinerea* 45, 262
○ *Morchella costata* 299
○ *elata* 54, 298
○ *rotunda* 298
○ *vulgaris* 54, 298
Morel 298
　False 292
○ *Mucidula mucida* 673
Mushroom, Caesar's 385
　Field 366
　Parasol 617
　Saint George's 863
○ *Mutinus caninus* 243, 880
○ *Mycena acicula* 131, 661
○ *cena adonis* 131, 662
○ *aetites* 132, 660
○ *alcalina* 659
○ *alphitophora* 665
○ *amicta* 119, 650, 656
○ *arcangeliana* 118, 650
○ *aurantiomarginata* 129, 650
○ *belliae* 94, 668
○ *bulbosa* 664
○ *capillaripes* 116, 646
○ *chlorantha* 127, 649
○ *clavicularis* 118, 656
○ *corticola* 663
○ *corynephora* 112, 665
○ *crocata* 130, 653
○ *elegans* 650
○ *epipterygia* 119, 654
○ *erubescens* 118, 655
○ *filopes* 118, 657
○ *flavoalba* 117, 660
○ *galericulata* 115, 641
○ *galopus* 119, 651
○ *haematopus* 131, 652
○ *inclinata* 115, 642
○ *leptocephala* 118, 658
○ *maculata* 115, 643
○ *meliigena* 130, 662
○ *metata* 658
○ *olivaceomarginata* 649
○ *oregonensis* 662
○ *pelianthina* 134, 645
○ *phyllogena* 658
○ *polygramma* 115, 641
○ *pterigena* 121, 648
○ *pura* 134, 643, 645
○ *renatii* 131, 648
○ *rorida* 89, 669
○ *rosea* 133, 644, 645
○ *rosella* 130, 647
○ *roseofusca* 647
○ *rubromarginata* 646
○ *sanguinolenta* 131, 652
○ *seynii* 133, 645
○ *silvae-nigrae* 659
○ *speirea* 667
○ *speirea* var. *camptophylla* 128
○ *stipata* 650
○ *stylobates* 111, 664
○ *tenerrima* 111, 665
○ *thymicola* 650
○ *vitilis* 119, 659
○ *vulgaris* 656
○ *zephyrus* 643
○ *Mycenastrum corium* 240, 876
○ *Mycenella bryophila* 119, 640
○ *Mycoacia aurea* 319
○ *fuscoatra* 62, 319
○ *uda* 62, 319
○ *Myriostoma coliforme* 241, 872

N

○ *Naucoria escharoides* 485
○ *phaea* 485
○ *Nectria cinnabarina* 39, 253
○ *galligena* 254
○ *Neobulgaria pura* 43, 268
○ *Nidularia farcta* 242, 878
○ *Nyctalis agaricoides* 117, 864
○ *asterophora* 864
○ *parasitica* 114, 865

O

○ *Octospora rutilans* 49, 275
Old Man of the Woods 408
○ *Oligoporus caesius* 67, 343
○ *subcaesius* 343
○ *Omphalina barbularum* 95, 814
○ *galericolor* 96, 811
○ *hepatica* 96, 812
○ *obscurata* 813
○ *oniscus* 97, 815
○ *pseudomuralis* 96, 814
○ *pyxidata* 96, 812
○ *rickenii* 810
○ *rivulicola* 96, 813
○ *umbellifera* 815
○ *Omphalotus olearius* 103, 684
○ *Onygena equina* 256
○ *corvina* 256
　Orange Peel Fungus 283
○ *Orbilia xanthostigma* 42, 262
○ *Otidea bufonia* 50, 285
○ *onotica* 50, 284
○ *Oudemansiella longipes* 673
○ *mucida* 112, 673
○ *pudens* 673
○ *radicata* 137, 672

P

○ *Panaeolina foenisecii* 120, 407
○ *Panaeolus acuminatus* 404
○ *ater* 135, 405
○ *dunensis* 134, 406
○ *fimicola* 407
○ *obliquoporus* 135, 405
○ *rickenii* 121, 403
○ *semiovata* 407
○ *sphinctrinus* 135, 404
○ *Panellus mitis* 686
○ *serotinus* 87, 686
○ *stipticus* 87, 686
　Panther Cap 387
　Parasol, Shaggy 618
○ *Paxillus atrotomentosus* 88, 682
○ *filamentosus* 682
○ *involutus* 105, 681
○ *panuoides* 85, 683
○ *Peniophora aurantiaca* 57, 308
○ *eriksonii* 309
○ *laeta* 309
　Penny Bun 422
○ *Peziza ammophila* 51, 290
○ *ampelina* 50, 288
○ *badia* 52, 286
○ *badioconfusa* 287
○ *cerea* 51, 288
○ *fimeti* 50, 289
○ *gerardii* 51, 287

902 Index

- merdae 44, 289
- petersii 51, 287
- phyllogena 52, 286
- pseudoammophila 291
- succosa 50, 290
- varia 289
- vesiculosa 289
- Phaeocollybia arduennensis 126, 529
- christinae 531
- lugubris 157, 530
- Phaeolepiota aurea 165, 679
- Phaeomarasmius erinaceus 147, 531
- Phaeotellus griseopallidus 97, 815
- rickenii 94, 810
- Phallus hadriani 243, 879
- impudicus 242, 879
- hippophaecola 69, 329
- robustus 330
- Phellodon niger 64, 322
- Pholiota alnicola 152, 794
- apicrea 794
- curvipes 147, 792
- destruens 795
- flammans 151, 792
- graminis 142, 793
- jahnii 151, 791
- lenta 136, 795
- mutabilis 789
- salicicola 794
- squarrosa 151, 790
- tuberculosa 792
- aporos 402
- arrhenii 129, 402
- hadrocystis 402
- vestita 152, 403
- Phylloporus pelletieri 419
- rhodoxanthus 169, 419
- Phyllotopsis nidulans 683
- Physisporinus vitraeus 66, 326
- Phytoconis ericetorum 97, 815
- Piptoporus betulinus 69, 340
- Pisolithus arrhizus 239, 869
- Pleurotus atricapillus 689
- cornucopiae 88, 688
- eryngii 88, 688
- nebrodensis 688
- ostreatus 87, 687
- Plicaturopsis crispa 84, 318
- Pluteus aurantiorugosus 182, 693
- cervinus 181, 689
- chrysophaeus 695
- leoninus 182, 691
- luteomarginatus 691
- lutescens 694
- minutissimus 692
- pellitus 691
- petasatus 180, 690
- phlebophorus 181, 694
- podospileus 693
- romellii 181, 694
- salicinus 690
- seticeps 182, 692
- thomsonii 181, 695
- tricuspidatus 690
- umbrosus 181, 692
- Poculum firmum 43, 266
- petiolorum 267
- sydowianum 267
- Poison Pie 487
- Polypore, Birch 340
- Many-zoned 341
- Polyporus badius 345
- durus 71, 345
- forquignonii 344
- lepideus 72, 345
- picipes 345
- sbrumalis 346
- slentus 344
- sleptocephalus 345
- smori 344
- squamosus 69, 344
- tuberaster 344
- varius 345
- Porphyrellus porphyrosporus 228, 409
- Psalliota ammophila 377
- meleagris 376
- Psathyrella ammophila 132, 448
- artemisiae 122, 453
- badiophylla 454
- badiophylla var. neglecta 123, 454
- candolleana 116, 450
- conopilus 122, 448
- coprobia 122, 449
- cotonea 113, 446
- gracilis 123, 447
- hirta 449
- hydrophila 451
- lacrymabunda 140, 444
- leucotephra 148, 450
- lutensis 123, 452
- maculata 149, 446
- piluliformis 149, 451
- pygmaea 437
- pyrotricha 140, 445
- spadicea 149, 452
- squamosa 453
- subatrata 448
- Pseudocraterellus sinuosus 359
- undulatus 80, 359
- Pseudohydnum gelatinosum 63, 304
- Psilocybe chionophila 134, 803
- pratensis 134, 803
- squamosa 130, 802
- thrausta 803
- Pterula multifida 77, 350
- Ptychoverpa bohemica 297
- Pulcherricium caeruleum 57, 313
- Pycnoporus cinnabarinus 67, 339
- Pyronema omphalodes 42, 272

R

- Ramaria abietina 356
- aurea 358
- botrytis 77, 357
- formosa 77, 356
- gracilis 76, 355
- largentii 77, 357
- ochraceovirens 76, 356
- stricta 76, 356
- Ramariopsis kunzei 76, 355
- Ramicola centunculus 148, 532
- centunculus fo. filopes 148
- Resinomycena saccharifera 664
- Resupinatus applicatus 84, 806
- striatulus 807
- trichotis 807
- Rhizina undulata 52, 292
- Rhodocybe caelata 132, 558
- fallax 89, 558
- gemina 100, 557
- mundula 558
- popinalis 100, 557
- truncata 558
- Rhodotus palmatus 150, 559
- Rhytisma acerinum 42,

- *Rickenella fibula* **95**, 816
- *swartzii* **94**, 816
- *Ripartites helomorphus* 832
- metrodii 832
- strigiceps 832
- tricholoma **93**, 832
- rocaerulea 807
- Rooting Shank 672
- *Rosellinia aquila* 38
- *Rozites caperatus* **165**, 456
- *Rugosomyces carneus* **141**, 862
- ionides **144**, 862
- *Russula acetolens* 729
- acrifolia **204**, 700
- aeruginea **205**, 715
- alnetorum **216**, 740
- amara **215**, 730
- amarissima **214**, 753
- amethystina 730
- amoena **214**, 714
- amoenicolor 714
- amoenolens **203**, 707
- artesiana **215**, 745
- atropurpurea 739
- aurea **210**, 723
- aurora **209**, 713
- badia **213**, 750
- betularum **208**, 735
- cavipes **214**, 742
- cessans **213**, 728
- chloroides **202**, 702
- clariana 742
- claroflava **206**, 724
- cuprea 751
- cutefracta **207**, 708
- cyanoxantha **214**, 707
- decolorans **211**, 723
- delica **202**, 701, 702
- densifolia 701
- drimeia **213**, 747
- emetica **208**, 734
- exalbicans **207**, 744
- fageticola **208**, 736
- farinipes **203**, 703
- fellea **206**, 733
- foetens **204**, 703
- fragilis **215**, 738
- fragrans 706
- fragrantissima 706
- fuscorubra **213**, 749
- gracillima **207**, 743
- grisea 717
- helodes 747
- heterophylla **205**, 709
- illota **205**, 705
- integra **212**, 727
- ionochlora 717
- knauthii 739
- krombholzii **215**, 739
- laeta **211**, 725
- langei 708
- laricina 728
- laurocerasi **204**, 706
- lepida **209**, 712
- lepida var. ochroleucoides **203**, 713
- linnaei 753
- lundellii 752
- luteotacta **208**, 734
- medullata **207**, 717
- melitodes **212**, 726
- mesospora **210**, 752
- mustelina **205**, 711
- nana **207**, 737
- nauseosa 728
- nigricans **202**, 700
- nitida **211**, 722
- nobilis 737
- norvegica **215**, 740
- ochroleuca **206**, 732
- olivacea **203**, 731
- parazurea **207**, 716
- pectinata 703
- pectinatoides 707
- pelargonia **216**, 741
- persicina 734
- pseudopuellaris 719
- puellaris **212**, 718
- pumila 715
- queletii **213**, 746
- rhodopoda 747
- risigallina **211**, 729
- romellii 726
- rubroalba 726
- sanguinaria **209**, 746
- seperina 723
- sphagnophila 722
- subfoetens **204**, 704
- subrubens **210**, 720
- torulosa **212**, 748
- turci **211**, 730
- unicolor **212**, 718
- urens 752
- velenovskyi **210**, 721
- versicolor 719
- vesca **209**, 710
- veternosa **209**, 751
- violacea 742
- violeipes 715
- violeipes f. citrina **205**, 715
- virescens **206**, 710
- viscida 745
- vitellina 729
- xerampelina **210**, 719
- *Rustroemia maritima* 44
- firmum 266
- luteovirescens 266
- maritima 267

S

- *Saccobolus versicolor* **44**, 274
- *Sarcodon imbricatum* **64**, 322
- *Sarcodontia setosa* 319
- *Sarcoscypha austriaca* 291
- coccinea **48**, 291
- jurana 291
- macaronesia 291
- Scarlet Elf Cap 291
- Scarlet Hood 569
- *Scenidium nitidum* **71**, 343
- *Schizophyllum commune* **84**, 362
- *Schizopora flavipora* 327
- paradoxa **66**, 327
- radula 327
- *Scleroderma areolatum* 867
- bovista **240**, 867
- citrinum **239**, 866
- geaster **240**, 867
- polyrrhizon 867
- *Scutellinia crinita* 281
- kerguelensis **46**, 281
- setosa **46**, 281
- *Sericeomyces sericatellus* **194**, 609
- *Serpula lacrymans* 318
- *Setulipes androsaceus* 632
- quercophilus 633
- splachnoides 633
- *Simocybe centunculus* 532
- Slippery Jack 412
- *Sparassis crispa* **77**, 351
- laminosa 351
- *Spathularia alpestris* 268
- flavida 268
- neesii **74**, 267
- pectinata 268
- rufa 268
- *Sphaerobolus stellatus* **242**, 870
- *Sphaerosporella brunnea* **47**, 278
- hinnulea 278
- Split-gill 362
- *Steccherinum ochraceum* **62**, 320, 326
- *Stereum hirsutum* 59,

314
- *insignitum* 59, 315
- *ochraceoflavum* 315
- *rameale* 315
- *subtomentosum* 316

Stinkhorn 879
Dog 880
- *Strobilomyces strobilaceus* 228, 408
- *Stropharia aeruginosa* 797
- *caerulea* 161, 796
- *coronilla* 143, 798
- *cyanea* 796
- *ochrocyanea* 161, 797
- *pseudocyanea* 797
- *Suillus bellinii* 414
- *bovinus* 230, 414
- *collinitus* 229, 414
- *granulatus* 414
- *grevillei* 229, 413
- *luteus* 229, 412
- *tridentinus* 414
- *variegatus* 230, 415
- *viscidus* 414

T

Tar Spot 258
- *Tazzetta catinus* 47, 284
- *Tephrocybe palustris* 122, 863
- *rancida* 133, 864
- *Thelephora palmata* 78, 349
- *penicillata* 78, 349
- *terrestris* 78, 348
- *Trametes gibbosa* 342
- *hirsuta* 342
- *multicolor* 341
- *pubescens* 66, 342
- *versicolor* 67, 341
- *zonatella* 341
- *Trechispora farinacea* 56, 311
- *Tremella encephala* 303
- *foliacea* 60, 303
- *mesenterica* 60, 303
- *Tremellodon gelatinosum* 304
- *Tremiscus helvelloides* 60, 305
- *Trichaptum abietinum* 316
- *Trichoglossum hirsutum* 272
- *Tricholoma acerbum* 162, 838
- *album* 162, 836
- *argyraceum* 843
- *atrosquamosum* 172, 841
- *aurantium* 166, 855
- *auratum* 167, 848
- *basirubens* 842
- *bresadolanum* 840
- *bufonium* 849
- *caligatum* 170, 856
- *cingulatum* 171, 843
- *colossus* 855
- *columbetta* 162, 838
- *equestre* 167, 847
- *flammula* 857
- *focale* 855
- *fracticum* 170, 854
- *fucatum* 847
- *fulvum* 166, 851
- *gausapatum* 842
- *imbricatum* 166, 841
- *inamoenum* 837
- *joachimii* 847
- *lascivum* 837
- *myomyces* 842
- *orirubens* 842
- *ornata* 857
- *pardinum* 170, 844
- *pessundatum* 168, 852, 853
- *populinum* 853
- *portentosum* 171, 840
- *psammopus* 169, 851
- *pseudoalbum* 162, 837
- *pseudonictitans* 852
- *ramentaceum* 843
- *saponaceum* 161, 844
- *scalpturatum* 171, 842
- *sciodes* 840
- *sejunctum* 161, 845
- *squarrulosum* 841
- *sulfurescens* 837
- *sulfureum* 167, 848
- *terreum* 170, 841
- *ustale* 168, 853
- *ustaloides* 168, 853
- *vaccinum* 166, 850
- *vaccinum* var. *fulvosquamosum* 850
- *virgatum* 170, 839
- *viridifucatum* 847
- *viridilutescens* 161, 846
- *Tricholomopsis decora* 150, 856
- *rutilans* 173, 857

Truffle 299
- *Tubaria autochtona* 104, 532
- *conspersa* 104, 533
- *furfuracea* 533
- *hiemalis* 533
- *pellucida* 534
- *romagnesiana* 104, 534
- *Tuber aestivum* 300
- *uncinatum* 244, 299
- *Tubercularia vulgaris* 254
- *Tulostoma brumale* 241, 868
- *mammosum* 868
- *Tylopilus felleus* 231, 429
- *Typhula quisquilaris* 74, 353

V

- *Vascellum pratense* 238, 875
- *Verpa conica* 297
- *Volvariella bombycina* 182, 696
- *gloiocephala* 183, 698
- *hypopithys* 183, 696
- *loveiana* 697
- *parvula* 697
- *pusilla* 697
- *speciosa* 698
- *surrecta* 182, 697

W

Wax Cap, Meadow 563
Parrot 574
Weeping Widow 444
Witch's Butter 304
Wood Cup, Green 265
Woolly-foot, Wood 627

X

- *Xerocomus armeniacus* 231, 416
- *badius* 230, 418
- *chrysenteron* 231, 417
- *parasiticus* 236, 418
- *porosporus* 417
- *pruinatus* 417
- *rubellus* 230, 418
- *subtomentosus* 231, 416
- *versi* 418
- *Xylaria carpophila* 253
- *filiformis* 253
- *hypoxylon* 40, 252
- *longipes* 253
- *polymorpha* 40, 253